Nanotechnology for Water and Wastewater Treatment

Integrated Environmental Technology Series

Nanotechnology for Water and Wastewater Treatment

P. N. L. Lens, J. Virkutyte,
V. Jegatheesan, Seung-Hyun Kim
and S. Al-Abed

IWA Publishing
London • New York

Published by **IWA Publishing**
Alliance House
12 Caxton Street
London SW1H 0QS, UK
Telephone: +44 (0)20 7654 5500
Fax: +44 (0)20 7654 5555
Email: publications@iwap.co.uk
Web: www.iwapublishing.com

First published 2013

British Library Cataloguing in Publication Data
A CIP catalogue record for this book is available from the British Library

ISBN 9781780404585 (Paperback)
ISBN 9781780404592 (eBook)

Front cover photographs:
Top, right: Photocatalytic microreactor for water cleaning, containing immobilized TiO_2 nanoparticles (Courtesy: Aran H. C., Salamon D., Rijnaarts T., Mul G., Wessling M. and Lammertink R. G. H., Mesa + Institute for Nanotechnology, University of Twente, The Netherlands).
Top, left: Side view of the tetrameric *E. coli* aquaporin Z (AqpZ) complex consisting of four monomer backbone structures shown in deep teal, violet purple, pale green, and yellow. The rendering was generated using PyMol 1.5.0.2 using AqpZ PDB coordinates 2ABM (Courtesy: The Biomimetic Membrane Group, Department of Physics, Technical University of Denmark, Denmark).
Bottom: Selenium nanoparticles produced by anaerobic granular sludge treating selenite contaminated wastewater (Courtesy: Gonzalez Gil G. and Saikaly P., Water Desalination and Reuse Center, King Abdullah University of Science & Technology, Saudi Arabia).

Contents

Preface

In the last decade, application of nanotechnology in the production of modern day consumer products has enhanced the quality and the efficient usage of those products significantly. For example, the use of consumer products based on nanotechnology has increased over 20 fold between 2005 and 2010. Further, nanotechnology has been finding its way in various other applications such as targeted drug delivery in medicine, pollutant removal from water and wastewater, improvements to the properties of herbicides in agriculture, light and strong materials for space travel, and so on. Thus, it has opened new avenues for numerous technical improvements and more performing water production and wastewater treatment technologies. On the other hand, the presence of nanomaterials (NMs) and nanoparticles (NPs) in air, water and soil environments has become unavoidable. Therefore, clear understanding of the life cycle of manufactured NPs is essential in order to avoid adverse health effects on humans, plants and animals due to their exposure to NMs and NPs. The above was the thrust to publish a book on the application of nanotechnology in water and wastewater treatment. This book overviews nanotechnology applications and covers the important elements:

- applications of NMs and NPs in water and wastewater treatment, pollutant degradation and bioremediation, disinfection and biosensors
- the life cycle analysis of NMs and NPs
- their effects on the health of human and animals
- their fate in the aquatic environments
- manufacturing processes of NMs and NPs

This book is divided into seven parts; the introductory section deals with identifying the applications of nanotechnology in water and wastewater treatment and its effects on the health of human and the environment in three chapters. It also focuses on the life cycle assessment of nanomaterials in order to carry out risk/benefit analyses, product development, policy making and waste management for technologies based on nanomaterials. The second part contains two chapters which are dedicated to the characterisation of nanoparticles and their mobility and fate in the water environment. Aggregation of nanoparticles is one of the interesting topics that has been included in this section.

The third part contains five chapters where the applications of nanosorbents and nanocatalysts in the removal of pollutants, such as phosphates, heavy metals and organics (in particular polychlorinated biphenyls, PCBs) have been discussed extensively. Applications of nanosorbents such as magnesium hydroxide and calcium based layered double hydroxides as well as nanocatalysts based on palladium, activated carbon supported palladized iron nanoparticles and visible light active titanium doped with nitrogen and silver are presented. The above chapters open up great research opportunities to innovate application of nanotechnology to remove pollutants that are toxic and hard to treat through conventional treatment processes.

Two chapters in part IV are dedicated to the application of nanomaterials for disinfection. Microbially manufactured silver nanoparticles and electrospun nanofibrous membranes have a huge potential in disinfecting water. The chapters provide the state of the art towards improved disinfection efficiency and reduced formation of disinfection by-products using nanotechnology based water disinfection.

The two chapters in part V discuss how nanotechnology can aid in (bio)process engineering of water and wastewater treatment. Micro and nanoreactors are introduced and the enhancement of catalytic conversions when using microreactors for nitrogen removal is presented. In addition, the great opportunities nanotechnology gives for filtration are overviewed. In particular, the use of new types of membrane materials for water treatment is highlighted, with a particular focus on the design and potential use of bioinspired membranes.

Nanotechnology for biosensors is the topic of part VI. The ideal biosensor would offer the sensitivity and selectivity to detect single molecules in real time. Current biosensors can detect DNA and proteins at concentrations higher than 600 molecules per ml (atto-mole/L). In order to sense molecules at a concentration of 1 molecule per ml (zepto-mole/L), sophisticated methods should be employed and one of them is nanosensors. Graphene is considered as one of the candidates for nanosensors as it can be produced in sheets that are one atomic layer thick. When a molecule is adsorbed by the surface of a graphene sheet, the electrical properties of the surface will change significantly which would help to detect the molecule that is attached to the graphene sheet. A thorough review on the application of nanotechnology in the field of biosensors is given under this section.

The last two chapters are dedicated to nanomanufacturing of metal and metal oxide nanoparticles. Green sysnthesis routes for metallic nanoparticles such as Ag, Au, Pd, Pt, Fe, In_2O_3 and TiO_2 as well as bimetallic nanoparticles such as Au-Pd have been developed over the last years in line with pollution prevention and green chemistry approaches. These include natural reductants, such as glucose and plant extracts, as well as biocatalysis using bacteria and fungi. An extensive review has been included to evaluate the current state of plant-based manufacturing of nanoparticles.

This book covers the entire spectrum of nanotechnology through 18 chapters and each chapter contains more than 50 references to the scientific literature for further reading. The book is designed for scientists, engineers, tertiary educators as well as post and undergraduate students. However, readers from all walks of life can enjoy reading it to gain an insight on nanotechnology which is going to stay with us for a very long future.

The editors of this book sincerely thank all the contributing authors for their time in writing the valuable chapters. The editors are also grateful to IWA Publishing for providing an opportunity to publish this book.

February 2013
Piet Lens
Jurate Virkutyte
Veeriah Jegatheesan
Seung-Hyun Kim
Souhail Al-Abed

List of Contributors

CHAPTER 1

Giuseppe Laera

Istituto di Ricerca Sulle Acque
Consiglio Nazionale delle Ricerche
Viale F. De Blasio 5
70132 Bari
Italy
Email: giuseppe.laera@ba.irsa.cnr.it

Piet N. L. Lens

Department of Environmental Engineering and Water Technology
UNESCO-IHE Westvest 7
2611 AX Delft
The Netherlands
Email: P.Lens@unesco-ihe.org; Piet.Lens@wur.nl

CHAPTER 2

Babina Shrestha

The Institute of Environmental and Human Health
Texas Tech University
P.O. Box 41163
Lubbock, TX 79409
USA
Email: shrestha.babina@epa.gov

Jaclyn E. Cañas-Carrell

The Institute of Environmental and Human Health
Texas Tech University
P.O. Box 41163
Lubbock, TX 79409
USA

Shibin Li

United States Environmental Protection Agency Mid-Continent Ecology Division
6201 Congdon Blvd.
Duluth, MN 55804
USA

CHAPTER 3

Takuya Tsuzuki

Research School of Engineering
College of Engineering and Computer Science
Australian National University
Ian Ross Building 31, North Road

Canberra ACT 0200
Australia
Email: takuya.tsuzuki@anu.edu.au

CHAPTER 4

Erika J. Espinosa-Ortiz
Environmental Engineering and Water Technology Department
UNESCO-IHE Institute for Water Education
Westvest 7, 2611 AX, Delft
The Netherlands
Email: e.espinosaortiz@unesco-ihe.org

Mabel Vaca-Mier
Energy Department
Universidad Autónoma Metropolitana (Metropolitan autonomous university)
Av. San Pablo No. 180
Col. Reynosa Tamaulipas
02200, Delegación Azcapotzalco
Distrito Federal
México
Email: mvm@correo.azc.uam.mx

CHAPTER 5

Ujjwal Shrestha
School of Engineering
Deakin University
Geelong Waurn Ponds Campus
75 Pigdons Road
Waurn Ponds, VIC 3216
Australia
Email: mailujjwal93@yahoo.com

Veeriah Jegatheesan
School of Engineering
Deakin University
Geelong Waurn Ponds Campus
75 Pigdons Road
Waurn Ponds VIC 3216
Australia
Email: jega.j@deakin.edu.au; jega.jegatheesan@jcu.edu.au

Li Shu
School of Engineering
Deakin University
Geelong Waurn Ponds Campus
75 Pigdons Road
Waurn Ponds VIC 3216
Australia
Email: l.shu@deakin.edu.au

Jurate Virkutyte
Pegasus Technical Services Inc.
On-Site Contractor to the U.S. Environmental Protection Agency (US-EPA)
National Risk Management Research Laboratory (NRMRL)
EPA Center Hill Facility
5995 Center Hill Avenue
Cincinnati, OH 45216
USA
E-mail: Virkutyte.Jurate@epa.gov

CHAPTER 6

Ji Zhi Zhou
School of Environmental and Chemical Engineering
Shanghai University
No. 333 Nanchen Rd.
Shanghai 200444
P. R. China
E-mail: jizhi.zhou@shu.edu.cn

Quangren Qian
School of Environmental and Chemical Engineering
Shanghai University

No. 333 Nanchen Rd.
Shanghai 200444
P. R. China
E-mail: grqian@shu.edu.cn

Zhi Ping Xu

ARC Centre of Excellence for Functional Nanomaterials, Australian Institute for Bioengineering and Nanotechnology
The University of Queensland
Brisbane, QLD 4072
Australia
E-mail: gordonxu@uq.edu.au

CHAPTER 7

Weizhen Liu

State Key Laboratory of Structural Chemistry
Fujian Institute of Research on the Structure of Matter
Chinese Academy of Sciences
Fuzhou, Fujian, 350002
China
Email: wzhliu0824@gmail.com

Zhang Lin

State Key Laboratory of Structural Chemistry
Fujian Institute of Research on the Structure of Matter
Chinese Academy of Sciences
Fuzhou, Fujian, 350002
China
Email: zlin@fjirsm.ac.cn

Feng Huang

Key Laboratory of Optoelectronic Materials Chemistry and Physics
Fujian Institute of Research on the Structure of Matter
Chinese Academy of Sciences
Fuzhou, Fujian, 350002
China
Email: fhuang@fjirsm.ac.cn

CHAPTER 8

Ibrahim El Saliby

School of Civil and Environmental Engineering
University of Technology, Sydney
Post Box 129, Broadway, NSW 2007
Australia
Email: ibsaliby@hotmail.com

Ho Kyong Shon

School of Civil and Environmental Engineering
University of Technology, Sydney
Post Box 129, Broadway, NSW 2007
Australia
Email: Hokyong.Shon-1@uts.edu.au

Andrew McDonagh

School of Chemistry and Forensic Science
University of Technology, Sydney
P.O. Box 123, Broadway, NSW 2007
Australia
Email: Andrew.McDonagh@uts.edu.au

CHAPTER 9

Tom Hennebel

Civil and Environmental Engineering
University of California, Berkeley
207 O'Brien Hall
Berkeley, CA 94720-1720
USA
Email: hennebel@berkeley.edu

Jeffrey P. Fitts

Civil and Environmental Engineering
Princeton University

E430 Engineering Quad
Princeton, NJ 08544
USA
Email: fitts@princeton.edu

Sam Van Nevel

Laboratory for Microbial Ecology and Technology (LabMET)
Ghent University
Coupure links 653, 9000 Gent
Belgium
Email: Sam.VanNevel@Ugent.be

Nico Boon

Laboratory for Microbial Ecology and Technology (LabMET)
Ghent University
Coupure links 653, 9000 Gent
Belgium
Email: Nico.Boon@Ugent.be

CHAPTER 10

Hyeok Choi

Department of Civil Engineering
University of Texas at Arlington
Arlington, TX 76019-0308
USA
Email: hchoi@uta.edu

Prince Nfodzo

Department of Civil Engineering
University of Texas at Arlington
Arlington, TX 76019-0308
USA

Souhail R. Al-Abed

USEPA
National Risk Management Research Laboratory (NRMRL)
26 West Martin Luther King Drive
Mail Code: CHL
Cincinnati, OH 45268
USA
Email: al-abed.souhail@epa.gov

Shirish Agarwal

CH2M Hill
2812 Madison Rd.
Apt # 7 Cincinnati
OHIO 45209
USA

Dionysios D Dionysiou

Environmental Engineering and Science Program
School of Energy, Environmental, Biological and Medical Engineering
University of Cincinnati
Cincinnati, OH 45221
USA
Email: dionysios.d.dionysiou@uc.edu

CHAPTER 11

Liesje Sintubin

Laboratory for Microbial Ecology and Technology (LabMET)
Ghent University
Coupure Links 653 B-9000 Gent
Belgium
Email: Liesje.Sintubin@Ugent.be

CHAPTER 12

Eugene Thomas Cloete

Stellenbosch University
Private Bag X1
Matieland 7602
South Africa
Email: eugenecloete@sun.ac.za

Bert Klumperman

South African Research Chair in Advanced Macromolecular Architectures
Department Chemistry and Polymer Science
Stellenbosch University
Private Bag X1
Matieland 7602
South Africa
Email: bklump@sun.ac.za

Michele de Kwaadsteniet

Department Microbiology
Stellenbosch University
Private Bag X1
Matieland 7602
South Africa
Email: mdekwaad@sun.ac.za

Nonjabulo Prudence Gule

Department Chemistry and Polymer Science
Stellenbosch University
Private Bag X1
Matieland 7602
South Africa
Email: njabu@sun.ac.za

CHAPTER 13

Rob G. H. Lammertink

Soft matter, Fluidics and Interfaces
Mesa+ Institute for Nanotechnology
University of Twente
P.O. Box 217, 7500 AE Enschede
The Netherlands
Email: r.g.h.lammertink@utwente.nl

H. C. (CAN) Aran, M.Sc

Mesa+ Institute for Nanotechnology
University of Twente
P.O. Box 217, 7500 AE Enschede
The Netherlands
Email: h.c.aran@utwente.nl

CHAPTER 14

Claus Hélix-Nielsen

The Biomimetic Membrane Group, Department of Physics
Technical University of Denmark
Fysikvej 309, DK2800 Kgs. Lyngby
Denmark
Email: claus.helix.nielsen@fysik.dtu.dk; chn@aquaporin.dk

CHAPTER 15

Wenrong Yang

School of Life and Environmental Science
Deakin University
Geelong Waurn Ponds Campus
75 Pigdons Road
Waurn Ponds, VIC 3216
Australia
Email: wenrong.yang@deakin.edu.au

Collin Barrow

School of Life and Environmental Science
Deakin University
Geelong Waurn Ponds Campus
75 Pigdons Road
Waurn Ponds, VIC 3216
Australia
Email: collin.barrow@deakin.edu.au

CHAPTER 16

Ronald Kent

Department of Civil and Environmental Engineering
Center for the Environmental Implications of Nanotechnology (CEINT)

Virginia Tech Blacksburg, VA 24060
USA
Email: ronkent40@gmail.edu

Paramjeet Pati
Department of Civil and Environmental Engineering
Center for the Environmental Implications of Nanotechnology (CEINT)
Virginia Tech Blacksburg, VA 24060
USA
Email: param@vt.edu

Peter Vikesland
Department of Civil and Environmental Engineering
Center for the Environmental Implications of Nanotechnology (CEINT)
Virginia Tech Blacksburg, VA 24060
USA
Email: pvikes@vt.edu

CHAPTER 17

Jurate Virkutyte
Pegasus Technical Services Inc.
On-Site Contractor to the U.S. Environmental Protection Agency (US-EPA)
National Risk Management Research Laboratory (NRMRL)
EPA Center Hill Facility
5995 Center Hill Avenue
Cincinnati, OH 45216
USA
E-mail: Virkutyte.Jurate@epa.gov

Rajender Varma
USEPA
National Risk Management Research Laboratory (NRMRL)
26 West Martin Luther King Drive
Mail Code: CHL
Cincinnati, OH 45268
USA
Email: varma.rajender@epa.gov

CHAPTER 18

Jorge L. Gardea-Torresdey
Department of Chemistry
The University of Texas at El Paso
500 West University Ave.
El Paso, TX 79968
USA
Email: jgardea@utep.edu

Jose R. Peralta-Videa
Department of Chemistry
The University of Texas at El Paso
500 West University Ave.
El Paso, TX 79968
USA
Email: jperalta@utep.edu

Jie Hong
Environmental Science and Engineering Ph.D Program
The University of Texas at El Paso
500 West University Ave.
El Paso, TX 79968
USA
Email: jhong@miners.utep.edu

Chapter 1

Nanotechnology for water and wastewater treatment: potentials and limitations

G. Laera and P. N. L. Lens

1.1 INTRODUCTION TO NANOSCIENCE AND NANOTECHNOLOGY

Nanotechnology involves the manipulation of matter on a near atomic scale to produce new structures, devices, and materials. Particles that have at least one dimension in the range of 1–100 nm are defined as nanoscale materials (Morose, 2010).

As observed by Chaturvedi *et al.* (2012), as atoms and chemical bonds have sizes in the range of a few tenths of a nanometer, the smallest structures we will ever make will have the dimensions of a few nanometers. This is because as soon as a few atoms are placed next to each other, the resulting structure is in the nanometer size range.

Several examples of nanomaterials and nanodevices can be found in nature (e.g. DNA, proteins, micelles, viruses, minerals, clays, tiles on a butterfly wing, etc.) and obviously nanoscience and nanotechnology are learning and copying from nature. Also other concepts and features of nanotechnology, such as self-assembly and self-replication, can be easily recognised in several natural phenomena. For this reason, it is better to refer to engineered nanomaterials (ENM) and engineered devices (END), which exhibit specific properties and functions because of their nanoscale size. Nanotechnology also involves the integration and assembly of nanoscale structures into larger systems, and keeping control of the nanoscale construction is one of the most challenging aspects of this new technologic revolution.

Interface and colloid science has given rise to many materials which may be useful in nanotechnology, such as carbon nanotubes and fullerenes, various metallic and metal oxide nanoparticles and nanorods, organic dendrimers, … Two main approaches are used in their production: "bottom-up" and "top-down". In the "bottom-up" approach, materials and devices are built from molecular or atomic components which assemble themselves chemically by principles of molecular recognition (seeding, reverse micelle, enzyme-substrate interaction and similar biomimetic principles). In the "top-down" approach, nano-objects are constructed from larger entities without atomic-level control (nano-lithography, etching, etc.).

The properties of the nanosized materials can be consistently different compared to what they exhibit on a macroscale, due to the high surface area to volume ratio and quantum effects. These two factors can change reactivity and catalytic properties, mechanical strength and electrical characteristics (Chaturvedi *et al.*, 2012). For instance, opaque substances become transparent (copper); stable materials turn combustible (aluminium); solids turn into liquids at room temperature (gold); insulators become conductors (silicon). A material such as gold, which is chemically inert at normal scales, can serve as a potent chemical catalyst at nanoscales. These properties can be modulated by controlling size and shape of the nanomaterial as, for example, in quantum dots and their interaction with light (Figure 1.1).

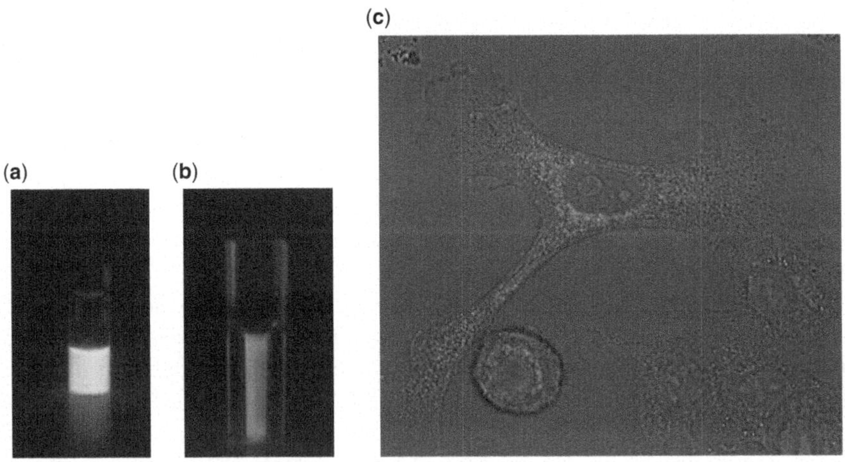

Figure 1.1 When irradiated with UV light, colloidal suspensions of CdSe nanocrystals emit different colours depending on the different size of the nanoparticles (a, b). This effect is called quantum confinement and predominates for crystal diameters below 10 nm. The quantum dots properties are maintained during interaction of the nanoparticles with living cells and can be used for diagnosis purposes, as shown in a confocal image of red quantum dots in A549 cells (c) (Courtesy of Pierpaolo Pompa, Virgilio Brunetti, Maria Ada Malvindi, Italian Institute of Technology).

The novel properties of nanomaterials offer great promise to provide new technological breakthroughs. At present, nanotechnology is already being used in hundreds of products across various industries, with a very rapidly expanding market, despite a certain concern regarding the health hazards of some nanoparticles (Morose, 2010).

The water sector also follows the current trend in nanoscience and nanotechnology and a great number of research activities are currently being run around the world for both aspects of nanomaterial usage, that is development of new applications and evaluation of safety and impacts for humans and the environment. This is reflected in the actions promoted by the European Commission, under the 6th and 7th Framework Programs, which include two major project clusters:

- nano4water – nanotechnology for water treatment (http://nano4water.eu), which groups 7 projects focused on the development of nanomembranes and on the use of nanomaterials in catalytic reactors.
- NanoSafety Cluster (http://nanosafetycluster.eu), which groups 39 projects addressing all aspects of nanosafety, including human and environmental toxicology, assessment of exposure, mechanisms and risks, standardization of tests and procedures.

Lots of projects related to the application of nanotechnology for water treatment have been funded in the USA, promoted by the National Nanotechnology Initiative (http://www.nano.gov/), the Environmental Protection Agency (http://www.epa.gov/nanoscience/), the National Science Foundation (http://www.nsf.gov/news/overviews/nano/index.jsp) and several state initiatives.

Also Japan is massively investing in nanotechnology research, including specific projects related to water treatment funded under the JST's CREST water research program (http://www.water.jst.go.jp/en/index.html) and general research initiatives.

South Africa has launched its National Nanotechnology Strategy in 2006, with new innovation centres for nanoscience, which also focus on nanoscience for water (e.g. the nanomembranes for water filtration studied at the University of Stellenbosch, see Chapter 12).

Brazil, Cuba, Saudi Arabia and Sri Lanka all host nanoscience centres working on this issue (Grimshaw, 2009). Also India, China and several developing countries are increasingly supporting nanoscience, including research into how nanotechnology can help deliver clean water.

1.2 NANOTECHNOLOGY FOR WATER AND WASTEWATER TREATMENT

Water scarcity problems are rapidly increasing due to population growth, pollution and climate change. As pointed out in a recent review, in the coming decades water scarcity may be a watchword that prompts action ranging from wholesale population migration to war, unless new ways to supply clean water are found (Shannon *et al.*, 2008). The same reviewers observed that, although conventional methods of water disinfection, decontamination and desalination can address many of these problems, these treatment methods are often chemically and energetically intensive, and require investments and engineering expertise that can preclude their use in many parts of the world.

On the other hand, the utilisation of alternative water sources like stormwater and reclaimed municipal wastewater is one of the most obvious and promising options in integrated water management. Major concerns about the safety of this exploitation route are connected to microbial and chemical contaminants occurring in wastewater, among which endocrine disrupters and pharmaceuticals are recognised as priority pollutants (PPs) due to their ubiquity and their capacity of affecting aquatic organisms also when present in trace concentrations. In this context, the development of new water treatment technologies is strongly needed in order to satisfy quality requirements for safe water reuse and environmental protection.

This opening chapter aims at exploring the latest developments of nanotechnology and nanomaterials with focus on potentials and limitations for their application to the treatment of water and wastewater. To this aim, a literature survey has been conducted by searching the Science Direct database for simple keywords in title and abstract of publications on scientific journals. Although the method was not very accurate and exhaustive, the results in Figure 1.2 are clear evidence that most of the research on nanotechnologies and nanomaterials for water treatment are applied to clean and potable water rather than wastewater treatment. When focusing on the applications to wastewater treatment, Figure 1.2b shows that the studies on nanofiltration technology (NF) have reached maturity and other nanotechnology application (such as catalysis, adsorption and disinfection) are rapidly growing during the last 7 years.

The following main topics could be selected from the literature survey:

- New materials for membrane filtration;
- Nanomaterials for catalysis and photocatalysis;
- Nanomaterials for water disinfection;
- Nanomaterials for adsorption of pollutants;
- Nanoscale zero-valent iron.

The applications of nanotechnology and nanomaterials in the main topics listed above attract most of the interests in the R&D sector and will be briefly discussed below, with particular emphasis to their application to water/wastewater treatment.

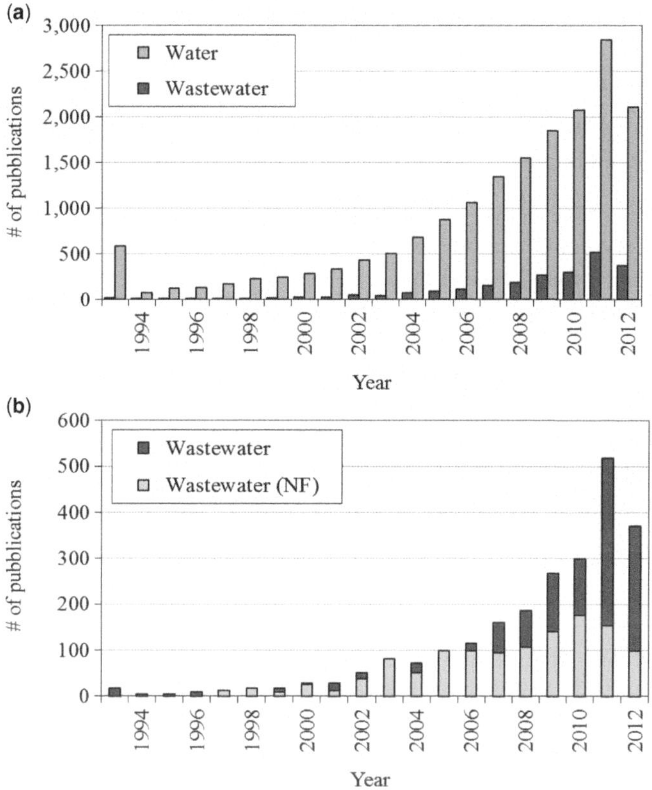

Figure 1.2 Literature scan for scientific journal publications containing nano- and water/wastewater in title, keywords and abstract (a) and specific focus on wastewater and nanofiltration (b) (Data from Science Direct, June 2012).

1.3 OVERVIEW OF EXISTING APPLICATIONS AND CURRENT TRENDS

1.3.1 New materials for membrane filtration

Many scientists believe that nanotechnology could possibly bring revolutionary advancements to the desalination industry, although the development of such membranes is only in the initial stages and many problems are yet to be overcome (Lee *et al.*, 2011). According to the National Geographic Magazine (Lange, 2010), three technologies promise to reduce the energy requirements for desalination of up to 30%:

- Forward osmosis (specific membranes are on the market and full scale processes are being tested at various locations in the world);

- Carbon nanotube composite membranes (expected on the market on 2013-15);
- Biomimetic membranes (expected on the market by the end of 2013).

Nanotechnology offers a wide range of solutions for novel membrane materials, including (Lee *et al.*, 2011; Ng *et al.*, 2013):

- Ceramic membranes for nanofiltration;
- New polymeric membranes with anti-fouling coating:
  - Organic brush-like coating
  - Membrane impregnated with nanoparticles
- New composite membranes:
  - Thin film composite membranes
  - Metal/metal oxides + Polymer
  - Carbon Nanotube + Polymer
  - Zeolites + Polymer
  - Aquaporin + Polymer

The ceramic membranes are mostly made from aluminum, silica, titanium, zirconium or a mixture of these materials. They are generally produced in multilayers with different pore sizes. They are characterized by very high resistance to extreme operating conditions (pH, temperature, flux, backwash intensity) and to the presence of oxidizing agents and UV light. Their surface can be modified with photocatalytic oxides and organic capping (for disinfection and fouling reduction). However, the cost of ceramic membranes is not sustainable for large scale water and wastewater treatments and they found application where polymeric membranes cannot be used (high temperature, radioactive/heavily contaminated feeds, and reactive environments).

Silver and titanium-based nanoparticles are ideal for incorporation into the membranes in order to reduce the biofouling of polymeric membranes (Ng *et al.*, 2013), and silver nanoparticles already found application in composite membranes for point-of-use water filtration/disinfection (as better described in the following sections).

Zeolites, carbon nanotubes and aquaporin promise to provide very high permeability. Although the increased permeability of zeolite thin film nanocomposite membranes is the lowest when compared with membranes incorporating carbon nanotubes (CNT) and aquaporins (AQP), they appear to be more readily adaptable to commercial use due to their similarity to current commercial RO membranes (Lee *et al.*, 2011). On the other hand, there are still several challenges in the production of composite membranes based on carbon nanotubes and aquaporin. In CNT-composites membranes, improvements are required both in tip-functionalization for more efficient salt rejection and preparation of aligned arrays of nanotubes. In the AQP based composite membranes, the rejection layer is formed by aquaporin proteins incorporated and immobilized into an ultrathin amphiphilic polymeric layer to mimic the natural membrane cell (see Chapter 14). A porous support layer is then added to one or both sides for mechanical support. Many practical issues need to be addressed to develop the AQP-composite, such as identification of appropriate support materials and optimization of the production and incorporation of the aquaporin protein. Moreover, for all these types of membranes, costs, specific rejection and long term operational stability are unknown (Lee *et al.*, 2011; Subramani *et al.*, 2011).

Pendergast and Hoek (2011) suggested that the new membranes based on nanotechnologies can be grouped in three categories: nanostructured ceramic, inorganic-organic and biologically inspired membranes. They also developed a ranking methodology to evaluate the new membranes over the current state-of-the-art, based on performance enhancement (permeability, selectivity, robustness) and commercial readiness (anticipated material costs, manufacturing scalability, and apparent time to commercialization). They concluded that biologically inspired membranes promise the greatest potential performance enhancements but are farthest from commercial reality, while zeolite membranes offer moderate performance enhancement and appear nearest to commercial viability (Figure 1.3). None of the membrane nanotechnologies fell in the optimal (upper right) quadrant of the chart, but this could change over time as biologically inspired membrane technology matures.

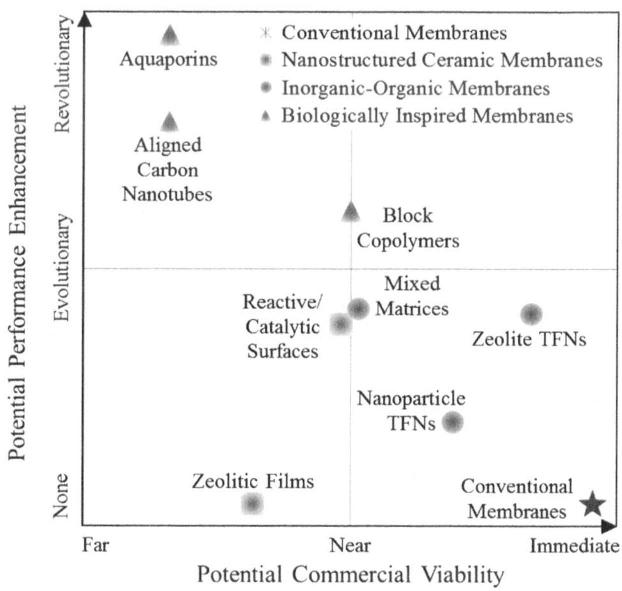

Figure 1.3 Comparison of the potential performance and commercial viability of nanotechnology-enabled membrane advances (adapted from Pendergast and Hoek, 2011).

The introduction of these new materials is turning in renewed interest for forward osmosis (Figure 1.4). Forward osmosis or simply said osmosis is a natural phenomenon based on the different chemical potential arising from the difference in solute concentrations across a selectively permeable membrane. As a result of this difference, a net movement of water across the membrane can be observed until osmotic and hydraulic pressures reach the equilibrium. In an engineered FO, a suitable draw solution allows to extract the water from a feed solution (e.g. sea water or treated wastewater). Then, conventional reverse osmosis (RO) or thermal processes are required to re-concentrate the dilute draw solution and recover the fresh water (Cath *et al.*, 2006). Figure 1.5 shows an example of an FO desalination process based on volatile draw solutes.

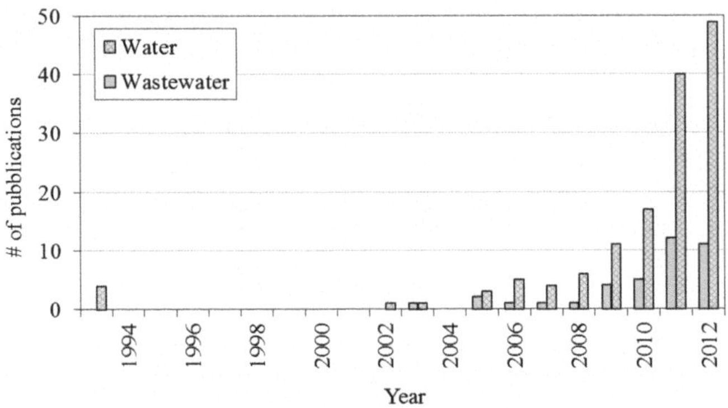

Figure 1.4 Number of publications related to the study of forward osmosis for water and wastewater purification (Data from Science Direct, June 2012).

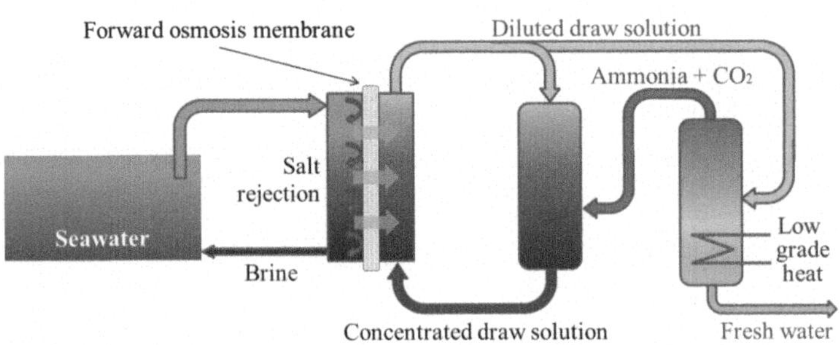

Figure 1.5 Example of application of forward osmosis for desalination with volatile draw solutes.

It has been estimated that the cost of these new treatment schemes can be considerably lower than that of conventional RO (Subramani *et al.*, 2011). Energy savings can be especially significant when low grade thermal energy is locally available (e.g. from industrial productions, thermal power plants and geothermal systems), however detailed cost studies are needed.

FO membranes can also be used for wastewater recovery and production of fresh water at a cost closer to that required for brackish water desalination. Figure 1.6 shows an example of the use of forward osmosis in a dual barrier system for desalination of treated effluents (Cath *et al.*, 2010).

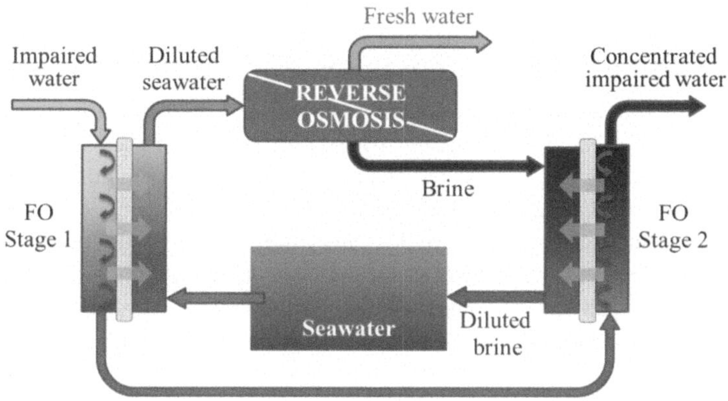

Figure 1.6 Dual barrier concept for desalination of treated effluents.

Besides its applications in desalination and water reuse, FO can be applied in pressure-retarded osmosis (PRO) for power generation (Achilli *et al.*, 2009). In a PRO process, the osmotic pressure difference between seawater/concentrated brines and fresh/recycled water is converted in hydrostatic pressure and then in electricity. The idea is not new (Loeb, 1975), but the commercialization of new efficient pressure exchangers and the development of new membranes are renewing the interest in PRO.

Further developments for wastewater treatment processes are the forward osmosis membrane bioreactor (Cornelissen *et al.*, 2008; Achilli *et al.*, 2009). In these systems, the biological treatment of the wastewater is directly coupled to the FO filtration for effluent withdrawal and recovery. As for the typical MBR based on micro and ultrafiltration membranes, different plant configurations depend on the membrane geometry, resulting in external or submerged FO membranes. Furthermore, the FO-MBR can be used for either clear water recovery or energy production. As in "conventional" MBRs, the main process drawbacks are related to membrane fouling control (Zhang *et al.*, 2012). Other aspects under investigation are salt back-diffusion and the effect of salinity accumulation in the bioreactor tank (Lay *et al.*, 2010).

Specific membranes need to be designed for application of forward osmosis and the two major practical challenges are the high cost of nano-structured materials, and the difficulty in scaling up nano-membrane manufacturing processes for commercial use. However, some products are commercially available (Hydration Technology Innovation, Albany, USA) and new developments are expected based on new available materials as reported above. As commercialization and standardization of FO membranes are expected to be developed in the coming years, capital costs can become competitive in short to medium horizon. This offers the best opportunity for the development and application of the osmotic technology for industrial practices.

1.3.2 Nanomaterials for catalysis and photocatalysis

Nano-materials are more effective than conventional catalysts for two reasons: their extremely small size (typically 10–80 nm, with consequent high surface area-to-volume ratio) and the higher reactivity related to the nanoscale size itself (Chaturvedi *et al.*, 2012). The study of this last aspect (i.e. understanding how decreasing the size of catalytic particles alters the intrinsic catalytic performance beyond simply expanding surface area), and the design/preparation of catalysts with the most effective size and structure are current objectives in catalysis research. This research – and heterogeneous catalysis in particular – has the potential to be one of the most important and productive areas of nanoscience and technology in the next decade (Nature, 2008).

Nanosized catalysts can be effectively used for chemical oxidation of organic and inorganic pollutants in water in advance oxidation processes (AOP). These processes are based on the in-situ formation of highly reactive radicals, which quickly react with target compounds or oxygen (to form hydroxyl radical and other oxygenated reactive species). The application of AOP is often limited by prohibitive costs resulting from the adoption of energy intensive technologies (UV lamps, ozonators, ultrasonicators, etc.) and the presence of hazardous chemicals (ozone and hydrogen peroxides).

An unaccountable number of nanomaterials have been developed for application in photocatalysis, thanks to the specific properties of nanosized materials. A photocatalytic reaction is activated by adsorption of a photon with energy greater than the band-gap energy of a semiconductor. When a photon illuminates the catalyst surface, an electron is transferred to the empty conduction band, leaving a positive hole in the valence band (Herrmann, 1999). The electron-hole pair can quickly recombine releasing thermal energy (and reducing the photoefficiency), or can be involved in redox reactions at the catalyst surface. Photons with wavelengths in the range of UV or visible light are usually required, depending on the bandgap energy that characterizes each photocatalyst. Great research efforts have been dedicated to the modification of the native bandgap of nanosized photocatalysts, by introducing metallic and non-metallic atoms into the crystal lattice (doping), in order to reduce the bandgap (and so shifting the adsorption spectrum towards the visible light range) or prevent the recombination of electron-hole pairs (Chong *et al.*, 2010).

The application of solar photo-catalytic processes based on newly developed nanomaterials can open new opportunities for the development of low-cost integrated processes suitable to reach the high quality required for stormwater and wastewater reclamation and reuse (see Chapter 8). Also in case of photocatalytic processes based on artificial UV light, the operation in mild conditions and the absence of chemicals other than the reusable photocatalyst make the technology attractive and environmental friendly. For these reasons, the research and testing of photocatalytic nanomaterials is turning in a great number of publications (Figure 1.7).

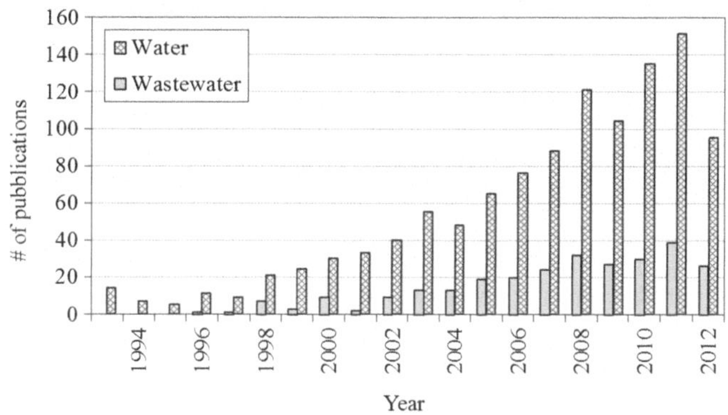

Figure 1.7 Number of publications related to the application of photocatalysis for water and wastewater treatment (Data from Science Direct, June 2012).

Two main process configurations are usually proposed for this heterogeneous catalytic process, which differs in the presence of suspended or immobilized nanoparticles (Chong *et al*., 2010). The adoption of slurries of suspended catalyst typically results in better light distribution and higher reactive surface, while the immobilized catalyst usually suffers from mass transfer limitations. The need for catalyst post-separation is the main drawback of the slurry reactors, although they offer greater flexibility and easier recovery/substitution of the catalyst.

A huge number of nanomaterials and preparation methods have been proposed for photocatalytic applications, employing different semiconductor catalysts (TiO_2, ZnO, Fe_2O_3, CdS, GaP and ZnS). However, common limitations have been found for the materials and the processes:

- Charge recombination (with consequent reduction of the light efficiency)
- Visible light transparency (although strong effort has been put in catalyst doping)
- Colloidal instability and fouling (potential occurrence and extent in real matrix is mostly unknown)

- Catalyst recovery (in case of slurry reactors)
- Low activity/light distribution (in case of supported catalysts).

Among the semiconductor catalysts, titanium dioxide (TiO_2) has received the greatest interest in R&D of photocatalysis technology, and successful removal of a broad range of organic molecules has been demonstrated (Chong *et al.*, 2010). Although TiO_2 has the disadvantage of being activated by UV light, in comparison with other nanosized photocatalysts, it offers the advantage of being an efficient catalyst, chemically and biologically inert, photocatalytically stable and easy to produce.

In order to improve visible light adsorption and develop solar photocatalytic processes, TiO_2 doping with heavy metals and non-metals has been largely investigated during the last decade (Han *et al.*, 2009), among which nitrogen doping seems to be the most promising.

In general, few studies are available that address the issue of colloidal instability due to salinity and the presence of humic substances or other dissolved organic and inorganic compounds that can be found in real water and effluents (Laera *et al.*, 2011; Kashif & Ouyang, 2009; Zhang *et al.*, 2009a).

Moreover, in applications for natural water and treated effluents, the influent turbidity is a limiting factor for any photocatalytic process. It is suggested that water with less than 5 NTU can be suitable for photocatalytic degradation, while 30 NTU is the limit for water disinfection (Chong *et al.*, 2010). The use of sonication instead of UV light has been recently proposed to overcome the turbidity issue in water and wastewater application (Pang *et al.*, 2011).

An interesting preparation of TiO_2 nanotubes was recently proposed (Okour *et al.*, 2010; Shon *et al.*, 2009), starting from sludge obtained during the flocculation of effluent from biological treatment of municipal wastewater (or directly municipal wastewater) with titanium salts, after drying and incineration. The above procedure could also be modified to produce thiourea-doped nanotubes, which showed high photocatalytic activity under visible light (Okour *et al.*, 2010).

Other interesting nanosized photocatalysts are based on zinc oxide (Chiu *et al.*, 2010), as this material is cheap and the active catalyst can be easily prepared in mild conditions and also as nanorods and nanofibers (with consequent simplified recovery). However, zinc oxide applications in water treatment could be limited by its easy dissolution and the environmental toxicity of dissolved zinc.

1.3.3 Nanomaterials for water disinfection

Several natural and engineered nanomaterials have been shown to have strong antimicrobial properties (Li *et al.*, 2008), including:

- chitosan
- silver nanoparticles (nAg)
- photocatalytic TiO_2

- fullerol,
- aqueous fullerene nanoparticles (nC60)
- carbon nanotubes (CNT)

As these antimicrobial nanomaterials are mild oxidants and are relatively inert in water, they are not expected to produce harmful disinfection by-products. Therefore, they have the potential to replace or enhance conventional disinfection methods, if properly incorporated into conventional treatment processes and in decentralized or point-of-use water treatment and reuse systems.

Several mechanisms have been proposed for the disinfection using nanoparticles, though in most cases the effective mode of action is still under investigation. In general, the disinfection with photocatalytic nanoparticles involves the production of hydroxyl radicals. Other disinfection mechanisms are related to direct damages to the cell membrane and DNA, or to interference with metabolic processes, that is interruption of electron transport and protein oxidation (Li *et al.*, 2008).

Silver nanoparticles are supposed to act by altering the membrane permeability, penetrating the cell and damaging the cell DNA, with significant dependence on size and shape of the nanoparticles. Furthermore, it is often suggested that a relevant antimicrobial action is related to the dissolution and release of silver ions.

The antibacterial activity of titanium dioxide is related to the production of hydroxyl radicals and peroxide. Enhancement of photocatalytic activity and visible light adsorption by doping also improves the inactivation of bacteria and viruses. In particular, doping TiO_2 with silver appears to be very promising for visible light disinfection.

Fullerenes and carbon nanotubes show good antibacterial activity, usually related to the induction of oxidative stress on the cell. However, they are only effective when properly dispersed in the aqueous media, by coating or surfactant utilization. They could find application for surface disinfection and fouling control, after their immobilization on membrane surfaces.

The main limitations identified for the application of nanomaterials in water and wastewater disinfection can be summarized as follows:

- the disinfection process often requires that the catalyst must be in contact with the cell membrane surface for the microbial inactivation to be successful;
- recovery of dispersed catalyst is difficult;
- no residual antimicrobial action is left in the water.

Possible promising applications of nanomaterials for water disinfection can be the use of supported catalyst in filters, providing a self-cleaning/self-sterilizing filtration unit for point-of-use applications. Membranes impregnated or coated with silver nanoparticles can already be found in commercial systems for home water purification (Li *et al.*, 2008). Applications of TiO_2 in enhanced UV disinfection have also been tested for improved virus removal and organic contaminant degradation.

Furthermore, the activation of TiO_2-based nanocatalyst by solar UV-A make this technology interesting for enhancing solar disinfection for emergency and point of use applications. Jafry *et al.* (2011) demonstrated that surface treatment of commercial TiO_2 (P25) with silica results in enhanced photocatalytic inactivation of viruses. Following these studies, a development of the SODIS method (the solar disinfection with plastic bottles promoted by WHO for point-of-use disinfection in developing countries) is currently being studied at the Rice University, with the goal of reducing the required exposure time from 6 hours (in bottles with no catalyst) down to 15 minutes when the modified photocatalyst is incorporated (Ruth, 2011).

1.3.4 Nanomaterials for pollutant adsorption

1.3.4.1 Carbon based nanomaterials

Adsorption of pollutants based on activated carbon is a traditional and widely diffused nanotechnology. Activated carbon (AC) is an amorphous material with microcrystalline structure and surface area in the order of 400–2000 m^2 per gram. The great porosity and specific surface area of AC is obtained in a two–step process, in which pyrolysis is followed by a thermal or chemical oxidative activation. Most of the active surface can be related to nanosized pores, with diameters in the range of 2–50 nm. Besides the presence of pores, the reactivity of AC is due to the presence of reactive species, such as resonating free radicals, oxygenated species and attached heteroatoms. The two main commercial forms of AC, that is powder (PAC) and granular (GAC), are extensively used in water treatment and are often considered best available technology (BAT) for both organic micropollutants and heavy metals.

A new class of carbon based adsorbent has been introduced with the discovery of carbon nanotubes (CNT) in the nineties. CNT can be synthesized as single- or multiple-walled nanotubes, consisting of graphitic shells arranged in tubes with diameters of a few nm and large length/diameter ratio. In comparison with AC, CNT offers similar high specific surface areas with the advantage of a well-defined and controlled structure. Indeed, the adsorption properties of the CNT can be modulated by controlling various structural factors (Ren *et al.*, 2011):

- The presence and amount of open end nanotubes can have a great influence, as these offer more adsorption sites, including the interior walls of the nanotubes and the internal axial sites.
- The presence of impurities, such as soot and other forms of carbon, typically reduce the adsorption performances.
- The presence of oxygenated functional groups, such as alcohol, carbonyl and carboxylic groups, usually results in enhanced adsorption performances. These functional groups can be formed during CNT preparation and purification or can be intentionally introduced with oxidizing post-treatments. The presence

of these polar moieties results in direct interaction with polar compounds as well as in increased wettability of the CNT.

- Doping with metal and non-metal atoms can change the CNT adsorption properties by influencing the electron density and potential of the carbon cage.

The adsorption of heavy metals from aqueous solutions mainly depends on pH of the solution and on the point of zero charge of the CNT (pH_{PZC} usually in the range of pH 4-6 for as-produced CNT). On the other hand, the adsorption of organic pollutants appears to be more influenced by CNT porosity and pore size distribution, and by the presence of functional groups that affect hydrophobic interactions and π–π bonds.

As also observed by the same authors (Ren *et al.*, 2011), it is important to note that the stability and properties of CNT suspensions are strongly influenced by the presence of natural organic matter, which can be adsorbed onto the CNT surface and change their physicochemical properties. As these substances are widely distributed in the environment, it is important to investigate the effectiveness of CNT adsorption processes in the real matrix. Furthermore, current research is mostly focused on the adsorption of a single solute, while complex mixtures of organic and inorganic contaminants coexist in environmental samples and wastewaters. This results in competition for the active sites on the CNT surface and interior. Hence, more research efforts are required in order to develop practical applications of these new materials.

1.3.4.2 Metal based nanomaterials

Nanosized metal oxides (NMO) have been widely explored as adsorbents for both heavy metals and organic contaminants, thanks to their large surface, capacity and selectivity (Hua *et al.*, 2012; Xu *et al.*, 2012). The adsorption properties of these NMO also depend on the size and shape and specific synthesis methods have been developed and optimized during the last decade.

Iron is one of the most diffused elements on the earth and is considered environmental friendly. Therefore, the use of iron oxide nanoparticles is considered a safe and cost-effective technology and different nanosized ferric oxides are under development. Ferric oxide nanomaterials show superior performances in the adsorption of heavy metals, and the adsorption process is often insensitive to the background electrolyte and to the ionic strength. Nanosized maghemite and magnetite can be easily separated by applying an external magnetic field and are characterized by high selectivity and good adsorption capacity for Cr(VI). Thanks to their easy post-separation, these nanosized ferric oxides are also proposed as magnetic cores for composite adsorbents. Efficient treatment of large volumes and fast separation via employing strong magnetic fields is also being explored for organic contaminants adsorption (Xu *et al.*, 2012), both as native nanoparticles and after surface modification (e.g. carbon coated Fe_3O_4).

Nanosized manganese oxides show a good adsorption potential for heavy metals, arsenate and phosphate. Mixed-valence manganese oxides form tunnel structures with a central cation (K^+, Mg^{2+}, Ca^{2+}), which act like ion exchangers. As different tunnel structures result in different selectivity towards metal ions, mixed-valence Mn oxides are considered a large class of selective ion exchangers and sorbents with potential interesting applications for the removal of heavy metals and radionuclides (Dyer *et al.*, 2000).

Similar to iron-NMO, nanosized zinc oxide is cheap, versatile and environmental friendly. Zinc oxide nanoparticles can be conveniently prepared in tailored morphologies, among which nanoplates show high removal efficiencies for heavy metals (Wang *et al.*, 2010).

Good performances in heavy metals removal are also shown by other nanosized metal oxides (based on aluminium, titanium, magnesium and cerium), with some nanomaterials showing also excellent removal efficiencies for Cr(VI) either by simple adsorption (magnesium-NMO) or by adsorption/reduction process (cerium-NMO). Furthermore, TiO_2 based nanomaterials have also been proposed as promising technology for arsenic removal based on oxidation/adsorption processes (Guan *et al.*, 2012).

1.3.5 Nanoscale zero-valent iron

The dosage of zero-valent iron in aqueous media invariably results in different corrosion reactions with dissolved oxygen and water, leading to the formation of ferrous and ferric iron, molecular hydrogen, hydroxyl ions and various precipitates. These corrosion reactions and the products thereof are then responsible for the reductive transformation and/or physical removal of the exposed chemical species (Crane & Scott, 2012). Depending on the specific contaminant chemistry, numerous removal mechanisms have been determined, including sorption, complexation, co-precipitation and surface-mediated chemical reductions.

Following the first application by Wang and Zhang (1997), the effectiveness of nanoscale zero-valent iron (nZVI) has been proved for a wide range of organic and inorganic contaminants, including chlorinated solvents and pesticides, azo-dyes, nitroamine and nitroaromatics, organophosphates, inorganic anions, arsenic, uranium and various metals (see Crane & Scott, 2012, for an exhaustive list of applications and references). Several research studies have been devoted to the improvement of dispersion (coatings and surfactants) and reactivity (bimetallic nanoparticles and thermal treatments).

In virtue of its relatively low cost and environmental compatibility, nZVI has been widely proposed for in-situ remediation of soil and groundwater. Several patents and full scale applications have already demonstrated the potential of these nanoparticles, although the performances obtained in lab-scale tests are usually found to overestimate the effectiveness in real environments.

Clean production, reactivity improvements, optimal dispersion and use of green dispersant, long term mobility and potential toxicity are among the aspects still open to the research.

On the other hand, the technology for in-situ application has been well developed, with different options depending on local geology and geochemistry. According to Tratnyek and Johnson (2006), the nZVI injection method can be adapted to the contamination mobility (Figure 1.8) and three approaches to application of Fe particles for groundwater remediation can be distinguished: (a) a conventional 'permeable reactive barrier' made with millimeter-sized granular iron, (b) a 'reactive treatment zone' formed by sequential injection of nZVI to form overlapping zones of particles adsorbed to the grains of native aquifer material and (c) treatment of nonaqueous phase liquid contamination by injection of mobile nanoparticles.

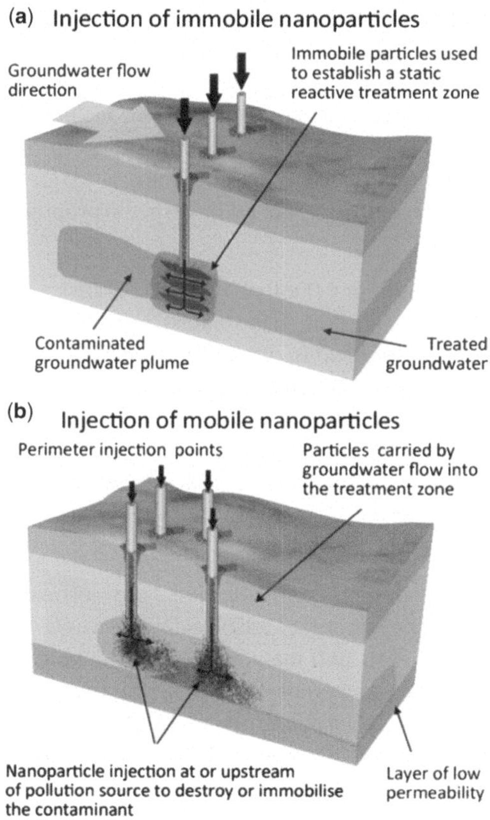

Figure 1.8 In-situ remediation technology with nZVI (adapted from Crane and Scott, 2012).

The use of nanoscale zero-valent iron also requires a detailed legislative framework, which has already been developed in North America and some European countries (Crane & Scott, 2012). The wide acceptability of this technology relies on a clear understanding of the nZVI behaviour in subsurface environmental systems. Further research is required for providing a sound body of scientific evidence upon which legislative decisions may be made.

1.4 PRACTICAL ASPECTS

1.4.1 Technical developments for direct applications of nanoparticles in water treatment

Nanomaterials promise excellent performances in catalysis, photocatalysis and adsorption processes for water and wastewater purification. This evaluation is mostly based on the results of fundamental research with stable dispersion of nanoparticles and in strictly controlled operating conditions, in which the unique chemical and physical properties of nanosized materials can be fully exploited. Most of the applications of NM for water and wastewater treatment are still at the early experimental or pilot stage.

For the development of practical applications, some technical bottlenecks need to be solved, among which difficult post-separation and performance losses due to particle agglomeration in real matrixes are common.

In photocatalysis, different approaches have been proposed to overcome the post-separation issue:

- use of filtration membranes (Doll & Frimmel, 2005);
- preparation of the catalyst in nanoscale fibrils (i.e. nanotubes, nanorods, nanofibers and nanowires), in which 1 dimension in the range of mm, allowing for easy post-separation by settling or conventional filtration (Chong *et al.*, 2010; Huang *et al.*, 2011);
- preparation of supported catalysts and nanocomposite materials, that is photocatalyst supported on glass, zeolites, clay, activated carbon (Chong *et al.*, 2010; Zhang *et al.*, 2009b). In most cases, these nanocomposites act both as adsorbant and photocatalyst.

Ultracentrifugation and membrane filtration are efficient techniques for post-separation of carbon nanotube adsorbants, however their application is limited by high operating costs and membrane pore blocking (Ren *et al.*, 2011).

Similarly, supported nanomaterials are often proposed for the application of nanosized metal oxide in adsorption processes, both for fast recovery from dispersed solutions and for reducing head losses in flow-through systems (Hua *et al.*, 2012). Hybrid adsorbents are obtained by impregnating or coating nanosized metal oxides on porous supports of larger sizes, including natural hosts (e.g. bentonite, zeolite, and sand), metal oxides (e.g. iron, aluminum and manganese oxides), polymeric materials (e.g. ion-exchange resins), activated carbon and sewage sludge (Hua *et al.*, 2012; Zhao *et al.*, 2011; Phuengprasop *et al.*, 2011). The adsorption capacities of the composite materials are always enhanced by the presence of the nanoparticles, although higher sensitivity to pH and ionic strength of the treated solution are often reported.

Another approach to solve both aggregation and post-separation problems is based on the production of magnetic composite sorbents with ferric oxides cores (Hua *et al.*, 2012; Xu *et al.*, 2012). Two different strategies can be identified, one in which the nanosized ferric oxide is coated with organic functional groups, such as chitosan, humic acid, polyacrilic acid and other polymers that can be further functionalized (e.g. with amino-groups). In the other approach, the nanosized ferric oxide is supported on porous materials, such as zeolite, mesopore molecular sieves, carbon nanotubes and graphene nanosheets.

Also in case of CNT adsorbants, magnetic separation is proposed as an effective solution and CNTs/magnetic composites are considered promising materials for the treatment of large amounts of water/wastewater (Ren *et al.*, 2011). In addition, many researchers are currently working on the surface modification of CNTs aimed at minimizing their aggregation in aqueous solutions, thus enhancing their adsorption properties. However, only plasma techniques are envisaged as a promising environmental friendly approach for surface modification and functional group introduction, as conventional chemical methods imply the usage of large amounts of chemicals, with consequent environmental pollution.

Finally, the regeneration of nanomaterials employed in adsorption of heavy metals is an important feature for evaluating their repeatability in use and the possibility of recovering valuable metals from the eluates. As evidenced by Hua *et al.* (2012), this kind of studies is very limited and more attention should be paid to the topic in order to promote practical application of these environmental nanotechnologies.

1.4.2 Costs and performance

The development of nano-composite membranes for reverse and forward osmosis is still in its early stages and the competition with commercial polymer membranes also passes through the production costs. As suggested by Lee *et al.* (2011), besides the very expensive production of zeolite nanoparticles, carbon nanotubes and aquaporin, additional cost gaps must be considered in relation to:

- energy for homogeneous dispersion of zeolites and CNT into the polymeric matrix
- cost of CNT tip functionalization
- packing density and supporting materials for aquaporin incorporation in biomimetic membranes

On the other side, these new membranes offer higher permeability and their application would result in reduced energy consumption and smaller plant footprint, with consequent reduction of operational and capital costs. Further capital cost savings could be obtained by designing single pass RO desalination, thanks to the better salt rejection performances, especially demonstrated by CNT and aquaporin composite membranes.

The cost of photocatalytic treatments is related to both the catalyst production and the specific energy required for the light source. Despite the great efforts in development of nanomaterials for solar photocatalysis, its application to wastewater treatment is contrasted by the huge surfaces required for the collection of sunlight. As a consequence, pilot and demonstrative applications have only been reported for specific locations, such as the Plataforma Solar de Almeria (Spain), where the surface availability is contrasted by the great distance from densely populated and industrial areas where the energy savings related to solar technologies could be more effective. On the other hand, photocatalytic processes based on nanomaterials could be conveniently applied in case of selective reactions for the removal of specific pollutants (e.g. pesticides and other toxic or refractory compounds) or in combination with existing technologies (e.g. for the improvement of conventional UV disinfection).

Similarly, the successful diffusion of adsorption technologies based on nanomaterials is mostly related to their specificity, rather than low cost of the nanoparticles production. As reviewed by Ren *et al.* (2011), much progress has been made over the last few years in adsorption applications of CNTs. The use of CNT nanosorbents may be advantageous in the future, as the higher adsorption capacities of CNTs compared to other media may offset their high cost. Besides CNT production, also their surface modification must be taken into account and the cost for plasma or conventional chemical techniques cannot be easily quantified at the moment.

As in many cases the technological development is still at the material development phase, it is hard to find treatment cost evaluations and comparative studies in which the performances of nanomaterials are compared with those of other conventional technologies. Hence, it is not clear at the moment which technologies and applications will find their way to the market.

1.4.3 Toxicity, fate and transport of nanomaterials

In almost all applications to water treatment, retention of the nanomaterials is critical not only because of the costs associated with their loss, but more importantly because of their potential impacts on human health and ecosystems (Li *et al.*, 2008).

Considering the use of TiO_2 nanoparticles in toothpaste and sunscreens, its potential ingestion is not of concern. However, it has been shown that TiO_2 nanoparticles could facilitate arsenic accumulation in fish by adsorbing the toxin and transporting it inside the animal (Sun *et al.*, 2007).

Similarly, no clear evidence of nano-Ag toxicity to humans has been described. However, their toxicity to a wide range of organisms has been extensively reported

(Levard *et al.*, 2012). Although the toxicity of Ag-NPs is partly explained by the release of Ag^{+} ions, it remains unclear if Ag-NPs are a direct cause of enhanced toxicity and to which extent the formation of silver sulphide at the nanoparticle surface contributes to mobility and toxicity reduction.

Some nanomaterials such as fullerenes and zinc oxide have shown to exhibit some toxicity to mammalian cells. However, the available information is not sufficient to determine the highest allowable concentration in drinking water. This could limit the application of these nanomaterials in large scale water treatment in the near future.

Most of the studies on catalytic properties and interactions with microbial activity have been conducted in simple and clean solutions (Sanchez *et al.*, 2011). In natural water the activity and environmental toxicity of nanoparticles could be limited by adsorption, coating and aggregation. On the other hand, important surface modifications can be induced by exposure to sunlight (Turco *et al.*, 2011; Aiken *et al.*, 2011).

As discussed by Brar *et al.* (2010), the dispersion of nanoparticles can be improved by surface modification with polyelectrolytes. Considering that the latter compounds find applications in a wide range of technologies, including in wastewater treatment plants, nanoparticles in wastewater can interact with polyelectrolytes and completely change their aggregation and reactivity, their potential dispersion in the effluent as well as their mobility and leaching from disposed sludge. Although there are various studies and clear indications that most of the nanoparticles entering a wastewater treatment plant will end up in the excess wasted sludge, the wastewater sludge contamination by nanoparticles is a completely neglected domain (Brar *et al.*, 2010). Furthermore, the stability and properties of both nanoparticle aggregates and nanoparticles adsorbed onto sludge flocs depend on ionic strength and specific environmental conditions. Considering the transfer among different sections of the wastewater treatment plant and to the soil and aquatic environment, the sludge and the nanoparticle aggregates may encounter different redox zones and solution chemistry. Hence, the fate of different specific nanoparticles can be very difficult to predict.

Nanoparticles-based technologies could be profitably adopted for the removal of micropollutants and other tertiary treatments aimed at improving the treated water quality for industrial and irrigational reuse. Also in this case, it is very important to minimize the loss of nanoparticles from the designed process and to understand the consequence of their release, as reused water can be an effective vehicle for diffusion of nanoparticles into soil and natural aquatic ecosystems.

Once nanoparticles end up in the soil, either via biosolids disposal or water reuse, the interaction with plants and crops can have positive or negative impacts. Although some studies refer of improved growth, possibly due to impacts on the nitrogen cycle, other toxicology studies reveal how the presence of nanoparticles can affect the root growth and sometimes also the germination phase (Brar *et al.*, 2010). However, most of the studies refer to virgin nanoparticles, whereas surface modifications and aggregates are expected to play a major role in real conditions. Hence, no concrete conclusions can be drawn with respect to mobility and impact of nanoparticles in soils.

There are no studies about the potential release of nanoparticles to the atmosphere during transport and disposal of dried sludge, or about the fate of nanoparticles in sludge disposed trough landfilling or incineration. Hence, it is neither clear if the nanoparticles can desorb and be released with the landfill leachate nor if they can be destroyed during incineration.

The lack of clear and sound knowledge of the behaviour and fate of nanoparticles in engineered and natural environmental conditions may result to be the main obstacle to the diffusion of nanoparticles-based technologies for water treatment and environmental remediation.

1.5 CONCLUDING REMARKS

After considerable efforts towards the application of nanomaterials in the treatment of water and wastewater, significant limitations still remain:

- More research is required to retain nanomaterials in the reactors. Better nanomaterials retention would result in reduced costs, minimized environmental impacts and limited harm for human health.
- Beside technical issues to be solved, better cost-benefit analysis are required for evaluating practical applicability to the water and wastewater treatment.
- Low cost and abundant materials should be preferred for long term and sustainable applications.
- Better knowledge is needed regarding the fate and effects of nanoparticles in real environmental conditions during water and wastewater treatment, as well as in soil and natural aquatic compartments.

REFERENCES

Achilli A., Cath T. Y., Marchand E. A. and Childress A. E. (2009). The forward osmosis membrane bioreactor: a low fouling alternative to MBR processes. *Desalination*, **238**, 10–21.

Aiken G. R., Hsu-Kim H. and Ryan J. N. (2011). Influence of dissolved organic matter on the environmental fate of metals, nanoparticles, and colloids. *Environ. Sci. Technol.*, **45**, 3196–3201.

Brar S. K., Verma M., Tyagi R. D. and Surampalli R. Y. (2010). Engineered nanoparticles in wastewater and wastewater sludge – evidence and impacts. *Waste Manage.*, **30**, 504–520.

Cath T. Y., Childress A. E. and Elimelech M. (2006). Forward osmosis: principles, applications, and recent developments. *J. Membr. Sci.*, **281**, 70–87.

Cath T. Y., Hancock N. T., Lundin C. D., Hoppe-Jones C. and Drewes J. E. (2010). A multi-barrier osmotic dilution process for simultaneous desalination and purification of impaired water. *J. Membr. Sci.*, **362**, 417–426.

Chaturvedi S., Dave P. N. and Shah N. K. (2012). Applications of nano-catalyst in new era. *J. Saudi Chem. Soc.*, **16**(3), 307–325.

Chiu W. S., Khiew P. S., Cloke M., Isa D., Tan T. K., Radiman S., Abd-Shukor R., Hamid M. A. Abd., Huang N. M., Lim H. N. and Chia C. H. (2010). Photocatalytic study of two-dimensional zno nanopellets in the decomposition of methylene blue. *Chem. Eng. J.*, **158**, 345–352.

Chong M. N., Jin B., Chow C. W. K. and Saint C. (2010). Recent developments in photocatalytic water treatment technology: a review. *Water Res.*, **44**, 2997–3027.

Cornelissen E. R., Harmsen D., de Korte K. F., Ruiken C. J., Qin J. J., Oo H. and Wessels L. P. (2008). Membrane fouling and process performance of forward osmosis membranes on activated sludge. *J. Membr. Sci.*, **319**, 158–168.

Crane R. A. and Scott T. B. (2012). Nanoscale zero-valent iron: future prospect for an emerging water treatment technology. *J. Hazard. Mater.*, 211–212, 112–125.

Doll E. T. and Frimmel H. F. (2005). Cross-flow microfiltration with periodical back-washing for photocatalytic degradation of pharmaceutical and diagnostic residues–evaluation of the long-term stability of the photocatalytic activity of TiO_2. *Water Res.*, **39**, 847–854.

Dyer A., Pillinger M., Newton J., Harjula R., Moller T. and Amin S. (2000). Sorption behaviour of radionuclides on crystalline synthetic tunnel manganese oxides. *Chem. Mater.*, **12**, 3798–3804.

Han F., Kambala V. S. R., Srinivasan M., Rajarathnam D. and Naidu R. (2009). Tailored titanium dioxide photocatalysts for the degradation of organic dyes in wastewater treatment: a review. *Appl. Catal. A: Gen.*, **359**, 25–40.

Herrmann J.-M. (1999). Heterogeneous photocatalysis: fundamentals and applications to the removal of various types of aqueous pollutants. *Catal. Today*, **53**, 115–129.

Hua M., Zhang S., Pan B., Zhang W., Lv L. and Zhang Q. (2012). Heavy metal removal from water/wastewater by nanosized metal oxides: a review. *J. Hazard. Mater.*, 211–212, 317–331.

Huang J., Cao Y., Deng Z. and Tong H. (2011). Formation of titanate nanostructures under different NaOH concentration and their application in wastewater treatment. *J. Solid State Chem.*, **184**, 712–719.

Grimshaw D. (2009). Nanotechnology for clean water: Facts and Figures. http://www.scidev.net/en/features/nanotechnology-for-clean-water-facts-and-figures.html (accessed 7 July 2012).

Guan X., Du J., Meng X., Sun Y., Sun B. and Hu Q. (2012). Application of titanium dioxide in arsenic removal from water: a review. *J. Hazard. Mater.*, 215–216, 1–16.

Jafry H. R., Liga M. V., Li Q. and Barron A. R. (2011). Simple route to enhanced photocatalytic activity of P25 titanium dioxide nanoparticles by silica addition. *Environ. Sci. Technol.*, **45**, 1563–1568.

Kashif N. and Ouyang F. (2009). Parameters effect on heterogeneous photocatalysed degradation of phenol in aqueous dispersion of TiO_2. *J. Environ. Sci.*, **21**, 527–533.

Laera G., Jin Bo., Zhu H. and Lopez A. (2011). Photocatalytic activity of TiO_2 nanofibers in simulated and real municipal effluents. *Catal. Today*, **161**, 147–152.

Lange K. E. (2010). The big idea. *Natl. Geogr. Mag.*, 15 March 2010, http://ngm.nationalgeographic.com/big-idea/09/desalination (accessed 7 July 2012).

Lay W. C. L., Yu L. and Fane A. G. (2010). Impacts of salinity on the performance of high retention membrane bioreactors for water reclamation: a review. *Water Res.*, **44**, 21–40.

Lee K. P., Arnot T. C. and Mattia D. (2011). A review of reverse osmosis membrane materials for desalination – development to date and future potential. *J. Membr. Sci.*, **370**, 1–22.

Levard C., Hotze E. M., Lowry G. V. and Brown G. E. Jr. (2012). Environmental transformations of silver nanoparticles: impact on stability and toxicity. *Environ. Sci. Technol.*, **46**, 6900–6914.

Li Q., Mahendra S., Lyon D., Brunet L., Liga M. V., Li D. and Alvarez P. J. J. (2008). Antimicrobial nanomaterials for water disinfection and microbial control: potential applications and implications. *Water Res.*, **42**, 4591–4602.

Morose G. (2010). The 5 principles of 'Design for safer nanotechnology'. *J. Clean. Prod.*, **18**, 285–289.

Mrayed S. M., Sanciolo P., Zou L. and Leslie G. (2011). An alternative membrane treatment process to produce low-salt and high-nutrient recycled water suitable for irrigation purposes. *Desalination*, **274**(1–3), 144–149.

Nature. (2008). Catalysis by design. *Editorial, Nat. Nanotechnol.*, **3(10)**575, doi:10.1038/nnano.2008.285.

Ng L. Y., Mohammad A. W., Leo C. P. and Hilal N. (2013). Polymeric membranes incorporated with metal/metal oxide nanoparticles: a comprehensive review. *Desalination*, **308**, 15–33.

Okour Y., Shon H. K., El Saliby I. J., Naidu R., Kim J. B. and Kim J.-H. (2010). Preparation and characterisation of titanium dioxide (TiO_2) and thiourea-doped titanate nanotubes prepared from wastewater flocculated sludge. *Bioresour. Technol.*, **101**, 1453–1458.

Pang Y. L., Bhatia S. and Abdullah A. Z. (2011). Process behaviour of TiO_2 nanotube-enhanced sonocatalytic degradation of rhodamine B in aqueous solution. *Sep. Purif. Technol.*, **77**, 331–338.

Phuengprasop T., Sittiwong J. and Unob F. (2011). Removal of heavy metal ions by iron oxide coated sewage sludge. *J. Hazard. Mater.*, **186**(1), 502–507.

Pendergast M. T. M. and Hoek E. M. V. (2011). A review of water treatment membrane nanotechnologies. *Energy Environ. Sci.*, **4**, 1946–1971.

Ren X., Chen C., Nagatsu M. and Wang X. (2011). Carbon nanotubes as adsorbent in environmental pollution management: a review. *Chem. Eng. J.*, **170**, 395–410.

Ruth D. (2011). Add dash of silicone for virus-free water. futurity – science and technology,. http://www.futurity.org/science-technology/add-dash-of-silicone-for-virus-free-water/ (accessed 29 June 2012).

Sanchez A., Recillas S., Font X., Casals E., Gonzalez E. and Puntes V. (2011). Ecotoxicity of, and remediation with, engineered inorganic nanoparticles in the environment. *Trends Anal. Chem.*, **30**(3), 507–516.

Shannon M. A., Bohn P. W., Elimelech M., Georgiadis J. G., Mariñas B. J. and Mayes A. M. (2008). Science and technology for water purification in the coming decades. *Nature*, **452**(20), 301–310.

Shon H. K., Vigneswaran S., Kandasamy J., Kim J.-B., Park H. J., Choi S. W. and Kim J.-H. (2009). Preparation of titanium oxide, iron oxide, and aluminium oxide from sludge generated from Ti-salt, Fe-salt and Al-salt flocculation of wastewater. *J. Ind. Eng. Chem.*, **15**, 719–723.

Subramani A., Badruzzaman M., Oppenheimer J. and Jacangelo J. G. (2011). Energy minimization strategies and renewable energy utilization for desalination: a review. *Water Res.*, **45**, 1907–1920.

Sun H., Zhang X., Chen Y., Qiu Q. and Crittenden J. C. (2007). Enhanced accumulation of arsenic in carp in the presence of titanium dioxide nanoparticles. *Water Air Soil Pollut.*, **178**, 245–254.

Tratnyek P. G. and Johnson R. L.(2006). Nanotechnologies for environmental cleanup. *Nanotoday*, **1**, 44–48.

Turco R. F., Bischoff M., Tong Z. H. and Nies L. (2011). Environmental implications of nanomaterials: are we studying the right thing? *Curr. Opin. Biotechnol.*, **22**, 527–532.

Wang C. B. and Zhang W. X. (1997). Synthesizing nanoscale iron nanoparticles for rapid and complete dechlorination of TCEs and PCBs. *Environ. Sci. Technol.*, **31**, 2154–2156.

Wang X. B., Cai W. P., Lin Y. X., Wang G. Z. and Liang C. H. (2010). Mass production of micro/nanostructured porous ZnO plates and their strong structurally enhanced and selective adsorption performance for environmental remediation. *J. Mater. Chem.*, **20**, 8582–8590.

Xu P., Zeng G. M., Huang L. D., Feng C. L., Hu S., Zhao M. H., Lai C., Wei Z., Huang C., Xie G. X. and Liu Z. F. (2012). Use of iron oxide materials in wastewater treatment: a review. *Sci. Tot. Environ.*, **424**, 1–10.

Zhang Y., Chen Y., Westerhoff P. and Crittenden J. (2009a). Impact of natural organic matter and divalent cations on the stability of aqueous nanoparticles. *Water Res.*, **43**, 4249–4257.

Zhang W., Zou L. and Wang L. (2009b). Photocatalytic TiO_2/adsorbent nanocomposites prepared via wet chemical impregnation for wastewater treatment: a review. *Appl. Catal. A: Gen.*, **371**, 1–9.

Zhang J., Lay W. C. L., Chou S., Tang C., Wang R. and Fane A. G. (2012). Membrane biofouling and scaling in forward osmosis membrane bioreactor. *J. Membr. Sci.*, 403–404, 8–14.

Zhao X., Lv L., Pan B., Zhang W., Zhang S. and Zhang Q. (2011). Polymer-supported nanocomposites for environmental application – a review. *Chem. Eng. J.*, **170**, 381–394.

Chapter 2

Environmental and human health effects of nanomaterials used in water and waste water treatment

J. E. Cañas, B. Shrestha and S. Li

2.1 INTRODUCTION

The use of nanotechnology in water and waste water treatment (WWWT) is emerging as a promising technology. Potential applications of nanomaterials in WWWT such as incorporation of nanostructures into water filtration membranes, water disinfection, nanoparticles as adsorbents for inorganic and organic contaminants, and nanocatalysts to break down pollutants (Smith, 2006) are currently under development. Different types of nanomaterials currently used in WWWT can be broadly categorized into four categories: carbon based, metal based, dendrimers, and zeolites (Savage & Diallo, 2005).

Current applications of carbon based nanomaterials in WWWT include water filtration membranes (Theron *et al.*, 2008) and nanosorbents (Savage & Diallo, 2005; Theron *et al.*, 2008; Theron *et al.*, 2010). Carbon nanotubes (CNTs) have been widely studied as water filters for bacterial pathogen removal (Srivastava *et al.*, 2004) and adsorbents of inorganic and organic contaminants (Li *et al.*, 2003; Peng *et al.*, 2003; Li *et al.*, 2004; Fugetsu *et al.*, 2004). Similarly, fullerenes have been found to be effective adsorbents for polycyclic aromatic hydrocarbons (Cheng *et al.*, 2004).

Many metal based nanomaterials have potential to be used in WWWT as photocatalysts, bioactive nanoparticles, nanosorbents, and nanostructures in water filtration membranes (Savage & Diallo, 2005; Theron *et al.*, 2008; Theron *et al.*, 2010). Among metal based nanomaterials, nano-sized titanium dioxide (nano-TiO_2) is the most widely studied and applied metal based nanomaterial for WWWT. Current applications of nano-TiO_2 in this sector include photocatalysts for degradation of organic contaminants as well as reduction of toxic metal ions (Chitose *et al.*, 2003; Kabra *et al.*, 2004) and bioactive nanoparticles for water disinfection (Theron *et al.*, 2008). Silver nanoparticles (nano-Ag) are another potential metal based nanomaterial currently being evaluated as bioactive nanoparticles in water disinfection (Theron *et al.*, 2008). Zero-valent iron nanoparticles and bimetallic zero-valent Fe nanoparticles (Pd/Fe, Ni/Fe etc) are another group of metal based nanomaterials that have demonstrated potential use in the degradation of chlorinated organic compounds and other organic contaminants (e.g. trichloroethylene), reduction of toxic heavy metal ions to less toxic forms (e.g. Cr(VI) to Cr(III)) and transformation of inorganic ions (e.g. nitrate to nitrogen gas) (Savage & Diallo, 2005; Theron *et al.*, 2008; Zhang & Elliott, 2006; Zhang, 2003). Moreover, potential applications of Fe nanoparticles in groundwater remediation are also gaining attention (Theron *et al.*, 2008). Besides these nanomaterials, other metal based nanoparticles that are currently studied for application in water and waste water treatments are ZnO (nanoscale photocatalyst), alumina nanoparticles (ultrafiltration membrane), and TiO_2/Al_2O_3 composites (photocatalyst) (DeFriend *et al.*, 2003; El-Kemary *et al.*, 2010; Yang & Li, 2008).

Besides carbon and metal based nanomaterials, nanocrystalline zeolites and dendritic polymers are also currently being studied for potential application in WWWT. Potential applications of nanocrystallines as sorbents and ion exchange media for metal ions have been reported (Theron *et al.*, 2008). Another emerging nanomaterial used in WWWT are dendritic polymers which are found to be effective for recovering metal ions, delivery of antimicrobial agents, removal of toxic metal ions, radionuclides, and inorganic anions (Diallo *et al.*, 2004; Diallo *et al.*, 2005). A few other nano-sized materials that have been proposed for use in WWWT are biopolymers (protein based nano-biomaterials), nanoporous ceramic oxide, and single enzyme nanoparticles (Theron *et al.*, 2008).

With the impending widespread application of nanomaterials in WWWT, an urgent need for evaluating their potential impact to the environment and humans has emerged. The uncertainty associated with environmental fate, transport and toxicity of nanomaterials (Colvin, 2003; Lecoanet *et al.*, 2004; Lecoanet & Wiesner, 2004) is an important issue to be considered prior to widespread utilization of nanomaterials in WWWT. In addition, there is a serious lack of promising techniques for removal of nanomaterials from water and waste water (Brar *et al.*, 2010; Theron *et al.*, 2008). Furthermore, there is a lack of information on nanomaterial stability in nanostructures and membranes produced for application in WWWT; there is also no safeguard against the potential release of nanomaterials from these nanostructures by weathering or other processes. Ultimately, nanomaterials used for WWWT could enter natural water bodies (Brar *et al.*, 2010). Also the direct application of nanomaterials for remediation, like groundwater treatment, by zero-valent iron nanoparticles (Zhang & Elliott, 2006) could result in direct entry of nanomaterials into the environment (Gottschalk & Nowack, 2011). Furthermore, aggregation and settling mechanisms will ultimately result in nanomaterials ending up in wastewater and sludge (Brar *et al.*, 2010). This could result in soil contamination during sludge application to agricultural fields. Hence, potential entry of nanomaterials into the environment from WWWT technologies cannot be ignored (Figure 2.1).

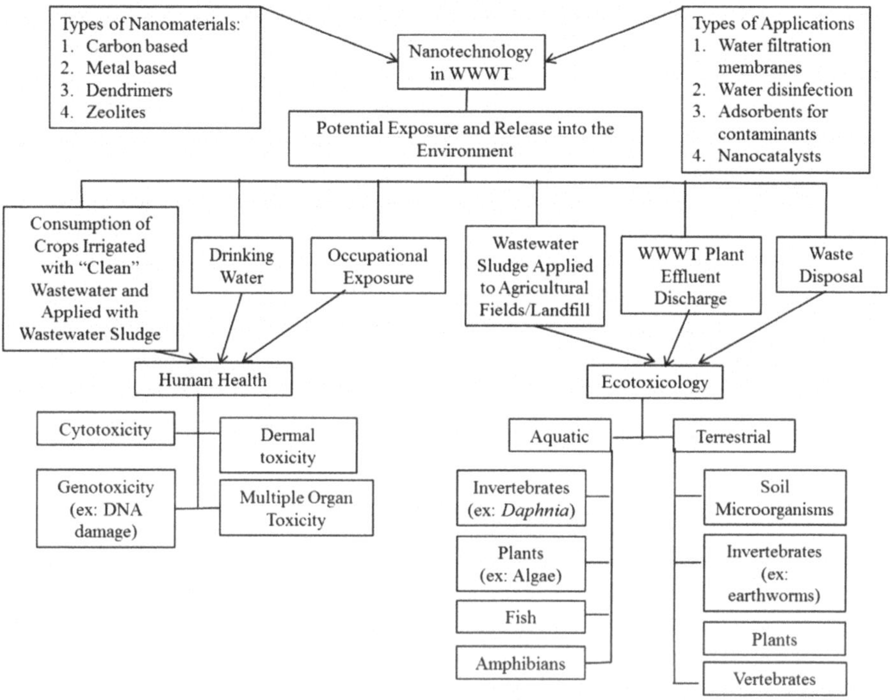

Figure 2.1 Nanotechnology in Water and Waste Water Treatment: Potential Release, Exposure and Effects on Human Health and Environment

Moreover, issues exist related to occupational human exposure, potential release of nanomaterials into the environment during the production and application of nanomaterials in secondary products, such as nanomembranes and nanosorbents, and generation of WWWT technology waste containing nanomaterials. Various studies have reported toxic effects of nanomaterials to humans, microorganisms, invertebrates, and plants (Aschberger *et al.*, 2011; Dinesh *et al.*, 2012; Kahru & Dubourguier, 2010; Klaine *et al.*, 2008; Navarro *et al.*, 2008a). Although, the application of nanomaterials in WWWT opens the door for new potential technologies for cleaning up drinking water for the growing world population and novel techniques for remediation to improve natural water quality and environmental protection, revisiting whether there are unforeseen hazards associated with nanomaterial applications in WWWT is a prerequisite. The remainder of this chapter will focus on toxicity studies with nanomaterials used in WWWT.

2.2 EFFECTS OF MANUFACTURED NANOMATERIALS ON HUMAN HEALTH AND THE ENVIRONMENT

2.2.1 Human health

Given the potential use of nanomaterials in WWWT and remediation, it is likely that these nanomaterials will end up in the environment posing a potential risk to humans. Humans can be exposed via several routes of exposure: ingestion, inhalation, and dermal. In the case of nanomaterials used in drinking water treatment without enough safeguards to prevent leaching of nanomaterials into the water, nanoparticles may exist in "clean" water ingested by humans (Wiesner *et al.*, 2006). Similarly, another potential source of human ingestion might be consumption of food (vegetables, fish, meat, etc.) that have accumulated nanoparticles from the environment (Wiesner *et al.*, 2006). In addition, humans can be subjected to occupational inhalational and dermal exposure while making or implementing the use of the nanomaterials in WWWT. As a result of these potential risks, the toxicity of all manufactured nanomaterials currently or potentially used in WWWT should be evaluated. This section will focus on toxicological studies that have been conducted with the most commonly used nanomaterials in WWWT.

2.2.1.1 Carbon based nanomaterials

2.2.1.1.1 In vitro toxicity

In vitro toxicity studies are often used as a rapid and effective method to determine the effect of chemicals, such as carbon based nanomaterials, on cells (known as cytotoxicity). These studies also often provide data on how chemicals interact with cells and help to inform further mechanistic studies. The increasing use and application of carbon nanotubes has raised concern regarding the safety of these nanomaterials, especially given their similarities to asbestos. Both Simate *et al.* (2012) and Zhao and Liu (2012) provide a review of in vitro toxicity studies with different types of CNTs (single walled carbon nanotubes (SWNTs) and multiple walled nanotubes (MWNTs) and mammalian cells, including human cell lines. A few examples will be provided here that resulted in important findings that represent the true potential effects of CNTs and highlight characteristics of CNTs that may affect toxicity.

Based on studies to date, CNT type, diameter size, and agglomeration state affect cytotoxicity. Single-walled carbon nanotubes were found to be more toxic to alveolar macrophages than MWNTs and fullerenes (Jia *et al.*, 2005). With increasing diameter size of CNTs, cytotoxicity decreased (Wang *et al.*, 2009b). Simon-Deckers *et al.* (2008) found CNTs to be more toxic to human pneumocytes than metal oxide nanomaterials (aluminum oxide and titanium dioxide) and CNT length and impurities did not influence toxicity. In a study with human lung cancer cell lines, toxicity increased from well-dispersed CNTs to agglomerated CNTs (Wick *et al.*, 2007).

These are just a few examples of studies that observed cytotoxicity following exposure to CNTs. Unfortunately, numerous contradictory studies have been conducted where no cytotoxicity was observed. All cytotoxicity studies were done with different biological models and types of CNTs (produced with different components or functionalization) which makes comparisons between studies almost impossible (Som *et al.*, 2011). Despite the conflicting data, the cytotoxicity studies have provided valuable information regarding which properties of CNTs are likely responsible for the observed toxicity (Simate *et al.*, 2012).

"CNT physical and chemical properties, such as their structure, length and size distribution, aspects ratio, surface area, shape, degree of dispersion and aggregation, agglomeration state, presence of amorphous impurities, extent of oxidation, bound functional group, method of manufacturing, metal impurities, coating, concentration and dose can affect the immobilization, cellular uptake or internalization and cytotoxic response of different cell types to CNTs (Auffan *et al.*, 2008; Helland *et al.*, 2008; Hussain *et al.*, 2009; Stella, 2011; Upadhyayula *et al.*, 2009)".

The fate and distribution of CNTs in the body is not well understood. At present CNTs are thought to be persistent, but recent studies have shown that certain peroxidases can break down CNTs (Allen *et al.*, 2008; Kagan *et al.*, 2010).

A few in vitro studies conducted with fullerenes (C_{60}) have concluded that there is little to no cytotoxicity of C_{60} in alveolar machrophages and human macrophage cells (Fiorito *et al.*, 2006; Jia *et al.*, 2005). In contrast, Sayes *et al.* (2004) found C_{60} to be highly toxic to human dermal fibroblasts and hepatocellular liver carcinoma cells. More recently, Song *et al.* (2012) observed that smaller C_{60} nanoparticles exhibited higher cytotoxicity in human lung adenocarcinoma cells than larger sized C_{60}. Given the conflicting data, more studies are warranted.

2.2.1.1.2 In vivo toxicity

While in vitro studies provide valuable data, Donaldson *et al.* (2009) highlight the potential dangers of only conducting in vitro testing. They state cells in culture do not encounter pathogenic effects likely to be observed in vivo as a result of translocation, toxicokinetics and coordinated tissue responses. A recent review article that included in vitro and in vivo toxicity studies discussed the interferences of the physical chemical properties of nanomaterials with in vitro assays (Dhawan & Sharma, 2010). In vivo studies are often used to observe the overall effects of a chemical on a living organism. Like in vitro studies, there are challenges in conducting in vivo studies with nanomaterials, such as dosimetry, dispersion, biodistribution and evaluation of interactions (Dhawan & Sharma, 2010). Hence, it is important to include both in vitro and in vivo studies of nanomaterials, in combination with imaging and characterization techniques, to fully assess the potential toxicity of nanomaterials (Arora *et al.*, 2012).

Numerous in vivo studies with CNTs have been conducted and toxic responses have been reported in multiple organ systems as outlined in Zhao and Liu (2012).

A few examples with different exposure routes will be included here to demonstrate the potential effects of CNTs. Animal models, such as mice and rats, are often used as surrogates to assess human exposure. Dermal exposure is a likely route of exposure for humans. In mice, exposure to unpurified CNTs generally caused oxidative stress, glutathione depletion, increased number of dermal cells, skin thickening, and alopecia (Koyama *et al.*, 2009; Murray *et al.*, 2009; Zhao & Liu, 2012). It is possible that impurities in CNTs may be the cause of these observed effects as mice implanted with high purity CNTs did not experience any hair loss (Zhao & Liu, 2012). Inhalation is the most likely route of exposure to CNTs. CNTs have been observed to penetrate alveolar macrophages and pleura in the lungs (Mercer *et al.*, 2010). Effects such as acute pulmonary toxicity, sub-pleural fibrosis and immune suppression have been observed in some studies while other studies have found contradicting results of no observed toxicity (Zhao & Liu, 2012). Even though changes are observed in vivo, it does not necessarily mean that overall CNT toxicity will be exhibited. CNT toxicity depends on numerous factors including dose, impurities, physical form, surface chemistry, and aggregation (Zhao & Liu, 2012).

In vivo studies with C_{60} have revealed that these materials can migrate throughout the mice and accumulate in the liver (Oberdörster *et al.*, 2006; Yamago *et al.*, 1995). Additional studies in human monocyte macrophages demonstrated accumulation of C_{60} in lysosomes, cytoplasm, along the plasma membrane, and inside the nucleus (Porter *et al.*, 2007; Porter *et al.*, 2006). A full understanding of fullerene toxicity is probably limited because of the different experimental conditions used and the complexity (size, shape, and charge) of C_{60} nanoparticles (Song *et al.*, 2012).

2.2.1.2 Metal based nanomaterials

2.2.1.2.1 Titanium dioxide (nano-TiO_2)

Titanium dioxide is stable in the environment. It is the most studied nanoparticle due to its widespread applications, especially in cosmetics and body care products (Som *et al.*, 2011). According to the recent research, nano-TiO_2 cannot be absorbed and cross the skin (Nanoderm: http://www.uni-leipzig.de/~nanoderm/Downloads/Nanoderm_Final_Report.pdf). Inhaled nano-TiO_2 has been found to cross the air-blood barrier and to be transported throughout the body. However, the amount of translocated particles is low relative to the initial dose (Geiser *et al.*, 2005; Kapp *et al.*, 2004). Overall most studies indicate little to no nano-TiO_2 toxicity (Som *et al.*, 2011). In general, in comparison to other nano-metal oxides, nano-TiO_2 is the least cytotoxic (Jin *et al.*, 2008; Lai *et al.*, 2008).

2.2.1.2.2 Nano-silver (nano-Ag)

Nano-silver has not received as much attention until very recently. To date, in vitro data indicate that nano-Ag may cause mitochondrial and DNA damage (Asharani *et al.*, 2008a; Hsin *et al.*, 2009; Hussain *et al.*, 2005). However, it is unclear if these observed effects were a result of free Ag^+ or nano-Ag. Nano-Ag also causes the generation of reactive oxygen species (ROS), which can ultimately cause cell damage, by disrupting ion and electron fluxes across the mitochondrial membrane (Almofti *et al.*, 2003; Arora *et al.*, 2009).

In in vivo mammalian studies, nano-Ag caused changes in rat lung function and increased inflammation markers (Sung *et al.*, 2008). An in vivo study indicated that nano-Ag can be distributed all over the mice body following inhalation (Takenaka *et al.*, 2001). Inhalational exposure to nano-Ag also affected liver and lung functioning (Ji *et al.*, 2007; Sung *et al.*, 2009; Sung *et al.*, 2008). Following oral exposure, silver was found in the liver (Cha *et al.*, 2008; Kim *et al.*, 2008; Sung *et al.*, 2008) and effects were seen in both the liver and kidney. Even though genotoxic effects (DNA damage) were observed in in vitro studies, no genotoxicity was observed in whole-animal studies.

2.2.1.2.3 Iron oxide nanoparticles (FeO-NPs)

Iron oxide nanoparticles are easy to synthesize and biocompatible which makes them ideal for numerous biomedical applications such as magnetic resonance imaging (MRI), magnetic cancer hyperthermia treatment or magnetically enhanced and target drug delivery (Hergt *et al.*, 2006; Namiki *et al.*, 2009; Soenen *et al.*, 2009a). At first, FeO-NPs were considered not cytotoxic, however recent studies have demonstrated that these nanoparticles can concentrate inside the cell (Rivera Gil *et al.*, 2010) and may be more difficult to efficiently clear from the body depending on their size (Sun *et al.*, 2008). Given the widespread biomedical use of FeO-NPs, it is important that their safety be assessed.

Iron oxide nanoparticles, depending on their coating, have demonstrated some cytotoxicity. However, conflicting in vitro studies have also shown that FeO-NPs are not cytotoxic. Only few studies have been conducted on cytotoxicity of FeO-NPs (Soenen *et al.*, 2011). Thus future cytotoxicity studies are needed to demonstrate the true cytotoxicity of FeO-NPs.

In a few in vitro studies, FeO-NPs generated and elevated levels of ROS (Arbab *et al.*, 2003; Arbab *et al.*, 2005; Soenen *et al.*, 2010a; Stroh *et al.*, 2004). However, contrasting studies indicate that FeO-NPs can actually reduce the amount of ROS present in the cell (Gao *et al.*, 2007; Huang *et al.*, 2009; Nel *et al.*, 2009). Thus, the relationship between ROS and cytotoxicity is still unclear. Numerous studies have indicated that FeO-NPs can affect the cellular cytoskeleton network and structure (Buyukhatipoglu & Clyne, 2011; Gupta & Curtis, 2004; Gupta & Gupta, 2005; Soenen *et al.*, 2009b, 2010b). Similar to ROS studies, studies on the effects of FeO-NPs on DNA damage and gene expression levels are conflicting. Some studies have found no effect of FeO-NPs on stem cell differentiation and no genotoxic effects (Arbab *et al.*, 2005; Auffan *et al.*, 2006). In contrast, inhibition of stem cell differentiation has been observed following exposure to FeO-NPs (Chen *et al.*, 2010; Kostura *et al.*, 2004). In addition,

FeO-NPs exhibited gentoxicity and disrupted cellular functionality (Kedziorek *et al.*, 2010; Pisanic *et al.*, 2007; Puppi *et al.*, 2011; Soenen *et al.*, 2010a). Based on in vitro studies to date and the conflicting results, it is very clear that the safety of FeO-NPs is unclear (Soenen *et al.*, 2011). In one in vivo study, FeO-NPs accumulated in the liver and spleen but slowly degraded over a 3 week period (Jain *et al.*, 2008). The FeO-NPs did not affect liver enzyme levels and organ tissue histology or cause oxidative stress. It is clear that additional in vitro and in vivo studies with FeO-NPs should be conducted, especially with the same types of FeO-NPs.

2.2.2 Ecotoxicological effects

2.2.2.1 Aquatic ecotoxicology

With antimicrobial properties, extremely small sizes, high surface-area-to-volume ratios and modifiable surfaces, various nanomaterials have been considered for the development of water treatment (Mauter & Elimelech, 2008; Musee *et al.*, 2011; Xu *et al.*, 2012). However, there is a possibility that nanomaterials will eventually enter various environmental compartments and pose a risk to ecosystems. As a major sink for contaminants, the aquatic environment is of particular concern.

Although toxicity studies of nanomaterials are still in their infancy, the toxicological impact of nanomaterials to various aquatic organisms has already been reported. Generally, four classes of nanomaterials have been evaluated or applied as functional materials for water treatment: carbon based nanomaterials, metal based nanoparticles, zeolites and dendrimers (Cloete, 2010; Li *et al.*, 2008; Sawai, 2003; Theron *et al.*, 2008). However, to date, toxicity studies with aquatic species and zeolites or dendrimers have not been conducted. Thus, in this section, the current knowledge related to aquatic toxicity of carbon based and metal based nanomaterials will be reviewed. The discussion will focus on algae, *Daphnia magna*, and several fish species as these are the major aquatic species that have been evaluated for nanomaterial toxicity (Zhao & Liu, 2012).

2.2.2.1.1 Carbon based nanomaterials

Routes of exposure. For aquatic organisms, ingestion has been the suggested or proposed route of exposure in most studies. Ingestion of SWNTs by a nematode (*Caenorhabditis elegans*) was reported (Oberdörster *et al.*, 2006). Also, both solubilized and unsolubilized SWNTs were detected in the fecal matter from the digestive tract of largemouth bass (Oberdörster *et al.*, 2006). It was recently reported that CNTs might interfere with normal gut processes if CNTs are not absorbed (Edgington, 2011). Jonker and Suijkerbuijk (2008) suggested that carbon based nanomaterials are taken up by aquatic organisms via direct ingestion or entry across epithelial boundaries.

Toxicity. Acute and reproductive toxicity of carbon based nanomaterials has been reported. When exposed to lysophophatidylcholine-coated SWNTs, there was 100% mortality of *D. magna* exposed to 20 mg/L for 96 h (Roberts *et al.*, 2007). MWNTs suspended in natural organic matter (2.0–18.5 mg/L dissolved organic carbon) were found to be toxic to *D. magna* with an average 96 h LC_{50} value of about 2 mg/L (Edgington, 2011). LC_{50} values represent the concentration at which 50% mortality is observed. When *D. magna* were exposed to 5 mg/L MWNTs suspended in natural organic matter, there was a 22% reduction in dry mass compared to the control (Alloy & Roberts, 2011). In contrast to the observed acute toxicity in *D. magna*, there was no observed acute mortality in juvenile rainbow trout (*Oncorhynchus mykiss*) exposed to SWNTs dissolved in sodium docedyl sulphate (Smith *et al.*, 2007).

Reproduction of *D. magna* were decreased in comparison to control when MWNT (Zhu *et al.*, 2006) exposure concentrations were higher than 0.24 mg/L (Alloy & Roberts, 2011). *Ceriodaphnia dubia* reproduction also significantly decreased relative to the controls at three exposure concentrations: 0.5, 2.5, and 5 mg/L (Alloy & Roberts, 2011). However, embryonic zebrafish (*Danio rerio*) development was not affected at SWNT concentrations of up to 360 mg/L (Cheng *et al.*, 2007).

Fullerene toxicity to aquatic organisms has also recently been studied. In a 48 h acute toxicity test, the LC_{50} for *D. magna* was around 0.35 mg/LC_{60} in water (Zhu *et al.*, 2006). There was no survival of fathead minnows (*Pimephales promelas*) when exposed to 0.5 mg/L fullerenes in tetrahydrofuran (THF) (Zhu *et al.*, 2006). However, in some cases, decomposition products of THF can be responsible for the toxicity if the fullerenes are induced by THF (Henry *et al.*, 2007). Hence, the high mortality observed might be confounded by the solvent rather than the fullerene itself. Furthermore, the toxicity of metal contaminants in carbon based materials cannot be ignored (Cheng *et al.*, 2007).

Mechanisms. Some studies have investigated the underlying mechanisms of the observed effects of carbon based nanomaterials. One study found increased glutathione levels in the gills and livers of juvenile *O. mykiss* (Smith *et al.*, 2007). Glutathione is a known free radical scavenger for contaminant detoxification and ROS (Jonker & Suijkerbuijk, 2008). Thus, increased levels might indicate exposure to SWNTs which might induce oxidative stress responses in aquatic organisms. Thomas *et al.* (2011) recently reported that MWNTs were responsible for ROS production when co-exposed with cobalt in *O. mykiss* hepatocyte cells. It has also been suggested that fullerenes could induce oxidative stress responses (Usenko *et al.*, 2007, 2008) in fathead minnows (Zhu *et al.*, 2006), *D. rerio* (Usenko *et al.*, 2008), adult goldfish (*Carassius auratus*) (Zhu *et al.*, 2008a), and goldfish larvae (Zhu *et al.*, 2008b). Caution should be taken however, since the exact mechanisms for oxidative effects of carbon based nanomaterials are still largely unknown (Jonker & Suijkerbuijk, 2008; Thomas *et al.*, 2011).

To date, the majority of studies have focused on the acute effects of carbon based nanomaterials on a single test species in the laboratory. There are limited studies that mimic a real ecosystem. However, one study investigated the effects of MWNT contaminated sediments on benthic macroinvertebrate communities. This study concluded that no negative effects of MWNTs on these communities are expected at environmentally relevant future CNT concentrations in aquatic sediments (Velzeboer *et al.*, 2011). More studies of MWNTs with environmental relevance are needed in the future.

In sum, there are limited studies estimating lethal concentrations of CNTs and fullerenes for *D. magna*. There are no lethal concentrations reported for fish exposed to carbon based nanomaterials. In addition, lethal dose estimation for CNTs and fullerenes varies highly for nanomaterials with different properties and dispersion methods (Handy *et al.*, 2011; Handy *et al.*, 2012). It is generally believed that carbon based materials induce oxidative stress in aquatic organisms. More studies are needed to examine the ecological effects of MWNTs at the population or community level.

2.2.2.1.2 Metal based nanomaterials

This section reviews the toxicity of metal based nanomaterials frequently used in WWWT. Compared with traditional contaminants, the toxicological profiles of metal based nanomaterials are still largely unknown with huge data gaps.

Titanium dioxide (nano-TiO_2). Many studies have investigated nano-TiO_2 acute toxicity to aquatic organisms (Aruoja *et al.*, 2009; Heinlaan *et al.*, 2008; Hund-Rinke & Simon, 2006; Wang *et al.*, 2009). Toxicity of nano-TiO_2 to aqueous organisms has been thoroughly reviewed in several papers (Baun *et al.*, 2008; Farre *et al.*, 2009; Handy *et al.*, 2008; Kahru & Dubourguier, 2010; Klaine *et al.*, 2008; Menard *et al.*, 2011).

Many nano-TiO_2 toxicity studies have been conducted with freshwater algae and invertebrates. In a study with algae (*Pseudokirchneriella subcapitata*), the EC_{50} value (concentration that causes an effect in 50% of exposed organisms) for nano-TiO_2 was 5.83 Ti mg/L, which was six times lower than that of bulk-TiO_2 (Aruoja *et al.*, 2009). This suggests that nano-TiO_2 (diameter = 25–70 nm) is more toxic than bulk-TiO_2. Similar conclusions were made in another study with nematodes (Wang *et al.*, 2009a). Several other studies have also investigated nano-TiO_2 toxicity values for *P. subcapitata* and determined EC_{50} values ranged from 16 to 241 mg/L (Aruoja *et al.*, 2009; Hartmann *et al.*, 2010; Warheit *et al.*, 2007). Another study found an inhibitory concentration by 50% (IC_{50}), a value similar to an EC_{50}, for *P. subcaptitata* exposed to nano-TiO_2 that was greater than 100 mg/L, which was within the range of that previously reported (Blaise *et al.*, 2008). Hund-Rinke and Simon (2006) observed that in green algae (*Desmodesmus subspicatus*) after 72 h, nano-TiO_2 with a smaller diameter

(25 nm, $EC_{50} = 44$ mg/L) was more toxic than larger nano-TiO_2 (100 nm, $EC_{50} = 50$ mg/L).

A study with invertebrates showed 60% mortality of *D. magna* at 20 g/L nano-TiO_2, while no toxicity was observed in fairy shrimp (*Thamnocephalus platyurus*) at the same concentration (Heinlaan *et al.*, 2008; Sharma, 2009). In another study, the 48 h EC_{50} value for *D. magna* was larger than 100 mg/L when exposed to Si/Al coated 140 nm TiO_2 (Warheit *et al.*, 2007). Exposure of *D. magna* to nano-TiO_2 could lead to an increase in mortality with increasing concentration (Lovern & Klaper, 2006). Hall *et al.* (2009) reported that when exposed to 10 nm nano-TiO_2, 96 h LC_{50} values for *C. dubia* and *Daphnia pulex* were 7.6 and 9.2 mg/L, respectively.

Similar to freshwater algae and invertebrates, it is almost impossible to compare nano-TiO_2 EC_{50} and LC_{50} values for fish based on different types of nano-TiO_2 and different durations of exposure (Menard *et al.*, 2011). Generally, very high LC_{50} values were reported (Menard *et al.*, 2011). For example, the 96 h LC_{50} for *O. mykiss* is larger than 100 mg/L (Warheit *et al.*, 2007), while no effects on *D. rerio* were found below 500 mg/L (Zhu *et al.*, 2008c). Sub-lethal effects of nano-TiO_2 in fish have also been investigated (Handy *et al.*, 2011). Federici *et al.* (2007) reported *O. mykiss* gill pathology with inhibition of bronchial Na^+K^+-ATPase after a 14 d treatment of 1 mg/L TiO_2 (24 nm size). Inhibition of brain Na^+K^+-ATPase activity was also observed during the exposure of 100 mg/kg TiO_2 (24 nm size) for eight weeks (Ramsden *et al.*, 2009). However, Scown *et al.* (2009) discovered that although 10–19% of the nano-TiO_2 (34 nm size) was detected in the kidney of *O. mykiss*, renal clearance was not affected. Limited studies have been conducted to investigate chronic exposure of TiO_2 to aquatic organisms. A chronic toxicological experiment reported a 25% decrease in growth at 1 mg/L of nano-TiO_2 (Hall *et al.*, 2009). More chronic exposure data are needed to better understand the risk posed by nano-TiO_2.

Other metal based nanomaterials. There are very limited studies on the aquatic toxicity of other metal based nanomaterials, besides nano-TiO_2, used in WWWT. Nano-Ag has been studied for application in water treatment because of its unique antimicrobial properties (Fabrega *et al.*, 2009; Garcia *et al.*, 2012). However, it is also believed that these same biocidal properties could pose a risk to other organisms (Kittler *et al.*, 2010). Toxicity of silver nanoparticles to various organisms in aqueous media has been reported, such as algae (Miao *et al.*, 2009; Navarro *et al.*, 2008b), aquatic invertebrates (Kim *et al.*, 2011a), and fish (Asharani *et al.*, 2008b; Laban *et al.*, 2010; Lee *et al.*, 2007). Similar as nano-TiO_2 studies, extremely variable toxicity was observed in these studies. Gagne *et al.* (2008) observed that aggregates of CdTe induced oxidative stress in gills and digestive gland tissues of freshwater mussels (*Elliptio complanata*). Cerium nanoparticles were extremely toxic to *D. magna*, causing inhibition higher than 80% at very low concentrations with a LC_{50} of 0.012 mg/ml (Garcia *et al.*, 2011). In the same study, toxicity of iron oxide nanoparticles increased with increasing FeO-NPs concentration, resulting in an EC_{50} value of 0.24 mg/mL (García *et al.*, 2011). In sum, future toxicological studies need to focus on other metal based nanomaterials, besides nano-TiO_2, used in WWWT.

2.2.2.2 *Terrestrial ecotoxicology*

Terrestrial ecotoxicology is related to the assessment of contaminant effects on soil nutrient cycling and functioning, plant biomass production, smaller invertebrates, amphibians, reptiles, birds, and mammals (CSTEE, 2000; Fairbrother & Hope, 2005; Garcia, 2004). Nanomaterials can enter the soil environment from disposal of waste containing nanomaterials and through the application of wastewater sludge, containing nanomaterials, on agricultural fields (Brar *et al.*, 2010). Although limited information is available on concentrations of nanomaterials in soil, the amount of nanomaterials that enter the environment will be higher in soil in comparison to air and water (Klaine *et al.*, 2008; Tiede *et al.*, 2009). Thus, it is important to re-evaluate the application of nanomaterials in large quantities. The current information known in the field of nanotoxicology regarding terrestrial ecotoxicology of nanomaterials is limited to soil microorganisms, smaller soil invertebrates, and plants.

2.2.2.2.1 Soil microorganisms

Soil microorganisms are an important part of the ecosystem as they are the major player for decomposition of organic matter and nutrient cycling dynamics in soil (Kennedy & Smith, 1995). Effects on soil microorganisms can limit the metabolic functioning of soil and ultimately affect nutrient turnover in the soil. Although it still has not been confirmed whether nanomaterials have toxic effects on soil microorganisms and metabolic functioning, a few published studies have reported nanomaterial toxicity to soil microorganisms. Nanomaterials can affect soil microorganisms by directly affecting cellular viability, by indirectly limiting availability of soil nutrients and increasing bioavailability of contaminants to have toxic effects (Simonet & Valcarcel, 2009). Possible mechanisms of nanomaterial toxicity to microorganisms include: ROS formation and accumulation in cytoplasm (Raghupathi *et al.*, 2011) and disruption of membranes, protein oxidation, genotoxicity, and energy transduction (Klaine *et al.*, 2008).

Microbial toxicity due to carbon based nanomaterials (Aoshima *et al.*, 2009; Fortner *et al.*, 2005; Kang *et al.*, 2008a; Kang *et al.*, 2008b; Lyon *et al.*, 2006; Lyon *et al.*, 2008; Sayes *et al.*, 2005) and metal nanoparticles (Adams *et al.*, 2006; Aruoja *et al.*, 2009; Baek & An, 2011; Dimkpa *et al.*, 2011; Gajjar *et al.*, 2009; Jaidev & Narasimha, 2010; Jiang *et al.*, 2009; Jones *et al.*, 2008; Rai *et al.*, 2009; Suresh *et al.*, 2010; Wang *et al.*, 2010) have been reported in cell cultures and water suspensions. Extrapolation of these results to the actual environment requires nevertheless consideration of different environmental factors, specifically in soil where direct contact between microbial cells and nanomaterials is limited by a decrease in bioavailability (Tong *et al.*, 2007). To date, only a few studies have been published on the effects of nanomaterials to soil microbial communities.

Carbon based nanomaterials. Microbial biomass and enzymatic activity related to soil nutrient cycling decreased in the presence of multi-walled carbon nanotubes in soil (Chung *et al.*, 2011). Another study observed a toxic effect on fast growing bacteria and observed a shift in bacterial diversity (Johansen *et al.*, 2008). In contrast, two other studies reported that there was no toxic effects of C_{60} fullerene to microbial communities in soil (1000 μg C_{60}/g soil dry weight) and anaerobic microbial communities in wastewater sludge (up to 50000 mg C_{60}/kg sludge dry weight) (Nyberg *et al.*, 2008; Tong *et al.*, 2007).

Metal based nanomaterials. Iron oxide magnetic nanoparticles stimulated carbon and nitrogen cycling and caused a shift in microbial communities in soil (He *et al.*, 2011). Nano-Ag was highly toxic to bacterial communities in a high altitude arctic soil and caused a shift in bacterial community structure (Kumar *et al.*, 2011). Ge *et al.* (2011) observed that nano-TiO_2 and nano-ZnO caused a significant reduction in microbial biomass, and community composition. The effect of nano-ZNO was higher than of nano-TiO_2 at 0.5 mg/kg concentration. Similar results were also observed in the Du *et al.* (2011) study in which nano- TiO_2 and nano-ZnO reduced soil protease, catalase, and peroxidase activities in soil. Soil dehydrogenase, β-glucosidase and acid phosphatase activities were affected by Zn and nano-ZnO during a pot study with *Cucumus sativus* plants (Kim *et al.*, 2011b). However, Shah and Belerozova (2009) found no effect on soil microbial communities following exposure to silicon, palladium, gold, and copper nanoparticles. Similarly, FeO-NPs showed no significant effect on dehydrogenase activity, hydrolase activity and ammonia oxidation potential in non-contaminated soil (Cullen *et al.*, 2011).

2.2.2.2.2 Soil invertebrates

Carbon based nanomaterials. Similar to studies with microorganisms, there are limited studies concerned with the assessment of nanoparticle toxicity to soil invertebrates. Double-walled carbon nanotubes (DWNT) affected the cocoon production in the earthworm *Eisenia veneta* at a dose above 37 mg DWNT/kg food (Scott-Fordsmand *et al.*, 2008). However hatchability, survival or mortality were not affected up to 495 mg DWNT/kg and 1000 mg C_{60}/kg In another study, *E. fetida* cocoon production was affected at high C_{60} concentrations (5% by weight of soil), but no effect on growth was observed at lower concentrations (1% by weight of soil) (Li & Alvarez, 2011).

Metal based nanomaterials. Hu *et al.* (2010) found that nano-TiO_2 and nano-ZnO bioaccumulated and caused oxidative stress, inhibited cellulase activity, and caused

DNA and mitochondrial damage in earthworms at concentrations above 1.0 g/kg in soil. Nano-ZnO and nano-TiO_2 inhibited *E. fetida* reproduction following a 4 week exposure in artificial soil (Canas *et al.*, 2011). Nano-sized ZnO, which also exhibited acute toxicity, was more toxic than nano-sized TiO_2. Drobne *et al.* (2009) found no effects of nano-TiO_2 on isopods (*Porcellio scaber*) at concentrations up to 1000 μg/g dry food. Other metal nanoparticles (ZnO, Al_2O_3 and TiO_2) inhibited growth and the reproductive capability of the nematode *Caenorhabditis elegans* (Wang *et al.*, 2009a).

2.2.2.2.3 Plants

Plants are a major producer of the ecosystem; hence any impact on plant biomass production can ultimately affect the whole ecosystem balance. Another major concern is the possible transfer of nanomaterials to and through the food chain creating the risk for organisms and humans (Rico *et al.*, 2011). Some available studies have observed uptake of some nanomaterials within different parts of plants, which further supports the need for studies on exposure of plants to nanomaterials and possible transfer along the food chain. Current studies on the ecotoxicological effects of nanomaterials on plants have resulted in three different scenarios of no observed effects, positive effects or negative effects. Since most phytotoxicity studies have been conducted in a water suspension and hydroponically, it is difficult to extrapolate these effects to actual soil environments. Despite the uncertainty in making definite conclusions regarding the effects of nanomaterials on plants, the negative effects observed in some studies cannot be neglected.

Uptake of nanomaterials into plants. Studies on plant uptake of nanomaterials are limited, especially given the lack of effective methods for quantification of nanomaterials (Rico *et al.*, 2011). However, studies on the uptake and translocation of nanomaterials are evolving. Magnetic carbon coated FeO-NPs penetrated the root, entered the vascular cylinder and were transported through xylem vessels to the aerial part of wheat, pea, sunflower and tomato plants within 24 h of exposure (Cifuentes *et al.*, 2010). Similar results were also observed with pumpkin plants exposed to magnetic carbon coated FeO-NPs (Corredor *et al.*, 2009). Khodakovskaya *et al.* (2011) found uptake and distribution of CNTs within different parts of tomato plants exposed to 50 μg/mL of CNTs in soil and growth medium. Uptake, transfer and accumulation of magnetic FeO-NPs in squash (*Cucurbita maxima*) plant tissues was observed following exposure to 0.5 g/L FeO-NPs in an aqueous suspension (Zhu *et al.*, 2008d).

Carbon based nanomaterials

Negative effects. There have been a few studies conducted to evaluate the effects of carbon nanotubes on plants which have resulted in negative results. A decrease in plant biomass was observed in zucchini plants hydroponically exposed to MWNTs for 15 d (Stampoulis *et al.*, 2009). Although MWNTs stimulated the growth of *Onobrychis arenaria* seedlings, it was found during the study that peroxidase activity was elevated and plants were stressed. Accumulation of MWNTs within the tissues of roots, stem and leaves was also observed (Smirnova *et al.*, 2012). The negative effects of MWNTs in sewage sludge from a wastewater treatment plant were demonstrated in a study conducted on the effect of MWNTs in sewage sludge on five plant species (*Lepidium sativum, Cucumis sativus, Raphanus sativus, Solanum lycopersicon, Sorghum saccharatum*). Except for *Sorghum saccharatum*, negative effects on the germination and root growth were observed. However, an increase in germination and root growth was observed in some of the plants tested, which the author explained as the adsorption of other toxic contaminants in sewage sludge by the MWNTs causing a decrease in toxicity of those contaminants (Oleszczuk *et al.*, 2011).

Positive effects. Two types of MWNTs (oxygen functionalized and non-functionalized) stimulated the growth and germination of mustard plants at lower concentrations, but a decrease in germination and growth was observed at higher concentrations (Mondal *et al.*, 2011). In tomato plants, MWNTs caused an increase in germination, water uptake and biomass (Khodakovskaya *et al.*, 2009). Fullerol induced longer hypocotyls in the *Arabidopsis thaliana* and no effect was observed in other physiological processes (Gao *et al.*, 2011).

Metal based nanomaterials

Negative effects. Nano-Ag caused cellular damage by breaking root cell walls and vacuoles of rice plants (*Oryza sativa*) at the highest concentration (1000 μg/mL) during a 12 d study (Majumdar & Ahmed, 2011). A decrease in plant biomass was observed in zucchini plants hydroponically exposed to nano-Ag for 15 d (Stampoulis *et al.*, 2009). In the same study, nano-Ag also reduced transpiration rates during a 17 d exposure while Cu nanoparticles caused a 90% decrease in biomass. Gubbins *et al.* (2011) observed inhibition in the growth of duckweed (*Lemna minor*) after exposure to both small (~20 nm) and large (~100 nm) nano-Ag at low concentrations (5 mg/L).

Zinc oxide nanoparticles accumulated in cellular and chromosomal modules of *Allium sepa* and affected root elongation following exposure to 5, 10, and 20 gm/L (Ghodake *et al.*, 2011). Velvet mesquite (*Prosopis juliflora velutina*) exposed to ZnO NPs showed an elevation in catalase and ascorbate peroxidase activities, which are related to detection of stress in plants. Also, accumulation of ZnO nanoparticles in the leaves, stem and roots were observed (Hernandez-Viezcas *et al.*, 2011). In a study by Lin and Xing (2007) on the effect of Zn and ZnO nanoparticles on six plant species, seed germination of ryegrass and corn was affected at exposure concentrations of 2000 mg/L of nano-Zn and nano-ZnO, respectively. Also root elongation was inhibited in most plant species at this concentration. Decreased biomass and increased oxidative stress was observed in *Cucumis sativus* exposed to CuO and ZnO NPs and the effects observed were dependent on the size, concentration and types of nanoparticles (Kim *et al.*, 2012).

In a study by Vochita *et al.* (2011), freshly germinated sunflower seeds injected with four different concentrations of aqueous suspensions of magnetic colloidal nanoparticles of Fe_3O_4, $CoFe_2O_4$, and $ZnFe_2O_4$ showed a diminished mitosis rate and increased chromosomal aberrations in root tips cells. Several other studies have observed inhibition in root elongation in some plants due to various nanoparticles (Canas *et al.*, 2008; Lin & Xing, 2007; Yang & Watts, 2005).

Positive effects. Several studies have found positive or no effects of nanomaterials on plant growth. Yang *et al.* (2007) found nano-TiO_2 enhanced photosynthesis and nitrogen fixation which ultimately increased spinach growth. A mixture of SiO_2 and TiO_2 increased the growth and germination of soybeans by increasing nitrate reductase activity (Lu *et al.*, 2002). Alumina nanoparticles increased root length, biomass and photosynthetic efficiency in *Lemna minor* (Juhel *et al.*, 2011). Nano-TiO_2 increased wheat germination and growth at lower concentrations (2 and 10 mg/kg); however, at higher concentrations there was no effect or only minor inhibitory effects observed (Feizi *et al.*, 2012).

No effects. No effects of nano-TiO_2 on willow tree seedling growth were observed (Seeger *et al.*, 2009). No effects were observed of alumina nanoparticles on kidney bean and rye grass at concentrations up to 17 mg/L (Doshi *et al.*, 2008). Lee *et al.* (2010) also observed no effects of Al_2O_3 nanoparticles on root elongation and growth of *Arabidopsis* plants. Despite concentration dependent growth inhibition of *Sorghum bicolor* and *Phaseolus radiatus* following exposure to nano-Ag in agar media, no effects on plant growth were found when the experiment was conducted in soil (Lee *et al.*, 2012).

2.3 CONCLUSION

Sufficient provision of clean drinking water is a current need of the growing world population. The development and application of nanotechnology has brought more efficient technologies for drinking water treatment and remediation of source water. Also, novel techniques using nanomaterials for wastewater treatment are bringing new possibilities for the prevention of contamination of receiving water bodies and supporting environmental sustainability goals. However, the possible threat to humans and the environment by these nanomaterials cannot be denied based on the outcome of numerous toxicity studies described in this chapter. It is obvious that more research is needed to be able to draw firm conclusions about the harm posed by the nanomaterials used in WWWT. As nanotoxicology research is still in its infancy, caution should be taken when using nanomaterials in WWWT.

REFERENCES

Adams L. K., Lyon D. Y. and Alvarez P. J. J. (2006). Comparative eco-toxicity of nanoscale TiO_2, SiO_2 and ZnO water suspensions. *Water Research*, **40**(19), 3527–3532.

Allen B., Kichambare P., Gou P., Vlasova I., Kapralov A., Konduru N. and Star A. (2008). Biodegradation of single-walled carbon nanotubes through enzymatic catalysis. *Nano Letter*, **8**(11), 3899–3903. doi: citeulike-article–id:3484103

Alloy M. M. and Roberts A. P. (2011). Effects of suspended multi-walled carbon nanotubes on daphnid growth and reproduction. *Ecotoxicology and Environmental Safety*, **74**(7), 1839–1843. doi: 10.1016/j.ecoenv.2011.06.020

Almofti M. R., Ichikawa T., Yamashita K., Terada H. and Shinohara Y. (2003). Silver ion induces a cyclosporine A-insensitive permeability transition in rat liver mitochondria and release of apoptogenic cytochrome C. *Journal of Biochemistry*, **134**(1), 43–49. doi: 10.1093/jb/mvg111

Aoshima H., Kokubo K., Shirakawa S., Ito M., Yamana S. and Oshima T. (2009). Antimicrobial activity of fullerenes and their hydroxylated derivatives. *Biocontrol Science*, **14**(2), 69–72.

Arbab A. S., Bashaw L. A., Miller B. R., Jordan E. K., Lewis B. K., Kalish H. and Frank J. A. (2003). Characterization of biophysical and metabolic properties of cells labeled with superparamagnetic iron oxide nanoparticles and transfection agent for cellular MR imaging. *Radiology*, **229**(3), 838–846. doi: 10.1148/radiol.2293021215

Arbab A. S., Yocum G. T., Rad A. M., Khakoo A. Y., Fellowes V., Read E. J. and Frank J. A. (2005). Labeling of cells with ferumoxides–protamine sulfate complexes does not inhibit function or differentiation capacity of hematopoietic or mesenchymal stem cells. *NMR in Biomedicine*, **18**(8), 553–559. doi: 10.1002/nbm.991

Arora S., Jain J., Rajwade J. M. and Paknikar K. M. (2009). Interactions of silver nanoparticles with primary mouse fibroblasts and liver cells. *Toxicology and Applied Pharmacology*, **236**(3), 310–318. doi: 10.1016/j.taap.2009.02.020

Arora S., Rajwade J. M. and Paknikar K. M. (2012). Nanotoxicology and in vitro studies: the need of the hour. *Toxicology and Applied Pharmacology*, **258**(2), 151–165. doi: 10.1016/j.taap.2011.11.010

Aruoja V., Dubourguier H. C., Kasemets K. and Kahru A. (2009). Toxicity of nanoparticles of CuO, ZnO and TiO_2 to microalgae *Pseudokirchneriella subcapitata*. *Science of the Total Environment*, **407**(4), 1461–1468. doi: 10.1016/j.scitotenv.2008.10.053

Aschberger K., Micheletti C., Sokull-Kluettgen B. and Christensen F. M. (2011). Analysis of currently available data for characterising the risk of engineered nanomaterials to the environment and human health – lessons learned from four case studies. *Environment International*, **37**(6), 1143–1156. doi: 10.1016/j.envint.2011.02.005

Asharani P. V., Low Kah Mun G., Hande M. P. and Valiyaveettil S. (2008a). Cytotoxicity and genotoxicity of silver nanoparticles in human cells. *ACS Nano*, **3**(2), 279–290. doi: 10.1021/nn800596w

Asharani P. V., Wu Y. L., Gong Z. and Valiyaveettil S. (2008b). Toxicity of silver nanoparticles in zebrafish models. *Nanotechnology*, **19**(25), 255102. doi: 10.1088/0957-4484/19/25/255102

Auffan M., Achouak W., Rose J., Roncato M. A., Chaneac C., Waite D. T. and Bottero J. Y. (2008). Relation between the redox state of iron-based nanoparticles and their cytotoxicity toward *Escherichia coli*. *Environmental Science and Technology*, **42**(17), 6730–6735. doi: 10.1021/es800086f

Auffan M., Decome L., Rose J., Orsiere T., De Meo M., Briois V. and Bottero J. Y. (2006). In vitro interactions between DMSA-coated maghemite nanoparticles and human fibroblasts: a physicochemical and cyto-genotoxical study. *Environmental Science and Technology*, **40**(14), 4367–4373. doi: 10.1021/es060691k

Baek Y. W. and An Y. J. (2011). Microbial toxicity of metal oxide nanoparticles (CuO NiO, ZnO, and Sb2O3) to *Escherichia coli, Bacillus subtilis, and Streptococcus aureus*. *Science of the Total Environment*, **409**(8), 1603–1608. doi: 10.1016/j.scitotenv.2011.01.014

Baun A., Hartmann N. B., Grieger K. and Kusk K. O. (2008). Ecotoxicity of engineered nanoparticles to aquatic invertebrates: a brief review and recommendations for future toxicity testing. *Ecotoxicology*, **17**(5), 387–395. doi: 10.1007/s10646-008-0208-y

Blaise C., Gagne F., Ferard J. F. and Eullaffroy P. (2008). Ecotoxicity of selected nano-materials to aquatic organisms. *Environmental Toxicology*, **23**(5), 591–598. doi: 10.1002/tox.20402

Brar S. K., Verma M., Tyagi R. D. and Surampalli R. Y. (2010). Engineered nanoparticles in wastewater and wastewater sludge – evidence and impacts. *Waste Management*, **30**(3), 504–520. doi: 10.1016/j.wasman.2009.10.012

Buyukhatipoglu K. and Clyne A. M. (2011). Superparamagnetic iron oxide nanoparticles change endothelial cell morphology and mechanics via reactive oxygen species formation. *Journal of Biomedical Materials Research Part A*, **96A**(1), 186–195. doi: 10.1002/jbm.a.32972

Canas J. E., Long M. Q., Nations S., Vadan R., Dai L., Luo M. X. and Olszyk D. (2008). Effects of functionalized and nonfunctionalized single-walled carbon nanotubes on root elongation of select crop species. *Environmental Toxicology and Chemistry*, **27**(9), 1922–1931. doi: 10.1897/08–117.1

Canas J. E., Qi B., Li S., Maul J. D., Cox S. B., Das S. and Green M. J. (2011). Acute and reproductive toxicity of nano-sized metal oxides (ZnO and TiO2) to earthworms (*Eisenia fetida*). *Journal of Environmental Monitoring*, **13**(12), 3351–3357.

Cha K., Hong H. W., Choi Y. G., Lee M. J., Park J. H., Chae H. K. and Myung H. (2008). Comparison of acute responses of mice livers to short-term exposure to nano-sized or micro-sized silver particles. *Biotechnology Letters*, **30**(11), 1893–1899. doi: 10.1007/s10529-008-9786–2

Chen R., Ratnikova T. A., Stone M. B., Lin S., Lard M., Huang G. and Ke P. C. (2010). Differential uptake of carbon nanoparticles by plant and mammalian cells. *Small*, **6**(5), 612–617. doi: 10.1002/smll.200901911

Cheng J., Flahaut E. and Cheng S. H. (2007). Effect of carbon nanotubes on developing zebrafish (*Danio rerio*) embryos. *Environmental Toxicology and Chemistry*, **26**(4), 708–716. doi: 10.1897/06–272r.1

Cheng X., Kan A. T. and Tomson M. B. (2004). Naphthalene adsorption and desorption from aqueous C_{60} fullerene. *Journal of Chemical and Engineering Data*, **49**(3), 675–683. doi: 10.1021/je030247m

Chitose N., Ueta S., Seino S. and Yamamoto T. A. (2003). Radiolysis of aqueous phenol solutions with nanoparticles. 1. Phenol degradation and TOC removal in solutions containing TiO_2 induced by UV, gamma-ray and electron beams. *Chemosphere*, **50**(8), 1007–1013. doi: Pii s0045-6535(02)00642-210.1016/s0045-6535(02)00642-2

Chung H., Son Y., Yoon T. K., Kim S. and Kim W. (2011). The effect of multi-walled carbon nanotubes on soil microbial activity. *Ecotoxicology and Environmental Safety*, **74**(4), 569–575. doi: 10.1016/j.ecoenv.2011.01.004

Cifuentes Z., Custardoy L., de la Fuente J. M., Marquina C., Ricardo Ibarra M., Rubiales D. and Perez-de-Luque A. (2010). Absorption and translocation to the aerial part of magnetic carbon-coated nanoparticles through the root of different crop plants. *Journal of Nanobiotechnology*, **8**, 26. doi: 10.1186/1477-3155-8-26

Cloete T. E. (2010). *Nanotechnology in Water Treatment Applications*. Caister Academic Press, Norfolk, UK.

Colvin V. L. (2003). The potential environmental impact of engineered nanomaterials. *Nature Biotechnology*, **21**(10), 1166–1170. doi: 10.1038/nbt875

Corredor E., Testillano P. S., Coronado M. J., Gonzalez-Melendi P., Fernandez-Pacheco R., Marquina C. and Risueno M. C. (2009). Nanoparticle penetration and transport in living pumpkin plants: in situ subcellular identification. *BMC Plant Biology*, **9**, 45(1–11). doi: 4510.1186/1471-2229-9-45

CSTEE (2000). Opinion on the available scientific approaches to assess the potential effects and risk of chemicals on terrestrial ecosystems; opinion in 19th CSTEE plenary meeting, http://ec.europa.eu/health/scientific_committees/environmental_risks/opinions/sctee/index_en.htm (accessed 21 April 2012)

Cullen L. G., Tilston E. L., Mitchell G. R., Collins C. D. and Shaw L. J. (2011). Assessing the impact of nano- and micro-scale zerovalent iron particles on soil microbial activities: particle reactivity interferes with assay conditions and interpretation of genuine microbial effects. *Chemosphere*, **82**(11), 1675–1682. doi: 10.1016/j.chemosphere.2010.11.009

DeFriend K. A., Wiesner M. R. and Barron A. R. (2003). Alumina and aluminate ultrafiltration membranes derived from alumina nanoparticles. *Journal of Membrane Science*, **224**(1–2), 11–28. doi: 10.1016/s0376-7388(03)00344-2

Dhawan A. and Sharma V. (2010). Toxicity assessment of nanomaterials: methods and challenges. *Analytical and Bioanalytical Chemistry*, **398**(2), 589–605. doi: 10.1007/s00216-010-3996-x

Diallo M. S., Christie S., Swaminathan P., Balogh L., Shi X. Y., Um W. and Johnson J. H. (2004). Dendritic chelating agents. 1. Cu(II) binding to ethylene diamine core poly (amidoamine) dendrimers in aqueous solutions. *Langmuir*, **20**(7), 2640–2651. doi: 10.1021/la036108k

Diallo M. S., Christie S., Swaminathan P., Johnson J. H. and Goddard W. A. (2005). Dendrimer enhanced ultrafiltration. 1. Recovery of Cu(II) from aqueous solutions using PAMAM dendrimers with ethylene diamine core and terminal NH2 groups. *Environmental Science and Technology*, **39**(5), 1366–1377. doi: 10.1021/es048961r

Dimkpa C. O., Calder A., Britt D. W., McLean J. E. and Anderson A. J. (2011). Responses of a soil bacterium, *Pseudomonas chlororaphis* O6 to commercial metal oxide nanoparticles compared with responses to metal ions. *Environmental Pollution*, **159** (7), 1749–1756. doi: 10.1016/j.envpol.2011.04.020

Dinesh R., Anandaraj M., Srinivasan V. and Hamza S. (2012). Engineered nanoparticles in the soil and their potential implications to microbial activity. *Geoderma*, **173–174**(0), 19–27. doi: 10.1016/j.geoderma.2011.12.018

Donaldson K., Borm P. J. A., Castranova V. and Gulumian M. (2009). The limits of testing particle-mediated oxidative stress *in vitro* in predicting diverse pathologies; relevance for testing of nanoparticles. *Particle and Fibre Toxicology*, **6**, 13(1–8). doi: 1310.1186/ 1743-8977-6-13

Doshi R., Braida W., Christodoulatos C., Wazne M. and O'Connor G. (2008). Nano-aluminum: transport through sand columns and environmental effects on plants and soil communities. *Environmental Research*, **106**(3), 296–303. doi: 10.1016/j.envres.2007.04.006

Drobne D., Jemec A. and Tkalec Z. P. (2009). In vivo screening to determine hazards of nanoparticles: nanosized TiO2. *Environmental Pollution*, **157**(4), 1157–1164. doi: 10.1016/j.envpol.2008.10.018

Du W., Sun Y., Ji R., Zhu J., Wu J. and Guo H. (2011). TiO_2 and ZnO nanoparticles negatively affect wheat growth and soil enzyme activities in agricultural soil. *Journal of Environmental Monitoring*, **13**(4), 822–828. doi: 10.1039/c0em00611d

Edgington A. J. (2011). Toxicity of Carbon Nanotubes to *Daphnia magna*. (Doctoral), Clemson University, South Carolina, USA.

El-Kemary M., El-Shamy H. and El-Mehasseb I. (2010). Photocatalytic degradation of ciprofloxacin drug in water using ZnO nanoparticles. *Journal of Luminescence*, **130**(12), 2327–2331. doi: 10.1016/j.jlumin.2010.07.013

Fabrega J., Fawcett S. R., Renshaw J. C. and Lead J. R. (2009). Silver nanoparticle impact on bacterial growth: effect of pH, concentration, and organic matter. *Environmental Science and Technology*, **43**(19), 7285–7290. doi: 10.1021/es803259g

Fairbrother A. and Hope B. (2005). Terrestrial ecotoxicology. In: *Encyclopedia of Toxicology*, P. Wexler and S. C. Gad (eds), Academic Press, San Diego.

Farre M., Gajda-Schrantz K., Kantiani L. and Barcelo D. (2009). Ecotoxicity and analysis of nanomaterials in the aquatic environment. *Analytical and Bioanalytical Chemistry*, **393**(1), 81–95. doi: 10.1007/s00216-008-2458-1

Federici G., Shaw B. J. and Handy R. D. (2007). Toxicity of titanium dioxide nanoparticles to rainbow trout (*Oncorhynchus mykiss*): gill injury, oxidative stress, and other physiological effects. *Aquatic Toxicology*, **84**(4), 415–430. doi: 10.1016/j.aquatox.2007.07.009

Feizi H., Moghaddam P. R., Shahtahmassebi N. and Fotovat A. (2012). Impact of bulk and nanosized titanium dioxide (TiO_2) on wheat seed germination and seedling growth. *Biological Trace Element Research*, **146**(1), 101–106. doi: 10.1007/s12011-011-9222-7

Fiorito S., Serafino A., Andreola F. and Bernier P. (2006). Effects of fullerenes and single-wall carbon nanotubes on murine and human macrophages. *Carbon*, **44**(6), 1100–1105. doi: 10.1016/j.carbon.2005.11.009

Fortner J. D., Lyon D. Y., Sayes C. M., Boyd A. M., Falkner J. C., Hotze E. M. and Hughes J. B. (2005). C_{60} in water: nanocrystal formation and microbial response. *Environmental Science and Technology*, **39**(11), 4307–4316. doi: 10.1021/es0498099n

Fugetsu B., Satoh S., Shiba T., Mizutani T., Lin Y. B., Terui N. and Watari F. (2004). Caged multiwalled carbon nanotubes as the adsorbents for affinity-based elimination of ionic dyes. *Environmental Science and Technology*, **38**(24), 6890–6896. doi: 10.1021/es049554i

Gagne F., Auclair J., Turcotte P., Fournier M., Gagnon C., Sauve S. and Blaise C. (2008). Ecotoxicity of CdTe quantum dots to freshwater mussels: impacts on immune system, oxidative stress and genotoxicity. *Aquatic Toxicology*, **86**(3), 333–340. doi: 10.1016/j.aquatox.2007.11.013

Gajjar P., Pettee B., Britt D., Huang W., Johnson W. and Anderson A. (2009). Antimicrobial activities of commercial nanoparticles against an environmental soil microbe, *Pseudomonas putida* KT2440. *Journal of Biological Engineering*, **3**(1), 9.

Gao J., Wang Y., Folta K. M., Krishna V., Bai W., Indeglia P. and Moudgil B. (2011). Polyhydroxy Fullerenes (Fullerols or Fullerenols): beneficial effects on growth and lifespan in diverse biological models. *PLoS ONE*, **6**(5), e19976. doi: 10.1371/journal.pone.0019976

Gao L., Zhuang J., Nie L., Zhang J., Zhang Y., Gu N. and Yan X. (2007). Intrinsic peroxidase-like activity of ferromagnetic nanoparticles. *Nature Nanotechnology*, **2**(9), 577–583. doi: 10.1038/nnano.2007.260

Garcia A., Delgado L., Tora J. A., Casals E., Gonzalez E., Puntes V. and Sanchez A. (2012). Effect of cerium dioxide, titanium dioxide, silver, and gold nanoparticles on the activity of microbial communities intended in wastewater treatment. *Journal of Hazardous Materials*, **199**, 64–72. doi: 10.1016/j.jhazmat.2011.10.057

Garcia A., Espinosa R., Delgado L., Casals E., Gonzalez E., Puntes V. and Sanchez A. (2011). Acute toxicity of cerium oxide, titanium oxide and iron oxide nanoparticles using standardized tests. *Desalination*, **269**(1–3), 136–141. doi: 10.1016/j.desal.2010. 10.052

Garcia M. V. (2004). *Effects of Pesticides on Soil Fauna: Development of Ecotoxicological Test Methods for Tropical Regions* (pp. 281). University of Bonn, Germany, Ecology and Development Series (No. 19).

Ge Y., Schimel J. P. and Holden P. A. (2011). Evidence for negative effects of TiO_2 and ZnO nanoparticles on soil bacterial communities. *Environmental Science and Technology*, **45**(4), 1659–1664. doi: 10.1021/es103040t

Geiser M., Rothen-Rutishauser B., Kapp N., Schurch S., Kreyling W., Schulz H. and Gehr P. (2005). Ultrafine particles cross cellular membranes by nonphagocytic mechanisms in lungs and in cultured cells. *Environmental Health Perspectives*, **113**(11), 1555–1560. doi: 10.1289/ehp.8006

Ghodake G., Seo Y. D. and Lee D. S. (2011). Hazardous phytotoxic nature of cobalt and zinc oxide nanoparticles assessed using *Allium cepa*. *Journal of Hazardous Materials*, **186** (1), 952–955. doi: 10.1016/j.jhazmat.2010.11.018

Gottschalk F. and Nowack B. (2011). The release of engineered nanomaterials to the environment. *Journal of Environmental Monitoring*, **13**(5), 1145–1155. doi: 10.1039/c0em00547a

Gubbins E. J., Batty L. C. and Lead J. R. (2011). Phytotoxicity of silver nanoparticles to *Lemna minor* L. *Environmental Pollution*, **159**(6), 1551–1559. doi: 10.1016/j.envpol.2011.03.002

Gupta A. K. and Curtis A. S. G. (2004). Lactoferrin and ceruloplasmin derivatized superparamagnetic iron oxide nanoparticles for targeting cell surface receptors. *Biomaterials*, **25**(15), 3029–3040. doi: 10.1016/j.biomaterials.2003.09.095

Gupta A. K. and Gupta M. (2005). Cytotoxicity suppression and cellular uptake enhancement of surface modified magnetic nanoparticles. *Biomaterials*, **26**(13), 1565–1573. doi: 10.1016/j.biomaterials.2004.05.022

Hall S., Bradley T., Moore J. T., Kuykindall T. and Minella L. (2009). Acute and chronic toxicity of nano-scale TiO_2 particles to freshwater fish, cladocerans, and green algae, and effects of organic and inorganic substrate on TiO_2 toxicity. *Nanotoxicology*, **3**(2), 91–97. doi: 10.1080/17435390902788078

Handy R. D., Cornelis G., Fernandes T., Tsyusko O., Decho A., Sabo-Attwood T. and Horne N. (2012). Ecotoxicity test methods for engineered nanomaterials: practical experiences and recommendations from the bench. *Environmental Toxicology and Chemistry*, **31**(1), 15–31. doi: 10.1002/etc.706

Handy R. D., Al-Bairuty G., Al-Jubory A., Ramsden C. S., Boyle D., Shaw B. J. and Henry T. B. (2011). Effects of manufactured nanomaterials on fishes: a target organ and body systems physiology approach. *Journal of Fish Biology*, **79**(4), 821–853. doi: 10.1111/j.1095-8649.2011.03080.x

Handy R. D., Henry T. B., Scown T. M., Johnston B. D. and Tyler C. R. (2008). Manufactured nanoparticles: their uptake and effects on fish-a mechanistic analysis. *Ecotoxicology*, **17**(5), 396–409. doi: 10.1007/s10646-008-0205-1

Hartmann N. B., Von der Kammer F., Hofmann T., Baalousha M., Ottofuelling S. and Baun A. (2010). Algal testing of titanium dioxide nanoparticles-testing considerations, inhibitory effects and modification of cadmium bioavailability. *Toxicology*, **269**(2–3), 190–197. doi: 10.1016/j.tox.2009.08.008

He S. Y., Feng Y. Z., Ren H. X., Zhang Y., Gu N. and Lin X. G. (2011). The impact of iron oxide magnetic nanoparticles on the soil bacterial community. *Journal of Soils and Sediments*, **11**(8), 1408–1417. doi: 10.1007/s11368-011-0415-7

Heinlaan M., Ivask A., Blinova I., Dubourguier H. C. and Kahru A. (2008). Toxicity of nanosized and bulk ZnO CuO and TiO_2 to bacteria *Vibrio fischeri* and crustaceans *Daphnia magna* and *Thamnocephalus platyurus*. *Chemosphere*, **71**(7), 1308–1316. doi: 10.1016/j.chemosphere.2007.11.047

Helland A., Wick P., Koehler A., Schmid K. and Som C. (2008). Reviewing the environmental and human health knowledge base of carbon nanotubes. *Ciencia and Saude Coletiva*, **13**(2), 441–452. doi: 10.1590/s1413-81232008000200019

Henry T. B., Menn F.-M., Fleming J. T., Wilgus J., Compton R. N. and Sayler G. S. (2007). Attributing effects of aqueous C_{60} nano-aggregates to tetrahydrofuran decomposition products in larval zebrafish by assessment of gene expression. *Environmental Health Perspectives*, **115**(7), 1059–1065.

Hergt R., Dutz S., Mueller R. and Zeisberger M. (2006). Magnetic particle hyperthermia: nanoparticle magnetism and materials development for cancer therapy. *Journal of Physics-Condensed Matter*, **18**(38), S2919–S2934. doi: 10.1088/0953-8984/ 18/38/s26

Hernandez-Viezcas J. A., Castillo-Michel H., Servin A. D., Peralta-Videa J. R. and Gardea-Torresdey J. L. (2011). Spectroscopic verification of zinc absorption and distribution in the desert plant *Prosopis juliflora-velutina* (velvet mesquite) treated with ZnO nanoparticles. *Chemical Engineering Journal*, **170**(2–3), 346–352. doi: 10.1016/j.cej.2010.12.021

Hsin Y. H., Chen C. F., Huang S., Shih T.-S., Lai P. S. and Chueh P. J. (2009). The apoptotic effect of nanosilver is mediated by a ROS- and JNK-dependent mechanism involving the mitochondrial pathway in NIH3T3 cells (vol 179, pg 130, 2008). *Toxicology Letters*, **185**(2), 142–142. doi: 10.1016/j.toxlet.2008.12.010

Hu C. W., Li M., Cui Y. B., Li D. S., Chen J. and Yang L. Y. (2010). Toxicological effects of TiO_2 and ZnO nanoparticles in soil on earthworm *Eisenia fetida*. *Soil Biology and Biochemistry*, **42**(4), 586–591. doi: 10.1016/j.soilbio.2009.12.007

Huang D. M., Hsiao J. K., Chen Y.-C., Chien L. Y., Yao M., Chen Y. K. and Chen Y.-C. (2009). The promotion of human mesenchymal stem cell proliferation by superparamagnetic iron oxide nanoparticles. *Biomaterials*, **30**(22), 3645–3651. doi: 10.1016/j.biomaterials.2009.03.032

Hund-Rinke K. and Simon M. (2006). Ecotoxic effect of photocatalytic active nanoparticles TiO_2 on algae and daphnids. *Environmental Science and Pollution Research*, **13**(4), 225–232. doi: 10.1065/espr2006.06.311

Hussain M. A., Sood A. K. and Kabir M. A. (2009). On the cytotoxicity of carbon nanotubes. *Current Science*, **96**(5), 664–673.

Hussain S. M., Hess K. L., Gearhart J. M., Geiss K. T. and Schlager J. J. (2005). In vitro toxicity of nanoparticles in BRL 3A rat liver cells. *Toxicology in Vitro*, **19**(7), 975–983. doi: 10.1016/j.tiv.2005.06.034

Jaidev L. R. and Narasimha G. (2010). Fungal mediated biosynthesis of silver nanoparticles, characterization and antimicrobial activity. *Colloids and Surfaces B-Biointerfaces*, **81**(2), 430–433. doi: 10.1016/j.colsurfb.2010.07.033

Jain T. K., Reddy M. K., Morales M. A., Leslie-Pelecky D. L. and Labhasetwar V. (2008). Biodistribution, clearance, and biocompatibility of iron oxide magnetic nanoparticles in rats. *Molecular Pharmaceutics*, **5**(2), 316–327. doi: 10.1021/mp7001285

Ji J. H., Jung J. H., Kim S. S., Yoon J.-U., Park J. D., Choi B. S. and Yu I. J. (2007). Twenty-eight-day inhalation toxicity study of silver nanoparticles in Sprague-Dawley rats. *Inhalation Toxicology*, **19**(10), 857–871. doi: 10.1080/ 08958370701432108

Jia G., Wang H. F., Yan L., Wang X., Pei R. J., Yan T. and Guo X. B. (2005). Cytotoxicity of carbon nanomaterials: single-wall nanotube, multi-wall nanotube, and fullerene. *Environmental Science and Technology*, **39**(5), 1378–1383. doi: 10.1021/es0487291

Jiang W., Mashayekhi H. and Xing B. (2009). Bacterial toxicity comparison between nano- and micro-scaled oxide particles. *Environmental Pollution*, **157**(5), 1619–1625. doi: 10.1016/j.envpol.2008.12.025

Jin C. Y., Zhu B. S., Wang X. F. and Lu Q. H. (2008). Cytotoxicity of titanium dioxide nanoparticles in mouse fibroblast cells. *Chemical Research in Toxicology*, **21**(9), 1871–1877. doi: 10.1021/tx800179f

Johansen A., Pedersen A. L., Jensen K. A., Karlson U., Hansen B. M., Scott-Fordsmand J. J. and Winding A. (2008). Effects of C_{60} fullerene nanoparticles on soil bacteria and protozoans. *Environmental Toxicology and Chemistry*, **27**(9), 1895–1903. doi: 10.1897/07–375.1

Jones N., Ray B., Ranjit K. T. and Manna A. C. (2008). Antibacterial activity of ZnO nanoparticle suspensions on a broad spectrum of microorganisms. *Fems Microbiology Letters*, **279**(1), 71–76. doi: 10.1111/j.1574-6968.2007.01012.x

Jonker M. T. O. and Suijkerbuijk M. P. W. (2008). Risk Assessment of Engineered Carbonaceous Nanoparticles in the Aquatic Environment. Masters thesis, Utrecht University, Netherland.

Juhel G., Batisse E., Hugues Q., Daly D., van Pelt F. N. A. M., O'Halloran J. and Jansen M. A. K. (2011). Alumina nanoparticles enhance growth of *Lemna minor*. *Aquatic Toxicology*, **105**(3–4), 328–336. doi: 10.1016/j.aquatox.2011.06.019

Kabra K., Chaudhary R. and Sawhney R. L. (2004). Treatment of hazardous organic and inorganic compounds through aqueous-phase photocatalysis: a review. *Industrial and Engineering Chemistry Research*, **43**(24), 7683–7696. doi: 10.1021/ie0498551

Kagan V. E., Konduru N. V., Feng W., Allen B. L., Conroy J., Volkov Y. and Shvedova A. A. (2010). Carbon nanotubes degraded by neutrophil myeloperoxidase induce less pulmonary inflammation. *Nature Nanotechnology*, **5**(5), 354–359. doi: 10.1038/nnano.2010.44

Kahru A. and Dubourguier H. C. (2010). From ecotoxicology to nanoecotoxicology. *Toxicology*, **269**(2–3), 105–119. doi: 10.1016/j.tox.2009.08.016

Kang S., Herzberg M., Rodrigues D. F. and Elimelech M. (2008a). Antibacterial effects of carbon nanotubes: size does matter. *Langmuir*, **24**(13), 6409–6413. doi: 10.1021/la800951v

Kang S., Mauter M. S. and Elimelech M. (2008b). Physicochemical determinants of multiwalled carbon nanotube bacterial cytotoxicity. *Environmental Science and Technology*, **42**(19), 7528–7534. doi: 10.1021/es8010173

Kapp N., Kreyling W., Schulz H., Hof V. I., Gehr P., Semmler M. and Geiser M. (2004). Electron energy loss spectroscopy for analysis of inhaled ultrafine particles in rat lungs. *Microscopy Research and Technique*, **63**(5), 298–305. doi: 10.1002/jemt.20044

Kedziorek D. A., Muja N., Walczak P., Ruiz-Cabello J., Gilad A. A., Jie C. C. and Bulte J. W. M. (2010). Gene expression profiling reveals early cellular responses to intracellular magnetic labeling with superparamagnetic iron oxide nanoparticles. *Magnetic Resonance in Medicine*, **63**(4), 1031–1043. doi: 10.1002/mrm.22290

Kennedy A. C. and Smith K. L. (1995). Soil microbial diversity and the sustainability of agricultural soils. *Plant and Soil*, **170**(1), 75–86. doi: 10.1007/bf02183056

Khodakovskaya M., Dervishi E., Mahmood M., Xu Y., Li Z., Watanabe F. and Biris A. S. (2009). Carbon nanotubes are able to penetrate plant seed coat and dramatically affect seed germination and plant growth. *ACS Nano*, **3**(10), 3221–3227. doi: 10.1021/nn900887m

Khodakovskaya M. V., de Silva K., Nedosekin D. A., Dervishi E., Biris A. S., Shashkov E. V. and Zharov V. P. (2011). Complex genetic, photothermal, and photoacoustic analysis of nanoparticle-plant interactions. *Proceedings of the National Academy of Sciences of the United States of America*, **108**(3), 1028–1033. doi: 10.1073/pnas.1008856108

Kim J., Kim S. and Lee S. (2011a). Differentiation of the toxicities of silver nanoparticles and silver ions to the Japanese medaka (*Oryzias latipes*) and the cladoceran *Daphnia magna*. *Nanotoxicology*, **5**(2), 208–214. doi: 10.3109/17435390.2010.508137

Kim S., Kim J. and Lee I. (2011b). Effects of Zn and ZnO nanoparticles and Zn^{2+}on soil enzyme activity and bioaccumulation of Zn in *Cucumis sativus*. *Chemistry and Ecology*, **27**(1), 49–55. doi: Pii 93354307610.1080/02757540.2010.529074

Kim S., Lee S. and Lee I. (2012). Alteration of phytotoxicity and oxidant stress potential by metal oxide nanoparticles in *Cucumis sativus*. *Water Air and Soil Pollution*, **223**(5), 1–8. doi: 10.1007/s11270-011-1067-3

Kim Y. S., Kim J. S., Cho H. S., Rha D. S., Kim J. M., Park J. D. and Yu I. J. (2008). Twenty-eight-day oral toxicity, genotoxicity, and gender-related tissue distribution of silver nanoparticles in Sprague-Dawley rats. *Inhalation Toxicology*, **20**(6), 575–583. doi: 10.1080/08958370701874663

Kittler S., Greulich C., Diendorf J., Koeller M. and Epple M. (2010). Toxicity of silver nanoparticles increases during storage because of slow dissolution under release of silver ions. *Chemistry of Materials*, **22**(16), 4548–4554. doi: 10.1021/cm100023p

Klaine S. J., Alvarez P. J. J., Batley G. E., Fernandes T. F., Handy R. D., Lyon D. Y. and Lead J. R. (2008). Nanomaterials in the environment: behavior, fate, bioavailability, and effects. *Environmental Toxicology and Chemistry*, **27**(9), 1825–1851. doi: 10.1897/08–090.1

Kostura L., Kraitchman D. L., Mackay A. M., Pittenger M. F. and Bulte J. W. M. (2004). Feridex labeling of mesenchymal stem cells inhibits chondrogenesis but not adipogenesis or osteogenesis. *NMR in Biomedicine*, **17**(7), 513–517. doi: 10.1002/nbm.925

Koyama S., Kim Y. A., Hayashi T., Takeuchi K., Fujii C., Kuroiwa N. and Endo M. (2009). *In vivo* immunological toxicity in mice of carbon nanotubes with impurities. *Carbon*, **47** (5), 1365–1372. doi: 10.1016/j.carbon.2009.01.028

Kumar N., Shah V. and Walker V. K. (2011). Perturbation of an arctic soil microbial community by metal nanoparticles. *Journal of Hazardous Materials*, **190**(1–3), 816–822. doi: 10.1016/j.jhazmat.2011.04.005

Laban G., Nies L. F., Turco R. F., Bickham J. W. and Sepulveda M. S. (2010). The effects of silver nanoparticles on fathead minnow (*Pimephales promelas*) embryos. *Ecotoxicology*, **19**(1), 185–195. doi: 10.1007/s10646-009-0404-4

Lai J. C. K., Lai M. B., Jandhyam S., Dukhande V. V., Bhushan A., Daniels C. K. and Leung S. W. (2008). Exposure to titanium dioxide and other metallic oxide nanoparticles induces cytotoxicity on human neural cells and fibroblasts. *International Journal of Nanomedicine*, **3**(4), 533–545.

Lecoanet H. F., Bottero J. Y. and Wiesner M. R. (2004). Laboratory assessment of the mobility of nanomaterials in porous media. *Environmental Science and Technology*, **38**(19), 5164–5169. doi: 10.1021/es0352303

Lecoanet H. F. and Wiesner M. R. (2004). Velocity effects on fullerene and oxide nanoparticle deposition in porous media. *Environmental Science and Technology*, **38** (16), 4377–4382. doi: 10.1021/es035354f

Lee C. W., Mahendra S., Zodrow K., Li D., Tsai Y. C., Braam J. and Alvarez P. J. J. (2010). Developmental phytotoxicity of metal oxide nanoparticles to *Arabidopsis thaliana*. *Environmental Toxicology and Chemistry*, **29**(6), 1399–1399. doi: 10.1002/etc.234

Lee K. J., Nallathamby P. D., Browning L. M., Osgood C. J. and Xu X.-H. N. (2007). In vivo imaging of transport and biocompatibility of single silver nanoparticles in early development of zebrafish embryos. *ACS Nano*, **1**(2), 133–143. doi: 10.1021/nn700048y

Lee W.-M., Kwak J. I. and An Y.-J. (2012). Effect of silver nanoparticles in crop plants *Phaseolus radiatus* and *Sorghum bicolor*: media effect on phytotoxicity. *Chemosphere*, **86**(5), 491–499. doi: 10.1016/j.chemosphere.2011.10.013

Li D. and Alvarez P. J. J. (2011). Avoidance, weight loss, and cocoon production assessment for *Eisenia fetida* exposed to C(60) in soil. *Environmental Toxicology and Chemistry* **30**, 2542–2545.

Li Q., Mahendra S., Lyon D. Y., Brunet L., Liga M. V., Li D. and Alvarez P. J. J. (2008). Antimicrobial nanomaterials for water disinfection and microbial control: potential applications and implications. *Water Research*, **42**(18), 4591–4602. doi: 10.1016/j.watres.2008.08.015

Li Q. L., Yuan D. X. and Lin Q. M. (2004). Evaluation of multi-walled carbon nanotubes as an adsorbent for trapping volatile organic compounds from environmental samples. *Journal of Chromatography A*, **1026**(1–2), 283–288. doi: 10.1016/j.chroma.2003. 10.109

Li Y. H., Ding J., Luan Z., Di Z., Zhu Y., Xu C. and Wei B. (2003). Competitive adsorption of Pb^{2+}, Cu^{2+} and Cd^{2+} ions from aqueous solutions by multiwalled carbon nanotubes. *Carbon*, **41**(14), 2787–2792. doi: 10.1016/s0008-6223(03)00392-0

Lin D. and Xing B. (2007). Phytotoxicity of nanoparticles: inhibition of seed germination and root growth. *Environmental Pollution*, **150**(2), 243–250. doi: 10.1016/j.envpol.2007. 01.016

Lovern S. B. and Klaper R. (2006). Daphnia magna mortality when exposed to titanium dioxide and fullerene (C_{60}) nanoparticles. *Environmental Toxicology and Chemistry*, **25**(4), 1132–1137. doi: 10.1897/05-278r.1

Lu C. M., Zhang C. Y., Wen J. Q., Wu G. R. and Tao M. X. (2002). Research of the effect of nanometer materials on germination and growth enhancement of *Glycine max* and its mechanisms. *Soyabean Science*, **21**(3), 168–172.

Lyon D. Y., Adams L. K., Falkner J. C. and Alvarez P. J. J. (2006). Antibacterial activity of fullerene water suspensions: effects of preparation method and particle size. *Environmental Science and Technology*, **40**(14), 4360–4366. doi: 10.1021/es0603655

Lyon D. Y., Brown D. A. and Alvarez P. J. J. (2008). Implications and potential applications of bactericidal fullerene water suspensions: effect of nC(60) concentration, exposure conditions and shelf life. *Water Science and Technology*, **57**(10), 1533–1538. doi: 10.2166/wst.2008.282

Majumdar H. and Ahmed G. U. (2011). Phytotoxicity effect of silver nanoparticles on *Oryza sativa*. *International Journal of ChemTech Research*, **3**(3), 1494–1500.

Mauter M. S. and Elimelech M. (2008). Environmental applications of carbon-based nanomaterials. *Environmental Science and Technology*, **42**(16), 5843–5859. doi: 10.1021/es8006904

Menard A., Drobne D. and Jemec A. (2011). Ecotoxicity of nanosized TiO_2. Review of in vivo data. *Environmental Pollution*, **159**(3), 677–684. doi: 10.1016/j.envpol.2010.11.027

Mercer R. R., Hubbs A. F., Scabilloni J. F., Wang L., Battelli L. A., Schwegler-Berry D. and Porter D. W. (2010). Distribution and persistence of pleural penetrations by multi-walled carbon nanotubes. *Particle and Fibre Toxicology*, **7**, 28(1–11). doi: 2810.1186/1743-8977-7-28

Miao A.-J., Schwehr K. A., Xu C., Zhang S. J., Luo Z., Quigg A. and Santschi P. H. (2009). The algal toxicity of silver engineered nanoparticles and detoxification by exopolymeric substances. *Environmental Pollution*, **157**(11), 3034–3041. doi: 10.1016/j.envpol. 2009.05.047

Mondal A., Basu R., Das S. and Nandy P. (2011). Beneficial role of carbon nanotubes on mustard plant growth: an agricultural prospect. *Journal of Nanoparticle Research*, **13**(10), 4519–4528. doi: 10.1007/s11051-011-0406-z

Murray A. R., Kisin E., Leonard S. S., Young S. H., Kommineni C., Kagan V. E. and Shvedova A. A. (2009). Oxidative stress and inflammatory response in dermal toxicity of single-walled carbon nanotubes. *Toxicology*, **257**(3), 161–171. doi: 10.1016/j.tox.2008.12.023

Musee N., Thwala M. and Nota N. (2011). The antibacterial effects of engineered nanomaterials: implications for wastewater treatment plants. *Journal of Environmental Monitoring*, **13**(5), 1164–1183. doi: 10.1039/c1em10023h

Namiki Y., Namiki T., Yoshida H., Ishii Y., Tsubota A., Koido S. and Tada N. (2009). A novel magnetic crystal-lipid nanostructure for magnetically guided in vivo gene delivery. [10.1038/nnano.2009.202]. *Nat Nano*, **4**(9), 598–606. doi: http://www.nature.com/nnano/journal/v4/n9/suppinfo/nnano.2009.202_S1.html (accessed 18 April 2012)

Nanoderm: Quality of skin as a barrier to ultra-fine particles – Final Report. (2007). Retrieved April 18, 2012, from http://www.uni-leipzig.de/~nanoderm/Downloads/Nanoderm_Final_Report.pdf

Navarro E., Baun A., Behra R., Hartmann N. B., Filser J., Miao A. J. and Sigg L. (2008a). Environmental behavior and ecotoxicity of engineered nanoparticles to algae, plants, and fungi. *Ecotoxicology*, **17**(5), 372–386. doi: 10.1007/s10646-008-0214-0

Navarro E., Piccapietra F., Wagner B., Marconi F., Kaegi R., Odzak N. and Behra R. (2008b). Toxicity of silver nanoparticles to *Chlamydomonas reinhardtii*. *Environmental Science and Technology*, **42**(23), 8959–8964. doi: 10.1021/es801785m

Nel A. E., Madler L., Velegol D., Xia T., Hoek E. M. V., Somasundaran P. and Thompson M. (2009). Understanding biophysicochemical interactions at the nano-bio interface. [10.1038/nmat2442]. *Nature Material*, **8**(7), 543–557.

Nyberg L., Turco R. F. and Nies L. (2008). Assessing the impact of nanomaterials on anaerobic microbial communities. *Environmental Science and Technology*, **42**(6), 1938–1943. doi: 10.1021/es072018g

Oberdörster E., Zhu S., Blickley T. M., McClellan-Green P. and Haasch M. L. (2006). Ecotoxicology of carbon-based engineered nanoparticles: effects of fullerene (C_{60}) on aquatic organisms. *Carbon*, **44**(6), 1112–1120. doi: 10.1016/j.carbon.2005.11.008

Oleszczuk P., Josko I. and Xing B. S. (2011). The toxicity to plants of the sewage sludges containing multiwalled carbon nanotubes. *Journal of Hazardous Materials*, **186**(1), 436–442. doi: 10.1016/j.jhazmat.2010.11.028

Peng X. J., Li Y. H., Luan Z. K., Di Z. C., Wang H. Y., Tian B. H. and Jia Z. P. (2003). Adsorption of 1,2-dichlorobenzene from water to carbon nanotubes. *Chemical Physics Letters*, **376**(1–2), 154–158. doi: 10.1016/s0009-2614(03)00960–6

Pisanic T. R., II, Blackwell J. D., Shubayev V. I., Finones R. R. and Jin S. (2007). Nanotoxicity of iron oxide nanoparticle internalization in growing neurons. *Biomaterials*, **28**(16), 2572–2581. doi: 10.1016/j.biomaterials.2007.01.043

Porter A. E., Gass M., Muller K., Skepper J. N., Midgley P. and Welland M. (2007). Visualizing the uptake of C_{60} to the cytoplasm and nucleus of human monocyte-derived macrophage cells using energy-filtered transmission electron microscopy and electron tomography. *Environmental Science and Technology*, **41**(8), 3012–3017. doi: 10.1021/es062541f

Porter A. E., Muller K., Skepper J., Midgley P. and Welland M. (2006). Uptake of C_{60} by human monocyte macrophages, its localization and implications for toxicity: studied by high resolution electron microscopy and electron tomography. *Acta Biomaterialia*, **2**(4), 409–419. doi: 10.1016/j.actbio.2006.02.006

Puppi J., Mitry R. R., Modo M., Dhawan A., Raja K. and Hughes R. D. (2011). Use of a clinically approved iron oxide MRI contrast agent to label human hepatocytes. *Cell Transplantation*, **20**(6), 963–975. doi: 10.3727/096368910x543367

Raghupathi K. R., Koodali R. T. and Manna A. C. (2011). Size-dependent bacterial growth inhibition and mechanism of antibacterial activity of zinc oxide nanoparticles. *Langmuir*, **27**(7), 4020–4028. doi: 10.1021/la104825u

Rai M., Yadav A. and Gade A. (2009). Silver nanoparticles as a new generation of antimicrobials. *Biotechnology Advances*, **27**(1), 76–83. doi: 10.1016/j.biotechadv.2008.09.002

Ramsden C. S., Smith T. J., Shaw B. J. and Handy R. D. (2009). Dietary exposure to titanium dioxide nanoparticles in rainbow trout, (*Oncorhynchus mykiss*): no effect on growth, but subtle biochemical disturbances in the brain. *Ecotoxicology*, **18**(7), 939–951. doi: 10.1007/s10646-009-0357-7

Rico C. M., Majumdar S., Duarte-Gardea M., Peralta-Videa J. R. and Gardea-Torresdey J. L. (2011). Interaction of nanoparticles with edible plants and their possible implications in the food chain. *Journal of Agricultural and Food Chemistry*, **59**(8), 3485–3498. doi: 10.1021/jf104517j

Rivera Gil P., Huhn D., del Mercato L. L., Sasse D. and Parak W. J. (2010). Nanopharmacy: inorganic nanoscale devices as vectors and active compounds. *Pharmacological Research*, **62**(2), 115–125. doi: 10.1016/j.phrs.2010.01.009

Roberts A. P., Mount A. S., Seda B., Souther J., Qiao R., Lin S. and Klaine S. J. (2007). In vivo biomodification of lipid-coated carbon nanotubes by Daphnia magna. *Environmental Science and Technology*, **41**(8), 3025–3029. doi: 10.1021/es062572a

Savage N. and Diallo M. S. (2005). Nanomaterials and water purification: opportunities and challenges. *Journal of Nanoparticle Research*, **7**(4–5), 331–342. doi: 10.1007/s11051-005-7523-5

Sawai J. (2003). Quantitative evaluation of antibacterial activities of metallic oxide powders (ZnO, MgO and CaO) by conductimetric assay. *Journal of Microbiological Methods*, **54**(2), 177–182. doi: 10.1016/s0167-7012(03)00037-x

Sayes C. M., Fortner J. D., Guo W., Lyon D., Boyd A. M., Ausman K. D. and Colvin V. L. (2004). The differential cytotoxicity of water-soluble fullerenes. *Nano Letters*, **4**(10), 1881–1887. doi: 10.1021/nl0489586

Sayes C. M., Gobin A. M., Ausman K. D., Mendez J., West J. L. and Colvin V. L. (2005). Nano-C60 cytotoxicity is due to lipid peroxidation. *Biomaterials*, **26**(36), 7587–7595. doi: 10.1016/j.biomaterials.2005.05.027

Scott-Fordsmand J. J., Krogh P. H., Schaefer M. and Johansen A. (2008). The toxicity testing of double-walled nanotubes-contaminated food to *Eisenia veneta* earthworms. *Ecotoxicology and Environmental Safety*, **71**(3), 616–619. doi: 10.1016/j.ecoenv.2008.04.011

Scown T. M., van Aerle R., Johnston B. D., Cumberland S., Lead J. R., Owen R. and Tyler C. R. (2009). High doses of intravenously administered titanium dioxide nanoparticles accumulate in the kidneys of rainbow trout but with no observable impairment of renal function. *Toxicological Sciences*, **109**(2), 372–380. doi: 10.1093/toxsci/kfp064

Seeger E. M., Baun A., Kaestner M. and Trapp S. (2009). Insignificant acute toxicity of TiO_2 nanoparticles to willow trees. *Journal of Soils and Sediments*, **9**(1), 46–53. doi: 10.1007/s11368-008-0034-0

Shah V. and Belozerova I. (2009). Influence of metal nanoparticles on the soil microbial community and germination of lettuce seeds. *Water Air and Soil Pollution*, **197**(1), 143–148. doi: 10.1007/s11270-008-9797-6

Sharma V. K. (2009). Aggregation and toxicity of titanium dioxide nanoparticles in aquatic environment – a review. *Journal of Environmental Science and Health Part a-Toxic/Hazardous Substances and Environmental Engineering*, **44**(14), 1485–1495. doi: 10.1080/10934520903263231

Simate G. S., Iyuke S. E., Ndlovu S., Heydenrych M. and Walubita L. F. (2012). Human health effects of residual carbon nanotubes and traditional water treatment chemicals in drinking water. *Environment International*, **39**(1), 38–49. doi: 10.1016/j.envint.2011.09.006

Simon-Deckers A., Gouget B., Mayne-L'Hermite M., Herlin-Boime N., Reynaud C. and Carriere M. (2008). In vitro investigation of oxide nanoparticle and carbon nanotube toxicity and intracellular accumulation in A549 human pneumocytes. *Toxicology*, **253** (1–3), 137–146. doi: 10.1016/j.tox.2008.09.007

Simonet B. M. and Valcarcel M. (2009). Monitoring nanoparticles in the environment. *Analytical and Bioanalytical Chemistry*, **393**(1), 17–21. doi: 10.1007/s00216-008-2484-z

Smirnova E., Gusev A., Zaytseva O., Sheina O., Tkachev A., Kuznetsova E. and Kirpichnikov M. (2012). Uptake and accumulation of multiwalled carbon nanotubes change the morphometric and biochemical characteristics of and *Onobrychis arenaria* seedlings. *Frontiers of Chemical Science and Engineering*, 1–7. **6**(2), 132–138. doi: 10.1007/ s11705-012-1290-5

Smith A. (2006). Opinion: nanotech – the way forward for clean water? *Filtration and Separation*, **43**(8), 32–33. doi: 10.1016/s0015-1882(06)70976-4

Smith C. J., Shaw B. J. and Handy R. D. (2007). Toxicity of single walled carbon nanotubes to rainbow trout, (*Oncorhynchus mykiss*): respiratory toxicity, organ pathologies, and other physiological effects. *Aquatic Toxicology*, **82**(2), 94–109. doi: 10.1016/j.aquatox. 2007.02.003

Soenen S. J., Rivera-Gil P., Montenegro J.-M., Parak W. J., De Smedt S. C. and Braeckmans K. (2011). Cellular toxicity of inorganic nanoparticles: common aspects and guidelines for improved nanotoxicity evaluation. *Nano Today*, **6**(5), 446–465. doi: 10.1016/j.nantod.2011.08.001

Soenen S. J. H., Himmelreich U., Nuytten N., Pisanic T. R., II, Ferrari A. and De Cuyper M. (2010a). Intracellular nanoparticle coating stability determines nanoparticle diagnostics efficacy and cell functionality. *Small*, **6**(19), 2136–2145. doi: 10.1002/smll.201000763

Soenen S. J. H., Nuytten N., De Meyer S. F., De Smedt S. C. and De Cuyper M. (2010b). High intracellular iron oxide nanoparticle concentrations affect cellular cytoskeleton and focal adhesion kinase-mediated signaling. *Small*, **6**(7), 832–842. doi: 10.1002/smll.200902084

Soenen S. J. H., Illyes E., Vercauteren D., Braeckmans K., Majer Z., De Smedt S. C. and De Cuyper M. (2009a). The role of nanoparticle concentration-dependent induction of cellular stress in the internalization of non-toxic cationic magnetoliposomes. *Biomaterials*, **30**(36), 6803–6813. doi: 10.1016/j.biomaterials.2009.08.050

Soenen S. J. H., Hodenius M. and De Cuyper M. (2009b). Magnetoliposomes: versatile innovative nanocolloids for use in biotechnology and biomedicine. *Nanomedicine*, **4**(2), 177–191. doi: 10.2217/17435889.4.2.177

Som C., Wick P., Krug H. and Nowack B. (2011). Environmental and health effects of nanomaterials in nanotextiles and facade coatings. *Environment International*, **37**(6), 1131–1142. doi: 10.1016/j.envint.2011.02.013

Song M., Yuan S., Yin J., Wang X., Meng Z., Wang H. and Jiang G. (2012). Size-dependent toxicity of nano-C_{60} aggregates: more sensitive indication by apoptosis-related bax translocation in cultured human cells. *Environmental Science and Technology*, **46**(6), 3457–3464. doi: 10.1021/es2039008

Srivastava A., Srivastava O. N., Talapatra S., Vajtai R. and Ajayan P. M. (2004). Carbon nanotube filters. *Nature Materials*, **3**(9), 610–614. doi: 10.1038/nmat1192

Stampoulis D., Sinha S. K. and White J. C. (2009). Assay-dependent phytotoxicity of nanoparticles to plants. *Environmental Science and Technology*, **43**(24), 9473–9479. doi: 10.1021/es901695c

Stella G. M. (2011). Carbon nanotubes and pleural damage: perspectives of nanosafety in the light of asbestos experience. *Biointerphases*, **6**(2), P1–P17. doi: 10.1116/1.3582324

Stroh A., Zimmer C., Gutzeit C., Jakstadt M., Marschinke F., Jung T. and Grune T. (2004). Iron oxide particles for molecular magnetic resonance imaging cause transient oxidative stress in rat macrophages. *Free Radical Biology and Medicine*, **36**(8), 976–984. doi: 10.1016/j.freeradbiomed.2004.01.016

Sun C., Lee J. S. H. and Zhang M. (2008). Magnetic nanoparticles in MR imaging and drug delivery. *Advanced Drug Delivery Reviews*, **60**(11), 1252–1265. doi: 10.1016/j.addr.2008.03.018

Sung J. H., Ji J. H., Park J. D., Yoon J. U., Kim D. S., Jeon K. S. and Yu I. J. (2009). Subchronic inhalation toxicity of silver nanoparticles. *Toxicological Sciences*, **108**(2), 452–461. doi: 10.1093/toxsci/kfn246

Sung J. H., Ji J. H., Yoon J. U., Kim D. S., Song M. Y., Jeong J. and Yu I. J. (2008). Lung function changes in Sprague-Dawley rats after prolonged inhalation exposure to silver nanoparticles. *Inhalation Toxicology*, **20**(6), 567–574. doi: 10.1080/08958370701874671

Suresh A. K., Pelletier D. A., Wang W., Moon J. W., Gu B., Mortensen N. P. and Doktycz M. J. (2010). Silver nanocrystallites: biofabrication using Shewanella oneidensis, and an evaluation of their comparative toxicity on gram-negative and gram-positive bacteria. *Environmental Science and Technology*, **44**(13), 5210–5215. doi: 10.1021/es903684r

Takenaka S., Karg E., Roth C., Schulz H., Ziesenis A., Heinzmann U. and Heyder J. (2001). Pulmonary and systemic distribution of inhaled ultrafine silver particles in rats. *Environmental Health Perspectives*, **109**, 547–551. doi: 10.2307/3454667

Theron J., Walker J. A. and Cloete T. E. (2008). Nanotechnology and water treatment: applications and emerging opportunities. *Critical Reviews in Microbiology*, **34**(1), 43–69. doi: 10.1080/10408410701710442

Thomas K. V., Farkas J., Farmen E., Christian P., Langford K., Wu Q. and Tollefsen K. E. (2011). Effects of dispersed aggregates of carbon and titanium dioxide engineered nanoparticles on rainbow trout hepatocytes. *Journal of Toxicology and Environmental Health-Part A-Current Issues*, **74**(7–9), 466–477. doi: Pii 93459870110.1080/15287394.2011.550557

Tiede K., Hassellov M., Breitbarth E., Chaudhry Q. and Boxall A. B. A. (2009). Considerations for environmental fate and ecotoxicity testing to support environmental risk assessments for engineered nanoparticles. *Journal of Chromatography A*, **1216**(3), 503–509. doi: 10.1016/j.chroma.2008.09.008

Tong Z., Bischoff M., Nies L., Applegate B. and Turco R. F. (2007). Impact of fullerene (C-60) on a soil microbial community. *Environmental Science and Technology*, **41**(8), 2985–2991. doi: 10.1021/es061953I

Upadhyayula V. K. K., Deng S., Mitchell M. C. and Smith G. B. (2009). Application of carbon nanotube technology for removal of contaminants in drinking water: a review. *Science of the Total Environment*, **408**(1), 1–13. doi: 10.1016/j.scitotenv.2009.09.027

Usenko C. Y., Harper S. L. and Tanguay R. L. (2007). In vivo evaluation of carbon fullerene toxicity using embryonic zebrafish. *Carbon*, **45**(9), 1891–1898. doi: 10.1016/j.carbon.2007.04.021

Usenko C. Y., Harper S. L. and Tanguay R. L. (2008). Fullerene C_{60} exposure elicits an oxidative stress response in embryonic zebrafish. *Toxicology and Applied Pharmacology*, **229**(1), 44–55. doi: 10.1016/j.taap.2007.12.030

Velzeboer I., Kupryianchyk D., Peeters E. T. H. M. and Koelmans A. A. (2011). Community effects of carbon nanotubes in aquatic sediments. *Environment International*, **37**(6), 1126–1130. doi: 10.1016/j.envint.2011.01.016

Vochita G., Creanga D. and Focanici-Ciurlica E. L. (2011). Magnetic nanoparticle genetic impact on root tip cells of sunflower seedlings. *Water, Air, and Soil Pollution*, 1–9. **223**(5), 2541–2549. doi: 10.1007/s11270-011-1046-8

Wang H., Wick R. L. and Xing B. (2009a). Toxicity of nanoparticulate and bulk ZnO, Al_2O_3 and TiO_2 to the nematode *Caenorhabditis elegans*. *Environmental Pollution*, **157**(4), 1171–1177. doi: 10.1016/j.envpol.2008.11.004

Wang X., Jia G., Wang H., Nie H., Yan L., Deng X. Y. and Wang S. (2009b). Diameter effects on cytotoxicity of multi-walled carbon nanotubes. *Journal of Nanoscience and Nanotechnology*, **9**(5), 3025–3033. doi: 10.1166/jnn.2009.025

Wang Z., Lee Y.-H., Wu B., Horst A., Kang Y., Tang Y. J. and Chen D. R. (2010). Anti-microbial activities of aerosolized transition metal oxide nanoparticles. *Chemosphere*, **80**(5), 525–529. doi: 10.1016/j.chemosphere.2010.04.047

Warheit D. B., Hoke R. A., Finlay C., Donner E. M., Reed K. L. and Sayes C. M. (2007). Development of a base set of toxicity tests using ultrafine TiO_2 particles as a component of nanoparticle risk management. *Toxicology Letters*, **171**(3), 99–110. doi: 10.1016/j.toxlet.2007.04.008

Wick P., Manser P., Limbach L. K., Dettlaff-Weglikowska U., Krumeich F., Roth S. and Bruinink A. (2007). The degree and kind of agglomeration affect carbon nanotube cytotoxicity. *Toxicology Letters*, **168**(2), 121–131. doi: 10.1016/j.toxlet.2006.08.019

Wiesner M. R., Lowry G. V., Alvarez P., Dionysiou D. and Biswas P. (2006). Assessing the risks of manufactured nanomaterials. *Environmental Science and Technology*, **40**(14), 4336–4345. doi: 10.1021/es062726m

Xu P., Zeng G. M., Huang D. L., Feng C. L., Hu S., Zhao M. H. and Liu Z. F. (2012). Use of iron oxide nanomaterials in wastewater treatment: a review. *Science of the Total Environment*, **424**(0), 1–10. doi: 10.1016/j.scitotenv.2012.02.023

Yamago S., Tokuyama H., Nakamura E., Kikuchi K., Kananishi S., Sueki K. and Ambe F. (1995). In vivo biological behavior of a water-miscible fullerene: 14C labeling, absorption, distribution, excretion and acute toxicity. *Chemistry & Biology*, **2**(6), 385–389. doi: 10.1016/1074-5521(95)90219-8

Yang F., Liu C., Gao F., Su M., Wu X., Zheng L. and Yang P. (2007). The Improvement of spinach growth by nano-anatase TiO_2 treatment is related to nitrogen photoreduction. *Biological Trace Element Research*, **119**(1), 77–88. doi: 10.1007/s12011-007-0046-4

Yang G. C. C. and Li C. J. (2008). Tubular TiO_2/Al_2O_3 composite membranes: preparation, characterization, and performance in electrofiltration of oxide-CMP wastewater. *Desalination*, **234**(1–3), 354–361. doi: 10.1016/j.desal.2007.09.104

Yang L. and Watts D. J. (2005). Particle surface characteristics may play an important role in phytotoxicity of alumina nanoparticles. *Toxicology Letters*, **158**(2), 122–132. doi: 10.1016/j.toxlet.2005.03.003

Zhang W. X. and Elliott D. W. (2006). Applications of iron nanoparticles for groundwater remediation. *Remediation Journal*, **16**(2), 7–21. doi: 10.1002/rem.20078

Zhang W. X. (2003). Nanoscale iron particles for environmental remediation: an overview. *Journal of Nanoparticle Research*, **5**(3–4), 323–332. doi: 10.1023/a:1025520116015

Zhao X. and Liu R. (2012). Recent progress and perspectives on the toxicity of carbon nanotubes at organism, organ, cell, and biomacromolecule levels. *Environment International*, **40**, 244–255. doi: 10.1016/j.envint.2011.12.003

Zhu S., Oberdorster E. and Haasch M. L. (2006). Toxicity of an engineered nanoparticle (fullerene, C-60) in two aquatic species, Daphnia and fathead minnow. *Marine Environmental Research*, **62**, S5–S9. doi: 10.1016/j.marenvres.2006.04.059

Zhu X., Zhu L., Lang Y. and Chen Y. (2008a). Oxidative stress and growth inhibition in the freshwater fish *Carassius auratus* induced by chronic exposure to sublethal fullerene aggregates. *Environmental Toxicology and Chemistry*, **27**(9), 1979–1985. doi: 10.1897/07–573.1

Zhu X.-S., Zhu L., Lang Y.-P., Li Y., Duan Z.-H. and Yao K. (2008b). Oxidative damages of long-term exposure to low level fullerenes (C-60) in *Carassius auratus*. *Huanjing Kexue*, **29**(4), 855–861.

Zhu X., Zhu L., Duan Z., Qi R., Li Y. and Lang Y. (2008c). Comparative toxicity of several metal oxide nanoparticle aqueous suspensions to Zebrafish (*Danio rerio*) early developmental stage. *Journal of Environmental Science and Health Part A-Toxic/Hazardous Substances and Environmental Engineering*, **43**(3), 278–284. doi: 10.1080/10934520701792779

Zhu H., Han J., Xiao J. Q. and Jin Y. (2008d). Uptake, translocation, and accumulation of manufactured iron oxide nanoparticles by pumpkin plants. *Journal of Environmental Monitoring*, **10**(6), 713–717. doi: 10.1039/b805998e

Chapter 3

Life cycle assessment of nanomaterials: towards green nanotechnology

T. Tsuzuki

3.1 INTRODUCTION

Nanotechnology is expected to revolutionize many aspects of our lives, in a way traditional science and technology could not deliver. In fact, the current and potential applications of nanotechnology are rapidly growing. These applications cover an extremely broad range of markets and industries including biomedical & cancer treatments, renewable energy, environmental protection, pharmaceuticals, personal care, surface coatings, plastics, textiles, food, building materials, electronics and the automotive industries (see Chapter 1). A number of nanoparticle-based consumer products are already available (Thomas *et al.*, 2006b). However, there is also a growing concern about the potential health and environmental impacts of nano-enabled products (Cobb & Macoubrie, 2004; Weisner & Colvin, 2005). In some cases, research in universities and product development in companies have been carried out without due consideration for the potential health and environmental impacts of nanomaterials.

Life cycle assessment (LCA) has recently emerged as a powerful holistic tool for the assessment of the environmental impacts of technology. LCA has already been successfully applied in many industries including agriculture, food, water and waste management (Horne *et al.*, 2009). Application of LCA to nanotechnology highlights both the positive and negative impacts of nanotechnology on the environment throughout the whole life cycle of nano-products. It also contributes to the development of "green nanotechnology" (Schmidt, 2007; Karn, 2008). This chapter gives an overview of LCA and discusses the applicability and challenges on the LCA of nanotechnology. Some alternative or complementary approaches are also discussed.

3.2 LIFE CYCLE ASSESSMENT (LCA)

3.2.1 What is LCA?

LCA is a systematic approach to assess the "environmental burden" of a product by analysing the flows of materials and energy during the "whole product life cycle". The life cycle includes literally all aspects of activities during the life of a product "from the cradle to the grave", such as the extraction of raw materials and resources, production processes and facilities, the usage of products, and post-use management of the product including recycling and disposal. The environmental burden covers a wide range of aspects including climate change, ozone-layer depletion, emission of green-house gas, release of hazardous substances, depletion of natural resources and the types of land and water use. Hence, LCA is often regarded as the most holistic approach to the assessment of environmental impacts. The United Nations Environmental Programme (UNEP) refers to LCA as a tool to reveal "the world behind the product" (Fava, 2002).

The history of LCA is relatively new (Hunt *et al.*, 1992). Resources and environmental profile analysis (REPA) was started by the Coca Cola Company in 1969. The company commissioned a study on the natural resources and environmental profile of their different beverage packaging materials. In the early 1970s, the oil shock triggered the use of REPA for energy analysis but the interest in REPA waned afterwards. In the late 1980s, there was increased public awareness towards the environment in the USA. This led to the resurgence of systematic multi-criteria inquiry on the environmental impacts of products and procedures. In 1990, the Society of Environmental Toxicology and Chemistry USA (SETAC) proposed to use the term "life cycle assessment" and embarked on the rapid development of LCA. In 2002, SETAC and UNEP formed a joint Life Cycle Initiative to develop LCA. Since 1994, the International Organization for Standardization (ISO) started working on creating the consensus amongst different schools of LCA. This was then crystallized as the ISO14000 series of industrial standards about environmental management systems.

LCA is different from risk analysis or Risk Assessment (RA). RA is defined as the "scientific process in which the risks posed by inherent hazards involved in the process or situations are estimated either quantitatively or qualitatively" (Fairman *et al.*, 1998). RA aims to ensure the safety of people and environment and hence RA can be a localised assessment on risks arisen from a certain activity in a specific location in a given time period (Kloepffer *et al.*, 2007). On the other hand, LCA is a more comprehensive non-localised assessment and can not only deal with risks but also with broader factors that impact environment.

3.2.2 Benefit of LCA

The objectives of an LCA are not limited to environmental protection. LCA can also be used for government policymaking as well as the development of business strategies in manufacturing and marketing. Some examples are given in the following list:

- Identifying the source of nano-waste and its environmental impacts (Allan, 2012) in order to form environmental regulatory systems (Bosso *et al.*, 2012).
- Identifying the "hot-spots" to cause environmental burden during the life cycle of the products or activity, in order to prevent adverse biological and environmental effects.
- Facilitating dialogue with customers and stakeholders on the risks and benefits of the products.
- Product development, including improvement or optimisation of production variants for existing products and designing of new products.
- Making choices between different technologies. For example, to analyse the benefits of renewable energy by analysing the production costs and environmental impacts not only during the use of the products but also in the phases of raw material generation, manufacturing and disposal.
- Making choices between different types of waste or risk management strategies (Bystrzejewska-Piotrowska *et al.*, 2009; Reijnders, 2006).
- Certifying "Eco-design" labels. For example, bamboo textiles have recently appeared in the clothing industry as eco-friendly products as opposed to cotton that requires a large amount of water and pesticides during growth. However, this eco-claim is only based on the raw materials. The manufacturing processes using toxic chemicals were not considered as part of the assessment until recently (Afrin *et al.*, 2010). LCA will resolve these types of issues with its holistic approach.

3.2.3 ISO14040 series

The ISO14040 series is the standard related to LCA. It defines the LCA process along with definitions of nomenclatures (Table 3.1). ISO14040:2006 defines LCA as the "compilation and evaluation of the inputs, outputs and potential environmental impacts of a product system throughout its life cycle".

ISO14040:2006 divides the LCA activity into 4 phases (Figure 3.1); goal and scope definition, life cycle inventory (LCI), life cycle impact assessment (LCIA) and life cycle impact interpretation.

In the first phase (goal and scope definition), the aims, intended application of the results and dissemination, along with scope (product system, system boundary, impact categories, data requirements, assumptions, etc.) are defined.

The second phase (LCI) involves the compilation and quantification of inputs and outputs for a product throughout its life cycle. The information includes the types and volumes of the materials used and released throughout the life cycle. The examples of inputs include energy, raw materials and consumables. The examples of outputs include the products, wastes and gas emission released to the environment.

Table 3.1 ISO14040 series of LCA standards.

Standard number	Title
ISO14040:2006	Environmental management – Life cycle assessment – Principles and framework
ISO14044:2006	Environmental management – Life cycle assessment – Requirements and guidelines
ISO/TR14047:2003	Environmental management – Life cycle impact assessment – Examples of application of ISO14042
ISO/TR14048:2002	Environmental management – Life cycle assessment – Data documentation format

Note that ISO14041:1998, 14042:2000 and 14043:2000 was replaced with the ISO14040:2006 and ISO14044:2006.

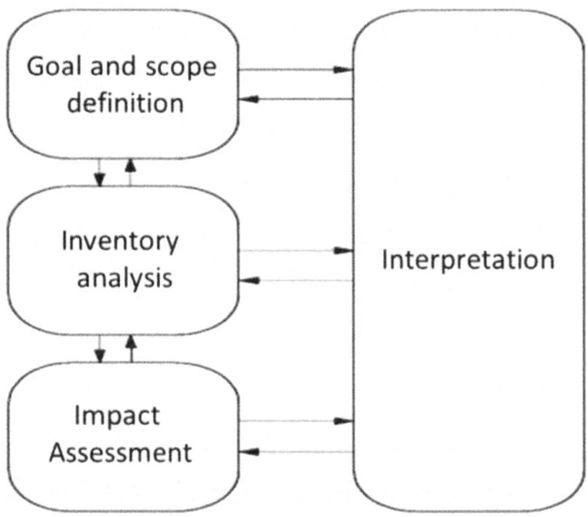

Figure 3.1 Life cycle assessment framework (ISO14040:2006).

The third phase (LCIA) aims at understanding and evaluating the magnitude and significance of the potential environmental impacts for a product system throughout the life cycle of the product. ISO14044:2006 defines four stages of the impact assessment: classification, characterization, normalization and weighting. The quantitative data of LCI is grouped into impact categories and category indicators are calculated to give the potential magnitude of the impacts.

In the fourth phase (life cycle interpretation), the findings at the previous stages are evaluated with relation to the defined goal and scope in order to reach conclusions and recommendations.

3.2.4 General limitation of LCA

By its own nature, LCA has many limitations. For the interpretation of LCA results, these limitations need to be carefully taken into consideration:

- Due to its holistic nature, LCA cannot be applied to assess localised impacts or temporary conditions.
- LCA is based on already known information and does not account for any technological breakthrough in the future.
- LCA is based on the flows of materials and products. It does not include the influence of and impacts on macro-economy, social and geopolitical factors.
- The accuracy of an LCA depends on the availability and accuracy of data. The examples of the required data include energy consumption, the types of production facilities, consumer behaviour, public health and safety, animal welfare, and so on.
- LCA takes its holistic approach by relying on assumptions and simplifications. The boundaries and limitations of the assumptions and simplification influence the results of an LCA.

3.3 LCA FOR NANOTECHNOLOGY

3.3.1 Nanotechnology and LCA

Nanotechnology encompasses a broad spectrum of technology, including nano-instrumentation, nano-electronics, nano-medicine and nanomaterials. To date, most of the health and environmental concerns are focused on the nanomaterials sector. Nanomaterials are a new class of industrial materials. Due to their extremely small size, nanomaterials exhibit new or enhanced properties compared with the same materials of larger dimensions (Tsuzuki, 2012a). Those unique attributes allow nanomaterials to be used in a broad range of applications that were not possible using traditional materials (Tsuzuki, 2012b).

Large quantities of nanoparticles are already in use in many commercial products. For example, the four major commercial markets of nanoparticles by volume are estimated as high as (i) 11500 tones in automotive catalysts, (ii) 9400 tones in chemical mechanical planarization slurry, (iii) 3100 tones in magnetic recording media and (iv) 1500 tones in sunscreens (Holister & Harper, 2002). The Woodrow Wilson International Centre for Scholars, Projects on Emerging Nanotechnology, counted nanomaterial-based consumer goods on the market in 2010 to be 1317 and the number is expected to reach 3400 by 2020 (Woodraw Wilson International Center for Scholars, 2011). The real number of nanomaterial-related products is estimated to be much higher. For example, integrated circuits used in computers, flush memories and mobile phones have nanostructures that are manufactured using nanoparticle polishing agents, but they are rarely counted as nano-enabled consumer-products. Also largely excluded in the count are the use of nano-SiO_2 in car tires, and magnetic nanoparticles in magnetic recording media such as video tapes.

However, recent escalation of the concerns towards the potential for adverse biological and environmental effects is impacting nanotechnology development both in academic research and commercial development (Nowack, 2009; Lu *et al.*, 2012). Normally, a precautionary approach is encouraged for the socially responsible development of nanotechnology (van Broekhuizen & Reijnders, 2011). While this approach sounds reasonable, it has drawbacks as it hampers the progress of nanotechnology and also creates unnecessary use of resources. LCA will assist in overcoming these drawbacks. As a proactive approach, LCA helps to identify the potential risks associated with nanomaterials and nano-enables products in advance. LCA can also clarify the risk-benefit balance during the entire life cycle of the nano-products (Sweet & Strohm, 2006).

Performing an LCA of nanomaterials has other benefits where it assists the development of "green nanotechnology" (Schmidt, 2007; Karn, 2008). One trend emerging among the nanotechnology community is to develop cleaner or greener methods during the life cycle of nanomaterials. Increasing research efforts have been made recently to develop new methods to synthesise nanomaterials without using raw materials containing scarce natural resources and hazardous substances, or new production methods that require low energy and low material consumption. Examples include the use of renewable raw materials (Tsuzuki *et al.*, 2011) and biosynthesis (Mohanpuria *et al.*, 2008). The concept of green nanotechnology applies not only to the manufacturing stage but also to the usage stage, in particular, in environmental applications (Weisner & Bottero, 2007). Caution is required so as not to consider the technology as "green', only because the applications are in the environmental sector. The potential environmental burden of the applications during manufacturing and disposal stages should be taken into account. LCA helps to guide the nanotechnology development towards true eco-friendliness, through the assessment of the *overall* risk-benefit balance during the production, usage and disposal stages of nano-products.

3.3.2 Challenges, limitations and obstacles specific to nanotechnology (Kloepffer *et al.*, 2007)

The LCA of nanotechnology should not only focus on the toxicology, but also include the following aspects (Sweet & Strohm, 2006):

- Exposure (amount, sites, frequency, transportability exposed party, etc.)
- Environmental fate (persistency, transformation, biological uptake, etc.)
- Product characteristics (design, chemical and physical stability, degradation characteristics, etc.)
- Mode of product usage (frequency, wear & tear characteristics, storage conditions, existence of accompanying chemicals on usage, etc.)

The assessment of nanotechnology in these aspects requires special consideration. In particular, it is important to recognise the limitations and obstacles arising from the following three unique characteristics of nanotechnology:

(i) Materials (Gavankar et al.*, 2012)*

The properties of nanomaterials differ greatly from those of their bulk counterparts. The unique properties of nanomaterials are associated with not only the physical size but also other factors including surface area, crystal structure, particle shape, aspect ratio, surface charge and surface chemistry. High surface areas result in high solubility, high reaction rates, high catalytic and photocatalytic activities. These physicochemical characteristics of nanomaterials are expected to largely influence the biological and environmental toxicology of nanomaterials (Oberdörster *et al.*, 2005; Wright & Jackson, 2012). Nanoscale coatings on nanomaterials will also govern the toxicology. The sheer number of nano-specific characteristics alone can complicate the set-up of the scopes and inputs-outputs analysis in an LCA. In addition, these nano-specific characteristics may collectively affect the toxicological profile.

The traditional dose-response relationship based on weight is not adequate to use for the assessment of these nano-specific characteristics. New dose-response relationship needs to be developed for each nano-specific characteristics.

(ii) Applications (Christensen, 2010)

Most of the nano-enabled products take advantage of a certain nano-specific characteristic of the raw nanomaterials. It is important to understand which nano-specific characteristics the product utilises and which characteristic is most significant in the nanomaterial ingredients. In this regard, the relationship between LCI and the key nano-specific functionality can be established only in the context of specific applications (Seager & Linkov, 2008). However, the information is difficult to come across, as they are often considered as commercially sensitive proprietary knowledge.

It is also important to recognise that not all nanotechnology applications end up as nanomaterial-containing products. Nanomaterials can be used as process tools such as chemical mechanical planarization agents in semiconductor industry or catalysts in methanol production. For the overall assessment of the safety and risks of nanotechnology, those nano-enabled products with no nanomaterials in them should also be analysed. However, it is difficult to identify such products as there is no regulation to declare the use of nanotechnology during the production process.

(iii) Uncertainty

Nanotechnology is an emerging and fast-developing science. The commercialization of nano-enabled products appeared only since the 1990s. This "newness" poses serious limitations to LCA in the following aspects, mainly as the lack of data and unpredictability:

- *Comparative analysis*: LCA is a comparative analysis. LCA normally assumes a functionally equivalent substitution of established technologies with new ones (Fleischer & Grunwald, 2008). For example, wind generation is compared against coal-fire power generation for the environmental impacts. However, some applications of nanotechnology are drastically new so that no substitutive technology can be found. For example, some nanomaterials such as buckminsterfullerenes, carbon nanotubes and graphenes are entirely new classes of materials and may be difficult to find suitable comparison with established technologies.
- *Production facilities and processes*: There are top-down and bottom-up approaches to the commercial scale production of nanomaterials (Tsuzuki, 2009). Bottom-up approaches are further classified into three types according to the medium (gas, liquid and solid) in which nanomaterials are formed. In the LCI phase (the second phase) of an LCA, it is essential to identify which production techniques are used, because the information is required to determine the inputs and outputs. The knowledge on the production methods is also critical, because the production methods largely influence the nano-specific characteristics of the products and, in turn, influence the health and environmental impacts of the nano-enabled products. However, many of the nanomaterial manufacturers are small to medium enterprises and their manufacturing processes and facilities, as well as the source of raw materials and waste management protocols are often proprietary. In addition, many commercial production facilities are still at the stage of scale-up, and the nature of manufacturing facilities and procedures will change as the production volume is increased. New technology is likely to quickly replace the existing capital equipment.
- *Products*: Nano-products, including nanomaterials as well as nano-enabled products that utilise nanomaterials, are relatively new in the market. The design of the products is also rapidly changing. Hence, there is not an established pattern in the way those products are handled by users and customers. As such, it is difficult to obtain the information about the mode of product usage including the frequency of use, wear & tear characteristics, storage conditions and likely end-of-life scenarios.
- *Scientific data*: Although nanotechnology has been an active research area since the 1980s, still many things about nanomaterials remain unknown. In particular, nano-toxicology is the youngest field of nanoscience and little is known about the biological and environmental toxicology of nanomaterials (Hoet *et al.*, 2004; Grieger *et al.*, 2012; Haase *et al.*, 2012). Most of the nano-toxicology research is confined to the laboratories; the behaviour and

environmental fate of nanomaterials in land and aquatic systems is not fully understood (Fent, 2010). Many techniques to detect/monitor nanomaterials in the environment are still under development (Hassellöv *et al.*, 2008; Wise & Brasuel, 2011). In the meantime, new types of nanomaterials and new properties of nanomaterials are discovered almost daily by scientists around the world.

Many new physicochemical properties of nanomaterials, an extremely wide breath of nano-applications, business and marketing characteristics of nano-products, and rapid progress in the science and commercialisation of nanomaterials – these factors all contribute to the unpredictability, uncertainty and lack of reliable information for the LCA of nanotechnology.

3.4 WATER RESEARCH AND LCA OF NANOMATERIALS

In order to assess the impacts of nanotechnology on aquatic systems, it is important to identify the possible routes for nano-products to enter into aquatic environments, through the life cycle of nano-products (Asmatulu *et al.*, 2012; Labille *et al.*, 2011). Examples of the entry routes are listed in Table 3.2.

Table 3.2 Possible entry routes of nano-related products into water.

Life cycle stage	Routes/scenarios
Production	• Process water (e.g. liquid-phase synthesis) • Cleaning of manufacturing facilities with water • Cleaning of package containers with water • Accidental spill • Disposal of rejected or quarantined batches
Use	• Wear & tear • Cleaning & laundering of nano-enabled products • Damage • Water-related applications (water treatment, water-splitting, desalination, etc.) • The use is accompanied with water (CMP slurries, etc.)
Disposal	• Leakage from landfill • Removal (e.g. sunscreens and other cosmetic products) with water • Cleaning of disposal facilities/apparatus with water

In the production stage of nanomaterials, the most likely entry routes are associated with the process water, that is, water used in the production process. As stated earlier in this chapter, there are 4 classes of production methods for commercial nanomaterial products. The use of process water in these production methods are characterised as follow:

(i) *Top-down approach*: In this approach, nanoparticles are produced from bulk materials through size reduction, often by mechanical grinding. Normally the grinding is performed under wet conditions, that is, in water or other solvents. The use of water is more common than organic solvents because of economic and safety reasons. Normally the final nanomaterial products need to be separated from process water. This can be achieved either by evaporation of water, nano-filtration or ultra-centrifugation. In the latter two methods, the residual nanomaterials in the removed process water should be closely monitored. In some cases, nanomaterials are sold in a slurry form in the presence of process water so that the removal of the process water is not necessary.

(ii) *Vapor-phase synthesis (bottom-up approach)*: Nanoparticles are synthesised by the rapid solidification of a liquid or vapour in vacuum or a gaseous medium. Process water is not required and the reaction chambers are normally well sealed.

(iii) *Liquid-phase synthesis (bottom-up approach)*: Nanoparticles are precipitated out from a solution that consists of a mixture of reagents. Liquid phase processes have been widely used in the chemical industry to make conventional micron-scale powders, and have been proven to be economical for the production of many materials. Water is a common process liquid in this technique. Normally the process water contains reaction by-products so that evaporation methods cannot be used to remove the process water from nanomaterials. Extensive rinsing of the nanomaterials with water may be carried out to avoid possible contamination from the reaction by-products.

(iv) *Solid phase synthesis (bottom-up approach)*: This technique uses high energy dry milling to mechanically induce chemical reactions to occur at low temperature in a ball mill. Milling of precursor powders leads to the formation of a nanoscale composite structure of the starting materials, which react with each other during milling or subsequent heat treatment to precipitate nanoparticles of the desired phase within a solid matrix. The reaction paths are designed so as to make the solid matrix water-soluble. The separation of nanoparticles from the matrix is carried out by washing the as-milled powder with water. Since the washing water contains the matrix phase material, evaporation methods cannot be used to remove the process water from nanomaterials.

Currently there are over 100 commercial companies that produce large quantities of inorganic nanoparticles around the world. According to the author's recent research (Tsuzuki, 2009), approximately 40% of the nanomaterial manufacturers use the liquid phase methods. About 10% of the companies use top-down approaches.

The volume of nanoparticles produced in the world is difficult to estimate, but Schmid and Riediker analysed that about 2500 tonnes of nanomaterials were used per year in the Swiss industry alone, during 2005 and early 2006 (Schmid & Riediker 2008). Thus, it is reasonable to assume that a considerable volume of water is being used as a process water to produce nanomaterials. However, the information about how the process water is treated in the manufacturing plants is difficult to obtain, as the details of the production process is often proprietary.

The nano-contamination is not limited to the process water used in the production of nanomaterials. The production of nano-enabled products, which contain engineered nanomaterials as their ingredients, may require the use of process water.

In the usage stage of nano-products, the major occurrence for the nanomaterials to enter into the aqueous environment is expected in the cases when the nanomaterials are used in water. The examples include the use of nanomaterials in water treatment applications, hydrogen-gas generation by water-splitting, and desalination. So far, those applications are still in the early stage of research and hence nano-waste has not yet become a serious environmental issue.

On the other hand, the use of nanomaterials in semiconductor industry is already happening. The construction of integrating circuits and high density memory storage devices involves the layer-by-layer formation of nanometer-thick multilayers. Lithography is applied to each layer to form micro and nano scale three dimensional electric circuits. Normally the surface of the deposited layers requires planarization to make the surface completely parallel to the substrate before the deposition of the next layer. This is critical because any curved or warped layer will compromise the device integrity. For the planarization of the nanometer-thick layers without making scratches, polishing compounds made of nanoparticles are used in a slurry form. The abrasive nanoparticles remove unnecessary parts of the layers by the synergistic effect of both physical and chemical forces and hence the process is called chemical-mechanical planarization (CMP). The consumption of nanoparticles in CMP slurries is expected to increase to thousands of tonnes per year (Holister & Harper, 2002). The treatment of used CMP slurry can be an important issue to be addressed in terms of possible environmental release of nanomaterials.

Once nanomaterials are released in an aquatic environment, the fate of nanomaterials needs to be identified. Table 3.3 lists some of the key considerations that should be taken into account for such an assessment. As discussed in the previous section of this chapter, there are many nano-specific physicochemical properties of nanomaterials and application-specific characteristics of nano-enabled products that influence the fate of nanomaterials in water. In addition, there are many properties of the aquatic environment such as pH and salinity. The combinations between these material-based characteristics and environment-based characteristics will ultimately govern the fate of nano-products and, in turn, their environmental impacts. Hence it is important to analyse all these combinations for the assessment of the impacts of nano-products. However, as shown in Table 3.3, the number of combinations can be very large, which could impede accurate and timely assessment of nano-products in aquatic systems.

Table 3.3 Key considerations in the fate and risks assessment of nanomaterials in water.

Subject	Consideration
Nanomaterials	• Size • Shape & aspect ratio • Solubility • Physical and chemical stability • Surface area • Surface charge • Surface chemistry • Catalytic activity • Chemical toxicity
Nano-enabled products	• Size (fragments, dust, etc.) • Characteristics of the matrix phase wherein the nanomaterials are embedded (liquid, solid, chemical and physical stability, ingredients, toxicity, solubility, etc.)
Water environment	• pH • Redox potential • Ionic strength • Flow rate • Characteristics of co-existing substances (ions, molecules, solids, etc.)
Behaviour of the nano-products in water	• Agglomeration/dispersion • Dissolution • Adhesion • Catalytic reaction • Transformation via chemical reactions • Shape change • Change in surface chemistry by adsorbing/absorbing other substances

3.5 OVERVIEW OF CASE STUDIES

To date, most of the LCA studies on nanotechnology have focused on commercially available *nanoparticulate* materials. This trend is reasonable, as the potential risks of nanoparticles are higher than nano-films, nano-porous bulk materials or other forms of nanomaterials. Some case studies on the release of engineered nanomaterials to aquatic environments are given in Chapters 1 and 5. The published studies of the

LCA of nanotechnology were reviewed by several authors (Meyer *et al.*, 2009; Theis *et al.*, 2011; Gavankar *et al.*, 2012). It was found that only a very small number of published studies deal with the LCA of nanomaterials in water (Meyer *et al.*, 2009). Theis *et al.*, (2011) analysed the number of published scientific articles on the LCA of nanotechnology. They found that, compared with nearly 1 million articles per annum on nanotechnology in general, the articles on the LCA of nanotechnology is considerably smaller, totalling less than 100 in the last 10 years.

Not only the small number of studies but also the limited scope of the studies reflects the fact that the LCA of nanotechnology, in particular for aquatic environments, is still in its infancy. Gavankar *et al.* (2012) analysed the published LCA studies of nanomaterials and concluded that none of them is ISO compliant. The impact categories of many studies are on the use of energy and materials only and the nano-specific factors are often ignored. The scarcity of raw materials such as rare earth elements and lithium, as well as the impacts from end-of-life activities, are often excluded. In addition, Theis *et al.* (2011) pointed out that, so far, little study has been conducted on the environmental impacts of the production processes of nanomaterials. When such studies exist, single metrics such as water usage, energy demand or usage of toxic substances, is used to evaluate the processes, but overall impacts on the environment were not properly addressed (Theis *et al.*, 2011).

The slow progress in the LCA of nanomaterials is due mainly to the lack of reliable data. In order to catch up with the rapid progress of nanotechnology commercialization, significant research efforts are required to fill the data gaps as a matter of urgency (Alvarez *et al.*, 2009).

3.6 NEW APPROACHES TO THE LCA OF NANOMATERIALS

The limitations and obstacles specific to nanotechnology put serious impediment on the LCA of nanomaterials and nano-enabled products. Considering the rapid expansion of fundamental nano-science and industrial nano-applications, it may not be possible to obtain complete data sets necessary for an LCA, for a long time. However, as large quantities of nano-enabled products are already on the market, the LCA of nanotechnology should not be postponed until complete data sets become available (Kloepffer *et al.*, 2007; Hischier & Walser, 2012). Hence it is prudent to device new interim strategies. In order to overcome the challenge, limitations and obstacles for the LCA of nanotechnology, many alternative or complementary approaches have been proposed. These include the following:

Life Cycle Thinking (Kloepffer et al., *2007)*

Life cycle thinking (LCT) is a platform to conduct broader assessment than LCA so as to include social and economic factors. For example, Samsung Electronics has developed an information technology system called EcoProduct Systems consisting of five factors: LCA, eco-design, green purchasing, environmental accounting and environmental customer treatment. BASF developed a system called Eco-Efficiency Tool for the analysis of economic, social and environmental impacts of their products, comprising six factors: consumption of resources and energy, emissions to air, emissions to water, emissions to soil, land use, and toxicology.

Screening approach (Kloepffer et al., *2007)*

In 2006, US Environmental Protection Agency (USEPA) organised a joint workshop with the Woodrow Wilson International Centre for Scholars the Project on Emerging Nanotechnologies, on "Nanotechnology and Life Cycle Assessment" (Kloepffer *et al.*, 2007). In the workshop, the limitations of applying a rigorous LCA to nanotechnology were highlighted and a Screening Approach (SA) was proposed as an alternative assessment process. SA was developed as a go/no-go decision making system and consists of the following five steps/gates:

(1) Traditional assessment of obvious harm. Check compliance with existing health, safety and environmental regulation in industry.
(2) Traditional full LCA on environmental impacts, excluding biological toxicology. Assess the risk/benefit balance.
(3) Traditional full LCA on biological toxicology and risk assessment, considering the possible exposure risks at each life cycle stage.
(4) Combined analysis of Steps 2 and 3, to give the assessment of the overall impacts and trade-offs.
(5) Scenario analysis to extrapolate the results of Step 4 to society-wide large scale use.

Risk scenario analysis & risk trigger analysis (Wardak et al., *2008)*

Wardak *et al.* (2008) discussed the application of earth system engineering management (Allenby, 1999) onto the early-stage analysis of potential risks of nanomaterials. The early-stage analysis is important for many reasons including prioritization of research, planning for product development and development of regulatory systems. Normally the early-stage analysis is hindered by the uncertainty and the lack of data availability. Wardak *et al.* (2008) proposed a new approach where expert elicitation is used to compensate the uncertainty and the lack of data availability. In this approach, the consultation to experts in the field is conducted to analyse the "use and disposal scenarios" and "risk triggers'. The "use and disposal scenarios" are the potential exposure scenarios for nanomaterials. For example, the exposure scenarios to humans include inhalation, skin penetration, ingestion, and entrance from follicles. The emission scenarios to the environment include the release into air, water and soil. The "risk triggers" are the causes of risks in each "use and disposal scenarios'. In this approach, the "risk triggers" are divided into in two types, namely, exposure-related risk triggers and hazard-related risk triggers. The examples of hazard-related risk triggers include photocatalytic activities, antibacterial activities and chemical toxicity. The exposure-related risk triggers include bioavailability and the numer of disposal pathways. Every "risk trigger" is scored from 1 = low to 5 = high risk by the experts. The mapping of the scores in the [exposure-related risk trigger – hazard-related risk trigger] space allows the identification of high risk "hot spots". An example of such analysis for the air freshener spray is shown in Figure 3.2.

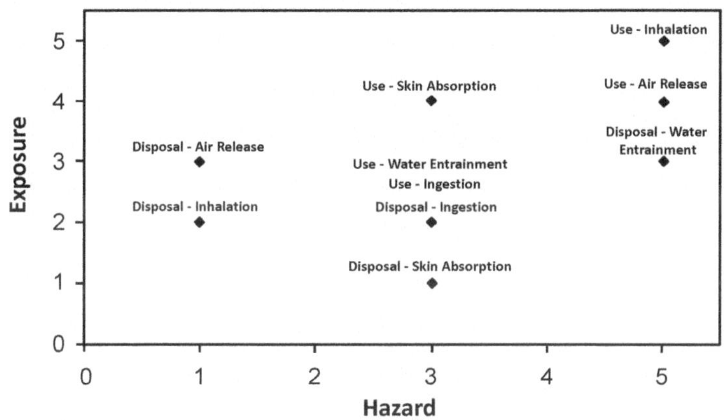

Figure 3.2 Scenario plot of hazard-related risk trigger versus exposure-related risk trigger for air freshener spray nano-product. Higher scores indicate greater risks (Wardak *et al.*, 2008). Reproduced with permission. Copyright 2008 Blackwell Publishing.

Screening Level Analysis (Gavankar et al.*, 2012)*

In order to overcome the lack of data availability, Gavankar *et al.* (2012) proposed a line of thinking based on a qualitative approach as illustrated in Figure 3.3. In this approach, when the characteristic size of nanomaterials is below 100 nm (in the "nano" range) but above the so-called scalability break-down point (i.e. the size below which nano-specific properties are manifested), then the LCA is carried out using the same approach as for traditional bulk materials. When the characteristic size of nanomaterials is below the scalability break-down point but no quantitative toxicological data is available, a qualitative assessment called "screening level assessment" is performed. The qualitative "screening level assessment" may follow the technique proposed by Bauer *et al.* (2008), where risk boundaries are drawn along the fixed/embedded or free/dispersed state of nanomaterials in a specific application/products.

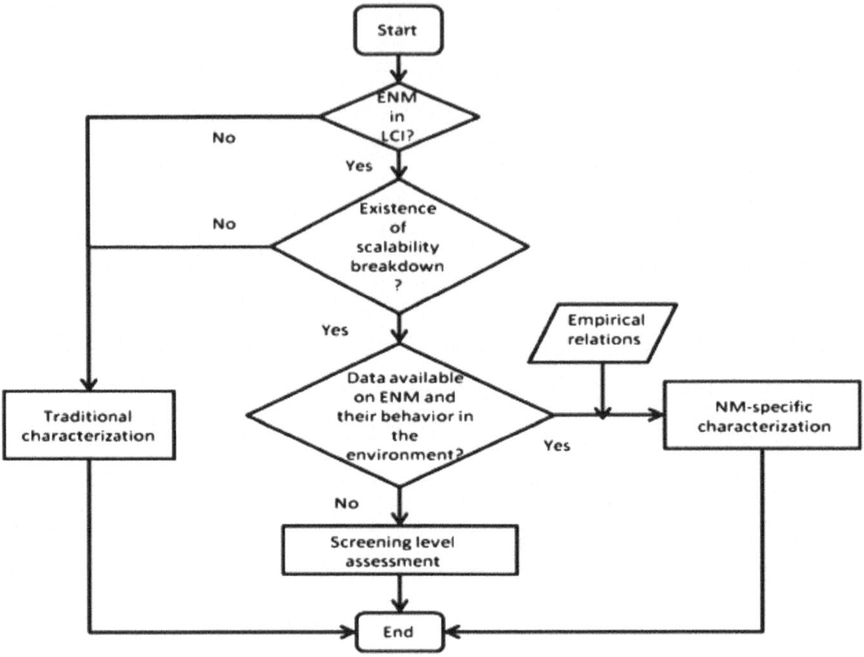

Figure 3.3 Screen-level approach for the LCA of nanomaterials (Gavanker *et al.*, 2012). Reproduced with permission. Copyright 2012 Springer.

Future Technology Analysis (Fleischer & Grunwald, 2008)

Future technology analysis (FTA) was proposed by Fleischer *et al.* (2008) to overcome the uncertainty and lack of data availability. FTA is a modified Technology Assessment (Genus, 2006) that serves as an early warning system for or against the development of new technologies in terms of social robustness. In FTA, an epistemological analysis is carried out on the future projections of the sustainability effects of nanotechnology, by uncovering and analysing the premises and presuppositions of the various potentials, promises and threats of nanotechnology. This will distinguish systematic foresight from plain speculation. In this manner, FTA can provide the *prospective knowledge* on the sustainability effects of nanotechnology in the production, use, and disposal stages in order to support decision-making activities.

Multi-Criteria Decision Analysis (Seager & Linkov, 2008)

Multi-Criteria Decision Analysis (MCDA) is a computational technique to facilitate the understanding of trade-offs and multiple perspectives in the impact assessment. Uncertainty in the data is quantified by combining scenario modelling with probabilistic representation of results. The MCDA problem is represented in the multi-dimensional decision space consisting of many criterion-axes. MCDA was recommended as one of the most promising tools for the risk assessment of nanotechnology (Roco, 2008).

Nano Risk Framework (Dupont) (Sellers, 2009)

Dupont and Environmental Defence (a non-profit group) developed a Nano Risk Framework (NRF) to assist decision-makings in the presence of knowledge gaps through the applications of assumptions. NRF does not consider resources input but focuses on potential environmental and health risks. The key aspect of NRF is that it uses "reasonable worst-case" default values or assumptions when actual data are not available.

XL Insurance Database Protocol (Sellers, 2009)

Rice University, Golder Associates and XL Insurance developed an assessment strategy based on the protocol that XL Insurance uses for calculating insurance premiums for chemical industries. The protocol is mainly used to perform risk assessment for the manufacture of nanomaterials, by focusing on the characteristics of the materials and production processes.

Many of these new approaches aim at overcoming the uncertainty and lack of information associated with nanotechnology, by adopting a number of assumptions. In this context, it is critical to clearly state the basis and limitations of the assumptions used in the LCA. This will satisfy transparency, credibility and accountability of the assessment and avoid misjudgement and misinterpretation of the outcomes of the LCA.

3.7 SUGGESTED IMPROVEMENT

In addition to the new approaches presented in the previous section, it is recommended that the following considerations be taken into account:

(i) Tandem LCA model

It is important to recognise that there are two types of nano-products: engineered nanomaterial products and nano-enabled products containing engineered nanomaterials. For example, Degussa P25 TiO_2 nanopowder is an engineered nanomaterial product, whereas a sunscreen or a toothpaste containing P25 is a nano-enabled product. The two types of nano-products have separate life-cycles. However, in the past, LCA of the two types of nano-products were often mixed up in a single assessment. This caused much confusion in the definition of "life cycle" in nanotechnology (Som *et al.*, 2010).

To avoid this confusion, it is more practical to separate LCA between the two types of nano-products, as shown in Figure 3.4. In this Tandem LCA model, a life cycle of an engineered nanomaterial starts from raw minerals/chemicals and ends with a particular type of nanomaterial products, such as TiO_2 nanopowder. Many different life cycles can be considered to produce the same type of nanomaterial products using different raw chemicals/minerals and different synthesis techniques (see Section 3.4). On the other hand, in the LCA of nano-enabled products, a particular type of nanomaterial products, such as TiO_2 nanopowder, is regarded as a raw material to produce various nano-enabled products. By separating the life-cycle of nanomaterial products from that of nano-enabled products, LCA can be conducted in a more simplified and systematic matter for a large number of nano-related products.

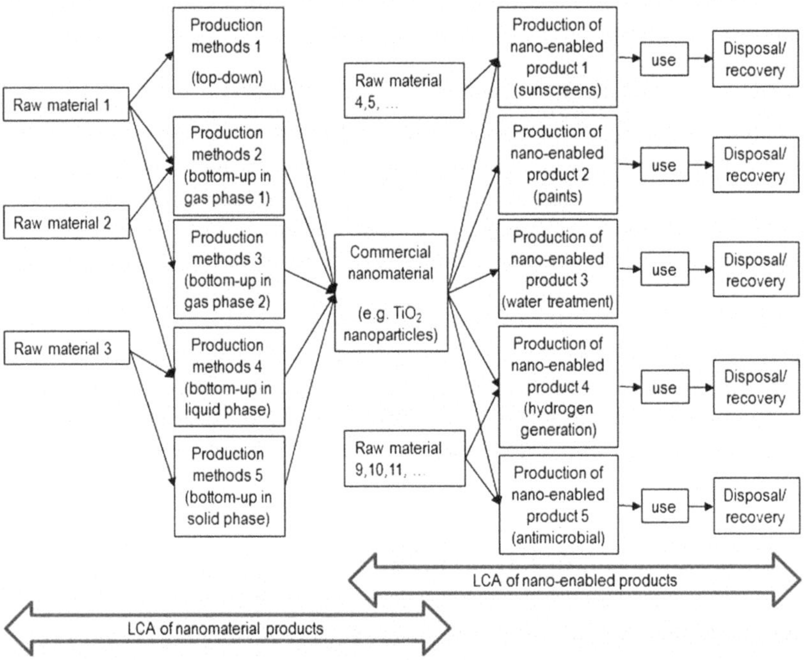

Figure 3.4 Tandem LCA model for nanomaterials and nano-enabled products (TiO_2 nanoparticles are used as an example).

It is also useful to create a world-wide database of the LCA of nanomaterial products, so that the LCA of a particular commercial nanomaterial product can be shared and utilised to perform the LCA of many nano-enabled products.

(ii) Comparative analysis with conventional technologies

LCA of nanotechnology should include the comparative analysis with conventional technologies. For example, in the LCA of nanoparticle sunscreens, a comparative analysis between conventional chemical-based (organic molecule-based) sunscreens and nano-enabled sunscreens should be included. This will give important decision-making information not only for the product development and regulatory framing but also for the overall risk-benefit analysis of nanotechnology.

(iii) Hot target identification

Due to the large number of nano-specific physicochemical properties of nanomaterials and equally large numbers of application/usage/disposal mode of nano-enabled products, it is a monumental task to complete the LCA of nanotechnology for all possible cases. In order to perform the LCA of nanotechnology fast enough to catch up with the rapid progress of nanotechnology commercialization, the LCA of nanotechnology needs focus and prioritization in an internationally coordinated manner.

The prioritization of nano-products for LCA should be made based on the magnitude of potential environmental and health risks. Risk is a function of hazards and exposure:

$$Risk = hazard \times exposure \tag{3.1}$$

However, the analysis of hazard and exposure are normally carried out on a single piece of a nano-product. The commercial production volume and the degree of product circulation/distribution are often ignored. Thus, the potential risks of nano-products should be assessed based on the modified formula:

$$Risk = hazard \times exposure\ (route) \times exposure\ (volume) \tag{3.2}$$

The exposure route and exposure volume should be scored according to the following criteria:

- *Exposure route*: the chance (threshold, frequency and locality) of exposure.
- *Exposure volume*: the production volume and distribution volume.

To score hazard, exposure routes and exposure volume, risk scenario analysis & risk trigger analysis, such as the ones Wardak *et al.* (2008) proposed, may be used. By mapping the scores in the hazard – exposure (chance) – exposure (volume) 3D space, the identification of "hot-target" and prioritization of nano-products for LCA studies will become possible.

3.8 INTERNATIONAL EFFORTS

A number of international organisations are actively making continued efforts to further develop the methodology of LCA for nanotechnologies (Thomas *et al.*, 2006a). Examples of such organisations include:

Society of Environmental Toxicology and Chemistry (SETAC) (www.setac.org)

SETAC is a non-profit international scientific organisation aiming to support the development of principles and practices for protection, enhancement and management of sustainable environmental quality and ecosystem integrity. SETAC was the first international body to coordinate the development of LCA. SETAC's nanotechnology advisory group was launched in 2007. Since then, they published several papers on the environmental assessment of nanomaterials.

International Organization for Standardization (ISO) (www.iso.org)

ISO has developed ISO14000 series of industrial standards associated with environmental management systems that include ISO14040 series for the standards related to LCA. There is no ISO standard specific to the LCA of nanotechnology. Nonetheless, ISO formed a technical committee TC229-Nanotechnologies in 2005, aiming to develop new standards relevant to nanotechnology, including ISO/TS13278-2011 Nanomaterial risk evaluation.

Woodrow Wilson International Center for Scholars (www.nanotechproject.org)

The Woodrow Wilson International Center for Scholars and The Pew Charitable Trusts launched an initiative "the Project on Emerging Nanotechnologies" to help business, government and the public to manage environmental and health impacts of nanotechnology. The Project also makes a catalogue of nanomaterial-containing consumer goods.

European Commission (ec.europa.eu/nanotechnology)

The European Commission advocates the importance of LCA in many policy and communication documents. The European Commission funds a project "the European Platform on Life Cycle Assessment" to establish a European Reference Life Cycle Data System. It also funds the Coordination Action for Innovation in Life-Cycle Analysis for Sustainability (CALCAS).

NanoImpactNet (www.nanoimpactnet.eu)

NanoImpactNet is an European network on the health and environmental impacts of nanomaterials. Their mission is to create a scientific basis to ensure the safe and responsible development of engineered nanomaterials and nano-enabled products. They also support the standardization, regulatory measures and implementation of legislation in Europe. Their core activities include RA and LCA. They organize annual international conference on the topic.

3.9 CONCLUSIONS

LCA is a holistic approach to assess the overall environmental impacts of technologies and human activities. LCA analyses the flows of materials and energy during the whole product life cycle including raw materials and resources, production processes and facilities, the usage of products, and post-use management of the product. It is anticipated that LCA will greatly help the development of nanotechnologies, in particular, for the analysis of risk-benefit balances, identification of research foci, product development, policy making, development of regulation and waste management strategies.

However, performing an LCA of nanotechnology faces many challenges. The major obstacles include the uncertainty due to the immature nature of the technology and markets, lack of risk-associated information, a high number of nano-specific properties to be considered, and the fast development pace of technology commercialization.

Despite these challenges, many case studies have started to appear in the scholarly literature. Many alternative, complementary or improved approaches to conventional LCA have been proposed for the assessment of nanotechnology. Many international organisations are engaging with activities to develop and apply LCA for nanotechnology. Since the late 2000's, many international conferences on nanotechnology have started to include at least one session on the environmental and health impacts of nanomaterials. In addition, many conferences solely dedicated to nano-safety and nano-LCA have been initiated around the world. As a result, LCA is emerging as a powerful tool for the sustainable development of nanotechnology.

GLOSSARY

- CALCAS: Coordination Action for Innovation in Life-Cycle Analysis for Sustainability
- FTA: Future technology analysis
- ISO: International Organization for Standardization
- LCA: Life cycle assessment
- LCI: Life cycle inventory
- LCIA: Life cycle impact assessment
- LCT: Life cycle thinking
- MCDA: Multi-criteria decision analysis
- RA: Risk assessment
- REPA: Resouces and environmental profile analysis
- SA: Screening approach
- SETAC: Society of Environmental Toxicology and Chemistry
- UNEP: United Nations Environmental Programme

REFERENCES

Afrin T., Tsuzuki T. and Wang X. (2010). Bamboo: a distinctive green fibre. Proceedings of the 1st International Conference on Textile and Apparel, 29–30, January 2010, Dhaka, Bangladesh, pp. 14–19.

Allan J. (2013). Managing nanowaste – concepts and challenges for nanomanufacturers, Chapter 11. In: *Nanotechnology Commercialization*, T. Tsuzuki (ed.), Pan Stanford Publishing, Singapore.

Allenby B. (1999). Earth system engineering: the role of industrial ecology in an engineered world. *Journal of Industrial Ecology*, **2**(3), 73–93.

Alvarez P. J. J., Colvin V., Lead J. and Stone V. (2009). Research priorities to advanceeco-responsible nanotechnology. *ACS Nano*, **3**(7), 1616–1619.

Asmatulu E., Twomey J. and Overcash M. (2012). Life cycle and nano-products: end-of-life assessment. *Journal of Nanoparticle Research*, **14**(3), 720 (8 pages).

Bauer C., Buchgeister J., Hischier R., Poganietz W. R., Schebek L. and Warsen J. (2008). Towards a framework for life cycle thinking in the assessment of nanotechnology. *Journal of Cleaner Production*, **16**(8–9), 910–926.

Bosso C., Isaacs J. and Walker W. (2012). Life cycle assessment in regulatort decision-making. Proceedings of the IEEE International Symposium on Sustainable systems and Technology, 16–18 May 2012, Boston, USA, pp. 1–4.

Bystrzejewska-Piotrowska G., Golimowski J. and Urban P. L. (2009). Nanoparticles: their potential toxicity, waste and environmental management. *Waste Management*, **29**, 2587–2595.

Christensen F. M. (2010). Reflections from an OECD workshop on environmental benefits of nanotechnology. *International Journal of Life Cycle Assessment*, **15**, 137–138.

Cobb M. D. and Macoubrie J. (2004). Public perceptions about nanotechnology: risks, benefits and trust. *Journal of Nanoparticle Research*, **6**, 395–405.

Fairman R., Mead C. D. and Williams W. P. (1998). *Environmental Risk Assessment: Approaches, Experiences and Information Sources*. Environmental Issues Series No. 4. European Environment Agency, Copenhagen.

Fava J. A. (2002). Life cycle initiative: a joint UNEP/SETAC partnership to advance the life-cycle economy. *International Journal of LCA*, **7**(4), 196–198.

Fent K. (2010). Ecotoxicilogy of engineered nanoparticles, Chapter 11. In: *Nanoparticles in the Water Cycle*, F. H. Frimmel and R. Nießner (eds), Springer-Verlag, Berlin, pp. 183–205.

Fleischer T. and Grunwald A. (2008). Making nanotechnology developments sustainable: a role for technology assessment. *Journal of Cleaner Production*, **16**(8–9), 889–898.

Gavankar S., Suh S. and Keller A. F. (2012). Life cycle assessment at nanoscale: review and recommendations. *International Journal of Life Cycle Assessment*, **17**, 295–303.

Genus A. (2006). Rethinking constructive technology assessment as democratic, reflective, discourse. *Technological Forecasting and Social Change*, **73**(1), 13–26.

Grieger K. D., Linkov I., Hansen S. F. and Baun A. (2012). Environmental risk analysis for nanomaterials: review and evaluation of frameworks. *Nanotoxicology*, **6**, 196–212.

Haase A., Tentschert J. and Luch A. (2012). Nanomaterials: a challenge for toxicological risk assessment? *Molecular, Clinical and Environmental Toxicology*, **101**, 219–250.

Hassellöv M., Readman J. W., Ranville J. F. and Tiede K. (2008). Nanoparticle analysis and characterizatio methodologies in environmental risk assessment of engineered nanoparticles. *Ecotoxicology*, **17**, 344–361.

Hiscgier R. and Walser T. (2012). Life cycle assessment of engineered nanomaterials: state of the art and strategies toovercome existing gaps. *Science of the Total Environment*, **425**, 271–282.

Hoet P. H. M., Brüske-Hohlfeld I. and Salata O. V. (2004). Nanoparticles – known and unknown health risks. *Journal of Nanobiotechnology*, **2**, 12 (15 pages).

Holister P. and Harper T. E. (2002). *The Nanotechnology Opportunity Report Vols. 1 & 2*. CMP Cientificia, Ilkley, West Yorkshire, UK.

Horne R., Grant T. and Verghese K. (2009). *Life Cycle Assessment, Principles, Practice and Prospects*. CSIRO Publishing, Collingwood, Australia.

Hunt R. G., Sellers J. D. and Franklin W. E. (1992). Resources and environmental profile analysis: a life cycle environmental assessment for products and procedures. *Environmental Impact Assessment Review*, **12**, 245–269.

Kloepffer W., Curran M. A., Frankl P., Heijungs R., Koehler A. and Olsen S. I. (2007). Nanotechnology and life cycle assessment. Proceedings of the Workshop on Nanotechnology and Life Cycle Assessment, Washington, DC, 2–3 October 2006, a joint publication by European Commission DG Research & the Woodrow Wilson International Center for Scholars.

Karn B. (2008). The road to green nanotechnology. *Journal of Industrial Ecology*, **12**(3), 263–266.

Labille J., Hendren C. O., Masion A. and Weisner M. R. (2011). Life cycle models and risk assessment, Chapter 17. In: *Nanoethics and Nanotoxicology*, P. Houdy, M. Lahmani and F. Marano (eds), Springer, Heidelberg, pp. 397–417.

Lu L. Y. Y., Lin B. J. Y., Liu J. S. and Yu C. Y. (2012). Ethics in nanotechnology: what's being done? What's missing? *Journal of Business Ethics*, **109**, 583–598.

Meyer D. E., Curran M. A. and Gonzalez M. A. (2009). An examination of existing data for the industrial manufacture and use of nanocomponents and their role in the life cycle impact of nanoproducts. *Environmental Science & Technology*, **43**(5), 1256–1263.

Mohanpuria P., Rana N. K. and Yadav S. K. (2008). Biosynthesis of nanoparticles: technological concepts and future applications. *Journal of Nanoparticle Research*, **10**, 507–517.

Nowack B. (2009). Is anything out there? What life cycle perspectives of nano-products can tell us about nanoparticles in the environment. *Nano Today*, **4**, 11–12.

Oberdörster G., Oberdörster E. and Oberdörster J. (2005). Nanotoxicology: an emerging discipline evolving from studies of ultrafine particles. *Environmental Health Perspectives*, **113**(7), 823–839.

Reijnders L. (2006). Cleaner nanotechnology and hazard reduction of manufactured nanoparticles. *Journal of Cleaner Production*, **14**, 124–133.

Roco M. C. (2008). Possibilities for global governance of converging technologies. *Journal of Nanoparticle Research*, **10**, 11–29.

Schmid K. and Riediker M. (2008). Use of nanoparticles in Swiss industry: a targeted survey. *Environmental Science and Technology*, **42**(7), 2253–2260.

Schmidt K. F. (2007). *Green Nanotechnology: It's Easier than you Think.* Woodraw Wilson International Center for Scholars & the Pew Charitable Trusts, available from www.wilsoncenter.org/nano (last accessed on 30 August 2011).

Seager T. P. and Linkov I. (2008). Coupling multicriteria decision analysis and life cycle assessment for nanomaterials. *Journal of Industrial Ecology*, **12**(3), 282–285.

Sellers K. (2009). Balancing the risks and rewards, Chapter 11. In: *Nanotechnology and the Environment*, K. Sellers, C. Mackay, L. L. Bergeson, S. R. Clough, M. Hoyt, J. Chen, K. Henry and H. Jane (eds), CRC Press, Boca Raton, pp. 249–263.

Som C., Berges M., Chaudhry Q., Dusinska M., Fernandes T. F., Olsen S. I. and Nowack B. (2010). The importance of life cycle concepts for the development of safe nanoproducts. *Toxicology*, **269**, 160–169.

Sweet L. and Strohm B. (2006). Nanotechnology – life cycle risk management. *Human and Ecological Risk Assessment*, **12**, 528–551.

Theis T. L., Bakshi B. R., Durham D., Fthenakis V. M., Gutowski T. G., Isaacs J. A., Seager T. and Wiesner M. R. (2011). A life cycle framework for the investigation of environmentally benighn nanoparticles and products. *Physica Status Solidi RRL*, **5**(9), 312–317.

Thomas K., Aguar P., Kawasaki H., Morris J., Nakanishi J. and Savage N. (2006a). Research strategies for safety evaluation of nanomaterials, Part VIII: International efforts to develop risk-based safety evaluations for nanomaterials. *Toxicological Science*, **92**(1), 23–32.

Thomas T., Thomas K., Sadrieh N., Savage N., Adair P. and Bronaugh R. (2006b). Research strategies for safety evaluation of nanomaterials, part vii: evaluating consumer exposure to nanoscale materials. *Toxicological Science*, **91**(1), 14–19.

Tsuzuki T. (2013a). Properties of nanoparticulate materials, Chapter 1. In: *Nanotechnology Commercialization*, T. Tsuzuki (ed.), Pan Stanford Publishing, 8 Temasek Boulevard, Singapore.

Tsuzuki T. (2013b). Applications and market opportunities of nanoparticles, Chapter 2. In: *Nanotechnology Commercialization*, T. Tsuzuki (ed.), Pan Stanford Publishing, Singapore.

Tsuzuki T., Rana R., Liu Q., Zhang L. and Wang X. (2011). Production of Green Nanomaterials, Proceedings of the International Conference on Nanoscience and Nanotechnology 2010, Sydney, Australia, 22–26th February 2010, pp. 150–153.

Tsuzuki T. (2009). Commercial scale production of inorganic nanoparticles. *International Journal of Nanotechnology*, **6**(5/6), 567–578.

van Broekhuizen P. and Reijnders L. (2011). Building blocks for a precautionary approach to the use of nanomaterials: positions taken by trade unions and environmental NGOs in the European Nanotechnologies Debate. *Risk Analysis*, **31**(10), 1646–1657.

Wardak A., Gorman M. E., Swami N. and Deshpande S. (2008). Identification of risks in the life cycle of nanotechnology-based products. *Journal of Industrial Ecology*, **12**(3), 435–448.

Wise K. and Brasuel M. (2011). The current state of engineered nanomaterials in consumer goods and waste streams: the need to develop nanoproperty-quantifiable sensors for monitoring engineered nanomaterials. *Nanotechnology, Science and Applications*, **4**, 73–86.

Woodraw Wilson International Center for Scholars. (2011). Project on Emerging Nanotechnologies, retrieved 30 August 2011, from http://www.nanotechproject.org/inventories/consumer/analysis_draft/.

Weisner M. R. and Colvin V. L. (2005). Environmental implications of emerging nanotechnologies, Chapter 4. In: *Environmentalism and the Technologies of Tomorrow: Shaping the Next Industrial revolution*, R. Olson, D. Rejeski (eds), Island Press, Washington, DC, pp. 41–52.

Weisner M. R. and Bottero J. Y. (2007). *Environmental Nanotechnology: Applications and Impacts of Nanomaterials*. McGraw–Hill Professional, New York.

Wright P. F. A. and Jackson N. R. C. (2013). Safety of engineered nanomaterials and OH&S issues for commercial scale production, Chapter 10. In: *Nanotechnology Commercialization*, T. Tsuzuki (ed.), Pan Stanford Publishing, 8 Temasek Boulevard, Singapore.

Chapter 4

Physical and chemical analysis of nanoparticles

E. J. Espinosa-Ortiz and M. Vaca-Mier

4.1 INTRODUCTION

The use of nano-scale (at least one dimension ~1–100 nm) materials has rapidly increased over the last decades, currently there exist over 2800 nanoproducts (Nanowerk Nanomaterial Database Inventory, 2012). Nano-sized materials show enhanced physical and chemical properties compared to their bulk materials, which is the reason why these materials have been used in different industrial applications, consumer products and health care technologies. The term nanomaterials may refer to nanofilms, nanowires, nanotubes and nanoparticles. For general purposes, the term nanoparticles will be used in this chapter.

There are no simple methods to characterize nanoparticles in a complex matrix (e.g. water). The inability to determine the presence and/or chemical structure of nanoparticles accurately severely hinders efforts to evolve treatment technologies that use them or to investigate the impact of these materials once they are released into an uncontrolled environment. So, it highlights the importance to not only characterize but also quantify nanoparticles in environmental samples (e.g. water, wastewater).

A priority set of properties to characterize nanoparticles in environmental samples proposed by different authors include: size distribution, shape and morphology, surface charge (ζ), as well as elemental/chemical composition and particle concentration in the sample (Stone *et al.*, 2010). Nanoparticles can be characterized either physically or chemically under a set of basic methodologies. This chapter provides a summary of the most common methods to characterize nanoparticles in environmental samples (Figure 4.1).

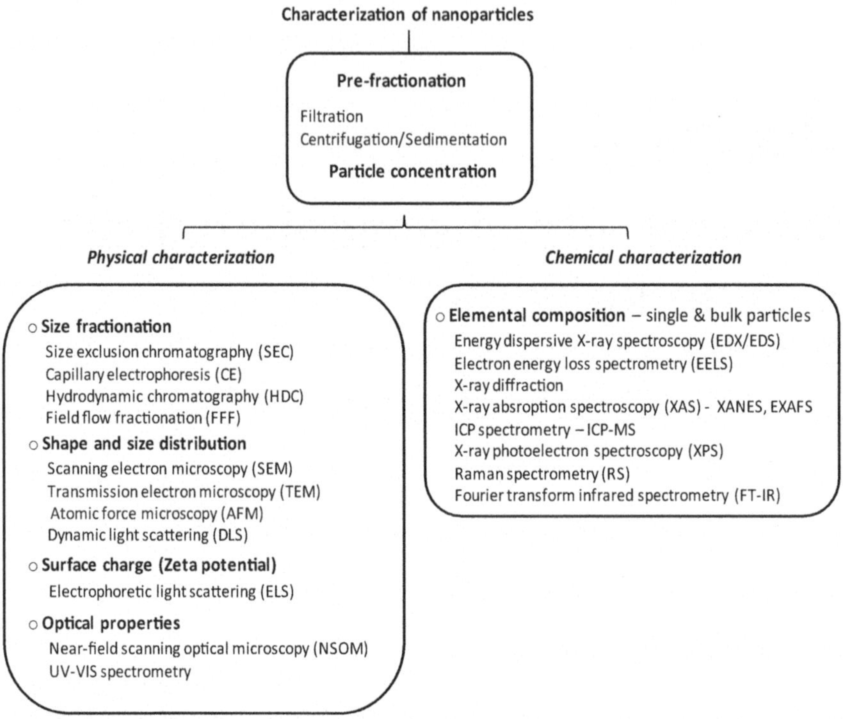

Figure 4.1 Physical and chemical methods to characterize nanoparticles in environmental samples.

4.2 SAMPLE PREPARATION - PREFRACTIONATION

The main purpose of the pretreatment of a sample is to reduce the volume of the mixture of nanoparticles or to extract them from the matrix. Filtration and centrifugation are the most common traditional techniques for preparative fractionation of samples. These analytical tools have the advantage of being simple, fast and cheap, and they do not require a major sample preparation. Table 4.1 shows a summary of the prefractionation techniques.

4.2.1 Filtration

The filtration technique relies on a series of filtration steps with filters with decreasing pore diameters, which allows generating fractionated particles. Traditional membrane filtration (0.2–1 μm) and microfiltration (>0.1 μm) are usually considered in the prefractionation. Filtration is characterized by its simplicity, field-portability, low-cost, low-sample perturbation and preparation, and capacity to treat high volumes. Aside from the benefits mentioned before, conventional filtration (0.45–0.40 μm) offers three particular advantages (Morrison & Benoit, 2001):

- Removal of suspended sediment and large organic matter fraction, which could influence the colloidal characteristics of the sample.
- Removal of biological agents, which could cause biological transformation in the sample.
- Filtration is the most widely known and used technique, which allows having comparison standards.

Table 4.1 Methods applied for the prefractionation of nanoparticles (Hassellöv *et al.*, 2008; Tiede *et al.*, 2008; Ferreira da Silva *et al.*, 2011; Handy *et al.*, 2008; Weinberg *et al.*, 2011).

Method	Approximate size range (nm)	Limit of detection	Advantages	Disadvantages/ Limitations
Pretreatment-sample preparation – Prefractionation				
Filtration	Microfiltration: 100 to >1000 Ultrafiltration: 1–30	Detection method dependent	• Simplicity • Low-cost • Low sample perturbation • High volume • Fast and little sample preparation	• Not suitable for complex, heterogeneous samples • Charging effects and clogging • Binding with ultrafiltration membrane, aggregation • Poor size resolution
Centrifugal sedimentation techniques	10 to >1000	Detection method dependent	• Size fractionation • Low-cost • High volume • Minimum perturbation	• Can suffer from size gradient • Not suitable for complex, heterogeneous samples • Does not provide direct information on size distribution • Low size resolution

This traditional technique is, however, also prone to artifacts such as nanoparticle deposition in the membrane, concentration polarization and pore clogging (Buffle *et al.*, 1992; Morrison & Benoit, 2001). Filtration is as well not suitable for complex samples and it does not provide direct information on size distribution (Hoque *et al.*, 2012).

4.2.2 Centrifugal-sedimentation techniques

The centrifugal-sedimentation technique consists of a series of differential centrifugation techniques at high speeds to generate fractions containing different particle sizes. Centrifugation is mainly considered to be a prefractionation technique, although it can also be applied as the main nanoparticle fractionation method when used as ultracentrifugation.

The principle of centrifugation is quite simple and most people in the laboratories are familiar with it. Besides, the equipment required (high speed centrifuges) to apply this technique is commonly available in every lab.

Centrifugation allows working with a broad range of particle sizes (10 to >1000 nm). Other advantages include minimum perturbation of the sample as well as little sample preparation. In fact, the only requirement for this technique is that particles are in suspension (Handy *et al.*, 2008). Apart from all the benefits provided by this technique, centrifugation can induce agglomeration between particles and a size gradient, which makes it a low size resolution method. Moreover, heterogeneous or complex samples are not suitable for this technique. Attention must be paid to the medium used for centrifugation since its composition could alter the colloidal properties arising errors (Tiede *et al.*, 2008; Hoque *et al.*, 2012).

4.3 METHODS FOR DETERMINING BULK PARTICLE CONCENTRATION

Due to the diversity and differences in physical and chemical properties of nanoparticles, there is no specific method that can be broadly applied to measure the nanoparticle concentration.

One novel and suited laser-based technique that allows determining the bulk concentration of nanoparticles and their average equivalent particle size is the laser-induced breakdown detection (LIBD). Particles in a liquid medium are exposed to a pulsed laser beam, which generates selective dielectric breakdowns on the particles. The colloid concentration and particle size is a function of the number of breakdown events produced per total number of laser pulses and their spatial distribution in the laser focus (Boller & Kaegi, 2011). A measurement usually takes several minutes, there is no required pretreatment of the sample and it is meanly non-invasive (Table 4.2). Kaegi *et al.* (2008) demonstrated that the combination of LIBD and microscopic methods (AFM, TEM) is a promising analytical approach for characterizing nanosized particles in water. In this study, the authors showed that LIBD was able to detect nanoparticles down to the 10 nm range, determining the mean diameter for particle mixtures of different size.

Disadvantages of LIBD include the incapability to distinguish between different particles (organic, inorganic, biological) and the requirement of particle-specific size calibration (Bundschuh *et al.*, 2001a, b).

Table 4.2 Method for determining bulk nanoparticle concentration (Adapted from Ferreira da Silva *et al.*, 2011; Handy *et al.*, 2008; Hassellöv *et al.*, 2008).

Method	Approximate size range (nm)	Limit of detection	Advantages	Disadvantages/ Limitations
Method for determining bulk particle concentration				
Laser induced breakdown detection (LIBD)	5 to >1000	ppt	• Extremely sensitive (ng/dm^3) • Estimation of particle size as well as concentration • No required sample preparation • Non-invasive • Provides both the concentration and the particle size	• Cannot discriminate between different types of nanoparticles • Not possible to use one set of calibration standards for different types of nanoparticles • Equipment still not commercially available • External calibrations are required for particle-size measurements

As an example of application, Kaegi *et al.* (2008) characterized nanosized particles in drinking water from a lake water source after conventional treatment and UF membrane treatment. Morphological information was obtained with AFM and TEM, whereas LIBD was used to determine the total number of particles down to 10 nm as well as the average particle diameter in the water samples. Results show differences in average particle diameter under conventional and UF-membrane treatment at different stages of the treatment. Results from LIBD were in good agreement compared to image analysis data. Figure 4.2 shows what it is typically obtained with LIBD, the particle size distribution of particles, from which then the average particle size can be estimated (Kaegi *et al.*, 2008).

Even though UV-VIS spectroscopy is mainly used to obtain information on the optical and structural properties of the nanoparticles (see Section 4.4.4.2), it is also commonly applied for the estimation of the bulk particle concentration in a solution. This method is particularly suitable for measurement of metallic nanoparticles, for example Au nanoparticles (Haiss *et al.*, 2007). Accurate measurement of particle concentration using UV-VIS spectroscopy is challenging, it requires the exact value of the extinction coefficient (see Section 4.4.4.1), which is a function of the size, shape, surface coatings and particle composition (Khlebtsov, 2008).

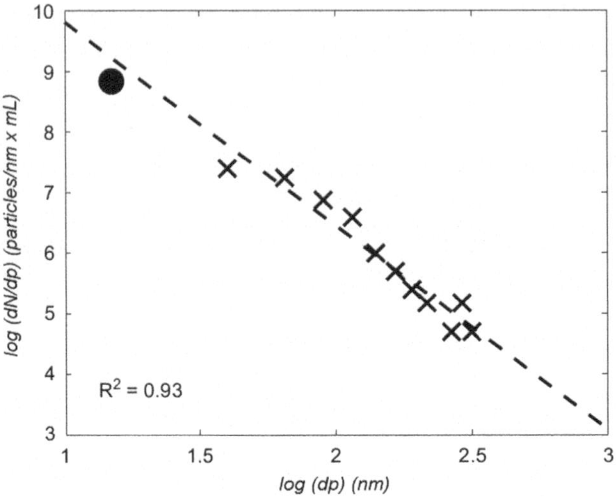

Figure 4.2 Log–log plot of the particle size distribution of particles (x). The average particle size and concentration derived from LIBD measurements is indicated by the black dot (Kaegi *et al.*, 2008)[1].

Novel methods have also been recently developed for the determination of nanoparticles concentration. Stuart *et al.* (2012) used the anodic particle coulometric method to determine the concentration of metal nanoparticles (Ni and Ag) in aqueous solution by fitting with the integrated flux equation. This method was shown to be useful to determine the unknown concentration of both aggregating and non-aggregating nanoparticles. Li *et al.* (2011) developed a simple and fast combined method consisting of electrospray differential mobility analysis with a statistical analysis of droplet-induced oligomer formation. Based on the droplet size generated from a particular electrospray source, this method estimates the absolute concentration of particles. Producing similar results to those obtained with UV-VIS measurements, the electrospray differential mobility analysis was shown to be a suitable and reliable method for the measurement of the concentration of monodispersed Au nanoparticles.

4.4 PHYSICAL CHARACTERIZATION

4.4.1 Separation techniques

The analysis of nanoparticles in environmental samples usually requires extraction or separation from the matrix. Thus, high pre-concentration factors are required

[1]Reprinted from Water Research, 42, Kaegi *et al.*, Size, number and chemical composition of nanosized particles in drinking water determined by analytical microscopy and LIBD, 2778–2786, 2008, with permission from Elsevier.

to meet detector sensitivity limitations. Chromatographic techniques to separate nanoparticles have been widely used: size-exclusion chromatography (SEC), capillary electrophoresis (CE), hydrodynamic chromatography (HDC) and field flow fractionation (FFF). These techniques are characterized for being sensitive, non-destructive, and for allowing to further characterize nanoparticles when coupled/combined with other techniques. Table 4.3 summarizes the characteristics and opportunities of these separation techniques.

4.4.1.1 Size exclusion chromatography

Size exclusion chromatography (SEC) is one of the most common and well known size separation techniques for nanoparticles. In this method, particles are passed through a column which is packed with porous beads (stationary phase) with a distribution of pore sizes in the range of the particles to be fractionated (Barth & Boyes, 1992). Particles are retained in the pores depending on their size and shape, thus generating chromatographic fractions according to their molecular weight. To ensure the efficiency of this separation technique, the minimum size of the pores of the packing material in the column has to be in accordance to the size of the solutes injected. Once the nanoparticles have been isolated, they can be further characterized when SEC is coupled or combined with different detection techniques, that is voltametry, ICP-MS or DLS (Weinberg *et al.*, 2011).

Advantages of using SEC include a good separation efficiency, combination/coupling with detection methods and high resolution molecular weight distributions (Tiede *et al.*, 2008; Weinberg *et al.*, 2011). Most important, this separation technique allows obtaining information regarding the size distribution: since particles are separated according to their molecular weight, plots of size distributions can be generated (Barth & Boyes, 1992). Disadvantages include unwanted interactions between the column and the solute leading to irreversible adsorption of nanoparticles onto the stationary phase, risks of sample loss in the column and molar mass calibration can be problematic. SEC can only be used when particles in the sample are in suspension, obtaining better results when the size range of the particles in the suspension is below 0.2 μm (Handy *et al.*, 2008). To avoid clogging issues, it has been proven that the addition of surfactants to the mobile phase might be a good option. Wei and Liu (1999) showed that separation of 5.3 nm and 38.3 nm citrate-functionalized gold nanoparticles improves in the presence of an anionic surfactant (sodium dodecylsulfate, SDS), which seems to be able to electrostatically repel the negatively charged particles (citrate-Au nanoparticles) from the stationary phase. SEC has been proven to be a useful size characterization tool of quantum dots (CdSe – Krueger *et al.*, 2005), metallic nanoparticles (nAu – Song *et al.*, 2004) and carbon nanotubes (Duesberg *et al.*, 1998).

Table 4.3 Separation techniques for the characterization of nanoparticles (Adapted from Hassellöv *et al.*, 2008; Tiede *et al.*, 2008; Ferreira da Silva *et al.*, 2011; Handy *et al.*, 2008; Weinberg *et al.*, 2011).

Method	Information obtained	Approximate size range (nm)	Limit of detection	Advantages	Disadvantages/Limitations
Separation techniques					
Size exclusion chromatography (SEC)	Size distribution	0.5–10	Detection method dependent	• Good separation efficiency • Allows combination with detection methods • Chromatography equipment is generally available • High resolution molecular weight distributions	• Unwanted interactions between the column and the solute • Risk of sample loss in the column • Molar mass calibration can be problematic
Capillary electrophoresis (CE)	Zeta potential	Not defined	Not defined	• Well suited to seize-based separation of a mixture of water soluble and charge particles • Direct data on charge • Faster than electron microscopy • Low consumption of sample and reagents	• Interpretation of migration times can be cumbersome • Lack of sensitivity of on-line detectors when low sample volumes • High concentration is required • Mobile phase interactions cannot be excluded
Hydrodynamic chromatography (HDC)	Size distribution	5–1200	Detection method dependent	• Robust method for the analysis of nanoparticles in environmental samples	• Poor peak resolution • Poor separation efficiency

				• Reduction of interactions with other particles • Allows combination with detection methods • Independent of particle size or density	
Field flow fractionation (FFF)	Diffusion coefficient and size distribution	Flow FFF: 1–1000 Sed FFF: 50–1000	Detection method dependent	• Reduction of sample complexity • Versatile in terms of separation range, selectivity and resolution • High resolution • Analysis on subfractions with defined particle size is possible • Versatility in carrier composition	• Requires pre-concentration of sample and standards to calibrate the FFF equipment • Lowest volume accepted 20 μL • Needs to be coupled to a detector that responds to the nanoparticle number or mass concentration (UV absorbance, MALLS, ICP-MS) • Errors may arise due to particle shape or density
Asymmetric-flow FFF (AsFlFFF)	Size distribution	1–500	Not defined	• High resolution, applicability to a large size range • Allows for on-channel pre-concentration of particulate samples	

4.4.1.2 Capillary electrophoresis

Capillary electrophoresis (CE) is designed to separate species and particles in different solutions according to their differential mobilities, based on their charge and size distribution, in the interior of hair thin capillaries filled with an electrolyte, while a high voltage is applied. Separation of nanoparticles by capillary zone electrophoresis is very suitable because of the electrical double layers formed around nanoparticles when exposed to an electrical field. This technique thus provides useful information about the zeta potential (ζ) of the nanoparticles, since ζ depends on the electrophoretic mobility of the nanoparticles (Weinberg *et al.*, 2011).

CE has emerged as a very useful separation technique, and it has been recognized as a major advancement in the separation of nanoparticles by size (Surugau & Urban, 2009). It is considerably faster than electron microscopy, and compared to other techniques (i.e. SEC), CE presents less surface effects. Another advantage of this separation technique is the low amount of samples and reagents that are consumed during the analysis.

Due to the fact that this technique is not solely based on particle size, interpretation of the data (migration times) can be very complex. Other limitations of CE include the lack of sensitivity of on-line detectors when working with low sample volumes, the necessity of working with high concentrations and the inevitability of mobile phase interactions (Tiede *et al.*, 2008; Fabrega *et al.*, 2011; Weinberg *et al.*, 2011).

An extensive review on electrophoretic methods was made by Surugau and Urban (2009). The authors found there was a consistent correlation between electrophoretic mobility and nanoparticle size (diameter Φ). Capillary electrophoresis has been used to separate nanoparticles of Ag (Liu *et al.*, 2005), Au and Au/Ag (Liu *et al.*, 2007), quantum dots (Oszwaldowski *et al.*, 2011) and carbon nanotubes (Lopez-Pastor *et al.*, 2008).

4.4.1.3 Hydrodynamic chromatography

Hydrodynamic chromatography (HDC) is a size exclusion chromatographic technique based on the nanoparticle hydrodynamic radius. Unlike SEC which employs a packed porous bed, HDC uses a column with non-porous packing material forming capillary routs where particles are separated according to size by flow velocity and the velocity gradient across the particle. Within the column there are regions of slow or fast flow velocities; small particles tend to be near to the column wall which is a slow flow velocity region, whereas larger particles remain in the centerline with faster velocities. Thus, larger particles appear first in the column effluent, reflecting this as an early peak in the chromatogram (McGowan & Langhorst, 1982; Fischer, 1999).

HDC columns allow separating particles in a wide range from 5 to 1200 nm (Weinberg *et al.*, 2011), according to the column length, which makes HDC a robust method for the analysis of nanoparticles in environmental samples (Tiede *et al.*, 2008). This chromatographic technique also offers the advantage of reducing solid-phase interactions since the stationary phase is not porous. Moreover, HDC can be combined with other detection methods, the most common being the UV-VIS detector (Williams *et al.*, 2002).

Tiede *et al.* (2009) developed a combined HDC-ICP-MS system for the analysis of metal-based nanoparticles in environmental samples. The developed system offers several benefits: rapid analysis time (<10 min), extended sizing range (5–300 nm) and limited sample pre-treatment required. The HDC-ICP-MS system was then used in a later study to detect and characterize silver nanoparticles in activated sludge, showing their efficiency to study the occurrence and behavior of engineered nanoparticles in complex natural environments (Tiede *et al.*, 2010). The challenge of working with HDC is the poor separation efficiency and poor peak resolution (Hassellöv *et al.*, 2008).

4.4.1.4 Field flow fractionation

Similar to chromatographic techniques, Field-Flow Fractionation (FFF) is a high resolution size-fractionating method. Contrary to chromatographic techniques, FFF does not utilize a stationary phase (Giddings, 1993; Giddings *et al.*, 1976). Particle separation is merely a physical phenomenon in an unobstructed channel. Particles are separated when interacting with an external field (no interaction with a solid-phase) which controls the particle transport velocity according to the particle diffusion coefficients when placed in different laminar flows in a very thin channel (Giddings, 1993; Hassellöv *et al.*, 2008). There are four common types of fields that can be applied when using FFF:

- Sedimentation (SdFFF) – force field is induced by centrifugation (40,000 g), 50 nm–1 µm.
- Thermal (ThFFF) – a temperature gradient is applied across the channel, 1 nm–50 nm.
- Hydraulic (Flow) – two crossed flow streams are superimposed on the same channel, 1 µm–1 nm. Typically, two types of Flow-FFF (Figure 4.3) are used: symmetrical (FIFFF) and asymmetrical (AsFIFFF).
- Electrical (EIFFF) – an electrical potential is applied across the channel.

According to the applied field and the operation mode, FFF allows separating and characterizing particles ranging from 300 Da to 100 mm (Schimpf *et al.*, 2000).

FFF has been described as the most versatile technique in terms of separation range, selectivity and resolution (Dubascoux *et al.*, 2008). It offers high resolution, reduction of sample complexity, direct relation between time and size, versatility in carrier composition and the possibility of analysis on subfractions with defined particle size. Rameshwat *et al.* (2006) demonstrated not only that Flow-FFF is a suitable method for the size characterization of water-soluble nanoparticles (Au, ZnS, ZnS-Mn^{2+}) and quantum dots (CdSe, CdSe-DNA), but also that compared to other conventional techniques (SEM, TEM and DLS), FFF is easier and more convenient. Although a fair amount of studies have been published about the use of FFF to characterize nanoparticles, still information on how to apply this technique to engineered nanoparticles in environmental samples is missing.

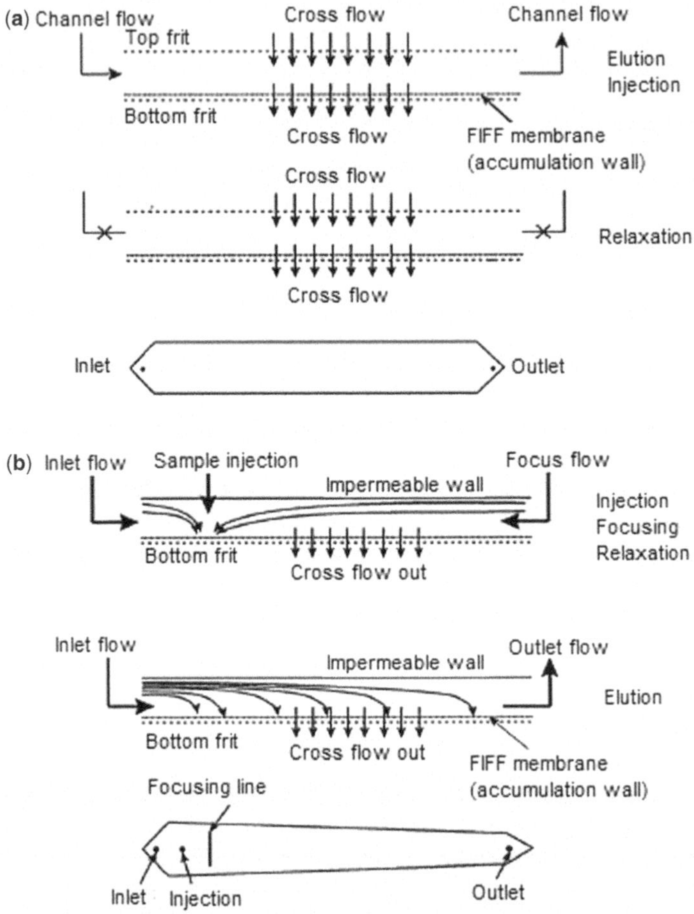

Figure 4.3 Schematic representation of a) symmetric FIFFF Flow scheme and channel geometry and b) typical asymmetrical FIFFF channel[2].

The most important shortcomings of the FFF-technique include the errors that might arise due to particle shape or density, the necessity of standards to calibrate

[2]Reprinted from J of Chromatography A, 1218, Baalousha *et al.*, Flow field-flow fractionation for the analysis and characterization of natural colloids and manufactured nanoparticles in environmental systems: A critical review, 4078–4103, 2011, with permission from Elsevier.

the FFF equipment and the fact that it needs to be coupled to a detector. This mild fractionation technique requires to be combined with some kind of detector. Common detectors coupled with FFF include: UV absorbance, light scattering (von der Kammer *et al.*, 2005) and ICP-MS (Hassellöv *et al.*, 1999). ICP-MS has been extensively used as a detector for FFF, due to its high sensitivity and selectivity, large dynamic range and capacity for multi-element monitoring. Dubascoux *et al.* (2010) made an extensive review of the benefits, limitations and applications of FFF coupled to ICP-MS. Poda *et al.* (2011) showed that the coupled system Flow-FFF-ICP-MS is a sensitive method for the detection and characterization of silver nanoparticles.

A special case of FFF is the Asymmetric-flow FFF (AsFIFFF), which has become in the most widely used and currently the most instrumentally developed technique based on Flow-FFF (Hoque *et al.*, 2012). AsFIFFF construction is simpler when compared to the symmetric FIFFF. Moreover, when using AsFIFFF there is the possibility to visualize the sample through a transparent upper wall (Wahlund and Giddings, 1987). This technique is suitable for the fractionation of nanoparticles because of its high resolution, wide operational size range (1–500 nm), and its ability to be coupled off-line with diverse detectors (Weinberg *et al.*, 2011). Most typically, AsFIFFF is coupled with UV-VIS spectroscopy and ICP-MS (Hoque *et al.*, 2012).

Silver nanoparticles were size characterized and quantified by AsFIFFF coupled to ICP-MS (Bolea *et al.*, 2011). Isaacson and Bouchard (2010) developed a size separation method for aqueous C_{60} fullerene using AsFIFFF coupled to a dynamic light scattering (DLS) in flow-through mode. An analytical system coupling AsFIFFF-MALS-DLS-IPC-MS was established by Schmidt *et al.* (2011) to separate and quantify the size and mass concentration of gold nanoparticles in aqueous suspensions.

4.4.2 Methods for assessing the shape, size distribution and surface structure of nanoparticles

Shape and size distribution are important nanoparticle properties. Size does not only define and classify nanoparticles, but it is also one of the primary properties to describe nanoparticle transport behavior. Affinity or accessibility of nanoparticles is determined by their shapes. The most common methods for nanoparticle visualization are based on electron and scanning probe microscopy: scanning electron microscopy (SEM), transmission electron microscopy (TEM) and atomic force microscopy (AFM). These analytical tools not only provide visual characterization of nanoparticles below the nm range, but also offer information about aggregation, size, structure, shape and dispersion of nanoparticles (Mavrocordatos *et al.*, 2004). Table 4.4 shows a summary of the methods to assess shape and size.

Table 4.4 Techniques to assess size and shape of nanoparticles (Adapted from Hassellöv *et al.*, 2008; Tiede *et al.*, 2008; Ferreira da Silva *et al.*, 2011; Handy *et al.*, 2008; Weinberg *et al.*, 2011).

Method	Information obtained	Approximate size range (nm)	Limit of detection	Advantages	Disadvantages/Limitations
Methods for assessing the size, shape and morphology of nanoparticles					
Scanning electron microscopy (SEM)	Visualization. Particle size in two dimensions, shape, structure and aggregation	1–1000	ppb-ppm	• High resolution • Suitable for elemental analysis by complementary techniques	• Samples may required coating and maintained under vacuum to avoid secondary electrons-gas molecules interaction • Destructive • Incompatible with wet or liquid samples • Requires extensive sample preparation • No information on particle concentration
Transmission electron microscopy (TEM)	Visualization. Size, shape and structural information	TEM-SAED: 1 nm TEM/HR-TEM: 1 to >1000	ppb-ppm	• Very high resolution • Direct visual information • When coupled to specific detectors (X-EDS, EELS), chemical information can be obtained	• Destructive • Charge effects • Incompatible with wet or liquid samples • Could be improved by moving to STEM • Time consuming and demanding. Requires extensive sample preparation • Does not provide information on particle concentration

Environmental scanning electron microscopy (ESEM)	Visualization. Size and elemental composition information when coupled with EDX	30–50	Not defined	• No sample preparation • No charging effects	• Loss in resolution • Atmospheric pressure and imaging under fully wet conditions not possible
Atomic force microscopy (AFM)	Visualization. Size, shape, particle aggregation and crystal structure	0.5 to >1000	ppb-ppm	• High resolution. • Provides 3D surface plots • Sub-nm resolution • Dry, moist or liquid samples	• Poor lateral nanoparticle dimensions accuracy • Affected by tip coating and loading • Limited for liquid samples due to particle movement • Artifacts by movement of particles
Dynamic light scattering (DLS)	Size distribution (particle size in the whole dispersion)	3 to >1000	ppm	• Instruments are available • Can be combined with FFF for more complex characterization • Rapid and simple • Non-invasive • Gives an indication of the aggregation state	• Not very suitable for analyzing complex heterogeneous matrices • Cannot be used to characterize suspensions at low concentrations • Size biased towards large particles

4.4.2.1 Scanning electron microscopy

Scanning electron microscopy (SEM) is a versatile technique that allows analyzing the microstructure morphology and chemical composition of nanoparticles. SEM is based on the exposure of the sample to a high energy focused beam of electrons to scan across the surface of the specimen, measuring secondary electrons and producing large amounts of signals, which are converted to a visual signal displayed on a cathode ray tube (Zhou *et al.*, 2006).

This high resolution technique not only provides 2 dimensional visualization images of particles (3D is also possible), but also useful particle information such as size, size distribution, shape, structure and aggregation (Tiede *et al.*, 2008; Hassellöv *et al.*, 2008).

Fukui (2007) compiled a series of examples in which SEM is shown to be a suitable tool to evaluate different types of nano-composite particles: core-shell (Fukui *et al.*, 1998), porous body (Fukui *et al.*, 2001) and internal dispersion composite structure (Fukui *et al.*, 1997). SEM has a broad operational sizing range (1 nm – 1 μm) and when coupled with complementary techniques (e.g. energy dispersive X-ray spectroscopy, EDS), elemental analysis of the particles is possible (Ferreira da Silva *et al.*, 2011). For example, Tiede *et al.* (2009) applied a wet scanning electron microscopy (WetSEM^{TM}-allows scanning observations of wet samples) combined with EDS to visualize and elementally identify Au, TiO_2, ZnO and Fe_2O_3 nanoparticles, under fully liquid conditions in distilled water, lake water and soil suspension. Compared to traditional SEM, combination of WetSEM^{TM}/EDS showed to be a promising tool for in situ investigation of engineered nanoparticles and their aggregates in natural matrices.

Extensive sample preparation (dehydration, cryofixation, embedding) is required when using SEM. Due to the fact that SEM has to be operated under vacuum conditions, dry samples are required, which may lead to sample alteration and dehydration artifacts (Mavrocordatos *et al.*, 2007). Hydrated samples can be exceptionally analyzed when using environmental SEM (ESEM). In some cases, samples may require coating (Au or graphite) and be maintained under ultrahigh vacuum in order to avoid undesired electron-gas molecules interactions (Hassellöv *et al.*, 2008). This technique may also be time consuming, particularly when trying to obtain representative and robust statistics.

4.4.2.2 Transmission electron microscopy

Originally invented in 1930s, transmission electron microscopy (TEM) operates on the principle that an electron beam of uniform current density interacts strongly with a very thin sample. Interaction between electrons and the sample atoms occurs by elastic and inelastic scattering, which makes a requirement to have very thin samples (5–100 nm). Once the electrons are transmitted through the sample, the electrons that were not absorbed are then focused onto an imaging detector (fluorescence screen or charged coupled device camera) (Reimer & Kohl, 2008).

Absorbance of the electrons (contrast image) in the sample material depends on the elemental composition and density of the sample. Sample preparation plays a very important role; special preparation techniques such as cryofixation, embedding and electropolishing are required. Obtaining good image contrast is also crucial in this technique, thus, samples containing light elements are regularly stained using a heavy metal solution (Hassellöv *et al.*, 2008; Reimer & Kohl, 2008).

TEM provides a number of advantages over other techniques for the characterization of nanoparticles: capability of determining crystallographic structure, size and shape of particles, in addition to an extreme high visual resolution. Chemical information of nanoparticles can also be obtained when TEM is combined with specific detectors (X-EDS, EELS) (Baalousha *et al.*, 2011). TEM has been applied to detect nanoparticles (Ag, Au) biogenically produced by bacteria (Zaki *et al.*, 2011; He *et al.*, 2007), as well as to characterize nanoparticle dispersions (ZnO, CNTs) in ecotoxicological studies (Brayner *et al.*, 2006; Smith *et al.*, 2007) .

High resolution TEM images might be extremely complicated to interpret, even for the most experienced lab technicians. Aside from being time consuming and demanding, extensive sample preparation to deal with wet or liquid samples is also a disadvantage for this technique (Hassellöv *et al.*, 2008; Domingos *et al.*, 2009). Typical TEM samples are less than 200 nm thick, which might represent a limiting factor for some samples.

Modern transmission electron microscopes include two high-resolution imaging modes: high-resolution transmission electron microscopy (HR-TEM) and scanning transmission electron microscopy (STEM). Being essentially a phase contrast technique, HR-TEM has become a routine technique to characterize the atomic structure of nanoparticles. Morphology of different nanoparticles has been characterized with HR-TEM (e.g. Ag and Cu nanoparticles; Peña *et al.*, 2007). STEM is a combination of SEM and TEM. STEM offers very high resolution images and with the help of dark field microscopy, high contrast images can be obtained without staining. This technique offers information on size, shape and structure of nanoparticles, and it is also complementary for elemental analysis with EDS. Ponce *et al.* (2012) extensively reviewed the STEM methods for the analysis of nanoparticles.

4.4.2.3 Atomic force microscopy

Atomic force microscopy (AFM) is a high resolution method in the family of scanning probe microscopy (Balnois *et al.*, 2007). AFM uses an oscillating cantilever with a silicon nitride tip to scan the surface of the sample. Changing electrostatic forces (10^{-7}–10^{-12} N) between the tip and the sample change as well as the angle of the cantilever, then images are created based on the separating distance between the tip and the sample. A real topographical image of the surface of the sample with atomic resolution (<1 nm) can then be obtained

(Mikayama, 2007). AFM allows particle structure visualization, producing 3D surface plots with height resolutions of ~0.5 nm. It also offers information on size, shape, size distribution and aggregation of the particles (Tiede *et al.*, 2008; Ferreira da Silva *et al.*, 2011).

Unlike TEM, AFM has the capability of analyzing not only dry but also moist and liquid samples. Nevertheless, the use of liquid samples may produce some artifacts disturbing the images. Under liquid conditions, particles are not attached to a substrate, thus they move around eventually sticking to the cantilever, provoking smearing effects and changes in the oscillation move of the cantilever (Hassellöv *et al.*, 2008; Tiede *et al.*, 2008). One of the main disadvantages of AFM is the poor lateral nanoparticle dimensions accuracy. The geometry of the tip is considerably larger than the particles in the sample. Thus, when the tip is scanning the sample, it usually recognizes the particle before its center has approached the outermost part of the particle. This leads to accurate height measurements, but overestimated lateral dimensions. Samples need to be applied onto a substrate (holder) which can be made of different materials such as mica, graphite or silicon. Application of samples needs to be carefully done. Sample preparation is also required, different preparation techniques include ultracentrifugation, drop deposition and adsorption (Balnois & Wilkinson, 2002).

AFM has been widely used to characterize nanoparticles and it has several applications. For example, Rao *et al.* (2007) demonstrated the use of AFM to characterize TiO_2, ZrO_2 and Al_2O_3 nanoparticles, while comparing the efficiency of different substrates (particle deposition) such as mica, clean and chemically treated silicon. Zhang *et al.* (2011) used AFM to measure interaction forces between *E.coli* cells and hematite (α-Fe_2O_3) or corundum (α-Al_2O_3) nanoparticles immobilized on surfaces in an aqueous environment.

4.4.2.4 Dynamic light scattering

Dynamic light scattering (DLS), also referred as photon correlation spectroscopy (PCS) or quasielastic light scattering (QELS), is a well-known technique for measuring size and size distribution of particles in the range of 3 to 1000 nm. Traditionally, DLS is applied to the characterization of particles dispersed or dissolved in a liquid. Diffusion of particles in suspension (Brownian motion) causes laser light to be scattered at different intensities. Intensity is then measured in short periods of time; the correlation time-intensity yields the velocity of the Brownian motion and thus the hydrodynamic radius distribution of particles using the Stokes-Einstein relationship. It is important to highlight that DLS only approximates the particle size of the whole dispersion in as much as this technique is based on the measurement of scattering intensity weighted diffusion coefficients and not the direct determination of the particle size (Ford, 1985; Filella *et al.*, 1997; Hassellöv *et al.*, 2008).

DLS is particularly useful for fast *in situ* and real-time nanoparticle sizing and physical characterization of aggregated nanoparticles (Ledin *et al.*, 1994). DLS is usually combined with SEC, as well as voltammetry, multi-angle laser light scattering (MALLS) and ICP-MS.

This technique is not suitable for analyzing complex heterogeneous matrices. However, when DLS is coupled with separation analytical tools such as FFF, flow-FFF or AsFIFFF, more complex characterization of size fractions can be achieved (Weinberg *et al.*, 2011). Isaacson and Bouchard (2010) applied AsFIFFF size separation coupled with in-line dynamic light scattering (DLS). Another disadvantage of DLS is that it cannot be used to characterize suspensions at low concentrations, so its capability might be limited when analyzing polydisperse samples (Hoque *et al.*, 2012). Moreover, the interpretation of the results based on intensity weighted sizes is complex, and results tend to be size based towards large particles.

4.4.3 Methods for assessing the surface charge (Zeta potential) of nanoparticles

Particles in the nanometer range are characterized by their large specific surface area, which means that the particle-liquid interface has a very strong influence on the physical properties of nanoparticles, such as dispersibility and stability. So, the behavior of nanoparticles is related to the surface structure and interfacial phenomena. Commonly used to characterize the nanoparticle's surface (charge) is the zeta potential, which is known as the electric potential at the boundary of the double layer of particles. The latter is the difference in potential between the layer at the location of the notional boundary in the surface of the particle (slipping plane) and a point in the dispersion medium. It should be highlighted that although the zeta potential is not exactly the surface potential (surface charge), it determines the interparticle forces, thus being a good index of the magnitude of interaction between nanoparticles as well as their stability and tendency to aggregate (Morrison & Ross, 2002).

The formation or breakage of aggregates between nanoparticles and water components is determined by surface properties (e.g. surface charge), which is in turn affected by pH and natural organic matter in environmental waters (Baalousha, 2009). It has been proven that natural organic matter controls the aggregation of nanoparticles in natural aquatic systems (Wilkinson *et al.*, 1997) by creating a surface coating on nanoparticles, thus changing their surface charge (zeta potential value), enhancing their stability and reducing their aggregation rate (Chen & Elimelech, 2007). The zeta potential is one of the most important parameters to characterize nanoparticles, offering valuable information about the potential impact of commercial nanoparticles on the environment and their effects on human health (Zhang *et al.*, 2008).

The surface charge of nanoparticles in liquid media can be assessed by electrophoretic light scattering (ELS), also known as laser Doppler electrophoresis or laser Doppler velocimetry (Table 4.5). This technique is used to determine the electrophoretic mobility of the particles, which can be performed via the determination of the Doppler shifts in the scattered light. Electrophoresis occurs when charged particles in dispersion move toward the electrode of opposite polarity when an electric field is applied to them. A shift in the frequency can be observed when a laser beam passes through the sample undergoing electrophoresis, as the laser light is scattered by the moving particles (Xu, 2000).

ELS has been extensively used in the last decades for zeta potential measurements. This technique allows measuring the complete mobility spectrum, while offering faster measurements compared to conventional manual techniques. ELS offers zeta potential measurements for a size range from a few nanometers to ~30 μm, with the advantage of handling low volumes samples (30 μL). Determination of the zeta potential requires samples to be highly diluted, which might lead to any kind of contamination, affecting the ionic strength or pH conditions of the system (Guimaraes & Ré, 2011).

Although the use of ELS appears to be fairly simple, it requires careful optimization to obtain usable signals. Moreover, there has been little systematic research to assess the reproducibility of zeta-potential measurements of dispersed nanoparticles. Tantra *et al.* (2010) reported the effect of the nanoparticle concentration on zeta potential measurements at dilute concentrations. Zeta potential values of silica, multi-walled carbon nanotubes (bamboo-shaped and hollow nanotubes) and gold nanoparticles at various concentrations were determined with a laser Doppler electrophoresis instrument. Results showed there is a concentration range within which the zeta potential and particle size are not affected by the nanoparticle concentration. Consistent results depended on the nature of the sample under study and ranged between 10^{-2} and 10^{-4} wt%. The zeta potential values shifted to negative values below this concentration range, accompanied by an increase in particle size (Tantra *et al.*, 2010).

4.4.4 Optical properties of nanoparticles

4.4.4.1 UV-VIS spectrometry

UV (ultraviolet)-VIS (visible) spectrometry is a well-established and versatile technique in nanomaterial research, which provides information on the reflection, transmission and absorption properties of nanoparticles. It has also been used to confirm the presence of nanoparticles in a suspension; particularly effective for metal nanoparticles characterization (Sullivan *et al.*, 2012). This technique is based on the principle that particles absorb UV or visible light, and so it measures the wavelength and intensity of absorption of near ultraviolet (200–400 nm) and visible light (400–800 nm) by a sample. UV-VIS spectrometry follows the Lambert-Beer equation (Eq. 4.1). According to the Beer's law, absorbance is

Table 4.5 Techniques to assess the surface charge and optical properties of nanoparticles (Adapted from Hassellöv *et al.*, 2008; Tiede *et al.*, 2008; Ferreira da Silva *et al.*, 2011; Handy *et al.*, 2008).

Method	Information obtained	Approximate size range (nm)	Limit of detection	Advantages	Disadvantages/ Limitations
Method for measuring surface charge					
Electrophoretic light scattering (ELS)	Zeta potential	Few-30,000	Not defined	• Handle low volume (30 μL) • Fast measurements	• Extreme caution with any kind of contamination • Interferences (e.g. natural organic matter)
Methods for measuring optical properties					
UV-VIS spectrometry	Reflection, transmission and absorption properties; concentration and some structural information	Not defined	ppb-ppm	• Equipment is generally available • Non-invasive • Simple, quick, inexpensive • No sample preparation required • Particularly effective for metal nanoparticles • High reproducibility	• Signal depends on concentration and absorptivity • Interferences from turbidity and dissolved organic matter
Near-field scanning optical microscopy (NSOM)	Topology, optical properties, sizing and visualization to the 30 nm level	50–100	Not defined	• Every surface • Optical properties • Improved spatial resolution • Simultaneous topographic image	• Low resolution • Sensitivity • Only surface features can be imaged • Performance depends on the probe material • Requires thin samples <200 nm

directly proportional to the path length (b) and the concentration (c) of the absorbing particles. Lambert's law states that the fraction of incident radiation absorbed by a medium is independent of the intensity of the incident radiation (I_0) (Yadav, 2005).

$$\log_{10}\frac{I_0}{I} = A = \varepsilon bc \qquad \text{(Eq. 4.1)}$$

where ε is the absorptivity (also known as extinction coefficient, it is a constant of proportionality) and I is the intensity of radiation transmitted through the sample. Thus, the concentration of the particles in a solution can be determined by measuring the absorbance at some wavelength.

This spectroscopic method offers a series of advantages. It is inexpensive compared to other techniques, it is simple, quick and it does not require an extensive sample preparation. Besides determination of the bulk particle concentration, UV-VIS spectroscopy might also indicate the shape of the nanoparticles, although confirmation via a visualization method (i.e. TEM) might be required. UV-VIS spectroscopy is usually applied as a first approximation to determine the presence of nanoparticles in a sample. However, turbidity and dissolved organic matter might interfere with the results, which are in turn dependent on the concentration of the nanoparticles and their extinction coefficient. For more detailed information on UV-VIS spectroscopy and its applications, see Perkampus (1992).

4.4.4.2 Near-field scanning optical microscopy

Near-field scanning optical microscopy (NSOM) is a scanning probe microscopic technique. It allows the direct correlation between nanostructures and their optical properties, providing simultaneous measurements of the topography and optical properties (fluorescence) of the particles. This technique, firstly introduced by Synge in 1928, uses a light source as the imaging mechanism, transmitting light to the nanometer scale regions on the surface, thus creating a pattern of various photosensitive materials. NSOM images can be obtained when the tip of the fiber optic probe is placed very close to the surface of the sample (distance much smaller than wavelength of light λ), and scans the surface in the near-field regime. The imaging resolution depends then on the tip diameter and the placement, offering a better resolution than the diffraction limit (Kim & Song, 2007).

NSOM provides information on optical properties of nanoparticles, offering simultaneous topographic images of the particles and so allowing for sizing and visualization to the 30 nm level (Table 4.5). It can be used for every surface without any special requirement. Nevertheless, the performance of NSOM depends on the probe material and only surface features of samples can be imaged. Moreover, this technique has low resolution and sensitivity (Betzig *et al.*, 1986).

Besides being a powerful imaging and spectroscopy tool, NSOM has shown great potential for fabricating various structures at the nanoscale. Due to its low cost and technical potential NSOM has proven to be a useful technique for: (a) material modification, including conventional photoresists and non-conventional electro-magneto-optical materials as well as ferroelectric and self-assembled monolayers; (b) material removal or etching, using NSOM as a source of energy to either induce a chemical etching process or ablate/melt materials; and (c) material addition or deposition (Tseng, 2007). For a more extensive review about the recent developments in nanofabrication using NSOM see Tseng (2007), whereas Kim and Song (2007) give more detailed information about the most recent advances on nanotechnology with NSOM.

4.5 CHEMICAL CHARACTERIZATION

4.5.1 Methods for measuring the elemental composition of single nanoparticles

The chemical composition of single particles affects the stability, as well as the optical and magnetic properties of the nanoparticle systems. Spectroscopic/ spectrometric techniques are useful analytical tools to determine the chemical composition of nanoparticles. Table 4.6 overviews the techniques to assess the elemental composition of single and bulk nanoparticles.

4.5.1.1 Energy dispersive X-ray spectroscopy

Energy dispersive X-ray spectroscopy (EDX or EDS) is an analytical technique used for the semi-quantitative elemental analysis of single particles. When a solid sample is bombarded with a high-energy beam of charged particles like electrons or X-rays, an X-ray spectrum is emitted, which is then measured with an energy dispersive spectrophotometer. Then, the observed energy peaks correspond to the elements present in the sample. Elements with atomic numbers between 4 and 92 can be theoretically detected; however, the sensitivity is not very good for light elements (atomic number <10). So, in practice this technique works best for heavier elements of the particles and associated heavy metals with high concentrations (Hassellöv *et al.*, 2008). EDX is characterized for being fast and accurate, for major elements accuracy is usually $\pm 1\%$, but the overall analytical accuracy is $\sim 20\%$ (Mavrocordatos *et al.*, 2004; 2007).

EDX is generally found on scanning electron microscopes (SEM-EDX) for elemental analysis. The main disadvantage of SEM-EDX is that it yields results as the average chemical composition of only a specific analyzed part of the sample surface, in contrast to other analytical methods such as laser-ablation-ICP-MS (Coedo *et al.*, 2007). EDX has been applied to determine the elemental composition of biosynthesized nanoparticles. EDX was applied to characterize selenium nanoparticles biogenically produced by *Pseudomonas alcaliphila*; the

Table 4.6 Techniques to assess the elemental composition of single or bulk nanoparticles (Adapted from Hassellöv *et al.*, 2008; Tiede *et al.*, 2008; Ferreira da Silva *et al.*, 2011; Handy *et al.*, 2008; Weinberg *et al.*, 2011).

Method	Information obtained	Approximate size range (nm)	Limit of detection	Advantages	Disadvantages/ Limitations
Methods for measuring the elemental composition of single nanoparticles					
Energy dispersive X-ray (EDX)	Semi-quantitative elemental composition for single particles	<10	ppm	• Accurate and fast • Single particle elemental information • Spatial resolution can be even <10 nm • Sensitivity is best for heavier elements • When microscopy methods are fitted with EDX, the elemental composition can be obtained	• Semi-quantitative information • Measurement uncertainty is generally $\sim$20% • Biased towards heavy metals
Electron energy loss spectrometry (EELS)	Elemental and chemical information	$\sim$1	Not defined	• Provides chemical information of single particle • Requires thin samples • Also applicable for light elements	• Time consuming • Changes in the nanoparticle structure and aggregation are possible • Complex and expensive
X-ray diffractometry (XRD)	Elemental information of major elements, crystal and crystallite size	0.5 to >1000	Dry powder	• Detailed crystallographic information • Elemental composition of major elements can be obtained	• Powdered samples • Low sensitivity compared to other methods (ICP-MS) • Requires a large amount of sample 1–3% wt limit • Cumbersome for low-level engineered nanoparticles

X-ray absorption spectrometry (XAS) – XANES – EXAFS	Oxidation state, elemental composition, structure	Not defined	ppm	• Non destructive • Solid and liquid samples can be used directly	• Lack of accuracy to identify the scattering of the nearest neighboring atoms
Vibrational spectroscopies					
Fourier transform infrared spectroscopy (FTIR)	Structural and chemical composition, chemical bond types, vibrational energies of molecules	Not defined	Not defined	• All kind of samples can be measured • Small amount of sample required ($<$1 mg) • Non-destructive • Faster and more sensitive than traditional IR spectroscopy	• Aqueous solutions are difficult to analyze • Complex mixtures in sample produces complex spectra • Artifacts
Raman spectroscopy	Structural and chemical composition, vibrational and rotational information	Not defined	ppt	• All kind of samples can be measured • Small amount of sample required • Relative lack of interferences • No sample pretreatment • Non-destructive	• Not very sensitive • Fluorescence interference can bias the results • Raman effect is weak, only 1 in 108 incident photons are Raman scattered

(*Continued*)

Table 4.6 Techniques to assess the elemental composition of single or bulk nanoparticles (Adapted from Hassellöv *et al.*, 2008; Tiede *et al.*, 2008; Ferreira da Silva *et al.*, 2011; Handy *et al.*, 2008; Weinberg *et al.*, 2011) (*Continued*).

Method	Information obtained	Approximate size range (nm)	Limit of detection	Advantages	Disadvantages/ Limitations
Methods for measuring the elemental composition of bulk nanoparticles					
Inductively coupled plasma mass spectrometry (ICP-MS)	Elemental composition	5–25	ppt	• Simultaneous elemental composition of tens of elements • Better selectivity than optical emission spectroscopy with the use of collision cells to eliminate interferences • Can produce isotopic composition	• Requires an acid digestion of the nanomaterial isolated from the sample • Destructive • Equipment is expensive
X-ray photoelectron spectroscopy (XPS)	Shape, size, elemental composition and oxidation state	1–10	Not defined	• Atomic composition of layers	• Control of surface contamination, adsorption or oxidation to avoid measurement errors • No resolution for single particles

EDX spectra indicated that the formed nanoparticles were entirely composed of selenium (Zhang *et al.*, 2011). Castro-Longoria *et al.* (2011) confirmed the formation of alloy-type Au/Ag bimetallic nanoparticles by SEM-EDX, when exposing $AgNO_3$ and $HAuCl_4$ to the filamentous fungus *Neurospora crassa*. Figures 4.4 and 4.5 show the EDX maps and spectra for the dried biomass of *N. crassa* exposed to ionic solutions.

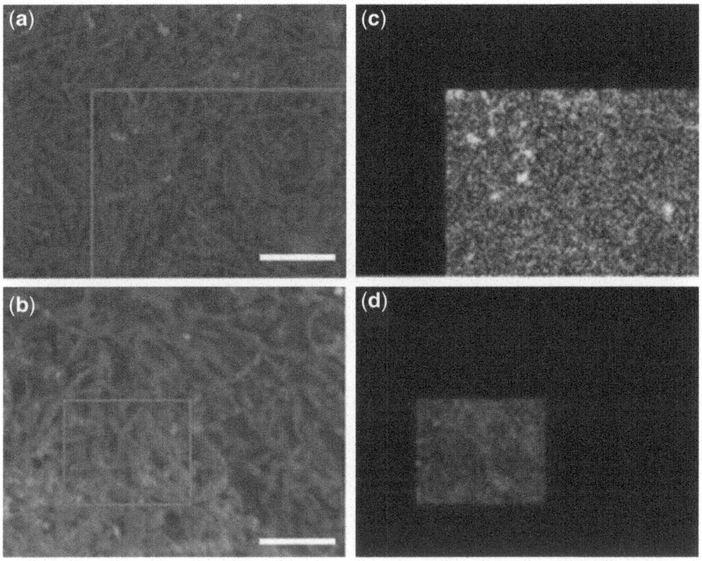

Figure 4.4 Dry biomass of *Neurospora crassa* analyzed by SEM-EDS. a) biomass exposed to $AgNO_3$; b) biomass exposed to $HAuCl_4$; c) EDS map indicating detection of Ag from the selected area in a; d) EDS map indicating detection of Au from the selected area in b. Scale bars: A-30 μm, B-20 μm [3].

4.5.1.2 Electron energy loss spectrometry

Another analytical tool that allows the determination of the elemental composition of nanoparticles is electron energy loss spectroscopy (EELS). This technique involves the bombardment of the sample with an energetic beam of electrons. When passing through the particles, electrons lose energy; such changes in kinetic energy of electrons are specific for each element. Hence, the composition of the sample can be determined directly or indirectly. This technique provides structural and chemical information with a spatial resolution down to the atomic level in favorable cases. EELS is generally used with transmission electron

[3]Reprinted from Collids and Surfaces B: Biointerfaces, 83, Castro-Longoria *et al.*, Biosynthesis of silver, gold and bimetallic nanoparticles using the filamentous fungus Neurospora crassa, 42–48, 2011, with permission from Elsevier.

microscopy (TEM), which offers a very high spatial resolution and capability to analyze extremely small volumes of material (Egerton, 2009).

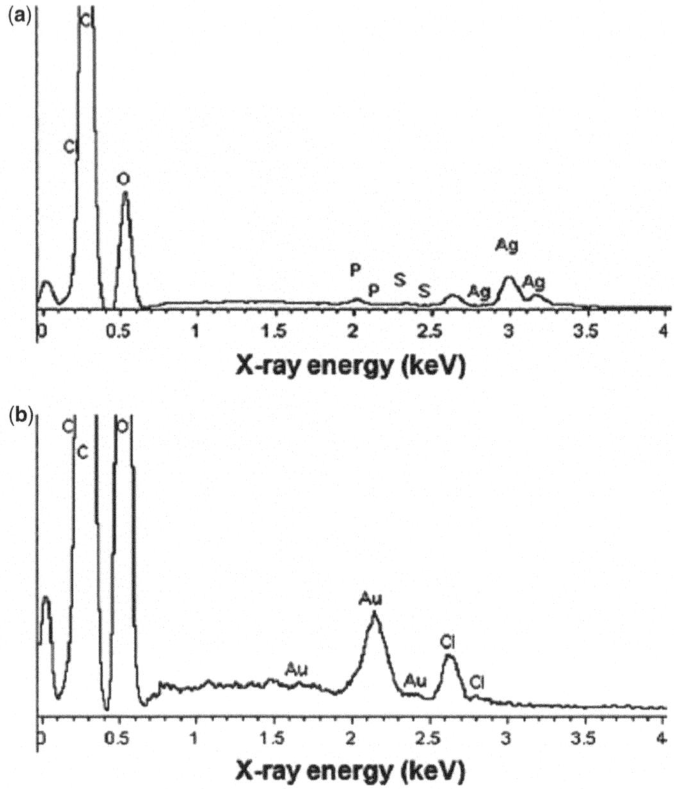

Figure 4.5 EDS spectra recorded from *N. crassa* dried biomass after exposure to ionic solutions. a) EDS from biomass exposed to $AgNO_3$, b) EDS from biomass exposed to $HAuCl_4$ [4].

EELS provides information on the elemental, chemical and dielectric characteristics of single particles. Crystalline properties have also been reported to be obtained with EELS (Mavrocordatos *et al.*, 2004). This technique offers high sensitivity to most elements, and unlike EDX, EELS has high detection efficiency for light elements (low atomic number) (Kaneko, 2007). The accuracy of EELS is ~10%. One of the main limitations of EELS is the complexity of

[4]Reprinted from Collids and Surfaces B: Biointerfaces, 83, Castro-Longoria *et al.*, Biosynthesis of silver, gold and bimetallic nanoparticles using the filamentous fungus Neurospora crassa, 42–48, 2011, with permission from Elsevier.

interpreting EELS spectra, besides being time consuming, complex and expensive (Ferreira da Silva *et al.*, 2011).

Estradé *et al.* (2012) characterized the structure and chemical properties of bi-magnetic MnO_x (antiferromagnetic)/MnO_y (ferromagnetic) and FeO_x (soft-ferromagnetic)/MnO_x (hard-ferromagnetic) by means of a combined TEM-EELS system. The local composition of nanoparticles was determined by EELS. Results allowed to distinguish the core from the shell, as well as to assess the nature of the involved manganese oxides.

4.5.1.3 X-ray diffraction

X-ray diffraction (XRD) allows determining the crystalline structure and the chemical composition of the material under investigation (Naumov, 2012). The technique is based on the principle that each crystalline solid possesses a unique characteristic X-ray pattern. The way the beam of X-ray is scattered by the atoms when passing through the sample depends on the particular crystalline structure of the materials analyzed. This non-destructive technique is the main method for crystal structure characterization for bulk, nano and thin film materials (Sharma *et al.*, 2012), allowing the estimation of the phase content and changes in phase structure of the samples. Although information about the elemental composition is obtained, it is biased towards major elements, and the sensitivity is considerably low compared to other elemental analysis methods (IPC-MS, AES). Sample preparation can be challenging; dry samples are required and they must be placed as a thin film on a ceramic support. XRD also requires a larger amount of sample (1–3%) compared to TEM-EELS. For crystal structure characterization, XRD offers considerably lower sensitivity and spatial resolution when compared to TEM-EELS. As an example of XRD application, Nurmi *et al.* (2005) used a combined methodology including XRD and XPS in order to characterize zero-valent Fe nanoparticles which were used in remediation.

4.5.1.4 X-ray absorption spectroscopy

X-ray absorption spectroscopy (XAS) is an element specific method to investigate the electronic and geometric structure of materials. Specifically applied to the study of nanoparticles, XAS provides information about the oxidation state, the elemental composition and the structure of the particles (Tiede *et al.*, 2008). This technique is based on the concept that each element has unique electron binding energies, which makes their X-ray absorption edges to occur at unique energies. More detailed information on the fundamentals of XAS techniques can be found in Schulze and Bertsch (1995). In XAS, the sample is irradiated by a monochromatic X-ray beam (produced by synchrotron radiation) which interacts with the electrons bound in an atom, leading to the scattering of the X-ray photons or to their absorption by the material in the sample (Figure 4.6).

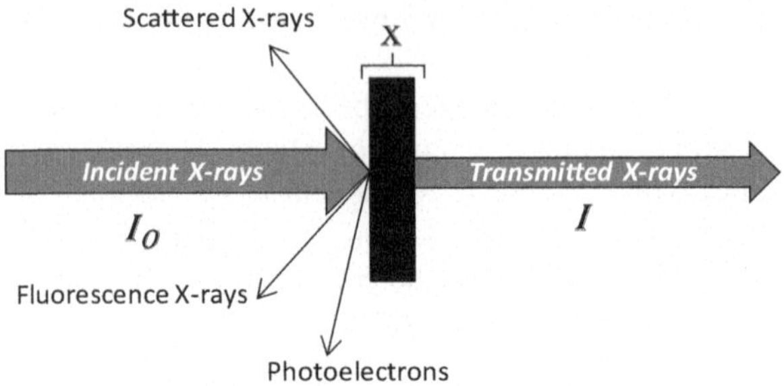

Figure 4.6 XAS principle (adapted from Jalilehvand, 2000).

The intensity I of X-rays transmitted through an object follows the Lambert-Beer equation (Attwood, 1999):

$$I(E) = I_o(E)\exp(-\mu(E) * x) \quad \text{(Eq. 4.2)}$$

where I_0 is the linear absorption coefficient, E is the photon energy and x the thickness of the sample material. When the sample is targeted with the X-ray energy, there are some energy areas where an increase in the absorption cross-section occurs. Such energies, denominated as absorption edges, correspond to the binding energies of the inner-core shell electrons (K, L, M) of the absorbing elements. The absorption edge does not solely indicate an abrupt increase in absorption, it provides information about the structure in the neighboring area of the edge jump and well above the edge (Penner-Hahn, 2003). The most common division for the XAS spectrum (Figure 4.7), based on the energy range of the X-ray beam compared to the absorption edge, includes:

- *Pre-edge*. Right before the absorption edge, there is no ionization, only transition to other orbits. $E < E_0$, where E_0 is the ionization energy.
- *X-ray Near Edge Structure (XANES)*. Region right in the vicinity of the edge, where $E = E_0 \pm 10$ eV.
- *Near Edge X-ray Absorption Fine Structure (NEXAFS)*. Region between $E = E_0 \pm 10$ eV and $E = E0 \pm 50$ eV.
- *Extended X-ray Absorption Fine Structure (EXAFS)*. Region where $E > E_0 \pm 50$ eV.

Some authors divide the XAS spectrum only in two parts: XANES below the edge and up to 30–50 eV, and EXAFS above 30–50 eV.

XAS studies in the XANES region allow obtaining information about the valence state, three dimensional geometry and coordination environment of the element in the sample. The size of nanoparticles can also be estimated using XANES (Bazin *et al.*, 1997; Bazin & Rehr, 2003). In the EXAFS range, information about the average local atomic environment and nearest neighboring atoms to the atom of interest can be obtained (Koningsberger & Prins, 1988). Different nanoparticles have been characterized by EXAFS and XANES, including Co (Cheng *et al.*, 2004), Fe^0 (Sun *et al.*, 2006), TiO_2 (Wu *et al.*, 2002), among others. Biogenically produced nanoparticles have also been characterized using XANES and EXAFS. For example, Gardea-Torresdey *et al.* (2002) applied XAS to characterize gold nanoparticles synthesized by alfalfa plants from a Au^{3+} rich agar-based solid media. Results from XANES and EXAFS confirmed the presence of Au^0 nanoparticles in the growth medium. It was also demonstrated that alfalfa plants were able to take up the Au^0 nanoparticles (2–18 nm).

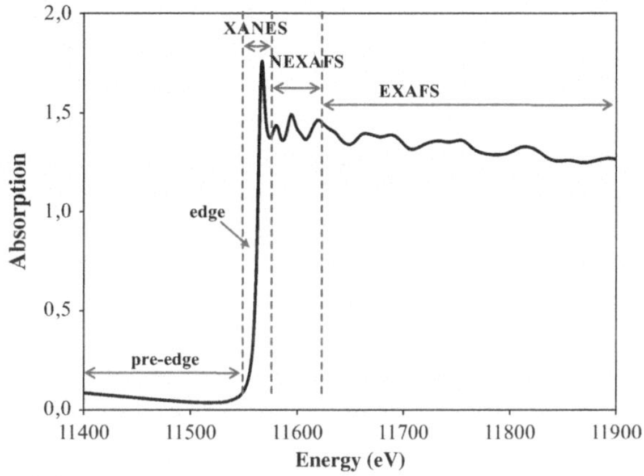

Figure 4.7 Schematic illustration of the different regions of XAS spectrum (Figure taken with permission from Jalilehvand, 2000).

XAS techniques are non-destructive (sample can be preserved), and they do not require an extensive sample preparation, thus solid or even liquid samples can be used directly. One of the most important drawbacks of XAS is the lack of accuracy to identify the scattering of the nearest neighboring atoms. For instance, XAS in the EXAFS region does not distinguish scattering atoms that are nearest neighboring atoms on the periodic tables such as sulfur or phosphorous (Penner-Hahn, 1999). The use of XAS is usually complemented with X-ray diffraction; XAS uses the structural analysis of compounds produced by XRD to estimate parameters of different samples (Koningsberger & Prins, 1988). The main difference between XRD and XAS is the ability of XAS to obtain local structural information, based on both the crystalline and the amorphous parts of the samples (Bienenstock, 1997).

4.5.1.5 Vibrational spectroscopy

V*ibrational spectroscopies* include infrared (IR) and Raman spectroscopy (RS), which measures the vibration spectra of groups of atoms. Both techniques are commonly used to characterize nanoparticles, providing information on the structural and chemical composition of the particles.

IR spectroscopy is based on the principle that atoms vibrate at a unique frequency when they are exposed to an incident infrared light, absorbing IR vibration in a specific wavelength. The unabsorbed light passes through the sample to a detector, which yields the IR spectrum (absorbed energy versus frequency). The Fourier transform infrared spectroscopy (FT-IR) has displaced, in most of the cases, the traditional IR spectroscopy. FT-IR consists of irradiating the sample with a wider range of frequencies. The obtained results are then mathematically converted to a typical spectrum through the use of *Fourier transform* (Larkin, 2011). For more detailed information on the principles and applications of FT-IR see Larkin (2011) and Smith (2011). FT-IR provides information on the specific molecular components and structures, and it allows quantitative analysis such as the determination of the concentration of a compound in the characteristic region of the spectra, since the strength of the absorption is proportional to the concentration (Reig *et al.*, 2002). Due to its simplicity, sensitivity and speed of analysis, FT-IR is considered one of the most powerful techniques for chemical analysis as well as for characterization of nanoparticles. This technique has been used to characterize nanocrystalline materials (Pejova *et al.*, 2000; Song *et al.*, 2005) and nanoparticles (Signorell *et al.*, 2000; Lemine, 2009). FT-IR is a non destructive technique, and it allows analyzing all kind of samples, although aqueous solutions might be difficult to analyze since water strongly absorbs IR. Disadvantages of FT-IR include the complexity of spectra when analyzing complex mixtures and analytical artifacts, for example the presence of water vapor and carbon dioxide (Smith, 2011).

In contrast to IR spectroscopy, Raman (scattering) spectroscopy uses an inelastic scattering of a monochromatic excitation source, usually from a laser source in the visible, near-infrared, or near-ultraviolet range (Le Ru & Etchegoin, 2009). When the monochromatic light interacts with the material, some of the photons of the laser light are absorbed by the atoms in the sample and some others pass through, creating a shift in the frequency of the reemitted photons compared with the original monochromatic frequency, which is known as the Raman effect (Larkin, 2011; Le Ru & Etchegoin, 2009). Raman spectroscopy can provide a fingerprint spectrum (molecular identity) as well as information on the structural composition and changes of the sample. It is a non- destructive, non-invasive technique, which allows analysis of all kind of samples with small amount of sample required and no necessary pretreatment. Although Raman spectroscopy has relative little analysis interferences, it is not very sensitive. Its main constrain consists on fluorescence interference, which can bias the results. Moreover, the Raman effect is inherently weak, only 1 in 10^8 incident photons are Raman scattered (Le Ru & Etchegoin, 2009). Raman spectroscopy has proven to be a useful technique to characterize nanofibers, nanocrystalline materials, nanowires, CNTs, quantum dots and nanoparticles, providing information on structural changes and micromechanical behavior. The use of Raman spectroscopy to characterize nanomaterials has been extensively reviewed elsewhere. A complete book on Raman spectroscopy for nanomaterials characterization has been edited by Kumar (2012). As well, Gouadec and Colomban (2007) present a detailed description on how Raman spectra relate to disorder, particle size and mechanical properties of nanomaterials.

A variant of the traditional Raman scattering, is the Raman microspectroscopy/microscopy, in which the material is analyzed not only by a Raman spectrometer, but also by a conventional optical microscope. Raman microscopy requires a very small sample volume ($<1\ \mu m^3$) and provides higher resolution and sensitivity (Turrel & Corset, 1996). Ultrasensitive chemical analysis and single molecule detection can be achieved when surface-enhancement (surface enhanced Raman scattering, SERS, see chapter 17) is used (Kneipp *et al.*, 1997). Dietz *et al.* (2012) discusses in more detailed the fundamentals and significance of confocal surface-enhanced Raman spectroscopic (SERS) microscopy, a proven method to obtain optical and geometrical information of nanosize metallic structures.

4.5.2 Methods for measuring the elemental composition of bulk nanoparticles

4.5.2.1 Inductively coupled plasma spectrometry

ICP spectrometry is traditionally used in the nanotechnology field, since it is a powerful analytical tool to determine the total element concentration as well as the elemental composition of the colloidal solutions (Beermann *et al.*, 2007). In addition, recent studies have approached the capability of ICP for size characterization of nanoparticles. ICP is generally combined with some kind of detection technique, mass spectrometry and optical emission spectrometry detection being the most common ones. When coupled with such detectors, ICP becomes one of the most powerful tools for elemental analysis with high sensitivity and robustness (Scheffer *et al.*, 2008).

In the family of mass spectrometry analysis, inductively coupled plasma mass spectrometry (ICP-MS) is one of the most conventional and widely used techniques for elemental determinations.

There are different configurations of ICP-MS, however all of them work under the same principle and have common elements: nebulizer, spray chamber, plasma torch and detector. Samples (usually required to be in liquid form) are introduced with the help of a peristaltic pump into a nebulizer, where particles are then transformed into a fine aerosol with argon. In the spray chamber the aerosol droplets are separated from larger droplets, and then the fine aerosol exits the spray chamber and is transported into the plasma torch, where it is ionized. Unlike ICP-Optical emission spectrometry (ICP-OES), the plasma torch in ICP-MS is placed horizontally, which is used to generate positively charged ions.

The massive production and detection of these ions is what gives ICP-MS the capability of detecting parts per trillion (Thomas, 2004).

ICP-MS has several advantages over other techniques: multielement determination, speed of analysis, detection limit (ppt) and isotopic capability. This technique also allows determining the concentration of nanoparticles. Compared to ICP-OES, ICP-MS is more sensitive (~3 orders of magnitude), which makes it the most suitable option when maximum detection power is required (Scheffer *et al.*, 2008). One of the main disadvantages of ICP-MS is that sample preparation is quite exhaustive; it requires an acid digestion of the sample, which makes it a destructive technique. In addition, the equipment can be highly expensive (Ferreira da Silva *et al.*, 2011).

ICP-MS is one of the most commonly used techniques to assess the concentrations of nanoparticles in environmental samples. For example, Kaegi *et al.* (2011) investigated the behavior of silver nanoparticles in a pilot wastewater treatment plant fed with municipal wastewater. Concentrations of silver were determined at different stages of the wastewater treatment plant using ICP-MS; predictions based on a mass balance matched the results obtained with ICP-MS.

4.5.2.2 X-ray photoelectron spectroscopy

X-ray photoelectron spectroscopy (XPS) is a powerful analytical tool to determine the elemental composition of nanoparticles at the near-surface area (10 nm) (Fadley, 1978). This technique measures the ejected photoelectrons from the sample surface when irradiated by a low-energy X-ray beam. The XPS spectrum shows the correlation of the number of electrons ejected and their binding energies, hence elements on the sample surface can be determined from the integration of peak areas and the intensity factor of every element (Lu, 2013). XPS does not have the capability of assessing the spatial resolution to analyze single particles, but it is usually applied to analyze collections of particles. This technique has been widely used for the characterization of the composition of the surface of nanoparticles and nanostructured bulk materials. However, attention must be paid to control surface contamination, adsorption and oxidation of the surface to avoid errors using this technique.

4.6 SUMMARY

Appropriate physical and chemical characterization of nanoparticles is imperative to predict the bioavailability and ecological impacts of such nano-sized materials in the environment, as well as to determine the best remediation technologies and treatment processes. Characterization of nanoparticles in environmental samples is challenging. For instance, we can expect to find nanoparticles in water sources (e.g. spill in a river), sediments or sludge (i.e. as sinks of nanoparticles). The assessment of nano-sized materials in such complex matrices might be difficult due to the presence and interference of natural nanoparticles (e.g. colloids, carbon black, iron oxides). As mentioned at the beginning of this chapter, the inability to accurately determine the presence and physicochemical properties of nanoparticles severely hinders efforts to evolve treatment technologies that use them or to investigate the potential impact of these nano-sized materials once they are released into the environment. Therefore, a comprehensive research about how nanoparticles behave under different circumstances and in media is essential to have a better understanding of the likely fate, behavior, uptake and eco-toxicity of nanoparticles under environmental conditions.

Given the nature of nanoparticles, size, structure and composition are important for characterization of bulk and individual particles. Separation, isolation and concentration of nanoparticles are equally critical to their detection. The ability to describe nanomaterials in unaltered environments (i.e. wet versus dry samples) is as well imperative when performing separation and analysis (Ferreira da Silva *et al.*, 2011). Ideal analytical methods for detection and characterization of nanoparticles in environmental matrices have to be able to deal with heterogeneous samples, minimize sample alteration to avoid artifacts and provide as much information as possible, because most characterization techniques are destructive and, therefore, samples often cannot be analyzed twice or used for more than one technique (Tiede *et al.*, 2008).

Comprehensive reviews have been published providing a detailed and full overview of analytical techniques suitable for nanoparticle analysis in complex media, see for example Tiede *et al.* (2008), Hassellöv *et al.* (2008) and Weinberg *et al.* (2011).

Characterization of nanoparticles involves the prefractionation and size fractionation of the particles as a first step. For nanoparticle separation, techniques like ultracentrifugation, nanofiltration and cross-flow filtration are commonly used. A further step is to assess the shape and size/size distribution of the nanoparticles. Microscopy-based techniques (SEM, TEM and AFM) provide information on nanoparticle aggregation, dispersion, size, structure and shape. Techniques to assess size distribution analysis of polydispersed samples include field flow fractionation (FIFFF and AsFIFFF), hydrodynamic chromatography (HDC) and size-exclusion chromatography (SEC). Coupled to several different types of detectors like ICP-MS, UV–visible spectroscopy or LIBD, more specific information can be obtained, such as the nanoparticles concentration in the sample. X-ray absorption spectroscopy techniques (XANES and EXAFS) provide significant information about the oxidation state, elemental composition and structure of the nanoparticles.

REFERENCES

Attwood D. (1999). *Soft X-rays and Extreme Ultraviolet Radiation: Principles and Applications*. Cambridge University Press, London, UK.

Baalousha M. (2009). Aggregation and disaggregation of iron oxide nanoparticles: influence of particle concentration, pH and natural organic matter. *Science of the Total Environment*, **407**, 2093–2101.

Baalousha M., Stolpe B. and Lead J. R. (2011). Flow field-flow fractionation for the analysis and characterization of natural colloids and manufactured nanoparticles in environmental systems: a critical review. *Journal of Chromatography A*, **1218**, 4078–4103.

Balnois E., Papastavrou G. and Wilkinson K. J. (2007). Force microscopy and force measurements of environmentalcolloids. In: *Environmental Colloids and Particles: Behavior, Structure and Characterization*, K. J. Wilkinson and J. R. Lead (eds), Wiley, Chichester, pp. 405–468.

Balnois E. and Wilkinson K. J. (2002). Sample preparation techniques for the observation of environmental biopolymers by atomic force microscopy. *Colloids and Surfaces A*, **207**, 229–242.

Barth H. G. and Boyes B. E. (1992). Particle size analysis. *Analytical Chemistry*, **67**, 257R–272R.

Bazin D. and Rehr J. J. (2003). Limits and advantages of x-ray absorption near edge structure for nanometer scale metallic clusters. *Journal of Physical Chemistry B*, **107**, 12398.

Bazin D., Sayers D., Rehr J. J. and Mottet C. (1997). Numerical simulation of the platinum L-III edge white line relative to nanometer scale clusters. *Journal of Physical Chemistry B*, **101**, 5332–5336.

Beermann B., Carillo-Nava E., Scheffer A., Buscher W., Jawalekar A., Seela F. and Hinz H. J. (2007). Association temperature governs structure and apparent thermodynamics of DNA-gold nanoparticles. *Biophysical Chemistry*, **126**, 124–131.

Betzig E., Lewis A., Harootunian A., Isaacson M. and Kratschmer E. (1986). Near-field scanning optical microscopy (NSOM) Development and biophysical applications. *Biophysical Journal*, **49**, 269–279.

Bienenstock A. (1997). *X-rays and the Structures of Amorphous Materials and Liquids.* Roentgen Centennial, World Scientific Publishing. Printed in Singapore, pp. 682–693.

Bolea E., Jimenez-Lamana J., Laborda F. and Castillo J. R. (2011). Size characterization and quantification of silver nanoparticles by asymmetric flow field-flow fractionation coupled with inductively coupled plasma mass spectrometry. *Analytical and Bioanalytical Chemistry*, **401**(9), 2723–2732.

Boller M. and Kaegi R. (2011). Characterization of aquatic particles. In: *Handbook on Particle Separation Processes*, A. van Nieuwenhuijzen and J. van der Graaf (eds), IWA Publishing, London, UK.

Brayner R., Ferrari-Iliou R., Brivois N., Djediat S., Benedetti M. F. and Fievet F. (2006). Toxicological impact studies based on *E.coli* bacteria in ultrafine ZnO nanoparticles colloidal medium. *Nano Letters*, **6**(4), 866–870.

Buffle J., Perret J. and Newman J. (1992). The use of filtration and ultrafiltration for size fractionation of aquatic particles, colloids and macromolecules. In: *Environmental Particles I*, J. Buffle and H. P. van Leeuwen (eds), Lewis, Chelsea, pp. 171–230.

Bundschuh T., Hauser W., Kim J. I., Knopp R. and Scherbaum F. J. (2001a). Determination of colloid size by 2-D optical detection of laser induced plasma. *Colloids and Surfaces A: Physicochemical and Engineering Aspects*, **180**, 285–293.

Bundschuh T., Knopp R. and Kim J. I. (2001b). Laser-induced breakdown detection (LIBD) of aquatic colloids with different laser systems. *Colloids and Surfaces A: Physicochemical and Engineering Aspects*, **177**, 47–55.

Castro-Longoria E., Vilchis-Nestor A. R. and Avalos-Borja M. (2011). Biosynthesis of silver, gold and bimetallic nanoparticles using the filamentous fungus *Neurospora crassa*. *Colloids and Surfaces B: Biointerfaces*, **83**, 42–48.

Chen K. L. and Elimelech M. (2007). Influence of humic acid on the aggregation kinetics of fullerene (C_{60}) nanoparticles in monovalent and divalent electrolyte solutions. *Journal of Colloid Interface Science*, **309**, 126–34.

Cheng G., Carter J. D. and Guo T. (2004). Investigation of Co nanoparticles with EXAFS and XANES. *Chemical Physics Letters*, **400**, 122–127.

Coedo A. G., Dorado T., Padilla I. and Farinas J. C. (2007). Study of heterogeneities in steels and determination of soluble and total aluminum and titanium concentration by laser ablation inductively coupled plasma mass spectrometry. *Talanta*, **71**(5), 2108–2120.

Dietz H., Sandmann G., Anders A. and Plieth W. (2012). Confocal surface-enhanced Raman microscopy at the surface of noble metals. In: *Raman Spectroscopy for Nanomaterials Characterization*, C. S. S. R. Kumar (ed.), Springer-Verlag, Berlin-Heidelberg, pp. 167–190.

Domingos R. F., Baalousha M., Ju-Nam Y., Reid M., Tufenkji N., Lead J. R., Leppard G. G. and Wilkinson K. J. (2009). Characterizing manufactured nanoparticles in the environment: multimethod determination of particle sizes. *Environmental Science Technology*, **43**(19), 7277–7284.

Dubascoux S., Le Hécho I., Hassellöv M., Von Der Kammer F., Potin Gautiera M. and Lespes G. (2010). Field-flow fractionation and inductively coupled plasma mass spectrometer coupling: history, development and applications. *Journal of Analytical Atomic Spectrometry*, **25**, 613–623.

Dubascoux S., Von Der Kammerb F., LeHécho I., Pautiera M. P. and Lespesa G. (2008). Optimization of asymmetrical flow field flow fractionation for environmental nanoparticlesseparation. *Journal of Chromatography*, **1206**, 160–165.

Duesberg G. S., Burghard M., Muster J., Philipp G. and Roth S. (1998). Separation of carbon nanotubes by size exclusion chromatography. *Chemical Communications*, **3**, 435–436.

Egerton R. F. (2009). Electron energy-loss spectroscopy in the TEM. *Reports on Progress in Physics*, **72**(1), 1–22.

Estradé S., Yedra Ll., López-Ortega A., Estrader M., Salazar-Alvarez G., Baró M. D., Nogués J. and Peiró F. (2012). Distinguishing the core from the shell in MnO_x/MnO_y and FeO_x/MnO_x core/shell nanoparticles through quantitative electron energy loss spectroscopy (EELS) analysis. *Micron*, **43**, 30–36.

Fabrega J., Luoma S. N., Tyler C. R., Galloway T. S. and Lead J. R. (2011). Silver nanoparticles: behavior and effects in the aquatic environment. *Environment International*, **37**, 517–531.

Fadley C. S. (1978). Basic concepts of x-ray photoelectron spectroscopy. In: *Electron Spectroscopy: Theory, Techniques and Applications*, C. R. Brumble and A. D. Baker (eds), Academic Press, London, **2**, pp. 1–156.

Ferreira da Silva B., Pérez S., Gardinalli P., Singhal R. K., Mozeto A. A. and Barceló D. (2011). Analytical chemistry of metallic nanoparticles in natural environments. *Trends in Analytical Chemistry*, **30**(3), 528–540.

Filella M., Zhang J., Newman M. E. and Buffle J. (1997). Analytical applications of photon correlation spectroscopy for size distribution measurements of natural colloidal suspensions: capabilities and limitations. *Colloids and Surfaces A: Physicochemical and Engineering Aspects*, **120**(1–3), 27–46.

Fischer C. H. (1999). Chromatography of colloidal inorganic nanoparticles. In: *Interfacial Phenomena in Chromatography. Surfactant Science Series 80*, E. Pefferkorn (ed.), CRC Press, NY, USA, pp. 173–226.

Ford N. C. Jr. (1985). Chapter 2 – Light scattering apparatus. In: *Dynamic Light Scattering – Applications of Photon Correlation Spectroscopy*, R. Pecora (ed.), Plenum, London, UK, pp. 7–19.

Fukui T. (2007). Chapter 1. Basic Properties and measuring methods of nanoparticles – composite structure. In: *Nanoparticle Technology Handbook*, M. Hosokawa, K. Nogi, M. Naito and T. Yokogama (eds), Elsevier, Amsterdam, The Netherlands, pp. 23–28.

Fukui T., Okawa H., Hotta T., Naito M. and Yokoyama T. (2001). Synthesis of CoO/Ni composite powder for molten carbonate fuel cells. *Journal of American Ceramic Society*, **84**, 233–235.

Fukui T., Ohara S. and Mukai K. (1998). Long-term stability of Ni-YSZ anode with a new microstructure prepared from composite powder. *Electrochemical and Solid-State Letters*, **1**, 120–122.

Fukui T., Oobuchi T., Ikyhara Y., Ohara S. and Kodera K. (1997). Synthesis of (La,Sr) MnO_3-YSZ composite particles by spray pyrolysis. *Journal of American Ceramic Society*, **80**, 261–263.

Gardea-Torresdey J. L., Parsons J. G., Gomez E., Peralta-Videa J., Troiani H. E., Santiago P. and Yacaman M. J. (2002). Formation and growth of Au nanoparticles inside live alfalfa plants. *Nano Letters*, **2**, 397–401.

Giddings J. C. (1993). Field-flow fractionation: analysis of macromolecular, colloidal, and particulate materials. *Science*, **260**, 1456–1465.

Giddings J. C., Yang F. J. and Myers M. N. (1976). Flow field-flow fractionation: a versatile newseparation method. *Science*, **193**, 1244–1245.

Guimaraes K. L. and Ré M. I. (2011). Chapter 5: Lipid nanoparticles as carriers for cosmetic ingredients: the first (SLN) and the second generation (NLC). In: *Nanocosmetics and Nanomedicines – New Approaches for Skin Care*, R. Beck, S. Guterres and A. Pohlmann (eds), Springer-Verlag, Berlin-Heidelberg, pp. 101–122.

Gouadec G. and Colomban P. (2007). Raman spectroscopy of nanomaterials: how spectra relate to disorder, particle size and mechanical properties. *Progress in Crystal Growth and Characterization of Materials*, **53**, 1–56.

Haiss W., Thanh N. T. K., Aveyard J. and Fernig D. G. (2007). Determination of size and concentration of gold nanoparticles from UV-Vis spectra. *Analytical Chemistry*, **79** (11), 4215–4221.

Handy R. D., von der Kammer F., Lead J. R., Hassellöv M., Owen R. and Crane M. (2008). The ecotoxicology and chemistry of manufactured nanoparticles. *Ecotoxicology*, **17**, 287–314.

Hassellöv M., Lyven B., Haraldsson C. and Sirinawin W. (1999). Determination of continuous size and trace element distribution of colloidal material in natural water by on-line coupling of flow field-flow fractionation with ICP-MS. *Analytical Chemistry*, **71**, 3497–3502.

Hassellöv M., Readman J. W., Ranville J. F. and Tiede K. (2008). Nanoparticle analysis and characterization methodologies in environmental risk assessment of engineered nanoparticles. *Ecotoxicology*, **17**, 344–361.

He S., Guo Z., Zhang Y., Zhang S., Wang J. and Gu N. (2007). Biosynthesis of gold nanoparticles using the bacteria *Rhodospseudomonas capsulata*. *Materials Letters*, **61** (18), 3984–3987.

Hoque M. E., Khosravi K., Newman K. and Metcalfe C. D. (2012). Detection and characterization of silver nanoparticles in aqueous matrices using asymmetric-flow field flow fractionation with inductively coupled plasma mass spectrometry. *Journal of Chromatography A*, **1233**, 109–115.

Isaacson C. W. and Bouchard D. (2010). Asymmetric flow field flow fractionation of aqueous C60 nanoparticles with size determination by dynamic light scattering and quantification by liquid chromatography atmospheric pressure photo-ionization mass spectrometry. *Journal of Chromatography A*, **1217**(9), 1506–1512.

Jalilehvand F. (2000). Structure of Hydrated Ions and Cyanide Complexes by X-ray Absorption Spectroscopy. *Doctoral thesis*, Department of Chemistry, Royal Institute of Technology, Stockholm. http://www.chem.ucalgary.ca/research /groups/faridehj/ (accessed 15 January 2013)

Kaegi R., Wagner T., Hetzer B., Sinnet B., Tzvetkov G. and Boller M. (2008). Size, number and chemical composition of nanosized particles in drinking water determined by analytical microscopy and LIBD. *Water Research*, **42**, 2778–2786.

Kaegi R., Voegelin A., Sinnet B., Zuleeg S., Hagendorfer H., Burkhardt M. and Siegrist H. (2011). Behavior of metallic silver nanoparticles in a pilot wastewater treatment plant. *Environmental Science Technology*, **45**(9), 3902–3908.

Kaneko K. (2007). Chapter 5: characterization methods for nanostructure of materials – analytical TEM (AEM). In: *Nanoparticle Technology Handbook*, M. Hosokawa, K. Nogi, M. Naito and T. Yokoyama (eds), Elsevier, Amsterdam, The Netherlands, pp. 306–310.

Kim J. and Song K. B. (2007). Recent progress of nano-technology with NSOM. *Micron*, **38**, 409–426.

Khlebtsov N. G. (2008). Determination of size and concentration of gold nanoparticles from extinction spectra. *Analytical Chemistry*, **80**, 6620–6625.

Kneipp K., Wang Y., Kneipp H., Perelman L. T., Itzkan I., Dasari R. and Feld M. S. (1997). Single molecule detection using surface-enhanced Raman scattering (SERS). *Physical Review Letters*, **78**(9), 1667–1670.

Koningsberger D. C. and Prins R. (1988). Catalysis. In: *X-ray Absorption: Principles, Applications, Techniques of EXAFS, SEXAFS and XANES*, D. C. Koningsberger and R. Prins (eds), John Wiley & *Sons. Inc*., New York, pp. 321–372.

Krueger K. M., Al-Somali A. M., Falkner J. C. and Colvin V. L. (2005). Characterization of nanocrystalline CdSe by size exclusion chromatography. *Analytical Chemistry*, **77**, 3511–3515.

Kumar C. S. S. R. (2012). *Raman Spectroscopy for Nanomaterials Characterization*. Springer, Berlin Heidelberg, XII.

Larkin P. (2011). *Infrared and Raman Spectroscopy. Principles and Spectral Interpretation*. Elsevier, San Diego, Calif, USA, pp. 1–35.

Le Ru E. C. and Etchegoin P. G. (2009). *Principles of Surface-Enhanced Raman Spectroscopy and Related Plasmonic Effects*. Elsevier, Amsterdam, The Netherlands, pp. 29–33.

Ledin A., Karlsson S., Duker A. and Allard B. (1994). Measurements *in-situ* of concentration and size distribution of colloidal matter in deep groundwaters by photon-correlation spectroscopy. *Water Research*, **28**, 1539–1545.

Lemine O. M. (2009). Microstructural characterization of α-Fe_2O_3 nanoparticles using XRD line profiles analysis, FE-SWM and FT-IR. *Superlattices and Microstructures*, **45**(6), 576–582.

Li M., Guha S., Zangmeister R., Tarlov M. and Zachariah M. R. (2011). Method for determining the absolute number concentration of nanoparticles from electrospray sources. *Langmuir*, **27**(24), 14732–14739.

Liu R. X., Lead J. R. and Baker A. (2007). Fluorescence characterization of cross flow ultrafiltration derived freshwater colloidal and dissolved organic matter. *Chemosphere*, **68**, 1304–1311.

Liu F. K., Ko F. H., Huang P. W., Wu C. H. and Chu T. C. (2005). Studying the size/shape separation and optical properties of silver nanoparticles by capillary electrophoresis. *Journal of Chromatography A*, **7**(1), 139–145.

Lopez-Pastor M., Dominguez-Vidal A., Ayora-Caada M. J., Simonet M., Lendl B. and Valcrcel M. (2008). Separation of single-walled carbon nanotubes by use of ionic liquid-aided capillary electrophoresis. *Analytical Chemistry*, **80**(8), 2672–2679.

Lu K. (2013). *Nanoparticulate Materials – Synthesis, Characterization, and Processing*. John Wiley & Sons, Inc., New Jersey, USA.

Mavrocordatos D., Pronk W. and Boller M. (2004). Analysis of environmental particles by atomic force microscopy, scanning and transmission electron microscopy. *Water Science and Technology*, **50**, 9–18.

Mavrocordatos D., Perret D. and Leppard G. G. (2007). Strategies and advances in the characterization of environmental colloids by electron microscopy. In: *Environmental Colloids and Particles: Behavior, Structure and Characterization*, K. J. Wilkinson and J. R. Lead (eds), Wiley, Chichester, pp. 345–404.

McGowan G. R. and Langhorst M. A. (1982). Development and application of an integrated, high-speed, computerized hydrodynamic chromatograph. *Journal of Colloid Interface Science*, **89**, 94–106.

Mikayama M. (2007). Characterization methods for nanostructure of materials – AFM. In: *Nanoparticle Technology Handbook*, M. Hosokawa, K. Nogi, M. Naito and T. Yokogama (eds), Elsevier, Amsterdam, The Netherlands, pp. 279–284.

Morrison M. A. and Benoit G. (2001). Filtration artifacts caused by overloading membrane filters. *Environmental Science Technology*, **35**, 3774–3779.

Morrison I. D. and Ross S. (2002). *Colloidal Dispersions: Suspensions, Emulsions, and Foams*. Wiley-Interscience, New York.

Naumov P. (2012). Chemical x-ray photodiffraction: principles, examples, and perspectives. In: *Advanced X-ray Crystallography*, K. Rissanen (ed.), Topics in Current Chemistry, Springer-Verlag, Berlin-Heidelberg, **315**, pp. 11–132.

Nanowerk Nanomaterial Database Inventory. (2012). Nanotechnology Commercial Organizations. November, 2012. http://www.nanowerk.com/nanotechnology/nanomaterial/commercial_all.php

Nurmi J. T., Tratnyek P. G., Sarathy V., Baer D. R., Amonette J. E., Pecher K., Wang C. M., Linehan J. C., Matson D. W., Penn R. L. and Driessen M. D. (2005). Characterization and properties of metalliciron nanoparticles: spectroscopy, electrochemistry, and kinetics. *Environmental Science Technology*, **39**, 1221–1230.

Oszwaldowski S., Zawistowska-Gibula K. and Roberts K. P. (2011). Capillary electrophoretic separation of nanoparticles. *Analytical and Bioanalytical Chemistry*, **399**, 2831–2842.

Pejova B., Kocareva T., Najdoski M. and Grozdanov I. (2000). A solution growth route to nanocrystalline nickel oxide thin films. *Applied Surface Science*, **165**, 271–278.

Penner-Hahn J. E. (1999). X-ray absorption spectroscopy in coordination chemistry. *Coordination Chemistry Reviews*, 190–192, 1101–1123.

Penner-Hahn J. E. (2003). Chapter 2.13 – X-ray absorption spectroscopy. In: *Comprehensive Coordination Chemistry II*, J. A. McCleverty and T. J. Meyer (eds.), Elsevier, Oxford, vol **2**, pp. 159–186.

Peña O., Cheang-Wong J. C., Rodríguez-Fernández L., Arenas-Alatorre J., Crespo-Sosa A., Rodríguez-Iglesias V. and Oliver A. (2007). Metal and metal oxide nanoparticles produced by ion implantation in silica: a microstructural study using HRTEM. *Nuclear Instruments and Methods in Physics Research B*, **257**, 99–103.

Perkampus H. H. (1992). *UV-VIS Spectroscopy and its Applications*. Springer Lab Manuals, Springer-Verlag, Berlin-Heidelberg.

Poda A. R., Bednar A. J., Kennedy A. J., Harmon A., Hull M., Mitrano D. M., Ranville J. F. and Steevens J. (2011). Characterization of silver nanoparticles using flow-field flow fractionation interfaced to inductively coupled plasma mass spectrometry. *Journal of Chromatography A*, **1218**(27), 4219–4225.

Ponce A., Mejía-Rosales S. and José-Yacamán M. (2012). Scanning transmission electron microscopy methods for the analysis of nanoparticles. In: *Nanoparticles in Biology and Medicine: Methods and Protocols in Molecular Biology*, **906**, M. Soloviev (ed.), Springer Protocols, Humana Press, NY, USA, pp. 453–471.

Rameshwat T., Samal S., Lee S., Kim S., Cho J. and Kim I. S. (2006). Determination of the size of water-soluble nanoparticles and quantum dots by field-flow fractionation. *Journal of Nanoscience and Nanotechnology*, **6**(8), 2461–2467.

Rao A., Schoenenberger M., Gnecco E., Glatzel T. and Meyer E. (2007). Characterization of nanoparticles using atomic force microscopy. *Journal of Physics: Conference Series*, **61**, 971–976.

Reig F. B., Fimeno-Adelantado J. V. and Moya-Moreno M. C. M. (2002). FTIR quantitative analysis of calcium carbonate (calcite) and silica (quartz) mixtures using the constant ratio method. Application to geological samples. *Talanta*, **58**, 811–821.

Reimer L. and Kohl H. (2008). *Transmission Electron Microscopy – Physics of Image Formation*. 5th edition, Springer series in Optical Sciences, Springer Science & Business Media (LLC), NY, USA, **36**, pp. 1–7.

Scheffer A., Engelhard C., Sperling M. and Buscher W. (2008). ICP-MS as a new tool for the determination of gold nanoparticles in bioanalytical applications. *Analytical and Bioanalytical Chemistry*, **390**, 249–252.

Schmidt B., Loeschner K., Hadrup N., Mortensen A., Sloth J. J., Koch C. B. and Larsen E. H. (2011). Quantitative characterization of gold nanoparticles by field-flow fractionation coupled online with light scattering detection and inductively coupled plasma mass spectrometry. *Analytical Chemistry*, **83**(7), 2461–2468.

Schimpf M. E., Caldwell K. and Giddings J. C. (2000). *FieldFlow Fractionation Handbook*. Wiley & Sons, Inc., USA.

Schulze D. G. and Bertsch P. M. (1995). Synchrotron x-ray techniques in soil, plant, and environmental research. *Advances in Agronomy*, **55**, 1–66.

Sharma R., Bisen D. P., Shukla U. and Sharma B. G. (2012). X-ray diffraction: a powerful method of characterizing nanomaterials. *Recent Research in Science and Technology*, **4**(8), 77–79.

Signorell R., Kunzumann M. K. and Suhm M. A. (2000). FTIR investigation of non-volatile molecular nanoparticles. *Chemical Physics Letters*, **329**, 52–60.

Smith B. C. (2011). *Fundamentals of Fourier Transform Infrared Spectroscopy*. 2nd edition, CRC Press, Taylor & Francis Group, FL, USA, pp. 8–16.

Smith C. J., Shaw B. J. and Handy R. D. (2007). Toxicity of single walled carbon nanotubes to rainbow trout (*Oncorhynchus mykiss*): respiratory toxicity, organ pathologies, and other physiological effects. *Aquatic Toxicology*, **82**, 94–109.

Song Y., Heien M. L. A. V., Jimenez V., Wightman R. M. and Murray R. W. (2004). Voltammetric detection of metal nanoparticles separated by liquid chromatography. *Analytical Chemistry*, **76**, 4911–4919.

Song W., Li G., Grassian V. H. and Larsen S. C. (2005). Development of improved materials for environmental applications: nanocrystalline NaY zeolites. *Environmental Science Technology*, **39**(5), 1214–1220.

Stone V., Nowack B., Baun A., Van den Brink N., Von der Kammer F., Dusinska M., Handy R., Hassellöv M., Joner E. and Fernandez T. F. (2010). Nanomaterials for environmental studies: classification, reference material issues, and strategies for physico-chemical characterization. *Science of the Total Environment*, **408**, 1745–1754.

Stuart E. J., Zhou Y. G., Rees N. V. and Compton R. G. (2012). Determining unknown concentrations of nanoparticles: the particle-impact electrochemistry of nickel and silver. *RSC Advances*, **2**, 6879–6884.

Sullivan T., Chapman J. and Regan F. (2012). Characterization of nano-antimicrobial materials. In: *Nano Antimicrobials: Progress and Prospects*, N. Cioffi and M. Rai (eds), Springer Berlin, Heidelberg, pp. 181–211.

Sun Y. P., Li X. Q., Cao J., Zhang W. X. and Wang P. H. (2006). Characterization of zero-valent iron nanoparticles. *Advances in Colloid and Interface Science*, **120**, 47–56.

Surugau N. and Urban P. L. (2009). Electrophoretic methods for separation of nanoparticles. *Journal of Separation Science*, **32**, 1889–1906.

Synge E. (1928). A suggested method for extending the microscopic resolution into the ultramicroscopic region. *Philosophical Magazine*, **6**, 356.

Tantra R., Shulze P. and Quincey P. (2010). Effect of nanoparticle concentration on zeta potential measurement results and reproducibility. *Particuology* **8**, 279–285.

Tiede K., Boxall A. B., Wang X., Gore D., Tiede D., Baxter M., David H., Tear S. P. and Lewis J. (2010). Application of hydrodynamic chromatography-ICP-MS to investigate the fate of silver nanoparticles in activated sludge. *Journal of Analytical Atomic Spectrometry*, **25**, 1149–1154.

Tiede K., Boxall A. B. A., Tiede D., Tear S. P., David H. and Lewis V. (2009). A robust size-characterization methodology for studying nanoparticle behavior in 'real' environmental samples, using hydrodynamicchromatography coupled to ICP-MS. *Journal of Analytical Atomic Spectrometry*, **24**, 964–972.

Tiede K., Boxall A. B. A., Tear S. P., Lewis J., David H. and Hassellöv M. (2008). Detection and characterization of engineered nanoparticles in food and the environment. *Food Additives & Contaminants*, **25**(7), 795–821.

Thomas R. (2004). *Practical Guide to ICP-MS*. Marcel Dekker, Inc., USA.

Tseng A. A. (2007). Recent developments in nanofabrication using scanning near-field optical microscope lithography. *Optics & Laser Technology*, **39**, 514–526.

Turrel G. and Corset J. (1996). *Raman Microscopy: Developments and Applications*. Academic Press, London.

Von der Kammer F., Baborowski M. and Friese K. (2005). Fieldflow fractionation coupled to multi-angle laser light scattering detectors: applicability and analytical benefits for the analysis of environmental colloids. *Analytica Chimica Acta*, **552**, 166–174.

Wahlund K. G. and Giddings J. C. (1987). Properties of an asymmetrical flowfield-flow fractionation channel having one permeable wall. *Analytical Chemistry*, **59**, 1332–9.

Wei G. T. and Liu F. K. (1999). Separation of nanometer gold particles by size exclusionchromatography. *Journal of Chromatography A*, **836**, 253–260.

Weinberg H., Galyean A. and Leopold M. (2011). Evaluating engineered nanoparticles in natural waters. *Trends in Analytical Chemistry*, **30**(1), 72–83.

Wilkinson K. J., Joz-Rolland A. and Buffle J. (1997). Different role of pedogenic fulvic acids and aquagenic biopolymers on colloid aggregation and stability in freshwaters. *Limnology and Oceanography*, **42**, 1714–1724.

Williams A., Varela E., Meehan E. and Tribe K. (2002). Characterization of nanoparticulate systems by hydrodynamic chromatography. *International Journal of Pharmaceutics*, **242**, 295–299.

Wu Z. Y., Zhang J., Ibrahim K., Abbas M. I., Li G., Tao Y., Hu T. D., Liu F. Q., Qian H. J., Xie Y. N., Zhang Q. H., Gao L. and Chen Z. Z. (2002). Structural determination of Titanium-oxide nanoparticles by x-ray absorption spectroscopy. *Acta Metallurgica Sinica (English Letters)*, **15**(1), 53–57.

Xu R. (2000). *Particle Characterization: Light Scattering Methods*. Particle technology series. Kluwer Academic Publishers, NY, USA.

Yadav L. D. S. (2005). *Organic Spectroscopy*. Kluwer Academic Publishers, MA, USA.

Zhang W., Chen Z., Liu H., Zhang L., Gao P. and Li D. (2011). Biosynthesis and structural characteristics of selenium nanoparticles by *Pseudomonas alcaliphila*. *Colloids and Surfaces B: Biointerfaces*, **88**, 196–201.

Zhang Y., Yang M., Portney N. G., Cui D., Budak G., Ozbay E., Ozkan M. and Ozkan C. S. (2008). Zeta potential: a surface electrical characteristic to probe the interaction ofnanoparticles with normal and cancer human breast epithelial cells. *Biomedical Microdevices*, **10**(2), 321–328.

Zaki S., El Kady M. F. and Abd-El-Haleem D. (2011). Biosynthesis and structural characterization of silver nanoparticles from bacterial isolates. *Materials Research Bulletin*, **46**(10), 1571–1576.

Zhou W., Apkarian R. P., Wang Z. L. and Joy D. (2006). Fundamentals of scanning electron microscopy. In: *Scanning Microscopy for Nanotechnology – Techniques and Applications*, W. Zhou and Z. L. Wang (eds), Springer Science Business Media, LLC, NY, USA, pp. 1–21.

Chapter 5

Mobility, fate and toxicity of nanomaterials/nanoparticles in water and wastewater

U. Shrestha, V. Jegatheesan, L. Shu and J. Virkutyte

5.1 INTRODUCTION

Nanotechnology is a technology that deals with and utilises material resources of nano-scale such as nanomaterials (NMs), nano-objects and nanoparticles (NPs) (Stone *et al.*, 2010). Since the introduction of the term "Nanotechnology" to describe ultra-fine machining by Professor Norio Taniguchi in 1974, there have been continuously new discoveries, innovations and developments in the Nanotechnology sector (Corbett *et al.*, 2000). With the advances in nanotechnology, its use for consumer products has also increased in geometric scale. According to an inventory, the number of consumer products based on nanotechnology has reached 1317 in 2010 from just 54 in 2005. Nanotechnology has been used to manufacture different categories of consumer products such as health and fitness, homes and gardens, automotive, food and beverage, cross cutting, electronics and computers, appliances and goods for children which signifies the importance of the nanotechnology in daily life (The Project on Emerging Nanotechnologies, 2011). Similarly, nanotechnology and related products have been used extensively for other purposes and in products such as environmental remediation, construction industry, nano fabrication, water and wastewater treatment, agricultural uses, industrial catalysts, space and astronomical applications (Aitken *et al.*, 2006; Corbett *et al.*, 2000; Reijnders, 2006). But the extensive use of nanotechnology based products has increased the prospects of the release of NMs and NPs into the environment through their use and consumption. Probable pathways for the release of the NMs/NPs into the environment are shown in Figure 5.1.

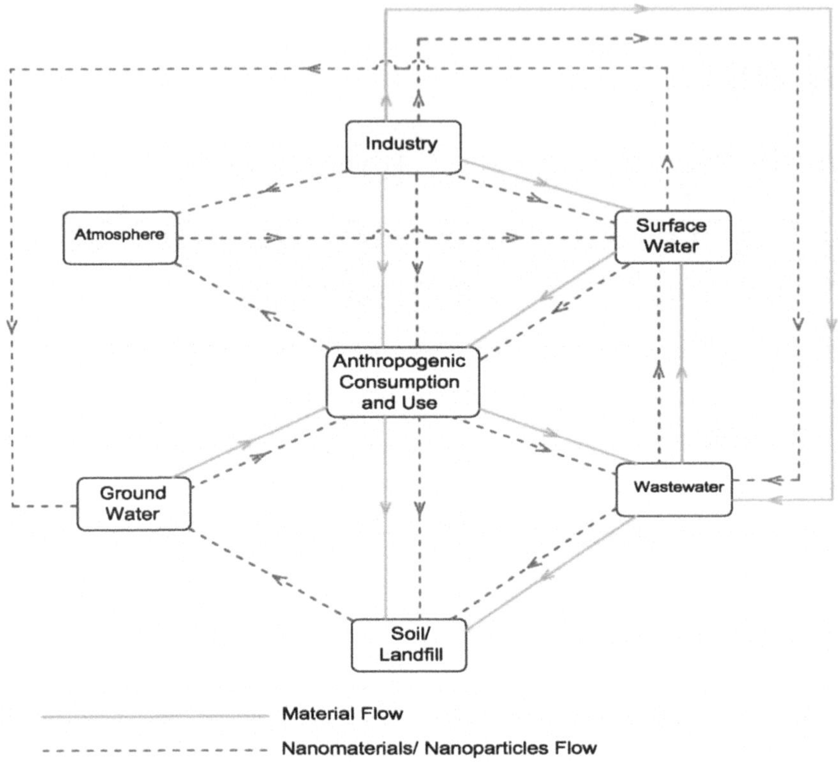

Figure 5.1 Mass flow diagram of nanomaterials/nanoparticles in environment.

5.2 RELEASE OF NANOMATERIALS/NANOPARTICLES INTO THE ENVIRONMENT

The sources of nanomaterials to the environment may be natural, incidental or engineered (Farré *et al.*, 2011). Natural colloids are the primary nano-scale materials present in nature, which supports sustainability and biodiversity through their activities and interactions with microorganisms (Jiménez *et al.*, 2011). Besides, natural colloids also contribute in pollutant control (Baalousha *et al.*, 2011). The natural nanomaterials generally evolve through different natural processes/activities/actions and are present in air, water and soil.

The incidental source of nanomaterials is unintentional unlike engineered/ manufactured nanomaterials (ENM/MNM). It may originate from combustion processes and industrial emissions (Farré *et al.*, 2011). The real concern is nanomaterials that are released through production, use and consumption of goods containing ENM/MNM as a result of the rapid development of nanotechnology as well as relatively lesser knowledge about their probable fate and toxicity. These ENMs/MNMs can be broadly classified into seven categories as given in Table 5.1. Applications of different types of ENP/MNP are given in Table 5.2.

Table 5.1 Classification of ENMs/MNMs.

Category	Examples
Metal oxides	Zinc oxide (ZnO), Titanium dioxide (TiO_2), Cerium dioxide (CeO_2), Silica (SiO_2), Aluminium oxide (Al_2O_3), Copper oxide (CuO), Iron Oxides (FeO, Fe_3O_4) and mixed oxides
Carbon products	Fullerene, Single-walled carbon nano-tubes (CNT), Multi-walled CNT, Amorphous carbon, Carbon black pigment
Metals	Silver (Ag), Iron (Fe), Gold (Au), Bimetallic NPs such as Iron-palladium (Fe-Pd), Iron-nickel (Fe-Ni), and Iron-silver (Fe-Ag)
Quantum dots and Semi-conductors	Cadmium-Tellurium (CdTe), Cadmium-selenium (CdSe), Zinc sulphide (ZnS), Lead Selenium (PbSe)
Nano-clays	Hectorites, Layered double hydroxides (LDHs)
Dendrimers	Polyphenylenedendrimers
Nano-emulsions/ Acrylic latex	Nano-emulsions of oil droplets in water

As shown in Table 5.2, various applications of the ENMs/MNMs are possible. These ENMs/MNMs may be released into the environment through various activities/products. Some examples of the release of ENM/MNM are summarised in Table 5.3. Correspondingly, the release of nanomaterials containing substances from different commercial products into wastewater are summarised in Table 5.4. Although Table 5.4 is based on usage of products in a European scenario, the authors Brar *et al.* (2010) emphasize that it is also relatively applicable to worldwide trends.

It is also necessary to know the flow pathways of NPs into the environment in order to quantify as well as reduce the release of NPs as desired. For example, the flow pathways of silver based NPs into the environment through the use of biocidal plastics and textile is elaborated in Figure 5.2 and the mass flow of silver from biocidal and other sources is shown in Figure 5.3. Similarly, mass flow of nano TiO_2 in environment with accumulation rates in tonne per year is shown in Figure 5.4. Figure 5.5 shows the mass flow and balance of the CeO_2 which is used in diesel fuel as additive.

Table 5.2 Applications of ENMs/MNMs.

	Nanomaterials																				
Application	**Fullerenes (carbon-60, carbon-70, carbon-80, derivatized)**	**Metallo-fullerenes (With endohedral metals)**	**Multi-wall nanotubes**	**Single-wall nanotubes**	**Rods**	**Fibres**	**Whiskers**	**Metals and metal oxides**	**Ceramics**	**Colloids**	**Quantum dots**	**Non-Quantum dot semi conducting materials**	**Silica**	**Polymers**	**Composites**	**Dendrimers**	**Inter-metallics**	**Ion-beam processed materials**	**Thin films (nm thickness)**	**Bio-molecules**	**Molecular motors**
Hydrogen storage	X		X	X											X						
Environmental remediation								X							X		X				
Catalysis		X						X			X				X		X				
Drug delivery	X	X	X	X				X							X	X				X	X
Medical imaging	X	X	X	X				X		X	X	X			X	X				X	X
Photovoltaics	X	X	X	X	X		X	X			X		X	X	X		X				
Textiles			X	X		X									X						
Therapeutics	X	X									X				X	X				X	X
Reinforced composites			X	X		X								X	X						
Electronics and electronic devices	X	X	X	X	X		X	X	X		X	X	X	X	X		X	X	X		
Optics and optical devices	X	X						X		X	X	X	X	X	X		X		X		
Coatings and pigments	X	X	X	X				X		X				X	X		X		X		

Cosmetics								X							X						
Ceramics applications									X												
Anti-oxidants									X												
Lubrication	X							X						X	X						
Sensors and sensing devices	X	X	X	X	X		X	X		X	X	X	X	X	X	X	X	X	X	X	
Absorbents								X		X				X	X		X				
Energetics and energetic materials			X	X				X							X		X				
Magnetics and magnetic devices								X							X		X				
Water purification and filtration media			X	X		X		X						X	X						
Air emissions reduction							X	X						X	X						
Natural and green products															X						
Quantum computing		X						X			X	X			X						
Masonry and building materials			X	X		X			X						X						
Photonics and photonic devices	X	X						X		X	X	X	X		X						
Surfactants																					

Source: Aitken *et al.* (2006), Copyright by Society of Occupational Medicine, Oxford University Press. Reprinted with permission.

Table 5.3 Examples of ENM/MNM release to the environment.

Activity/Product	Release Mechanism	Types Released	Receiving Technosphere	Reference
Construction Industry	Construction waste disposal, Aerosolization of manufactured nanomaterials, Abrasion and corrosion of civil infrastructure	Carbon Based Nanomaterials, metal oxide NPs, Metal NPs	Air, water and soil	Lee *et al.* (2010)
Nanotextile	Direct Contact, Impact during washing process, Unintentional release from geotextile	Carbon Based Nanomaterials, Metal oxide NPs, Metal NPs	Soil, water	Som *et al.* (2011)
Facade Coatings	Weathering (rain, heat, UV-light, and abrasion) or Maintenance Activities (cleaning or sanding processes), Urban Runoff	Metal oxide NPs, Metal NPs	Air, water and soil	Som *et al.* (2011); Kaegi *et al.* (2008)
Commercialized Sunscreens	Washing	Metal oxide (nano TiO_2)	Water	Botta *et al.* (2011); Johnson *et al.* (2011)
Biocidal Plastics and textiles	Through wastewater and solid waste	Silver based NPs	Air, water and soil	Blaser *et al.* (2008)
Vehicle tires	Wear, Interaction between tires and road pavement	SiO_2	Air	Reijnders (2009)

Table 5.4 Release of nanomaterials containing substances from different commercial products into wastewater.

Product type	Emission (g/pc/d)	Comments	References
Antiperspirant	0.35	Based on 50% population use of the product at dose = 0.7 g/d	Loretz *et al.* (2005, 2006)
Body lotion	1.2	20% population use of the product at dose = 6.0 g/d	Loretz *et al.* (2005, 2006)
Body wash	0.32	Home chemical usage survey in Denmark	Eriksson *et al.* (2003)
Cleaners	0.3	Home chemical usage survey in Denmark	Eriksson *et al.* (2003)
Deodorants	0.08	Home chemical usage survey in Denmark	Eriksson *et al.* (2003)
Face cream	1.64	20% population use of the product at dose = 8.2 g/d	Loretz *et al.* (2005, 2006)
Hair conditioner	0.47	Home chemical usage survey in Denmark	Eriksson *et al.* (2003)
Hair styling products	0.10	Home chemical usage survey in Denmark	Eriksson *et al.* (2003)
Lime deposit removers	0.11	Home chemical usage survey in Denmark	Eriksson *et al.* (2003)
Paint	0.09–0.36 ml/pc/yr	Data on usage in the UK and release of 1–4% during cleaning	European Commission (2004)
Laundry detergents	10.1–20.5	Data for USA, Sweden, Denmark, Finland and Norway	Eriksson *et al.* (2002)
Oral hygiene products	0.7	Survey of home chemical usage in Denmark	Eriksson *et al.* (2003)
Perfume	0.05	Dose = 0.5 g/d and 10% of population use product	Loretz *et al.* (2005, 2006)
Shampoo	1.83–6.30	Data for Sweden and Denmark	Eriksson *et al.* (2002, 2003)
Shaving foam	0.07	Survey of home chemical usage in Denmark	Eriksson *et al.* (2003)
Soap	2.5	Data for Sweden	Eriksson *et al.* (2002)
Skin care products	1.3	Survey of home chemical usage in Denmark	Eriksson *et al.* (2003)
Softeners	16.4	Date based on use in Europe	Eriksson *et al.* (2002)
Sunscreen	3.0	Dose = 30 g/d and 10% of population use product	Loretz *et al.* (2005, 2006)
Window cleaners	0.03	Survey of home chemical usage in Denmark	Eriksson *et al.* (2003)

Source: Brar *et al.* ((2010)), Copyright by Elsevier Limited. Reprinted with permission.

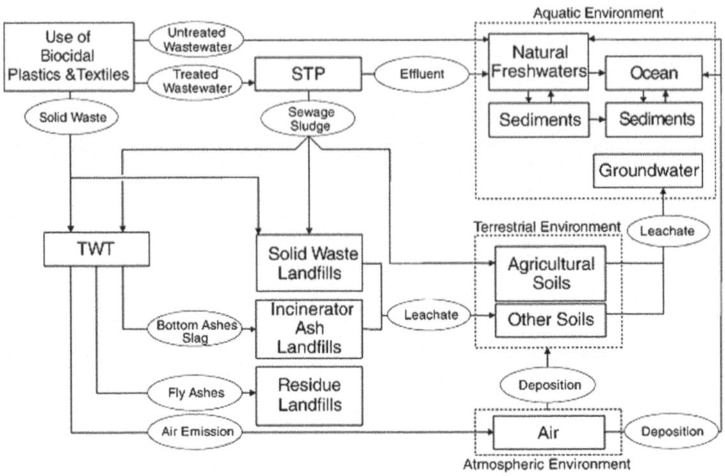

Figure 5.2 Release of silver based NPs into the environment from uses of biocidal plastics and textile. *Source*: Blaser *et al.* (2008). Copyright by Elsevier BV. Reprinted with permission.

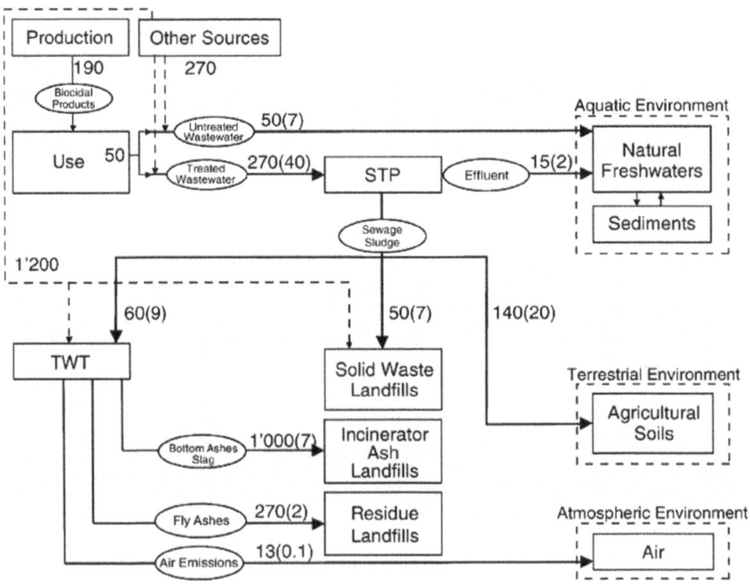

Figure 5.3 Mass flow of silver into the environment from the use of biocidals and other sources in tonnes per year. Flows caused by biocidal plastics and textiles (⟶), silver flows from other sources (– – →), combined flow of biocidal products and other sources(———→). The figure in brackets represents the contribution of biocidals in total flow. *Source*: Blaser *et al.* (2008), Copyright by Elsevier BV. Reprinted with permission.

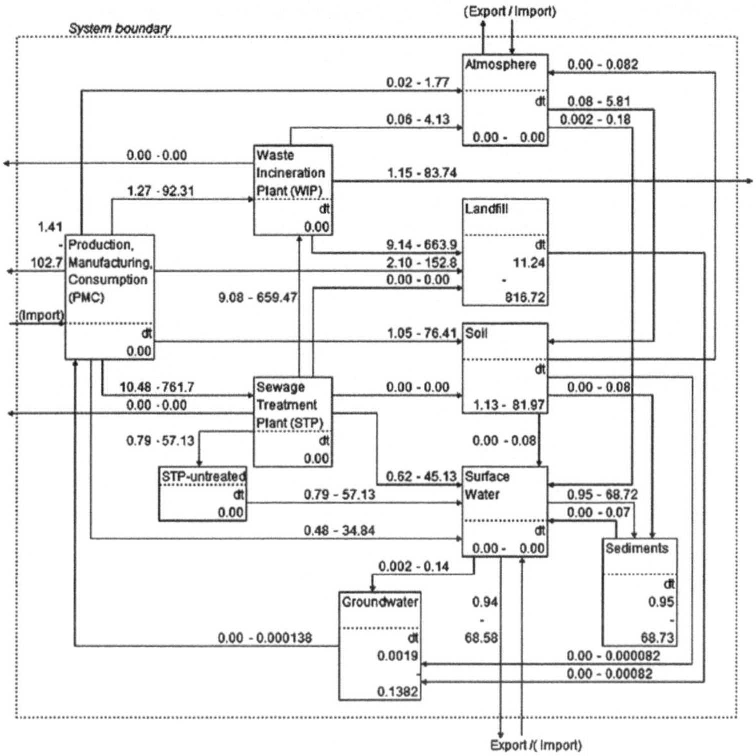

Figure 5.4 Mass flow of nano TiO_2 in the environment with accumulation rates in tonne per year. *Source*: Gottschalk *et al.* (2010). Copyright by Pergamon. Reprinted with permission.

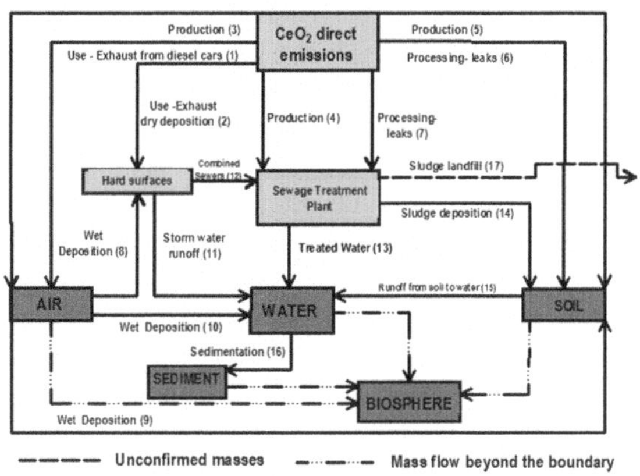

Figure 5.5 Mass flow and balance of CeO_2 (a diesel fuel additive). Reprinted with permission from Hansen *et al.* (2011).

5.3 PROPERTIES AND CHARACTERIZATION OF NANOMATERIALS/NANOPARTICLES

5.3.1 Properties

Size is one of the important properties of NMs as their nano-scale size ascribes dissimilar properties compared to substances which are chemically the same but larger in size. The decrease in size of particles generally results in greater surface area relative to the mass and hence the NMs, including NPs, have comparatively larger surface area and higher reactivity. Besides, shape is another primary property influencing the agglomeration of particles (Sellers, 2009). Penetrability of the NPs into cell membranes/organs is also affected by particle size and shape (Brar & Verma, 2011; Sellers, 2009). Primary particle size distribution (PPSD) is also an important parameter related to size, and aggregation of the NPs is affected by the PPSD as well (Peng *et al.*, 2012).

In terms of size, engineered nano-scale materials with one of the dimension less than 100 nm are defined as nanomaterials, whereas NPs have all dimensions less than 100 nm. The natural colloids which are naturally occurring nano-scale materials have dimensions of about 1 to 1000 nm (Jiménez *et al.*, 2011). The extent of the size for different ENMs varies with their type. Engineered oxides and quantum dots are metals containing ENP with dimensions of less than 10 nm, whereas combustion by products and engineered soot comprising of carbon based ENPs may reach the upper limit of 100 nm in size. Engineered metals may vary from 5 nm to greater than 100 nm (Weinberg *et al.*, 2011; Westerhoff *et al.*, 2008).

NPs in general have lower zeta potential values than the actual surface charge and may have negative zeta potential in natural water as the surface charge is affected by existing natural organic matter (NOM) (Westerhoff *et al.*, 2008).

5.3.2 Analysis and characterization of nanoparticles

Different methods are available for analysis and characterization of nanomaterials (see Chapter 4). A combination of more than one analytical method may be required to conclude on the physico-chemical properties of nanomaterials (Weinberg *et al.*, 2011; Domingos *et al.*, 2009). In general, the methods which are used for the analysis of classical pollutants are being utilized for the analysis of nanomaterials as well (Weinberg *et al.*, 2011). The major techniques that are used for analysis and characterization can be classified as microscopic techniques, scattering and spectroscopic techniques, interfacing separation and characterization, and sensor designs for analysis of ENPs (Farré *et al.*, 2011, Weinberg *et al.*, 2011).

Microscopic techniques include optical, electron and scanning probe microscopy. The limit of detection (LOD) of the optical microscopy is in higher range than the size of NPs and therefore electron and scanning probe microscopy are preferred for characterization, which includes size, shape, texture, structure, morphology and elemental composition (Hoyt, 2009; Farré *et al.*, 2011; Weinberg *et al.*, 2011).

Electron microscopy is particularly suitable for elements with high electron density (Tiede *et al.*, 2009). Transmission electron microscopy (TEM) and scanning electron microscopy (SEM) are two electron microscopy methods which are most commonly used and sample surfaces are exposed to electron beams in both methods. The measurement of backscattered electrons or secondary electrons emitted from the surface is taken when SEM is used. In TEM, the electrons which pass through the surface are focussed to an image detector for particle imaging (Hoyt, 2009; Hassellöv *et al.*, 2008). Higher image resolution is the advantage of TEM, whereas SEM provides larger area of focus permitting relatively higher amount of particles to be focussed. Environmental scanning electron microscopy (ESEM) is a relatively new method which can be used in different pressure and humidity but the image resolution is lower than TEM or SEM (Peters *et al.*, 2011; Hoyt, 2009).

Scanning probe microscopy techniques include atomic force microscopy (AFM) and scanning tunnelling microscopy (STM). AFM can be used for broader categories of materials and generates very high resolution and three dimensional images (Peters *et al.*, 2011). AFM can be used for analysis of particles in liquid as well (Hoyt, 2009; Farré *et al.*, 2011).

Field flow fractionation (FFF), hydrodynamic chromatography (HDC), size exclusion chromatography (SEC) and capillary electrophoresis (CE) are the techniques utilized for separation. SEC is the method that is used commonly and in which a porous packed column is used for separation of particles with respect to the hydrodynamic volume. Various force fields are applied to a sample flowing in a thin flow channel in FFF method to achieve separation of particles so that smaller particles travel faster causing fractionation (Hassellöv *et al.*, 2008; Farré *et al.*, 2011).

Light scattering methods are utilised in measurements of the particle size and determination of the particle size distribution. Some of the light scattering methods utilised in NPs characterization are static light scattering (STS), dynamic light scattering (DLS), laser induced breakdown detection (LIBD) and small angle X-ray scattering (SAXS) (Farré *et al.*, 2011; Weinberg *et al.*, 2011; Hassellöv *et al.*, 2008; Brar & Verma, 2011). DLS is the most common method among light scattering methods. It is applicable to broad categories of NPs and analysis is rapid too but is not preferable for characterization of polydispersed samples (Tiede *et al.*, 2009). But SAXS can be utilized for both monodispersed and polydispersed samples (Farré *et al.*, 2011). LIBD is a method which can be used to characterize NPS of the very small dimension up to 10 nm (Weinberg *et al.*, 2011).

Nuclear magnetic resonance (NMR) spectroscopy, Raman spectroscopy, flame atomic absorption spectrometry (FAAS), inductively coupled plasma mass spectrometry (ICP-MS), ultraviolet-visible spectroscopy (UV-VIS), electron energy loss spectroscopy (EELS), inductively couple plasma optical emission spectrometry (ICP-OES) atomic absorbance spectrometry (AAS) and X-ray spectroscopy which includes X-ray photoelectron spectrometry (XPS), X-ray fluorescence, X-ray diffraction (XRD), energy-dispersive X-ray spectroscopy (EDX) and X-ray absorption are some spectroscopic methods for characterization of NPs (Fabrega *et al.*, 2010; Farré *et al.*, 2011; Rezić, 2011; Weinberg *et al.*, 2011). X-ray spectroscopy is particularly suitable in determining the surface properties and crystallographic information (Farré *et al.*, 2011; Tiede *et al.*, 2009). EELS is generally used to determine the elemental and chemical information especially for the chemical information of single particles, but longer duration for analysis is a disadvantage. Bulk total elemental composition of multiple metals can be determined by processes like ICP-OES, ICP-MS, and AAS. Precise and rapid determination of elemental information of single particles can be carried out by EDX but is limited to semi quantitative information (Fabrega *et al.*, 2010).

Application of light scattering methods is particularly suitable for the measurement of NPs in liquid media. The application of DLS in combination with centrifugation and microscopic methods for analysis of NPs in studies of eco-toxicity and aggregation in water can cover broad categories of NPs. Hence combination of these methods can be highly suitable for characterisation of NPs in wastewater (Brar & Verma, 2011). Similarly, for the freshwater aquatic environments, electron microscopy and x-ray diffraction can be used to determine specific surface area (Klaine *et al.*, 2008). Analysis of CNTs in surface water can be carried out by combining CE techniques with either near infra-red (NIR)-fluorescence or Raman spectroscopy (Farré *et al.*, 2011).

5.4 FATE OF NANOMATERIALS/NANOPARTICLES IN WATER AND WASTEWATER

5.4.1 Fate in aqueous environment

5.4.1.1 Processes determining the fate of NMs/NPs

Properties such as size, size distribution, shape, surface area, surface charge, concentration, catalytic activity, purity and dissolution properties can influence the fate of the NPs in the aquatic media. Similarly, properties of the natural water system such as ion composition, ionic strength, pH and constituents such as natural organic material/dissolved organic material (NOM/DOM) also affect the fate of the NPs in aquatic systems (Lin *et al.*, 2010b; Weinberg *et al.*, 2011; Westerhoff *et al.*, 2008). Besides processes and activities such as dispersion/diffusion, aggregation and disaggregation, sedimentation, biotic and abiotic degradation, transformation and photoreaction/light parameter can modify the physical and chemical properties of NPs and consequently affect the fate of NPs in water (Baalousha *et al.*, 2011).

Nano-particles may be subjected to different fate in aquatic systems such as dissolution, deposition, sedimentation, agglomeration, coating, association, reaction and decomposition (Weinberg *et al.*, 2011). Among these, agglomeration of NPs is one of the processes in aquatic environments that has been prioritized by researchers. Past studies have shown that some of the nano-materials are very unstable in the water due to their readily agglomerating properties. For example, carbon nanotubes (CNT) of unadulterated and pure quality are highly hydrophobic and tend to agglomerate and get settled in water causing instability. Dispersants are needed to keep them in suspension (Lin, 2010a; Liu *et al.*, 2009). Similarly, unmodified fullerenes also aggregate quickly and dispersion aids are needed to hinder aggregation into large sizes (Ju-Nam & Lead, 2008). Hence this attribute may assist in their settling and removal in water or wastewater treatment processes.

5.4.1.2 Effect of NM/NP characteristics

Aggregation of NPs is influenced by various solution conditions. In case of fullerene particles, the pH of the solution greatly affects the aggregation with higher pH resulting in stability of fullerene. As the pH increases, the zeta potential increases negatively causing a decrease in the aggregation process (Chen & Elimelech, 2009). Another condition affecting aggregation as well as deposition of the NPs is the solution chemistry. The solution chemistry will be unfavourable or favourable depending on the dominant repulsive interaction. The presence of dominant repulsive interaction will impede the aggregation and deposition (Petosa *et al.*, 2010). The variation in PPSD also affects the structural formation of the aggregates of NPs in aqueous medium. Smaller PPSD variance generates netlike aggregated structures while the larger variances stimulate compact aggregated structures (Peng *et al.*, 2012).

Zeta potential is also an important parameter related to the aggregation of the NPs. The electric potential at slipping or shear plane located in the diffused layer away from the particle surface is known as the Zeta potential (Xu, 2008). Zeta potential fluctuates positively or negatively with changes in pH of the solution with respect to isoelctric point. When the pH falls below isoelctric point and gets further decreased, the zeta potential increases positively while the zeta potential increases negatively when the pH increases beyond the isoelctric point. As the aggregating behaviour of colloids is favoured when the pH is near the isoelctric point, it can also be inferred for NPs that values of the zeta potential near the value about zero (approximately within ± 5 mV) should be favourable for aggregation of NPs (Lin *et al.*, 2010b; Westerhoff *et al.*, 2008).

Particle concentration, shape, hydrophobicity/hydrophilicity are some other characteristics which also influence the agglomeration of the NPs (Brar *et al.*, 2010). Although many NPs in pure form may demonstrate readily agglomerating behaviour, functionalized NMs may not aggregate readily. In addition, there is a possibility of de-agglomeration of aggregated NMs hence these two aspects shall be given due importance (Som *et al.*, 2011).

The solubility of organic NPs in solution varies according to their type. Some organic NPs may be biodegradable (such as cellulose and starch NPs) while other organic NPs (such as fullerenes and carbon nanotubes) may be non biodegradable in aqueous environment (Kümmerer *et al.*, 2011).

The presence of natural organic material (NOM) can impede aggregation of NPs irrespective of ionic strength. Humic acid coated TiO_2 particles demonstrated a zeta potential which was similar to that of humic acid and hence has a higher resistance to salt induced aggregation. Consequently, deposition of NPs in aqueous media is highly affected by the presence of humic acid; the deposition behaviour of TiO_2 was observed to be in accordance with classical Derjaguin-Landau-Verway-Overbeek (DLVO) theory in the absence of humic acid, but no deposition was observed in the presence of humic acid (Thio *et al.*, 2011).

5.4.2 Fate in water treatment processes

As mentioned in previous sections, the probability of release of ENP/MNP into natural waters has increased with widespread trends of production and use of the commercial products containing NPs. It has become necessary to assess the efficiency of water treatment processes in the removal of NPs as well as to assess the necessary augmentation that may be needed to facilitate the removal of NPs. Water treatment processes such as coagulation, flocculation, sedimentation and filtration are key processes in drinking water production which may assist in the removal of the NPs from raw water. Although water treatment plants are capable of removing nano-scale bacteria and viruses, the efficiency of these plants in the removal of NPs is less known. But experiments conducted by Westerhoff *et al.* (2008) demonstrated that conventional processes like coagulation/flocculation and sedimentation can remove NPs to some extent, whereas advanced process like membrane filtration can remove NPs to a large extent.

Destabilization of the NPs can occur during coagulation and subsequent agglomeration will take place during flocculation. The agglomerated large particles are removed by sedimentation and particles escaping sedimentation can be filtered out by filtration processes. Straining action is the removal mechanism during membrane filtration whereas attachment to filter media is the removal mechanism during sand filtration (Westerhoff *et al.*, 2008).

The colloidal stability of NPs is the factor that will affect destabilization during coagulation. Generally, classical DLVO theory will be sufficient for assessing colloid stability in water, but extended DLVO with the incorporation of non-DLVO interactions might be essential for assessing the potential stability of NPs in water (Westerhoff *et al.*, 2008). Steric interactions, magnetic forces (for iron based NMs), hydration forces (Petosa *et al.*, 2010), hydrophilic and hydrophobic interactions are some important non-DLVO parameters with respect to NMs in aqueous media (Westerhoff *et al.*, 2008). Some of the NPs follow similar behaviour as predicted by the DLVO theory, while others may not conform to DLVO predictions. A study by Liu *et al.* (2009) showed that the boron nano-particle followed the DLVO and filtration model predictions in porous media experiments for quantifying aggregation and deposition kinetics of boron NPs. The experimental results for sonicated and stirred fullerene NPs conducted by Chen & Elimelech (2009) were consistent with the predictions using the DLVO theory in the fast aggregation regime. But minute variations in the predictions made by DLVO theory were observed for stirred fullerenes in the slow aggregation regime which was deemed to be caused by a discrepancy in the determination of the hydrodynamic and particle radius. The authors of that study emphasized that the discrepancy may have been aggravated by irregular shape and poly-dispersivity of the particles.

A laboratory study was conducted by Zhang *et al.* (2008) to evaluate the stability of various commercial metal oxide NPs such as TiO_2, Fe_2O_3, ZnO, NiO and SiO_2. The study demonstrated that the presence of electrolyte both induces destabilisation of the NPs as well as causes re-suspension of the aggregated NPs. This became evident by sedimentation experiments. Addition of electrolyte (0.1 M $MgCl_2$) to nanopure water with 10 mg/L of nano-particle suspensions and mixing for 8 hours yielded 40–80% removal, whereas 24 hr mixing did not bring any significant increase in the removal. The study also found that the conventional alum coagulation (with dosages between 20 and 60 mg/L), is not much effective in the removal of NPs as 20 to 80% remained in the effluent after coagulation and sedimentation. But the use of 0.45 μm membrane filtration was found very efficient with 100% removal of nano-particle aggregates of 500 nm.

In another study, the removal of NPs in porous media was carried out under different solution chemistry. Glass beads were used as filter medium in a 20 cm vertical column (Table 5.5). Table 5.5 shows that lower ionic strength as well as presence of organic matter increased the mobility of NPs in porous media. An increase in ionic strength increased agglomeration of NPs, thus reducing the mobility of the NPs investigated. Low flow rates increased the retention of NPs in the porous medium. Moreover, the presence of NOM further decreased the agglomeration when ionic strength was low (Ben Moshe *et al.*, 2009).

Table 5.5 Removal of NPs by glass beads acting as porous medium

Type of Nanoparticle	Percentage of NPs exiting the column		
	0.01 M NaCl	**0.10 M NaCl**	**0.01 M NaCl + 60 mg/L humic acid**
TiO_2	62	13	98
CuO	52	8.3	98
Fe_2O_3	16	6.2	62
ZnO	1.4	1.2	74

5.4.3 Fate in wastewater treatment processes

Extensive use and consumption of products containing nano-size materials have resulted in the contamination of domestic wastewater and storm water with NMs/NPs. The removal of NPs from wastewater happens through adsorption, aggregation and deposition, and biodegradation. But when surfactants are present in the wastewater, functionalization of the nano-materials might take place causing stabilization of NPs and preventing aggregation, even when coagulation is applied. Some NPs may get sorbed/adsorbed onto different substances present in the wastewater such as organic material, consumer products, and debris. Furthermore, natural organic matter may also bring changes in physical properties of nano-materials (Brar *et al.*, 2010).

5.4.3.1 Preliminary treatment

Generally preliminary treatment consists of bar screens and grit removal. Preliminary treatment is aimed at the removal of large solids, rags, debris as well as grit material consisting of sand, gravel, broken glass, egg shells, and other material with considerably higher settling velocity than the organic material (Mackenzie, 2010). Removal of a very small portion of nano-particles that are present in wastewater is anticipated in this stage of the treatment through sorption of NPs onto debris and other large particles (Brar *et al.*, 2010).

5.4.3.2 Primary treatment

Primary treatment is the process for the removal of organic and inorganic suspended solids through solid-liquid separation in primary settling tanks. The removal of nano-particles by plain sedimentation is not expected in primary treatment as the settling time required will be substantially higher than other removable particles. Very long detention times will be necessary to remove nano-size particles (Brar *et al.*, 2010). Figure 5.6 shows the settling velocities of different NPs with different densities with respect to increasing order of size. Figure 5.6 demonstrates that the settling velocities are too small for the removal of NPs by sedimentation. Similarly, Figure 5.7 shows the theoretical relationship between particle size and corresponding time required for diffusion or settlement. But if coagulation and flocculation processes are incorporated, they would enhance the removal of NPs by primary treatment due to the aggregation of NPs. Adsorption of NPs onto organic particles present in the wastewater is another mechanism which may assist the removal of NM/NP by primary treatment (Brar *et al.*, 2010).

5.4.3.3 Secondary treatment

Oxidation of readily biodegradable organic constituents and the removal of suspended solids escaping primary treatment are carried out in secondary treatment. Secondary treatment utilizes aerobic microorganisms either in

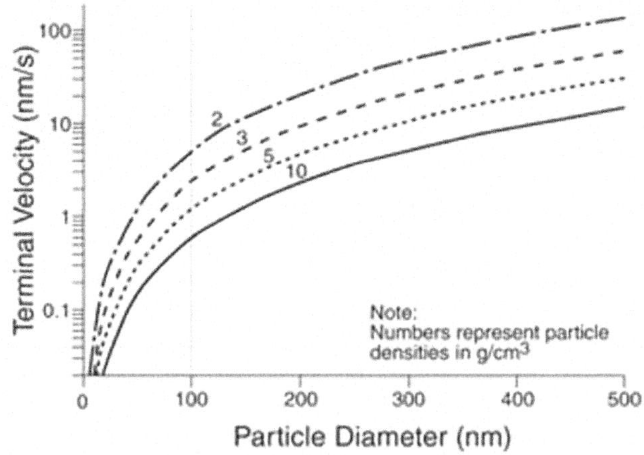

Figure 5.6 Velocities of various NPs of different diameter in order of increasing size. *Source*: Mackay *et al.* (2006). Copyright by Wiley Periodicals, Inc. Reprinted with permission.

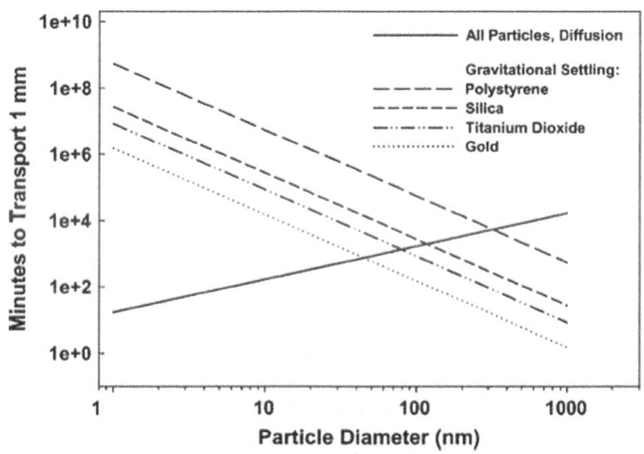

Figure 5.7 Theoretical relationship of particle size with time required for diffusion or settlement. *Source*: Teeguarden *et al.* (2006), Copyright by Oxford University Press. Reprinted with permission.

suspension or in biofilms attached to inert media to oxidize biodegradable constituents present in the wastewater. Mechanisms such as physical entrapment of insoluble particles into the floc, active cellular uptake (absorption), binding to extra cellular polymeric substances (EPS) and volatilisation are considered to be effective in the removal of metals in activated sludge. These mechanisms may also be equally applicable to the removal of NPs with physical entrapment of insoluble particles being the most effective mechanism.

Attachment/sorption to EPS are also found to be another significant mechanism for metal removal and also applicable to NPs. Attachment of NPs to the bound EPS (EPS that is attached to biomass) will result in the removal of NPs through settling along with the biomass, while attachment of NPs to the soluble EPS (EPS present in the supernatant) will result in stabilisation and non-removal. In the activated sludge process, the metal removal efficiency is enhanced by 10–20% when sludge age is increased, this can be inferred for NPs too (Boxall *et al.*, 2007).

The activated sludge process is one of the biological processes that is employed during secondary treatment. There have been some studies to determine the efficiency and applicability of the activated sludge process in removal of NPs. Kiser *et al.* (2010) studied about combined result of sorption of NPs onto different constituents of wastewater biomass which included adsorption to cell surfaces and EPS, and absorption into the cell. The combined net effect was termed biosorption. A fluorescent NP was prepared by doping the fluoresce in isothiocyanate (FITC) into silica NP and was utilised to investigate biosorption. Increased fluorescence was noticed in the resultant images with increased concentration of SiO_2-FITC mixed to the biomass, which indicated biosorption of NP in simulated activated sludge processes. The study further demonstrated that non functionalized Ag had the highest removal prospects through biosorption in biomass among the five NPs experimented followed by TiO_2, functionalized Ag, fullerol ($_nC_{60}(OH)_x$) and aq-$_nC_{60}$. The removal efficiency increased with the increase in the concentration of above mentioned NPs. The study also found out that the presence of NOM and EPS may not be favourable for the interaction between NPs and microorganisms in the biomass for some particular NPs such as aq-$_nC_{60}$ (a non functionalized fullerene suspension in water) (Kiser *et al.*, 2010).

Wang *et al.* (2012) also conducted a study about the removal of different NPs by biomass utilising a modified activated sludge process. When wastewater containing nanomaterials were treated with laboratory based sequencing batch reactors (SBR) with hydraulic retention time (HRT) and sludge retention time (SRT) that were resembling actual wastewater treatment plants, the results demonstrated that the efficiency of the SBR without biomass addition for the removal of the different nanomaterials can be different. The removal of silver NPs functionalized with polymer coating (fn-Ag) and fullerene (C_{60}) did not occur when microbial organisms were absent. But nanoscale TiO_2 removal was approximately 70% and aggregation of particles were provoked by the mild ionic strength in the absence of biomass. The experiment with biomass addition to SBR achieved removal of fn-Ag, C_{60}, TiO_2 as well as fullerol ($_nC_{60}(OH)_{24\text{-}y}Na_y$). The overall removal rate for fn-Ag was found to be 88%. But, the removal rate for TiO_2 with a biomass concentration of 1.3 ± 0.2 g TSS/L added to the SBR was $97 \pm 1\%$. The removal of fullerol and fullerenes were determined to be approximately 70% and greater than 95% respectively in short term SBR experiments with biomass. The researchers also studied the effect of extended time duration and variations in concentration of biomass for fullerene. The results showed that at higher $_nC_{60}$ concentration with low biomass concentration (0.6 g/L), the NP removal is relatively lower (83%) whereas at normal concentration of biomass (>1.5 g/L) the removal efficiency was 96%. The predominant removal mechanism of nanoscale TiO_2 was found to be aggregation and settlement, whereas sorption into biosolids was the predominant removal mechanism of fn-Ag. Thus, removal of certain NPs is effective in the activated sludge process.

5.4.3.4 Tertiary treatment/membrane filtration

The mechanism of NM/NP removal by membrane filtration is through straining for particles larger than the pore size of the membrane. However, adsorption of the NPs aids in the removal of NPs that are smaller than the pore size of the membrane but different NPs demonstrate different adsorption potential. In a study by Ladner *et al.* (2012), the removal of functionalized NPs by polymeric membranes of different pore sizes and diverse composition was investigated. The study demonstrated that negatively charged TiO_2 and Au have a higher adsorption affinity than negatively charged Ag-NPs and hence are retained in membranes to a larger extent. The rejection of negatively charged TiO_2, which has a similar adsorption potential as the negatively charged Au, was verified to be lower which was attributed to a higher NP-NP self-adsorption capacity. The study also experimented with some positively charged NPs and found that the adsorption mechanism is highly effective for the removal of positively charged TiO_2 and Au. These NPs were rejected readily by negatively charged membranes. In case of membranes with smaller pore sizes with respect to NPs, the retained NPs through adsorption tend to cause a decrease in pore size which further increases the rejection. But when the pore size is relatively larger than the NPs, breakthrough may occur although partial blockage by adsorption of particles may have occurred. Another finding of the study was that the mobility of the NPs seemed to be predominantly influenced by the properties of the NPs rather than by properties of the membranes used.

In another study, Kiser *et al.* (2009) evaluated the occurrence and fate of titanium nano-materials in wastewater with field as well as lab-based analyses. The removal efficiency was found to be 79% $\pm$ 23% for Ti during summer (June) when a field based research was conducted in a wastewater treatment plant comprising of a primary clarifier, secondary treatment (activated sludge process), tertiary filtration and disinfection. The efficiency was about 95% in winter (January) with overall average removal of 81% $\pm$ 21%. The measured average Ti concentration of the influent entering WWTP was 843 μg/L, while the average concentrations of effluent from the primary clarifier, secondary clarifier and tertiary filtration during summer were 99 μg/L, 35 μg/L, and 36 μg/L respectively. Comparison of these values to the average influent concentration at the primary clarifier (843 μg/L) shows all of the removal of Ti to occur in primary and secondary treatment. Hence it can be hypothesized that the tertiary filtration was unable to further polish the quality of the secondary effluent. The effluent samples from each treatment unit were also filtered with a filter of 0.7 μm pore size to find out the removal of the Ti which was filterable through a 0.7 μm mesh filter. The analysis showed that only the 42 ± 2% of the filterable Ti was removed which demonstrates this particular wastewater treatment process was not highly efficient for the removal of Ti aggregates and particles of a size less than 0.7 μm, as well as some portions of the Ti particles that may have remained unaggregated and stable. The removal efficiency was little higher during winter with average removal efficiency of 95%. In winter, concentrations were found to be 254 ± 15 μg/L and 12 ± 2 μg/L for influent and effluent, respectively. The analysis of the biosolids showed that the removed nanoscale Ti was in the form of small aggregates of single nano particles and aggregated spherical TiO_2. In the effluent, the Ti was present as aggregates with a size of a few hundred nanometres, comprising of primary TiO_x particles with some single TiO_x particles also present. Furthermore, the authors also reiterated that the source of these nano Ti found in the biosolids and effluent were originated from industrial/food applications of nano Ti as well as environment.

5.5 TOXICITY AND IMPLICATIONS

The NM/NP escaping with the effluent from the wastewater treatment process along with NM/NP removed during treatment process and associated with the sludge produced by the activated sludge process might be the major sources from where NM/NP could be released to the natural water, ground water and soil. Hence, environmental and health implications will arise due to inefficient performance of wastewater treatment processes. Besides, the NM/NP may also impede water and wastewater treatment as they might inhibit the microorganisms of biological treatment process of the wastewater treatment and also adversely affect the disinfection process during water treatment. The possible mechanisms through which NMs/NPs can become toxic are shown in Figure 5.8.

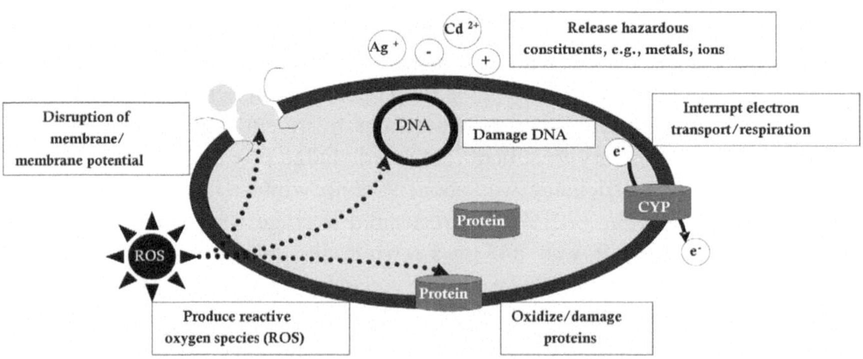

Figure 5.8 NM/NPs toxicity mechanisms to bacteria. *Source*: Klaine *et al.* (2008) Copyright by SETAC (SOCIETY). Reprinted with permission.

Garcia *et al.* (2012) found that the toxicity of NPs on microbial communities varied with different type of NPs that were utilised. The study was conducted to evaluate the effect of chosen NPs on the microorganisms such as heterotrophic organisms, ammonia oxidizing bacteria (AOB), and thermophilic and mesophilic anaerobic bacteria which are essential in wastewater treatment processes. CeO_2 NPs were highly toxic to these microbial communities and inhibited the biogas production almost completely. Ag-NP was less toxic than CeO_2 towards the microbial population, whereas Au-NPs and TiO_2-NPs had negligible negative effects on microbial communities (Garcia *et al.*, 2012).

Ag-NP which may have adverse effect on the microorganisms those are present in secondary wastewater treatment system. The presence of Ag-NP, which has anti-microbial properties, in wastewater can be destructive to microorganisms in the biofilms. Other microorganisms in the biofilms are not susceptible to Ag-NP exposure in general, but when loosely bound EPS are separated, then microorganisms can become vulnerable (Sheng & Liu, 2011).

The extensive use of zinc oxide (ZnO) NPs in industrial, medical and military applications has resulted in the discharge of waste containing ZnO NPs into wastewater treatment plants. Mu and Chen (2011) investigated the removal of ZnO by biological treatment and found that the activated sludge process was successful in the removal of ZnO NPs. The removed waste activated sludge which generally is sent to sludge treatment, with anaerobic digestion, contained the removed ZnO NPs. The study showed that the production of methane was not affected by lower doses of about 1 mg ZnO NPs/g-TSS, whereas at higher dosages of 30 and 150 mg/g-TSS, the production of methane was restrained through the inhibition of the hydrolysis and methanation. Furthermore, if the digested sludge after subsequent processing is dispatched for landfill, it can possibly be a pathway of releasing NPs to the soil.

Besides the probable toxicity of the NPs, they can assist the transportation of undesired substances. NPs have a large surface area compared to their sizes which enhance prospects of contaminants and metals getting attached to the NPs and transported together with NPs (Weinberg *et al.*, 2011; Farré *et al.*, 2011). Therefore, the possibility of transport of contaminants by NPs removed during the wastewater treatment cannot be ignored.

In a study by Zhang *et al.* (2008), the effect of storage of synthesized hematite NPs was investigated. The result showed that both the wet flocculated hematite as well as dry powder hematite aggregated to larger sizes after a month of storage. Furthermore, some quantity of hematite of both forms aggregated to irreversible aggregated NPs, as they remained intact even after sonication. The authors emphasized that when fate and toxicity of NPs are examined, the variability in size of NPs during the production, application in products and their ultimate release to the environment should be considered.

5.6 CONCLUSIONS

Understanding the fate and the removal of NPs is becoming increasingly important in order to manage their presence in the environment, which will minimise the health hazards posed by them to human and other forms of life. Aggregation of NPs plays a big role in their transport in the environment. Appropriate ionic strengths and pH of fresh water sources that contain NPs will assist in effective removal of NPs by water treatment processes. When wastewater contains NPs, primary and secondary treatment processes are the main processes in the removal of NPs. Tertiary treatment could become effective if the influent solution chemistry is altered to improve the removal of NPs. Toxicity of NPs to the microorganisms that are present in the activated sludge of an aeration system of a wastewater treatment plant must be managed carefully for its efficient functioning.

REFERENCES

Aitken R. J., Chaudhry M. Q., Boxall A. B. A. and Hull M. (2006). Manufacture and use of nanomaterials: current status in the UK and global trends. *Occupational Medicine*, **56**(5), 300–306.

Baalousha M., Lead J. and Nam Y. J. (2011). Natural colloids and manufactured nanoparticles in aquatic and terrestrial environment. *Treatise on Water Science*, **3**, 41.

Ben Moshe T., Dror I. and Berkowitz B. (2010). Transport of metal oxide nanoparticles in saturated porous media. *Chemosphere*, **81**(3), 387–393.

Blaser S. A., Scheringer M., MacLeod M. and Hungerbühler K. (2008). Estimation of cumulative aquatic exposure and risk due to silver: contribution of nano-functionalized plastics and textiles. *Science of the Total Environment*, **390**(2–3), 396–409.

Botta C., Labille J., Auffan M., Borschneck D., Miche H., Cabié M., Masion A., Rose J. and Bottero J. Y. (2011). TiO_2-based nanoparticles released in water from commercialized sunscreens in a life-cycle perspective: structures and quantities. *Environmental Pollution*, **159**(6), 1543–1550.

Boxall A. B. A., Chaudhry Q., Sinclair C., Jones A., Aitken R., Jefferson B. and Watts C. (2007). *Current and Future Predicted Environmental Exposure to Engineered Nanoparticles*. Central Science Laboratory, Sand Hutton, UK.

Brar S. K. and Verma M. (2011). Measurement of nanoparticles by light-scattering techniques. *TrAC Trends in Analytical Chemistry*, **30**(1), 4–17.

Brar S. K., Verma M., Tyagi R. D. and Surampalli R. Y. (2010). Engineered nanoparticles in wastewater and wastewater sludge – evidence and impacts. *Waste Management*, **30** (3), 504–520.

Chen K. L. and Elimelech M. (2009). Relating colloidal stability of fullerene (C_{60}) nanoparticles to nanoparticle charge and electrokinetic properties. *Environmental Science & Technology*, **43**(19), 7270–7276.

Corbett J., McKeown P., Peggs G. and Whatmore R. (2000). Nanotechnology: international developments and emerging products. *CIRP Annals – Manufacturing Technology*, **49** (2), 523–545.

Domingos R. F., Baalousha M. A., Ju-Nam Y., Reid M. M., Tufenkji N., Lead J. R., Leppard G. G. and Wilkinson K. J. (2009). Characterizing manufactured nanoparticles in the environment: multimethod determination of particle sizes. *Environmental Science & Technology*, **43**(19), 7277–7284.

Eriksson E., Auffarth K., Eilersen A.-M., Henze M. and Ledin A. (2003). Household chemicals and personal care products as sources for xenobiotic compounds in grey water. *Water SA*, **29**(2), 135–146.

Eriksson E., Auffarth K., Henze M. and Ledin A. (2002). Characteristics of grey water. *Urban Water*, **4**, 85–104.

European Commission. (2004). Towards a European strategy for nanotechnology (COM (2004). 338). Available from: http://www.cordis.lu/nanotechnology/src/communication.htm

Fabrega J., Luoma S., Tyler C., Galloway T. and Lead J. (2010). Silver nanoparticles: behaviour and effects in the aquatic environment. *Environment International*, **37**, 517–531.

Farré M., Sanchís J. and Barceló D. (2011). Analysis and assessment of the occurrence, the fate and the behavior of nanomaterials in the environment. *TrAC Trends in Analytical Chemistry*, **30**(3), 517–527.

García A., Delgado L., Torà J. A., Casals E., González E., Puntes V., Font X., Carrera J. and Sánchez A. (2012). Effect of cerium dioxide, titanium dioxide, silver and gold nanoparticles on the activity of microbial communities intended in wastewater treatment. *Journal of Hazardous Materials*, **199–200**(0), 64–72.

Gottschalk F., Scholz R. W. and Nowack B. (2010). Probabilistic material flow modeling for assessing the environmental exposure to compounds: methodology and an application to engineered nano-TiO_2 particles. *Environmental Modelling & Software*, **25**(3), 320–332.

Hansen S. F., Baun A., Tiede K., Gottschalk F., Meent D. V. D., Peijnenburg W., Fernandes T. and Riediker M. (2011). *Environmental Fate and Behaviour of Nanoparticles – Beyond Listing of Limitations*, Technical University of Denmark, Kgs. Lyngby.

Hassellöv M., Readman J. W., Ranville J. F. and Tiede K. (2008). Nanoparticle analysis and characterization methodologies in environmental risk assessment of engineered nanoparticles. *Ecotoxicology*, **17**(5), 344–361.

Hoyt M. (2009). Analysis of nanaoparticles in the environment. In: *Nanotechnology and the Environment*, K. Sellers, C. Mackay, L. L. Bergeson, S. R. Clough, M. Hoyt, J. Chen, K. Henry and H. Jane (eds) , CRC Press, Boca Raton.

Jiménez M. S., Gómez M. T., Bolea E., Laborda F. and Castillo J. (2011). An approach to the natural and engineered nanoparticles analysis in the environment by inductively coupled plasma mass spectrometry. *International Journal of Mass Spectrometry*, **307** (1–3), 99–104.

Johnson A. C., Bowes M. J., Crossley A., Jarvie H. P., Jurkschat K., Jürgens M. D., Lawlor A. J., Park B., Rowland P., Spurgeon D., Svendsen C., Thompson I. P., Barnes R. J., Williams R. J. and Xu N. (2011). An assessment of the fate, behaviour and environmental risk associated with sunscreen TiO_2 nanoparticles in UK field scenarios. *Science of the Total Environment*, **409**(13), 2503–2510.

Ju-Nam Y. and Lead J. R. (2008). Manufactured nanoparticles: an overview of their chemistry, interactions and potential environmental implications. *Science of the Total Environment*, **400**(1–3), 396–414.

Kaegi R., Ulrich A., Sinnet B., Vonbank R., Wichser A., Zuleeg S., Simmler H., Brunner S., Vonmont H., Burkhardt M. and Boller M. (2008). Synthetic TiO_2 nanoparticle emission from exterior facades into the aquatic environment. *Environmental Pollution*, **156**(2), 233–239.

Kiser M. A., Ryu H., Jang H., Hristovski K. and Westerhoff P. (2010). Biosorption of nanoparticles to heterotrophic wastewater biomass. *Water Research*, **44**(14), 4105–4114.

Kiser M. A., Westerhoff P., Benn T., Wang Y., Pérez-Rivera J. and Hristovski K. (2009). Titanium nanomaterial removal and release from wastewater treatment plants. *Environmental Science & Technology*, **43**(17), 6757–6763.

Klaine S. J., Alvarez P. J. J., Batley G. E., Fernandes T. F., Handy R. D., Lyon D. Y., Mahendra S., Mclaughlin M. J. and Lead J. R. (2008). Nanomaterials in the environment: behaviour, fate, bioavailability and effects. *Environmental Toxicology and Chemistry*, **27**(9), 27.

Kümmerer K., Menz J., Schubert T. and Thielemans W. (2011). Biodegradability of organic nanoparticles in the aqueous environment. *Chemosphere*, **82**(10), 1387–1392.

Ladner D. A., Steele M., Weir A., Hristovski K. and Westerhoff P. (2012). Functionalized nanoparticle interactions with polymeric membranes. *Journal of Hazardous Materials*, **211–212**, 288–295. doi: 10.1016/j.jhazmat.2011.11.051

Lee J., Mahendra S. and Alvarez P. J. J. (2010). Nanomaterials in the construction industry: a review of their application and environmental health and safety considerations. *ACSNANO*, **4**(7), 3580–3590.

Lin D., Liu N., Yang K., Xing B. and Wu F. (2010a). Different stabilities of multiwalled carbon nanotubes in fresh surface water samples. *Environmental Pollution*, **158**(5), 1270–1274.

Lin D., Tian X., Wu F. and Xing B. (2010b). Fate and transport of engineered nanomaterials in the environment. *Journal of Environment Quality*, **39**(6), 1896.

Liu X., Wazne M., Christodoulatos C. and Jasinkiewicz K. L. (2009). Aggregation and deposition behavior of boron nanoparticles in porous media. *Journal of Colloid and Interface Science*, **330**(1), 90–96.

Loretz L., Api A. M., Barraj L., Burdick J., Davis D.-A., Dressler W., Gilberti E., Jarrett G., Mann S., Pan Y. H. L., Re T., Renskers K., Scrafford C. and Vater S. (2006). Exposure data for personal care products: hairspray, spray perfume, liquidfoundation, shampoo, body wash, and solid antiperspirant. *Food and Chemical Toxicology*, **44**, 2008–2018.

Loretz L., Api A. M., Barraj L., Burdick J., Dressler W., Gettings S. D., Hsu H. H., Pan Y. H. L., Re T. A., Renskers K. J., Rothenstein A., Scrafford C. G. and Sewall C. (2005). Exposure data for cosmetic products: lipstick, body lotion, and face cream. *Food and Chemical Toxicology*, **43**, 279–291.

Mackay C. E., Johns M., Salatas J. H., Bessinger B. and Perri M. (2006). Stochastic probability modeling to predict the environmental stability of nanoparticles in aqueous suspension. *Integrated Environmental Assessment and Management*, **2**(3), 293–298.

Mackenzie L. D. (2010). *Water and Wastewater Engineering*. McGraw-Hill, New York.

Mu H. and Chen Y. (2011). Long-term effect of ZnO nanoparticles on waste activated sludge anaerobic digestion. *Water Research*, **45**(17), 5612–5620.

Peng Z., Doroodchi E. and Evans G. M. (2012). Influence of primary particle size distribution on nanoparticles aggregation and suspension yield stress: a theoretical study. *Powder Technology*, **223**(0), 3–11.

Peters R., Dam G. t., Bouwmeester H., Helsper H., Allmaier G., Kammer F. v., Ramsch R., Solans C., Tomaniová M., Hajslova J. and Weigel S. (2011). Identification and characterization of organic nanoparticles in food. *TrAC Trends in Analytical Chemistry*, **30**(1), 100–112.

Petosa A. R., Jaisi D. P., Quedevo I. R., Elimelech M. and Tufenkji N. (2010). Aggregation and deposition of engineered nanomaterials in aquatic environments: role of physiochemical reactions. *Environmental Science and Technology*, **44**(17), 18.

Reijnders L. (2006). Cleaner nanotechnology and hazard reduction of manufactured nanoparticles. *Journal of Cleaner Production*, **14**(2), 124–133.

Reijnders L. (2009). The release of TiO_2 and SiO_2 nanoparticles from nanocomposites. *Polymer Degradation and Stability*, **94**(5), 873–876.

Rezić I. (2011). Determination of engineered nanoparticles on textiles and in textile wastewaters. *TrAC Trends in Analytical Chemistry*, **30**(7), 1159–1167.

Sellers K. (2009). Nanoscale materials: definition and properties. In: *Nanotechnology and the Environment*, K. Sellers, C. Mackay, L. L. Bergeson, S. R. Clough, M. Hoyt, J. Chen, K. Henry and H. Jane (eds) , CRC Press, Boca Raton.

Sheng Z. and Liu Y. (2011). Effects of silver nanoparticles on wastewater biofilms. *Water Research*, **45**(18), 6039–6050.

Som C., Wick P., Krug H. and Nowack B. (2011). Environmental and health effects of nanomaterials in nanotextiles and façade coatings. *Environment International*, **37**(6), 1131–1142.

Stone V., Nowack B., Baun A., van den Brink N., von der Kammer F., Dusinska M., Handy R., Hankin S., Hassellöv M., Joner E. and Fernandes T. F. (2010). Nanomaterials for environmental studies: classification, reference material issues, and strategies for physico-chemical characterisation. *Science of the Total Environment*, **408**(7), 1745–1754.

Teeguarden J. G., Hinderliter P. M., Orr G., Thrall B. D. and Pounds J. G. (2006). Particokinetics *in vitro*: dosimetry considerations for *in vitro* nanoparticle toxicity assessments. *Toxicological Sciences*, **95**(2), 300–312.

The Project on Emerging Nanotechnologies. (2011). Inventories. Retrieved 04 December, 2011, from http://www.nanotechproject.org/inventories/consumer/analysis_draft/ (last accessed: 8 February 2013).

Thio B. J. R., Zhou D. and Keller A. A. (2011). Influence of natural organic matter on the aggregation and deposition of titanium dioxide nanoparticles. *Journal of Hazardous Materials*, **189**(1–2), 556–563.

Tiede K., Hassellov M., Breitbarth E., Chaudhry Q. and Boxall A. (2009). Considerations for environmental fate and ecotoxicity testing to support environmental risk assessments for engineered nanoparticles. *Journal of Chromatography A*, **1216**(3), 503–509.

Wang Y., Westerhoff P. and Hristovski K. D. (2012). Fate and biological effects of silver, titanium dioxide, and C_{60} (fullerene) nanomaterials during simulated wastewater treatment processes. *Journal of Hazardous Materials*, **201–202**(0), 16–22.

Weinberg H., Galyean A. and Leopold M. (2011). Evaluating engineered nanoparticles in natural waters. *TrAC Trends in Analytical Chemistry*, **30**(1), 72–83.

Westerhoff P., Zhang Y., Crittenden J. and Chen Y. (2008). Properties of commercial nanoparticles that affect their removal during water treatment. In: *Nanoscience and Nanotechnology: Environmental and Health Impacts*, V. H. Grassien (ed.), John Wiley and Sons, Inc., New Jersey.

Xu R. (2008). Progress in nanoparticles characterization: sizing and zeta potential measurement. *Particuology*, **6**(2), 112–115.

Zhang Y., Chen Y., Westerhoff P., Hristovski K. and Crittenden J. C. (2008). Stability of commercial metal oxide nanoparticles in water. *Water Research*, **42**, 2204–2212.

Chapter 6

Effective phosphate removal using Ca-based layered double hydroxide nanomaterials

J. Z. Zhou, Z. P. Xu and G. Qian

6.1 LAYERED DOUBLE HYDROXIDE

6.1.1 Introduction

Layered double hydroxides (LDHs), also known as hydrotalcite-like compounds, are a large group of natural and synthetic clay materials with high anion exchange capacity (Goh *et al.*, 2008; Zhang *et al.*, 2008). As shown in Figure 6.1, LDHs are layered materials consisting of positively charged metal hydroxide (brucite-like) sheets with octahedra sharing the edges and the interlayer anions. Their chemical composition can be generally expressed as $[M_{1-x}^{2+}M_x^{3+}(OH)_2][A^{n-}]_{x/n} \cdot yH_2O$, where M^{2+} and M^{3+} are divalent and trivalent cations, and A^{n-} an anion in the interlamellar space to balance the positive charge of the brucite-like layer with associated water molecules (Cavani *et al.*, 1991). Moreover, due to the relatively weak interlayer bonding, anions such as Cl^- and NO_3^-, intercalated in the LDHs, are readily exchanged with various organic and inorganic anions. Because of the anion exchange property, LDHs become a promising reservoir for diverse inorganic, organic, and biomolecular anions, as well as anionic polymers and drugs (Gu *et al.*, 2010; Palmer *et al.*, 2009; Rives & Ulibarri 1999; Zhang *et al.*, 2008).

For environmental applications, LDHs have shown great potential in removal of anionic contaminants from polluted waters through the anion-exchange process. For instance, there are many researches in using LDH materials to remove phosphates, arsenate, selenate, and chromate (Liu *et al.*, 2009; Triantafyllidis *et al.*, 2010; Wang *et al.*, 2009; Xu *et al.*, 2010; Zhou *et al.*, 2008) as well as organic anions, such as folate, nitrophenol, terephthalate, and so on (Greenwell *et al.*, 2010; Millange *et al.*, 2000; Xu *et al.*, 2008).

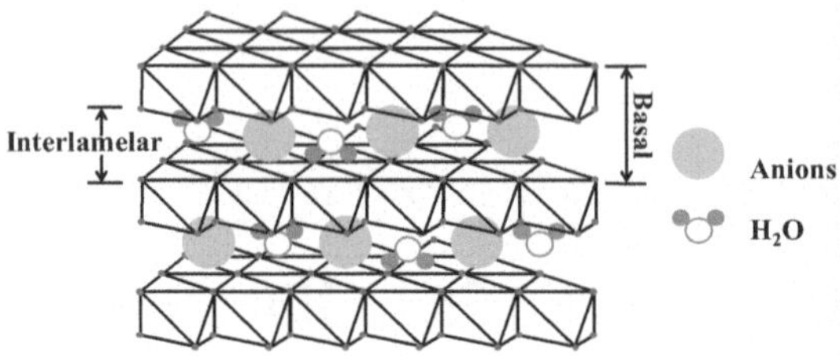

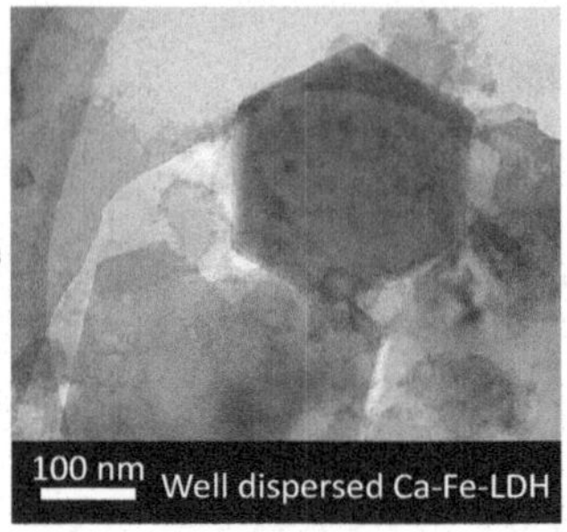

Figure 6.1 Structure schematic of LDHs and TEM image for Ca_2Fe-LDH.

Many methods have been developed to prepare LDH, such as coprecipitation, anion exchange within an LDH precursor, and preparation from the corresponding oxides or hydroxides. Among these methods, coprecipitation is the most common method (Braterman *et al.*, 2004). In coprecipitation, a steady addition of the solution containing all the cations in a ratio is mixed with a basic solution simultaneously under vigorous stirring. After aging for a period of time, LDH materials can be obtained via filtration and drying.

6.1.2 Composition variability

The composition of LDH is variable and tuneable. As in hydrotalctie ($Mg_6Al_2(OH)_{16}CO_3{\cdot}4H_2O$), a number of bivalent and trivalent metal cations have been reported to replace Mg and Al and form LDHs. For instance, LDHs can be composed of Cu, Fe(II), Ni, Zn, Co(II) as bivalent cations and Fe(III), Al, Cr, Mn as trivalent ones (Bravo-Suarez *et al.*, 2004). To form an LDH, the incorporated metal ion is required to have a radius close to that of Mg^{2+} or Al^{3+}. The radius suitable for metal ions in LDH is approximately 0.070 nm (Braterman *et al.*, 2004).

Besides the typical binary LDHs which involve only one bivalent and one trivalent metal, LDHs containing more than two metals, such as CuCo(II) FeAl-LDH and ZnCdCr-LDH (Cheng *et al.*, 2008; Guo *et al.*, 2005) were reported. Accordingly, some researchers have successfully prepared heavy metals incorporated LDHs such as ZnCuCr-LDH and ZnNiCr-LDH from an industrial wastewater (Wu *et al.*, 2009; Zhou *et al.*, 2010).

Apart from the type of metal cations, the ratio of trivalent to total metals (x) is tuneable as well. For most LDHs, the value of x ranges from 0.20 to 0.33, within which a pure LDH can be obtained (Khan & O'Hare, 2002). Otherwise, a mixture containing an LDH and a metal hydroxide, oxide, or other compound is formed (Zhou *et al.*, 2010). Therefore, the composition variability of LDH makes it possible to obtain multi-component LDHs that have various properties such as catalytic activity, aqueous stability and anionic pollutant removal capacity.

6.1.3 Structure features

Mg-based LDH is one kind of LDHs attracting much attention due to the efficient anion-exchange capacity with most guest anions (Chao *et al.*, 2008; Lv *et al.*, 2007; Toraishi *et al.*, 2002; Xu & Zeng, 2001). As a typical Mg-based LDH, hydrotalcite ($Mg_6Al_2(OH)_{16}CO_3{\cdot}4H_2O$) displays a series of strong XRD patterns conventionally indexed as 003, 006, and so on (Hibino & Tsunashima, 1997). These basal reflections illustrate the hexagonal (2H) or rhombohedral (3R) layer stacking sequence. Among the non-basal reflections (hkl, either h or k is not zero), the 110 reflection shows the information of cation distribution in the brucite-layer of LDH ($a = 2d_{110}$), which gives a hexagonal superlattice with $a' = 2a\sin60°$ for M(II)/M(III) = 2 or $a' = 2a$ for M(II)/M(III) = 3 (Braterman *et al.*, 2004).

On the other hand, Ca-based LDH, often observed in cement or concrete, is a different series of LDHs from Mg-based LDH. A typical Ca-based LDH is $Ca_2Al(OH)_6Cl{\cdot}nH_2O$ that is always known as hydrocalumite or Friedel's salt (Zhou *et al.*, 2008). Compared with that of Mg-based LDH, the presence of peak (020) is only indexed in the 002 system (Ca-Al-LDH). In the 030 system (Ca-Fe-LDH), the (020) peak is hardly observed, while strong (110) reflection, identified at about 31°, which reveals the different structure of target LDHs. CaFe-LDH also has a different structure from Mg-based LDH. The XRD pattern of Ca_2Fe-LDH displays a series of reflections such as 003, 006, 110 and so on in which the plane (110) locates at 31° (2θ). As a result, lattice parameter *a* is 0.58 nm, larger to that of the Mg-based LDH (often 0.31 nm).

This special structure of Ca-based LDH is attributed to the larger radius of Ca^{2+} (0.100 nm) than Mg^{2+} (0.072 nm) that results in the distorted close-stacking configuration of the Ca-O octahedron to some extent. This distortion leads Ca to be 7-coordinated while Al is kept 6-coordinated (Braterman *et al.*, 2004). Another significant difference is the higher solubility product of $Ca(OH)_2$ ($K_{sp} = 5.02 \times 10^{-6}$) than that of $Mg(OH)_2$ ($K_{sp} = 5.61 \times 10^{-6}$) (Lide, 2007), so the Ca-based LDH is instable in water due to dissolution of Ca-based LDH (Radha *et al.*, 2005). This aqueous instability of Ca-based LDH was also observed in the case of Ca-Fe-LDH (Seida & Nakano, 2002).

In terms of the controlled preparation, MgAl-LDH particle size can be readily tailored in the range of 60 to 300 nm with a relatively narrow particle size distribution (Xu *et al.*, 2006a, b), Ca-based LDH, such as $Ca_2Fe(OH)_6Cl{\cdot}xH_2O$, has a size of about 200–300 nm, as shown Figure 6.1b. Note that almost all LDH particles possess the typical hexagonal morphology, which can be attributed to the structural feature of the hydroxide layers and the stacking style of these layers (Braterman *et al.*, 2004).

6.2 REMOVAL OF PHOSPHATES BY LDH

LDHs have attracted much attention due to the composition and structure variability. Most metal cations can be introduced into the LDH structure, which makes these materials widely used in many applications. To remove anionic pollutants from water, the hydrotalcite-like LDH displays an efficient removal capacity due to the anion-exchange property. On the other hand, the aqueous instable Ca-based LDH also has potential to take up anionic pollutants through precipitating with Ca^{2+} dissolved from LDH, various phosphates in particular. Thus, this chapter briefly reviews recent investigations using Ca-based LDHs to remove phosphates.

6.2.1 Inorganic phosphate speciation

Inorganic phosphate is widely distributed in water, soil, and sediment. The species of inorganic phosphate includes monophosphate and condensed phosphates. Monophosphate (PO_4^{3-}, P1), known as orthophosphate, is the most common species of inorganic phosphorous (Huang *et al.*, 2005; Sannigrahi & Ingall, 2005). As an essential nutrition element for plant growth, monophosphate causes a fast growth of aquatic vegetation to consume dissolved oxygen, that is, eutrophication, thus deteriorating aquatic ecosystems (Baker *et al.*, 1998).

Diphosphate (P2, $P_2O_7^{4-}$) and triphosphate (P3, $P_3O_{10}^{5-}$) are two kinds of condensed phosphates, which are formed through linking tetrahedral [PO_4] units (Rashchi & Finch, 2000). The condensed phosphates are widely used in the industries and often found in the municipal or rural wastewaters (Halliwell *et al.*, 2001). The condensed phosphates have been thought to have low effect on the aquatic plant growth due to the low bioavailability. However, some researchers recently observed that in the presence of N and C species, P2 and P3 were absorbed by plant roots as nutrients like monophosphate (Sundareshwar *et al.*, 2001; Torres-Dorante *et al.*, 2006). Moreover, condensed phosphates can also be converted to nutrient monophosphate in a pH-dependent way. It is found that at $pH < 7.0$, the conversion of condensed phosphates is much faster than that at $pH > 8.0$ (Green, 1950; van Wazer *et al.*, 1955; van Wazer *et al.*, 1952; Vasudevan *et al.*, 1994). Thereby, the condensed phosphates, like monophosphate, are also contributing to eutrophication, and therefore it is necessary to eliminate P1, P2 and P3 from waters to keep healthy aquatic ecosystems (Xing *et al.*, 2008).

6.2.2 Removal isotherms

Mg-based LDH is used to remove various phosphates through the anion-exchange process as these phosphates are negatively charged in water and wastewaters (pH = 4–10) (Ookubo *et al.*, 1993). In comparison, the Ca-based LDH has been found to improve the removal of phosphates through dissolving Ca-LDH to release Ca^{2+} and then precipitating with phosphates (Seida & Nakano, 2002; Xu *et al.*, 2010). For instance, the Cl-intercalated Ca_2Fe-LDH was employed in P1 removal, as shown in Figure 6.2a (100 ml P1 solution with 1.0 g/L of LDH for 24 h reaction). The resultant removal sharply increased to the saturation amount (2.7 mmol(P)/g), when [P] increased from 0.0 to 0.1 mM. The practical maximum removal amount was 2.71 mmol(P)/g, which was much higher than the theoretical amount (1.10–1.65 mmol/g) if only the anion exchange process took place (Zhou *et al.*, 2011). Such a high removal capacity of Ca-based LDH could mostly be attributed to the precipitation of P1 with Ca^{2+} dissolved from Ca-LDHs.

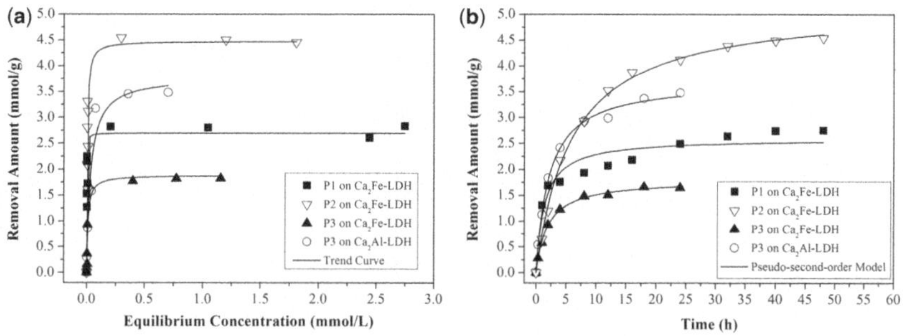

Figure 6.2 Isotherm (a) and kinetics (b) of phosphate removal by LDHs from solution.

Moreover, the isotherms of P2 and P3 removal over Ca_2Fe-LDH were similar to that of P1 (Figure 6.2a). Interestingly, the saturation removal amount of P2 was 4.5 mmol/g, which was significantly higher than that of P3 (1.8 mmol/g). It is likely related to the fact that the Ca-P3 precipitate is further complexed with P3 to be a soluble anionic species (Zhou & Carnali, 2000). In comparison, Ca_2Al-LDH removed up to 3.5 mmol/g of P3 (Figure 6.2a) under the same conditions. The low P3 removal amount by Ca_2Fe-LDH was probably due to the existence of Fe in the LDH that may partly associate with Ca in the LDH framework. This Ca is thus not available to precipitate P3 (see Section 6.3).

6.2.3 Removal kinetics

The removal kinetics of phosphates over Ca-LDHs are shown in Figure 6.2b. In general, in the first 2–3 h, the removal rate of phosphates was initially quick, and then the slow removal continued. For example, at the initial [P] = 4.83 mmol/L and the LDH dose = 1.0 g/L, the P1 removal amount of 1.6 mmol/g P was observed within 5 h which increased to its saturation amount (2.75 mmol/g) within 48 h. The P2 removal over Ca_2Fe-LDH (Figure 6.2b) was faster, with a much higher saturation removal amount at 48 h (4.50 mmol/g). On the contrary, the removal amount of P3 over Ca_2Fe-LDH (1.65 mmol/g) was much lower than that in the case of P2 removal. This difference implied that the aqueous stability of Ca-P precipitates varied in these three cases.

The kinetics of P2 and P3 removal fitted well with pseudo-first and pseudo-second order models, with $R^2 > 0.98$ (Table 6.1). As also shown in Table 6.1, the

P1 removal is not in good agreement with these two kinetics models, as $R^2 < 0.94$ (Goh *et al.*, 2009; Grover *et al.*, 2009). This indicates that the uptake process of P1 by Ca-Fe-LDH would be different from that of P2 and P3 removal.

Table 6.1 Kinetic parameters for phosphates removal by LDHs.

Phosphate	LDH	Model*	Parameters		
			K	Qe (mmol/L)	R^2
P1	Ca_2Fe-LDH	Pseudo-first order	0.570	2.38	0.846
		Pseudo-second order	0.279	2.60	0.927
P2	Ca_2Fe-LDH	Pseudo-first order	0.144	4.41	0.994
		Pseudo-second order	0.032	5.17	0.998
P3	Ca_2Fe-LDH	Pseudo-first order	0.419	1.57	0.990
		Pseudo-second order	0.278	1.80	0.997
P3	Ca_2Al-LDH	Pseudo-first order	0.385	3.24	0.982
		Pseudo-second order	0.123	3.72	0.997

* Pseudo-first order: $q_t = q_{m1}[1-\exp(-k_1t)]$;
Pseudo-second order: $q_t = k_2q_{m2}^2t/(1 + k_2q_et)$,
where t is time (h); q_t is the removal amount of phosphorous at time t (mmol/g); q_{m1} and q_{m2} are the maximum removal amount (mmol/g) in Pseudo-first order and Pseudo-second order models, respectively; k_1(1/h) and k_2(g/mmol $\times$ h) are the reaction rate constant, respectively.

6.2.4 Effect of absorbent composition

Al or Fe (M(III) is often incorporated in the Ca-based LDH as M(III), interacts with adjacent Ca cations and affects the leaching of Ca cations from the hydroxide layer. Meanwhile, Al-hydroxide is partially soluble in alkaline solution while Fe-hydroxide is not, and thus this difference further affects the amount of leached Ca cations from a Ca-LDH. Together these two aspects show significant influence on the phosphate removal effectiveness over Ca-based LDHs.

As clearly shown in Table 6.2, the amount of P1 and P3 removed by Ca_2Al-LDH was higher than that by Ca_2Fe-LDH. It is related to the different solubility of M(III) in LDH. Radha *et al.* (2005) supposed Ca^{2+} and $Al(OH)_4^-$ were leached out from Ca_2Al-LDH, which resulted in apatite ($Ca_5(PO_4)_3OH$) precipitation in the presence of PO_4^{3-}. This resulted in the high concentration of Al detected in the final solution at high pH (Table 6.2). In contrast, no Fe was determined in the final solution at the similar pH in the case of Ca_2Fe-LDH, which probably retains some Ca in the insoluble Fe oxide/hydroxide framework, leading to a lower phosphate removal.

Table 6.2 also indicates that in cases of P2 and P3, the soluble Ca concentration was much higher than that in the case of P1, likely relative to the formation of soluble Ca-P complexes, as discussed below.

Table 6.2 Elemental analysis of the components in final solution after P removal.

Case	Initial [P] (mmol/L)	LDH dosage (g/L)	Sample	Removal amount (mmol/g)	Final pH	Cations in Solution after P Removal (mmol/L)	
						Ca	Al (Fe)
P1	4.83	1.0	Ca_2Al-LDH	3.4	11.5	0.01	1.7
			Ca_2Fe-LDH	2.7	11.7	0.006	ND[a]
P2	4.83	1.0	Ca_2Fe-LDH	4.5	12.0	0.24	ND
P3	1.61	0.2	Ca_2Al-LDH	3.5	11.2	0.51	0.48
			Ca_2Fe-LDH	1.5	10.5	0.62	ND

[a]ND represents "not detected".

6.2.5 Effect of the phosphate concentration

The kinetics process of phosphate removal has been related to the initial P concentration. For instance, the kinetic process of P1 removal at a high initial P concentration was similar to that at slow initial P concentration within the first 0.5 h, but became much slower with the time at the high initial P concentration and took a longer time to reach the equilibrium (Xu *et al.*, 2010). In contrast, the kinetic process in the current three cases did not differ significantly, even considering that the initial P concentration in the case of P3 was 5 times smaller than in the cases of P1 and P2 (Table 6.2).

6.3 REMOVAL MECHANISM

6.3.1 Ion concentrations in treated solutions and Ca/P ratio in the collected solids

In Table 6.2, soluble Al was detected in the treated solution at $pH > 11.0$. The soluble species of Al is highly dependent on the solution pH. As Ca is leached out, the pH increases to 11.0 quickly (Xu *et al.*, 2010) and Al-hydroxide dissolves to form soluble $Al(OH)_4^-$. Such a conversion of Al-hydroxide to $Al(OH)_4^-$ was observed in the recent work where Ca_2Al-LDH was used for PO_4^{3-} removal (Xu *et al.*, 2010). In fact, for P1 to remove 3.4 mmol P, 5.77 mmol of Ca^{2+} is required to form apatite. This amount of Ca corresponds to 83% of Ca in 1 g of Ca_2Al-LDH ($Ca_{2.01}Al(OH)_{6.02}Cl_{0.97}(CO_3)_{0.015}{\cdot}2.5H_2O$, MW = 290). Taking into account the partially left LDH (Xu *et al.*, 2010), this comparison further demonstrates that almost all Ca cations in Ca-LDH are leached out and used to precipitate P1.

Furthermore, Fe-hydroxide was not dissolved to form any soluble Fe species (Table 6.2). This was also observed by Seida and Nakano (2002), who reported the concentration of Fe in the final solution was only from 0.15 to 0.2 mg/L in the P1 removal over Ca_2Fe-LDH regardless of the equilibrium P concentration varied from 0 to 1500 mg/L. The insolubility of Fe-hydroxide may affect Ca leaching and the subsequent P1 removal. For example, 2.7 mmol P1 would react with 4.58 mmol Ca^{2+} to form apatite, corresponding to 73% of Ca^{2+} in 1 g of Ca_2Fe-LDH, lower than that in Ca_2Al-LDH. A similar influence of M(III) was observed in P3 removal, indicating that both orthophosphate and condensed phosphate removal was contributed to the solubility of M(III) in LDH.

On the other hand, Ca concentrations in P2 and P3 were found higher than those in P1. The low concentration of soluble Ca in P1 indicated that ca. 99% of Ca was precipitated after P removal in the pristine LDH. In comparison, the high soluble Ca concentration revealed that ca. 37% of Ca in Ca_2Al-LDH and 48% in Ca_2Fe-LDH were not precipitated in the collected solid in the case of P3 removal. The percentage of soluble Ca in P2 removal was 3.7% (Ca_2Fe-LDH).

Accordingly, excluding the soluble Ca, the amount of Ca in the solid can be estimated by using the mass balance of Ca. Thus, the removed Ca^{2+}/P ratio (Ca^{2+}/P) refers to the practical amount ratio of the consumed Ca^{2+} to the removed P. Figure 6.3 illustrates the Ca^{2+}/P of the solid sample in cases of Ca_2Al-LDH and Ca_2Fe-LDH, respectively. It should be noted that Ca^{2+}/P (P1) over Ca_2Fe-LDH decreased from 1.75 to 1.3 with the initial P concentration increasing from 2.0 to 3.0 mmol/L in comparison to the constant Ca^{2+}/P ca. 1.8 in the case of Ca_2Al-LDH. Similar variety of the Ca^{2+}/P ratio was also observed in the removal of P2 on Ca_2Fe-LDH, in which Ca^{2+}/P decreased from 1.4 to 1.1. These observations indicate that more phosphate was removed by Ca_2Fe-LDH, which seems not to be attributed to the precipitation of apatite.

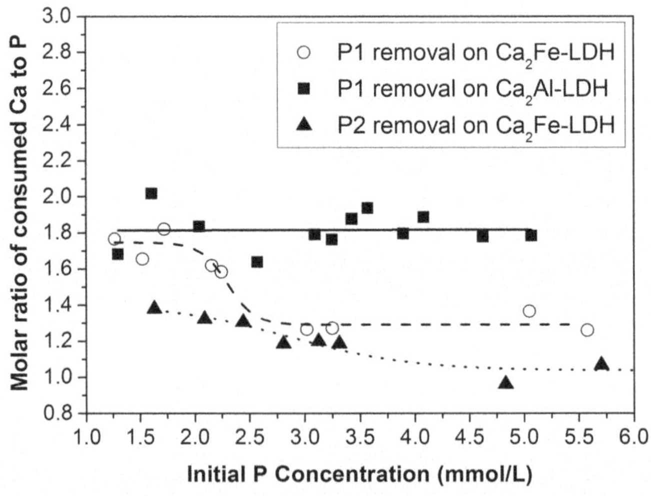

Figure 6.3 Molar ratio of consumed Ca to removed P (Ca^{2+}/P) in the precipitates after P1 removal compared with that after P2 removal over Ca_2Fe-LDH.

6.3.2 Formation of Ca-P precipitate

Figure 6.4a shows X-ray diffraction (XRD) patterns for the collected samples after phosphate removal compared with the pristine LDH. The XRD pattern of Ca_2Fe-LDH in the pristine LDH was indexed according to the diffraction pattern along (003), (006), (110), (116) and (316) planes. In contrast, only very weak XRD peaks for the layered structure in the products after phosphate uptake on LDH samples could be observed (Figure 6.4a). Instead, the hydroxyapatite ($Ca_5(PO_4)_3OH$) and $Ca_2P_2O_7{\cdot}2H_2O$ were clearly identified in the P1 and P2 removal, respectively, while no pattern for the Ca-P3 precipitate was observed after the P3 removal. The FTIR spectra of these samples in three cases (Figure 6.4b) contained the feature vibrations of phosphates in the range of 850–1300 cm^{-1}, indicating the uptake of phosphates (Gong, 2001; Guan *et al.*, 2005). These observations revealed that the dissolution of LDH and the precipitation of phosphate with Ca^{2+} occurred during the phosphate removal.

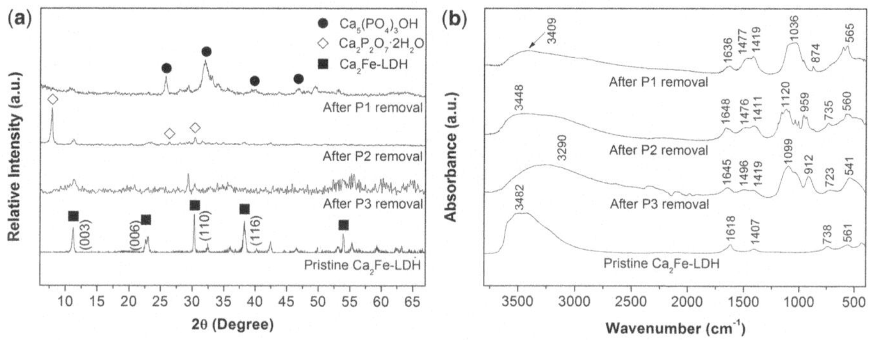

Figure 6.4 Structure characterization of P removed LDHs by XRD (a) and FTIR (b).

In particular, when mixing Ca^{2+} and P3, $Ca_5(P_3O_{10})_2{\cdot}nH_2O$ was amorphous and the dominative component, when the initial $[Ca^{2+}]/[P]$ ranged from 2.0 to 4.0 (Diamond & Grove, 1959; Zhou & Carnali, 2000). In the case of P3 removal on Ca_2Fe-LDH, the initial $[Ca^{2+}]/[P]$ ratio was about 2.6 if all Ca was leached out from the LDH. As a result, the precipitation of P3 with Ca^{2+} would form $Ca_5(P_3O_{10})_2{\cdot}nH_2O$ in spite of its amorphous structure. On the other hand, the Ca^{2+}/P ratio in the solid precipitate after P3 removal was about 2.0, excluding the soluble Ca in the case of Ca_2Fe-LDH (Table 6.2). It should be noted that it was much higher than the theoretical Ca/P in $Ca_5(P_3O_{10})_2{\cdot}nH_2O$ (Ca/P = 1.2). This further indicated that Ca remained partially in the Fe hydroxide frame after dissolution of LDH.

Moreover, degradation of condensed phosphate during phosphate removal was ruled out by the further characterization by X-ray photoelectron spectrometry

(XPS). Figure 6.5 displays the binding energy (BE) peaks of P 2p at 133.3 and 133.6 eV in the collected solids after P2 and P3 removal, respectively. Thus, P was predominant in the condensed phosphate form, much different from the BE of P 2p at 134.4 eV in the case of P1 removal. The stability of condensed phosphates was in agreement with other reports (Ramana *et al.*, 2007; Vasudevan *et al.*, 1994; Zhou & Carnali, 2000). In addition, XPS also exhibits a similar bond of Ca to P as the Ca 2p doublets (Ca 2p3/2 and Ca 2p1/2) were similar in all samples. These observations supported the assumption that the Ca-P precipitate ($Ca_5(PO_4)_3OH$, $Ca_2P_2O_7{\cdot}2H_2O$ or $Ca_5(P_3O_{10})_2{\cdot}nH_2O$) was the dominant component in the corresponding collected sample, consistent with the XRD pattern (Figure 6.4).

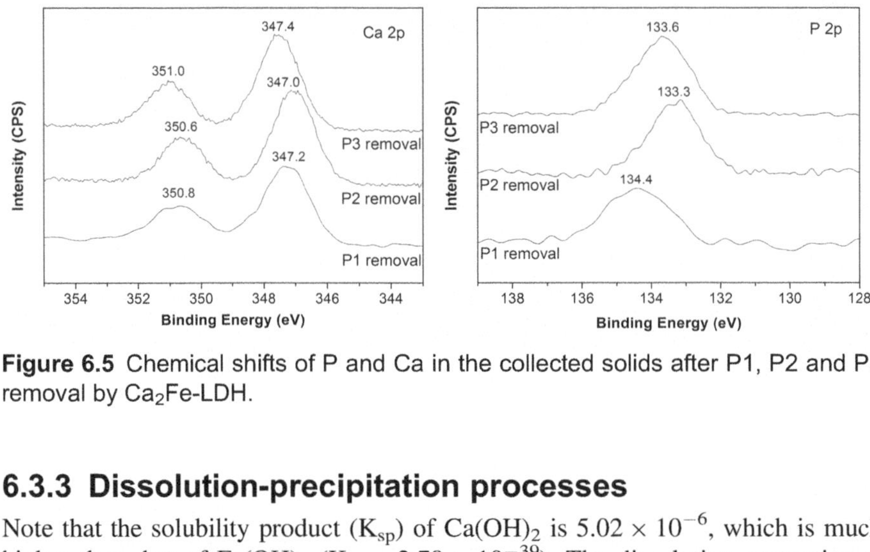

Figure 6.5 Chemical shifts of P and Ca in the collected solids after P1, P2 and P3 removal by Ca_2Fe-LDH.

6.3.3 Dissolution-precipitation processes

Note that the solubility product (K_{sp}) of $Ca(OH)_2$ is 5.02×10^{-6}, which is much higher than that of $Fe(OH)_3$ ($K_{sp} = 2.79 \times 10^{-39}$). The dissolution properties of these two hydroxides are mostly inherited in the Ca_2Fe-LDH, which leads most Ca to dissolve in water and yields a higher pH in the phosphate removal as described in the following equation (Zhou *et al.*, 2011):

$$Ca_2Fe(OH)_6Cl{\cdot}2H_2O \longrightarrow Ca^{2+} + Ca_xFe(OH)_{(3+2x)} + OH^- + H_2O + Cl^- \quad \text{(Eq. 6.1)}$$

As a result, the removal of phosphate (P1 to P3) over CaFe-LDH is attributed to the precipitation of phosphate with Ca^{2+}:

$$Ca^{2+} + PO_4{}^{3-} + OH^- \longrightarrow Ca_5(PO_4)_3OH \quad \text{(Eq. 6.2)}$$

$$2Ca^{2+} + P_2O_7{}^{4-} + 2H_2O \longrightarrow Ca_2P_2O_7{\cdot}2H_2O \quad \text{(Eq. 6.3)}$$

$$5Ca^{2+} + 2P_3O_{10}{}^{5-} + nH_2O \longrightarrow Ca_5(P_3O_{10})_2{\cdot}nH_2O \quad \text{(Eq. 6.4)}$$

Equation 6.1 suggests that Ca is partially retained in the framework as $Ca_xFe(OH)_{(3+2x)}$. The value of x ranged from 0.5–0.7 after the hydrolysis of LDH in a P free solution for 24 h at 25°C. Due to the high aqueous stability of Fe oxide/hydroxide, the collected sample contained $Ca_xFe(OH)_{(3+2x)}$ and Ca-P precipitates after the dissolution-precipitation process. As evidenced by TEM (Figure 6.6), these two solid compounds in the collected solid fraction in the case of P1 were identified. At spot A in Figure 6.6a, only Ca, P and O were detected, revealing $Ca_5(PO_4)_3OH$ was freshly formed on the top of a previous frame. However, at spot B in Figure 6.6a, Ca, Fe, P and O were all detected, indicating that spot A consisted of $Ca_xFe(OH)_{(3+2x)}$ but was topped with $Ca_5(PO_4)_3OH$.

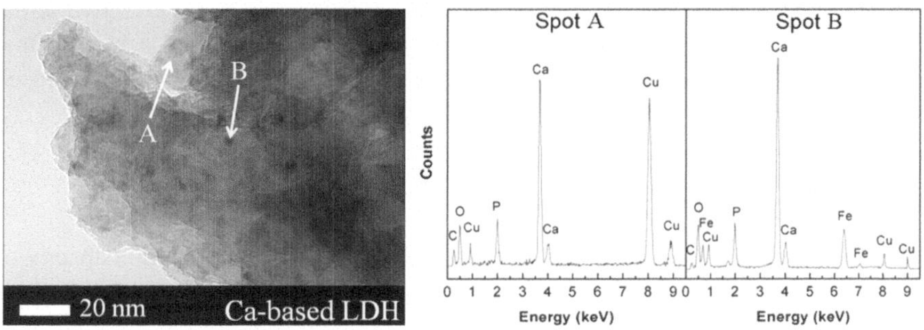

Figure 6.6 TEM image for P1-containing Ca_2Fe-LDH with the EDX spectra.

Obviously, the dissolution-precipitation was the dominant mechanism of P removal. A subsidiary process for P removal was also found in the case of Ca_2Fe-LDH. For instance, the reduction of the removed Ca^{2+}/P ratio from 1.8 to 1.3 in the case of P1 removal (Figure 6.3) suggested that more P was removed as no more Ca^{2+} consumed with the removal of P1. A similar situation was also observed in the case of P2 removal in which the ratio of Ca^{2+}/P decreased from 1.4 to 1.1. This reduction implied that P2 was removed partially via a subsidiary process other than precipitation of $Ca_2P_2O_7 \cdot 2H_2O$. The excess of P removal was attributed to the adsorption of P on the $Ca_xFe(OH)_{(3+2x)}$ as $Fe(OH)_3$ or FeO(OH) could efficiently adsorb P (Hesterberg & Liu, 2011; Ler & Stanforth, 2003). However, the obtained Ca^{2+}/P ratio in the collected solid was still higher than the theoretical ratio of Ca/P (=1.0) in $Ca_2P_2O_7 \cdot 2H_2O$, which could be due to the amount of Ca retained in $Ca_xFe(OH)_{(3+2x)}$.

In the case of P3 removal, the Ca^{2+}/P ratio in the collected solid was 0.8 (Zhou *et al.*, 2011), close to the Ca/P ratio in $Ca_5(P_3O_{10})_2 \cdot nH_2O$ (0.83), which demonstrated that the adsorption of P3 on $Ca_xFe(OH)_{(3+2x)}$ was insignificant.

It is worth mentioning that there was more soluble Ca in the final solution after the removal of P2 and P3 than that of P1 (Table 6.2), no matter how much P was left for precipitation (Eqs. 6.3 and 6.4). This can be attributed to the formation of Ca^{2+}-phosphate complexes at neutral or basic solution (Sawada *et al.*, 2003), which inhibits the Ca^{2+} precipitation:

$$Ca^{2+} + P_2O_7^{4-} \longrightarrow CaP_2O_7^{2-} \qquad \text{Log K1} = 6.50 \qquad \text{(Eq. 6.5)}$$

$$Ca^{2+} + HP_2O_7^{3-} \longrightarrow CaHP_2O_7^{-} \qquad \text{Log K2} = 4.30 \qquad \text{(Eq. 6.6)}$$

$$Ca^{2+} + P_3O_{10}^{5-} \longrightarrow CaP_3O_{10}^{3-} \qquad \text{Log K3} = 4.80 \qquad \text{(Eq. 6.7)}$$

$$Ca^{2+} + HP_3O_{10}^{4-} \longrightarrow CaHP_3O_{10}^{2-} \qquad \text{Log K4} = 8.41 \qquad \text{(Eq. 6.8)}$$

Obviously, the sequestration of Ca^{2+} with P2 or P3 was much effective while that with P1 was not observed. Therefore, most of the soluble Ca in the final solution after P removal contributed to the complexation of Ca^{2+} with condensed phosphate.

6.4 CONCLUDING REMARKS AND PERSPECTIVES

Compared with Mg-based LDH, Ca-based LDH shows a high removal capacity of all three inorganic phosphates. The predominant removal mechanism is dissolution-precipitation in which dissolved Ca^{2+} precipitates with phosphates (PO_4^{3-}, $P_2O_7^{4}$, and $P_3O_{10}^{5-}$) to form insoluble $Ca_5(PO_4)_3OH$, $Ca_2P_2O_7{\cdot}2H_2O$ and $Ca_5(P_3O_{10})_2{\cdot}nH_2O$, respectively. On the other hand, the excess removal amount of PO_4^{3-} and $P_2O_7^{4-}$ can be obtained over Ca_2Fe-LDH, and attributed to the adsorption of these two phosphates on $Ca_xFe(OH)_{(3+2x)}$ generated after hydrolysis of LDH. Unfortunately, the resultant high pH and high soluble Ca^{2+} or $Al(OH)_4^-$ in solution limits the application of Ca-based LDH in anion removal due to the secondary pollution.

Alternatively, an advanced strategy for anions removal over Ca-based LDH was developed recently. A ternary Ca-Mg-Fe-LDH (M(II)/M(III) = 2) gave a removal amount of 2.8 mmol/g, higher than that by Ca_2Fe-LDH (P3) when the Ca/Mg ratio was 1.54 in the ternary LDH (Zhou *et al.*, 2011). The final pH in the solution was about 9.3, which was significantly lower than that in the case of Ca_2Fe-LDH. The possible mechanism of P3 removal was the synergic effect of the precipitation of P3 with Ca^{2+} and the adsorption on Ca-Mg-Fe-LDH. Another benefit of ternary LDH is the high aqueous stability of the P3–containing product due to the lower P3 leaching in the water media than that in both cases of Ca_2Fe-LDH and Mg_2Fe-LDH. This observation indicates that some metals can be introduced into a stable LDH frame to make LDH more efficient in the high selective removal or separation of anions, and reservation of special drugs or chemicals.

REFERENCES

Baker M. J., Blowes D. W. and Ptacek C. J. (1998). Laboratory development of permeable reactive mixtures for the removal of phosphorus from onsite wastewater disposal systems. *Environ. Sci. Technol.*, **32**(15), 2308–2316.

Braterman P. S., Xu Z. P. and Yarberry F. (2004). Layered double hydroxides (LDHs). In: *Handbook of Layered Materials*, S. M. Auerbach, K. A. Carrado and P. K. Dutta (eds), Marcel Dekker, New York, 373.

Bravo-Suarez J. J., Paez-Mozo E. A. and Oyama S. T. (2004). Review of the synthesis of layered double hydroxides: a thermodynamic approach. *Quim. Nova*, **27**(4), 601–614.

Cavani F., Trifirò F. and Vaccari A. (1991). Hydrotalcite-type anionic clays: preparation, properties and applications. *Catal. Today*, **11**(2), 173–301.

Chao Y. F., Chen P. C. and Wang S. L. (2008). Adsorption of 2,4-D on Mg/Al-NO_3 layered double hydroxides with varying layer charge density. *Appl. Clay Sci.*, **40** (1–4), 193–200.

Cheng J., Yu J. J., Wang X. P., Li L. D., Li J. J. and Hao Z. P. (2008). Novel CH_4 combustion catalysts derived from Cu-Co/X-Al (X = Fe, Mn, La, Ce) hydrotalcite-like compounds. *Energy Fuels*, **22**(4), 2131–2137.

Diamond W. J. and Grove J. E. (1959). Composition of the solid phase in the $Na_5P_3O_{10}$-$CaCl_2$-H_2O System. *J. Phys. Chem.*, **63**(9), 1528–1529.

Goh K.-H., Lim T.-T. and Dong Z. (2009). Enhanced arsenic removal by hydrothermally treated nanocrystalline Mg/Al layered double hydroxide with nitrate intercalation. *Environ. Sci. Technol.*, **43**(7), 2537–2543.

Goh K. H., Lim T. T. and Dong Z. (2008). Application of layered double hydroxides for removal of oxyanions: a review. *Water Res.*, **42**(6–7), 1343–1368.

Gong W. Q. (2001). A real time *in situ* ATR-FTIR spectroscopic study of linear phosphate adsorption on titania surfaces. *Int. J. Miner. Process*, **63**(3), 147–165.

Green J. (1950). Reversion of molecularly dehydrated sodium phosphates. *Ind. Eng. Chem.*, **42**(8), 1542–1546.

Greenwell H. C., Jones W., Rugen-Hankey S. L., Holliman P. J. and Thompson R. L. (2010). Efficient synthesis of ordered organo-layered double hydroxides. *Green Chem.*, **12**(4), 688–695.

Grover K., Komarneni S. and Katsuki H. (2009). Uptake of arsenite by synthetic layered double hydroxides. *Water Res.*, **43**(15), 3884–3890.

Gu Z., Rolfe B. E., Xu Z. P., Thomas A. C., Campbell J. H. and Lu G. Q. M. (2010). Enhanced effects of low molecular weight heparin intercalated with layered double hydroxide nanoparticles on rat vascular smooth muscle cells. *Biomaterials*, **31**(20), 5455–5462.

Guan X. H., Liu Q., Chen G. H. and Shang C. (2005). Surface complexation of condensed phosphate to aluminum hydroxide: an ATR-FTIR spectroscopic investigation. *J. Colloid Interface Sci.*, **289**(2), 319–327.

Guo Y., Zhang H., Zhao L., Li G.-D., Chen J.-S. and Xu L. (2005). Synthesis and characterization of Cd-Cr and Zn-Cd-Cr layered double hydroxides intercalated with dodecyl sulfate. *J. Solid State Chem.*, **178**(6), 1830–1836.

Halliwell D. J., McKelvie I. D., Hart B. T. and Dunhill R. H. (2001). Hydrolysis of triphosphate from detergents in a rural waste water system. *Water Res.*, **35**(2), 448–454.

Hesterberg D. and Liu Y. T. (2011). Phosphate bonding on noncrystalline Al/Fe-hydroxide coprecipitates. *Environ. Sci. Technol.*, **45**(15), 6283–6289.

Hibino T. and Tsunashima A. (1997). Synthesis of paramolybdate intercalates of hydrotalcite-like compounds by ion exchange in ethanol/water solution. *Chem. Mater.*, **9**(10), 2082–2089.

Huang Q. H., Wang Z. J., Wang D. H., Wang C. X., Ma M. and Jin X. C. (2005). Origins and mobility of phosphorus forms in the sediments of lakes Taihu and Chaohu, China. *J. Environ. Sci. Health Part A-Tox./Hazard. Subst. Environ. Eng.*, **40**(1), 91–102.

Khan A. I. and O'Hare D. (2002). Intercalation chemistry of layered double hydroxides: recent developments and applications. *J. Mater. Chem.*, **12**(11), 3191–3198.

Ler A. and Stanforth R. (2003). Evidence for surface precipitation of phosphate on goethite. *Environ. Sci. Technol.*, **37**(12), 2694–2700.

Lide D. R. (ed.) (2007). *CRC Handbook of Chemistry and Physics*. CRC Press, Boca Raton, FL.

Liu R., Frost R. L. and Martens W. N. (2009). Absorption of the selenite anion from aqueous solutions by thermally activated layered double hydroxide. *Water Res.*, **43**(5), 1323–1329.

Lv L., He J., Wei M., Evans D. G. and Zhou Z. L. (2007). Treatment of high fluoride concentration water by MgAl-CO_3 layered double hydroxides: kinetic and equilibrium studies. *Water Res.*, **41**(7), 1534–1542.

Millange F., Walton R. I., Lei L. X. and O'Hare D. (2000). Efficient separation of terephthalate and phthalate anions by selective ion-exchange intercalation in the layered double hydroxide $Ca_2Al(OH)_6NO_3 2H_2O$. *Chem. Mater.*, **12**(7), 1990–1994.

Ookubo A., Ooi K. and Hayashi H. (1993). Preparation and phosphate ion-exchange properties of a hydrotalcite-like compound. *Langmuir*, **9**(5), 1418–1422.

Palmer S. J., Frost R. L. and Nguyen T. (2009). Hydrotalcites and their role in coordination of anions in Bayer liquors: anion binding in layered double hydroxides. *Coord. Chem. Rev.*, **253**(1–2), 250–267.

Radha A. V., Kamath P. V. and Shivakumara C. (2005). Mechanism of the anion exchange reactions of the layered double hydroxides (LDHs) of Ca and Mg with Al. *Solid State Sci.*, **7**(10), 1180–1187.

Ramana C. V., Ait-Salah A., Utsunomiya S., Morhange J. F., Mauger A., Gendron F. and Julien C. M. (2007). Spectroscopic and chemical imaging analysis of lithium iron triphosphate. *J. Phys. Chem. C*, **111**(2), 1049–1054.

Rashchi F. and Finch J. A. (2000). Polyphosphates: a review their chemistry and application with particular reference to mineral processing. *Miner. Eng.*, **13**(10–11), 1019–1035.

Rives V. and Ulibarri M. (1999). Layered double hydroxides (LDH) intercalated with metal coordination compounds and oxometalates. *Coordin. Chem. Rev.*, **181**(1), 61–120.

Sannigrahi P. and Ingall E. (2005). Polyphosphates as a source of enhanced P fluxes in marine sediments overlain by anoxic waters: evidence from P-31 NMR. *Geochem. T.*, **6**(3), 52–59.

Seida Y. and Nakano Y. (2002). Removal of phosphate by layered double hydroxides containing iron. *Water Res.*, **36**(5), 1306–1312.

Sundareshwar P. V., Morris J. T., Pellechia P. J., Cohen H. J., Porter D. E. and Jones B. C. (2001). Occurrence and ecological implications of pyrophosphate in estuaries. *Limnol. Oceanogr.*, **46**(6), 1570–1577.

Toraishi T., Nagasaki S. and Tanaka S. (2002). Adsorption behavior of IO_3^- by CO_3^{2-}- and NO_3^--hydrotalcite. *Appl. Clay Sci.*, **22**(1–2), 17–23.

Torres-Dorante L. O., Claassen N., Steingrobe B. and Olfs H. W. (2006). Fertilizer-use efficiency of different inorganic polyphosphate sources: effects on soil P availability and plant P acquisition during early growth of corn. *J. Plant Nutr. Soil Sci.*, **169**(4), 509–515.

Triantafyllidis K. S., Peleka E. N., Komvokis V. G. and Mavros P. P. (2010). Iron-modified hydrotalcite-like materials as highly efficient phosphate sorbents. *J. Colloid Interface Sci.*, **342**(2), 427–436.

Vasudevan T. V., Somasundaran P., Howiemeyers C. L., Elliot D. L. and Ananthapadmanabhan K. P. (1994). Interaction of pyrophosphate with calcium phosphates. *Langmuir*, **10**(1), 320–325.

Wang S. L., Liu C. H., Wang M. K., Chuang Y. H. and Chiang P. N. (2009). Arsenate adsorption by Mg/Al-NO_3 layered double hydroxides with varying the Mg/Al ratio. *Appl. Clay Sci.*, **43**(1), 79–85.

van Wazer J. R., Griffith E. J. and McCullough J. F. (1955). Structure and properties of the condensed phosphates. VII. Hydrolytic degradation of pyro- and tripolyphosphate. *J. Am. Chem. Soc.*, **77**(2), 287–291.

van Wazer J. R., Griffith E. J. and Mccullough J. F. (1952). Hydrolysis of condensed phosphates. *J. Am. Chem. Soc.*, **74**(19), 4977–4978.

Wu Y. Y., Bai H. M., Zhou J. Z., Chen C. X., Xu X., Xu Y. F. and Qian G. R. (2009). Thermal and chemical stability of Cu-Zn-Cr-LDHs prepared by accelerated carbonation. *Appl. Clay Sci.*, **42**(3–4), 591–596.

Xing K., Wang H. Z., Guo L. G., Song W. D. and Zhao Z. P. (2008). Adsorption of tripolyphosphate from aqueous solution by Mg-Al-CO_3-layered double hydroxides. *Colloid Surf. A*, **328**(1–3), 15–20.

Xu Y. F., Dai Y. C., Zhou J. Z., Xu Z. P., Qian G. R. and Lu G. Q. M. (2010). Removal efficiency of arsenate and phosphate from aqueous solution using layered double hydroxide materials: intercalation vs. precipitation. *J. Mater. Chem.*, **20**(22), 4684–4691.

Xu Z. P. and Zeng H. C. (2001). Abrupt structural transformation in hydrotalcite-like compounds $Mg_{1-x}Al_x(OH)_2(NO_3)_x{\cdot}nH_2O$ as a continuous function of nitrate anions. *J. Phys. Chem. B*, **105**(9), 1743–1749.

Xu Z. P., Jin Y. G., Liu S. M., Hao Z. P. and Lu G. Q. (2008). Surface charging of layered double hydroxides during dynamic interactions of anions at the interfaces. *J. Colloid Interface Sci.*, **326**(2), 522–529.

Xu Z. P., Stevenson G., Lu C. Q., Lu G. Q. Max, Bartlett P. F. and Gray P. P. (2006a). Stable suspension of layered double hydroxide nanoparticles in aqueous solution. *J. Am. Chem. Soc.*, **128**, 36–37.

Xu Z. P., Stevenson G., Lu C. Q. and Lu G. Q. Max. (2006b). Dispersion and size control of layered double hydroxide nanoparticles in aqueous solutions. *J. Phys. Chem. B*, **110**, 16923–16929.

Zhang F. Z., Xiang X., Li F. and Duan X. (2008). Layered double hydroxides as catalytic materials: recent development. *Catal. Surv. Asia*, **12**(4), 253–265.

Zhou J. Z., Qian G. R., Cao Y. L., Chui P. C., Xu Y. F. and Liu Q. (2008). Transition of Friedel phase to chromate-AFm phase. *Adv. Cement Res.*, **20**(4), 167–173.

Zhou J. Z., Wu Y. Y., Liu C., Orpe A., Liu Q., Xu Z. P., Qian G. R. and Qiao S. Z. (2010). Effective self-purification of polynary metal electroplating wastewaters through formation of layered double hydroxides. *Environ. Sci. Technol.*, **44**(23), 8884–8890.

Zhou J. Z., Xu Z. P., Qiao S., Liu Q., Xu Y. and Qian G. (2011). Enhanced removal of triphosphate by MgCaFe-Cl-LDH: synergism of precipitation with intercalation and surface uptake. *J. Hazard. Mater.*, **189**(1–2), 586–594.

Zhou Y. and Carnali J. O. (2000). Solid-state hydrolysis of calcium tripolyphosphate scales. *Langmuir*, **16**(11), 5159–5168.

Chapter 7

Recycling $Mg(OH)_2$ nanoadsorbents during the removal of heavy metals from wastewater using $Cr^{(VI)}$ as an example

W. Z. Liu, Z. Lin and F. Huang

7.1 INTRODUCTION

Nowadays, heavy metals have become one of the most important pollutants in industrial and municipal wastewater (Zarazua *et al.*, 2006; Wang *et al.*, 2011). They are reported as priority pollutants due to their mobility in natural ecosystems and their toxicity (Wang *et al.*, 2003; Demirbas, 2008). Heavy metal ions are not biodegradable, and can be accumulated in living organisms, resulting in persistent environmental contaminants and severe public health issues.

Recently, there has been a great interest about the use of inorganic materials (Pitcher *et al.*, 2004; Bhattacharyya & Gupta, 2008; Granados-Correa & Jimenez-Becerril, 2009), especially inorganic nanomaterials (Shin & Jang, 2007; Ruparelia *et al.*, 2008; Huang & Chen, 2009), as adsorbents to remove heavy metals from polluted wastewater due to their high effectiveness and low costs. Several studies have shown that inorganic particles can be potentially used in the removal of heavy metal ions as they have high exchange capacity, selectivity and fairly good radiation/temperature stability (Mishra *et al.*, 2007).

Using the typical heavy metal, $Cr^{(VI)}$, as an example, a recycling strategy of magnesium hydroxide [$Mg(OH)_2$] nanoadsorbent during treatment of heavy metal-containing wastewater was described (Liu *et al.*, 2011). Moreover, the practical applications of this carbonation strategy of $Mg(OH)_2$ in the treatment of heavy metal-containing $Mg(OH)_2$ nanowaste from traditional industry was explored. On the basis of research results, this chapter proposes the potential applications of this strategy for the treatment of other heavy metals by using Mg $(OH)_2$ nano-particles (NPs).

7.1.1 Environmental applications of $Mg(OH)_2$

In recent years, $Mg(OH)_2$ as a non-toxic and environmentally-friendly material, is of growing interest for environmental protection. The largest end-use for Mg $(OH)_2$ is in wastewater treatment and flue gas desulfurization (Schlag *et al.*, 2007). It has been widely used in the neutralization of acidic wastewater (Wu, 2002), removal of heavy metals (Moore & Anderson, 2007), colour removal from dye-containing wastewater (Lazaridis *et al.*, 2003), removal of phosphate and ammonium from municipal wastewater (Shin & Lee, 1998) and flue gas desulfurization (Sada *et al.*, 1983; Mondal, 2007).

$Mg(OH)_2$ has several main advantages in wastewater treatment: (i) it is more effective than other commonly used chemicals for pH adjustment of wastewater or sludge, because of its neutralization capacity and buffering intensity (Wu, 2002), (ii) because of its high activity and absorption capacity, $Mg(OH)_2$ can be used as adsorbent for removal of heavy metals, dye and nutrients from wastewater and (iii) $Mg(OH)_2$ is low-cost, non-corrosive, safe and easy to handle, making it an eco-friendly water treatment agent.

In addition to its capacity of removing heavy metal cations from wastewater by co-precipitation, $Mg(OH)_2$ has also the potential for the removal of highly toxic anionic pollutants, such as hexavalent chromium and arsenic. Generally, the $Cr^{(VI)}$ species may exist in the forms of dichromate ($Cr_2O_7^{2-}$), hydrogen chromate ($HCrO_4^-$) or chromate (CrO_4^{2-}) in aqueous solutions with different pH values (Li *et al.*, 2009). Furthermore, the most common species of arsenic in water are arsenate (AsO_4^{3-}) and arsenite (AsO_3^{3-}), leading to the well-known arsenicosis. Due to the positively charged surface of $Mg(OH)_2$ particles, these toxic anions can be firmly adsorbed onto $Mg(OH)_2$.

Over the past decade, the global consumption of $Mg(OH)_2$ in wastewater treatment exceeded 48,000 tons per year (Schlag *et al.*, 2007), which created a great amount of $Mg(OH)_2$ sludge. On the other hand, it has been recently recognized that actually some traditional industrial sludges are nanowastes with adsorbed heavy metals or organic pollutants (Cheremisinoff, 2003; Kent, 2003). For example, over 800 kilotons of nano-$Mg(OH)_2$-containing sludge are generated from chlor-alkali and chlorate plants per year in China (Viswanathan & Tilak, 1984; Li *et al.*, 2007). In order to solve these issues, this chapter reports some investigations related to the recycling of $Mg(OH)_2$ adsorbent and disposal of this hazardous pollutant-adsorbed $Mg(OH)_2$ nanowaste.

7.1.2 The risks of $Cr^{(VI)}$ and treatment of $Cr^{(VI)}$-containing wastewater

7.1.2.1 *Cr removal technologies*

Chromium is widely used in electroplating, steel fabrication, leather tanning, textile dyeing, paint and pigment industries (Kimbrough *et al.*, 1999). Therefore, large quantities of chromium are discharged into the environment. Chromium is usually present in the aqueous environment as $Cr^{(III)}$ or $Cr^{(VI)}$. $Cr^{(VI)}$ is much more toxic than $Cr^{(III)}$ due to its high water solubility and mobility. $Cr^{(VI)}$ is considered as an acute toxic substance (Cieslak-Golonka, 1996), which is strongly epithelial irritant and carcinogenic (Sedman *et al.*, 2006). It can easily penetrate the cell wall and exert its noxious influence to the cell. For the short-term exposure above the maximum contaminant level, $Cr^{(VI)}$ causes skin and stomach irritation or ulceration. For the long-term exposure above the maximum contaminant level, it causes dermatitis, liver damage, kidney circulation, nerve tissue damage and death in large doses (Owlad *et al.*, 2009).

Therefore, $Cr^{(VI)}$ is of particular concern, and it is necessary to remove $Cr^{(VI)}$ from industrial wastewater. Various techniques have been applied for the removal of $Cr^{(VI)}$, including reduction and precipitation (Qin *et al.*, 2005; Yang *et al.*, 2007), electrolysis (Vlyssides & Israilides, 1997; Korzenowski *et al.*, 2008), ion exchange (Rengaraj *et al.*, 2001; Galan *et al.*, 2005), membrane filtration (Pagana *et al.*, 2008), adsorption (Valix *et al.*, 2006; Alvarez-Ayuso *et al.*, 2007; Li *et al.*, 2009) and electrochemical precipitation (Golder *et al.*, 2011).

Among these techniques, the reduction and precipitation treatment is the most mature and applicable method, and is considered to be the most economical technique for wastewaters containing chromium at concentrations ranging from tens to hundreds of mg/L. When $Cr^{(VI)}$ is at low concentrations, for example less than 10 mg/L, plenty of reductants are needed, resulting in a great amount of sludge that needs further treatment.

The other methods, as ion exchange, electrolysis and membrane filtration, although usually effective, when treating only small volumes of wastewater require a high capital investment and running cost. For these reasons, adsorption (Alvarez-Ayuso *et al.*, 2007; Qu, 2008) becomes the most promising technique and a feasible alternative for $Cr^{(VI)}$ removal from contaminated water or wastewater, offering significant advantages, such as low cost, availability, profitability, ease of operation and efficiency, especially when dealing with low concentration of $Cr^{(VI)}$.

7.1.2.2 Cr adsorption technologies

Granular or powdered activated carbons (Dias *et al.*, 2007) are the most widely used adsorbents. However, these absorbents are not efficient in removing metallic substances unless they are chemically activated using mineral acids or transition metal salts. In this case, their use is usually limited due to the high cost (Wang *et al.*, 2003). Therefore, it is necessary to develop cost-effective alternative techniques or materials for removal of $Cr^{(VI)}$ from diluted solutions.

In recent years, a number of studies have been reported using different low-cost adsorbents for the removal of $Cr^{(VI)}$, including biomass (Baral *et al.*, 2006; Bhattacharya *et al.*, 2008; Vaghetti *et al.*, 2008) or biomaterial-based activated carbon (Babel & Kurniawan, 2004; Mohan & Pittman, 2006), activated clay

(Dursun *et al.*, 2008; Weng *et al.*, 2008), minerals such as boehmite (Granados-Correa & Jimenez-Becerril, 2009), zeolite (Basaldella *et al.*, 2007; Kocaoba *et al.*, 2007), iron oxides (Deng *et al.*, 1996), hydrotalcite-like compounds (Alvarez-Ayuso & Nugteren, 2005; Li *et al.*, 2009) and composite materials (Lazaridis & Charalambous, 2005; Qu, 2008).

The application of nanoadsorbents has emerged as a fast-developing and fascinating area of interest for treatment of pollutants (Sharma *et al.*, 2009). Because of their small size, they possess a large surface area and a great number of active sites, which results in high adsorption capacity and reactivity to pollutants (Aarthi & Madras, 2008; Zhang *et al.*, 2008). In comparison with commonly used adsorbents, the nanoadsorbents make the purification system for low concentrations of heavy metal ions to be more economically viable, especially if low-cost nanoadsorbents are used.

However, the nanoadsorbents become harmful substances after adsorbing pollutants. Currently, most efforts proposing the use of nanoadsorbents are still considering burial or storage of the nanoadsorbents after sequestration of contaminants (Hu *et al.*, 2008). Though the goal is to permanently immobilize the contaminants in the solid phase (Kanel *et al.*, 2005; Kugbe *et al.*, 2009), the actual stability of the resulting materials is poorly known. A very high possibility is that the pollutants can eventually be leached from the sludge into the groundwater again (Gilbert *et al.*, 2009). Therefore, for large-scale application of nanoadsorbents in wastewater treatment, the recycling of nanoadsorbents should be considered.

7.1.3 Recycling nanoadsorbents and the pre-concentration of heavy metals

To date, studies related to the detoxification or regeneration of nanoadsorbents are scarce. It has been reported that the adsorbed contaminants can be efficiently released from the surface of the nano-materials via adjusting the surface ions and pH (Banerjee & Chen, 2007; Shin & Jang, 2007). Therefore, the nanoadsorbents may be recycled.

It is well known that the enrichment of pollutants in water, like heavy metals, is the most important criterion to judge whether nanoadsorbents are successfully recycled. Especially for low (less than 10 mg/L) or very low (at μg/L level) concentrations of heavy metals, the pre-concentration is crucial for the determination and recovery of heavy metals (Baysal *et al.*, 2009; Kaur & Gupta, 2009). Due to the limitation of sensitivity, it is difficult to directly determine the toxic heavy metals at extremely low concentrations (less than 1 μg/L) by modern atomic spectroscopic methods, such as atomic absorption spectrometry (AAS) and inductively coupled plasma atomic emission spectrometry/mass spectrometry (ICP-AES/MS). Consequently, the pre-concentration of trace heavy metal ions from water is often required (Kaur & Gupta, 2009; Liang *et al.*, 2004). On the other hand, the pre-concentration is beneficial to the recovery and reuse of heavy metals, providing significant reductions in cost, time and energy to the industry. Moreover, the recovery of heavy metals can conserve natural resources and avoid environmental hazards, which is the ultimate goal of the treatment of heavy metals.

The pre-concentration of heavy metals by nanoadsorbents has the following requirements: (i) the nanoadsorbent itself should be non-toxic, moreover, it should have a high binding affinity and relatively high selectivity to the heavy metals at low concentration; (ii) the re-usage cycles of the nanoadsorbent should be as many as possible; (iii) the adsorbates can be removed from the surface of nanoadsorbent, and can be concentrated for further treatment or purification for reuse; (iv) for the convenience of the post-treatment of the adsorbates, the introduced chemical substance should be environmentally friendly and easy in handling. Ideally, in order to avoid creating other chemical byproducts during the treatment, the recycling loop should have no requirement for consumptive chemical substances.

In view of the above, treatment of wastewater with low concentrations of heavy metal ions by adsorbing the metal pollutants onto the low cost, non-toxic and effective nanoadsorbents combined with appropriate regeneration steps and pre-concentration of the pollutants is an ideal method. This solves the problem of sludge disposal and enriches the heavy metals for further recovery and reuse in the plant.

Another issue for the recycled reuse of nanoadsorbents is that the initial surface state and morphology of the nanoadsorbents may gradually disappear after several cycles due to the inherent thermodynamic instability of the nanophase (Gilbert *et al.*, 2009). For example, nano-particles (NPs) can grow or aggregate. This issue can be effectively resolved by creating a reversible transformation between the nanoadsorbent and the bulk material.

In this chapter, $Cr^{(VI)}$ is used as a typical example of a heavy metal and a reversible route between nano-$Mg(OH)_2$ and bulk-$MgCO_3{\cdot}3H_2O$ is described, which simultaneously realizes both recycling $Mg(OH)_2$ nanoadsorbent and upconcentrating of diluted $Cr^{(VI)}$ from wastewater.

7.2 RECYCLING $Mg(OH)_2$ NANO ADSORBENT DURING TREATMENT OF LOW CONCENTRATION OF $Cr^{(VI)}$

Scheme in Figure 7.1 shows the strategy of recycling $Mg(OH)_2$ nanoadsorbent, including three main steps as follows:

Step I: Adsorbing low concentration of $Cr^{(VI)}$ onto $Mg(OH)_2$ NPs.

Step II: Transforming $Cr^{(VI)}$-loaded $Mg(OH)_2$ NPs into $Cr^{(VI)}$-desorbed $MgCO_3{\cdot}3H_2O$ bulk material using CO_2 as the phase transformation medium.

Step III: Turning the $MgCO_3{\cdot}3H_2O$ bulk material back into the same $Mg(OH)_2$ NPs via calcination and hydration treatment.

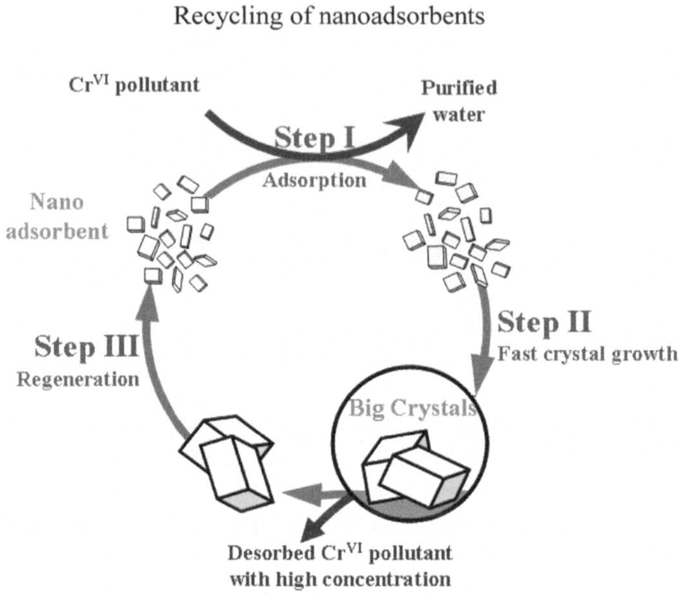

Figure 7.1 Strategy of recycling $Mg(OH)_2$ nanoadsorbent during the treatment of wastewater with low concentrations of $Cr^{(VI)}$ (Liu *et al.*, 2011).

7.2.1 Adsorption of $Cr^{(VI)}$ onto $Mg(OH)_2$ nanoparticles

$Mg(OH)_2$ NPs were synthesized by directly placing the heated MgO into $Cr^{(VI)}$-containing wastewater (Liu *et al.*, 2011), which combined "the preparation of $Mg(OH)_2$ NPs" and "the adsorption of $Cr^{(VI)}$" into one step. Briefly, 20 g of MgO was heated at 600°C for 1 or 2 hours, and then it was immediately quenched in 5 L of simulated $Cr^{(VI)}$-containing wastewater. The X-ray powder diffraction (XRD) patterns of raw MgO and $Mg(OH)_2$ products (Figure 7.2a) showed that most of the MgO could be completely converted to hexagonal Mg $(OH)_2$. Scanning electron microscope (SEM) images of raw MgO (Figure 7.2b) and $Mg(OH)_2$ products (Figure 7.2d) displayed that the size of granular MgO was about 100 to 200 nm, and the formed $Mg(OH)_2$ was hexagonal platelet with a thickness ranging from 10 to 20 nm and a width ranging from 50 to 500 nm.

Figure 7.2c shows that the adsorption equilibrium was obtained after 3 hours, and over 90% of $Cr^{(VI)}$ was removed. It indicated that the formed $Mg(OH)_2$ NPs had a good adsorption efficiency to the anionic $Cr^{(VI)}$. Further studies on the adsorption kinetics and thermodynamics of $Cr^{(VI)}$ onto the $Mg(OH)_2$ nanoadsorbent showed that the adsorption isotherm could be well described by the Langmuir model with a maximum adsorption capacity of 7.43 mg/g and equilibrium constant of 1.032 L/mg (unpublished data). These parameters reflected that the $Mg(OH)_2$ nanoadsorbent had a great affinity for $Cr^{(VI)}$ and was suitable for adsorption of low concentrations of $Cr^{(VI)}$. Furthermore, the adsorption enthalpy ($\Delta_{ad}H^{\theta} = 114.57$ kJ/mol) indicated that the adsorption process was chemisorption, and the adsorption could be improved by increasing the temperature (unpublished data).

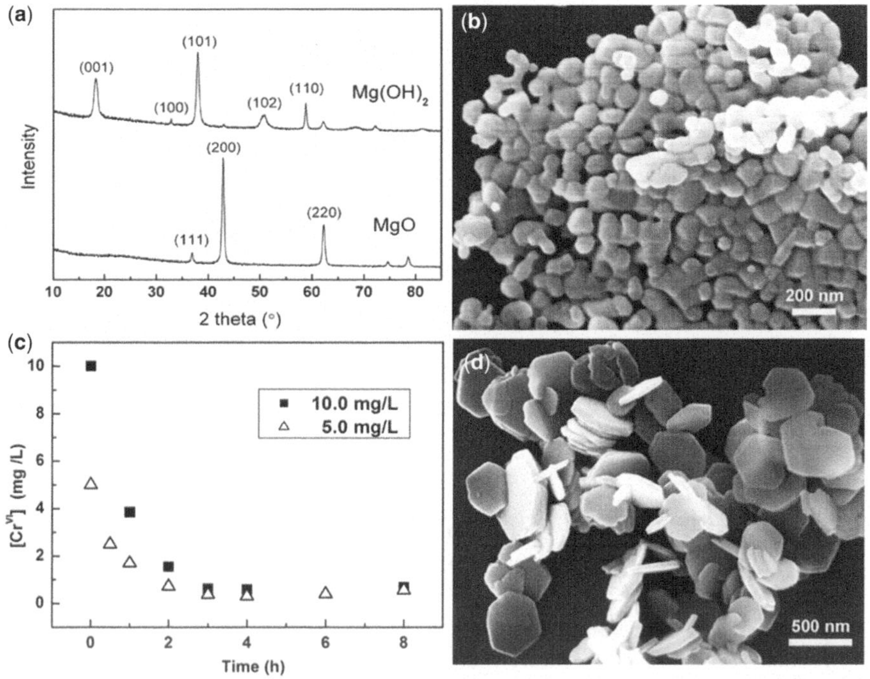

Figure 7.2 (a) XRD patterns of raw MgO and $Mg(OH)_2$ products, (b) SEM image of raw MgO, (c) Adsorption curves of $Cr^{(VI)}$ with initial concentrations of 5.0 and 10.0 mg/L (the dosage of MgO was 4.0 g/L), (d) SEM image of formed $Mg(OH)_2$ NPs.

7.2.2 Desorption and enrichment of $Cr^{(VI)}$

After adsorption, the $Cr^{(VI)}$-loaded $Mg(OH)_2$ NPs were filtered out for further carbonation treatment, which changed nano-$Mg(OH)_2$ into bulk-$MgCO_3{\cdot}3H_2O$ and facilitated the desorption of $Cr^{(VI)}$ from the solid surface.

XRD analysis revealed that $Mg(OH)_2$ was completely changed into $MgCO_3{\cdot}3H_2O$ crystals after exposure to 0.5 MPa of CO_2 for 12 hours (Figure 7.3a). SEM observations revealed that the produced $MgCO_3{\cdot}3H_2O$ crystal exhibited rod-like morphology, with an average width of 2–3 μm and length of 10–20 μm. Brunauer–Emmett–Teller (BET) measurements revealed that the specific surface area of $MgCO_3{\cdot}3H_2O$ was 18.6 m^2/g. As a result, over 90% of $Cr^{(VI)}$ was desorbed during the transformation process from nano-$Mg(OH)_2$ to bulk $MgCO_3{\cdot}3H_2O$. It is worth mentioning that the concentrated $Cr^{(VI)}$ supernatant could be easily extracted, since the size increase of magnesium compounds enhanced the settlement of the solids. Data showed that the $Cr^{(VI)}$ could be enriched over 50 times from step I to step II (Liu *et al.*, 2011).

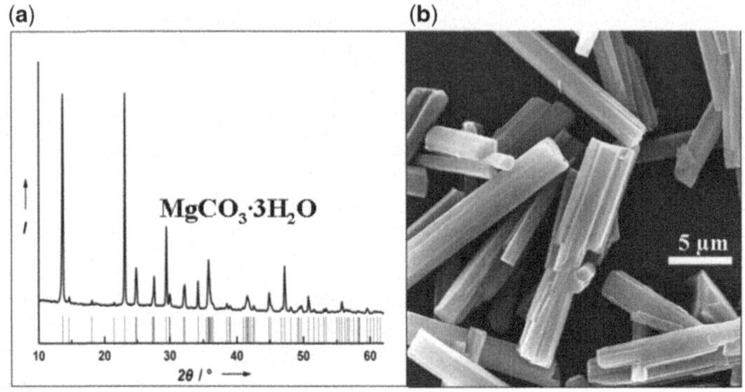

Figure 7.3 (a) XRD pattern of $MgCO_3 \cdot 3H_2O$, and (b) SEM image of $MgCO_3 \cdot 3H_2O$ generated from $Mg(OH)_2$ by carbonating at room temperature for 12 hours under 0.5 MPa of CO_2.

7.2.3 Recycle loop of the $Mg(OH)_2$ nanoadsorbent

Theoretically, the $Mg(OH)_2$ nanoadsorbent can be regenerated to adsorb $Cr^{(VI)}$ by calcining the desorbed $MgCO_3 \cdot 3H_2O$ crystals and then quenching the MgO product into $Cr^{(VI)}$-containing wastewater as described in Step I. Figure 7.4 shows that the recycled $Mg(OH)_2$ could still maintain the initial phase, size, morphology and $Cr^{(VI)}$ adsorption capacity during the repeating cycles. The SEM image of calcined $MgCO_3 \cdot 3H_2O$ (Figure 7.4a) revealed that the size and morphology of the formed MgO were similar to those of $MgCO_3 \cdot 3H_2O$. However, it had no effect on the size and morphology of the recycled $Mg(OH)_2$ (Figure 7.4b), suggesting that the hexagonal $Mg(OH)_2$ nanoplatelet was the most stable under this condition. Figure 7.4d shows that the $Cr^{(VI)}$ removal efficiency of the $Mg(OH)_2$ nanoadsorbent was maintained at a high level, and the enrichment of $Cr^{(VI)}$ was sustained over 40 times in each of the five cycles. These results implied that a stable recycling efficiency of $Mg(OH)_2$ nanoadsorbent could be achieved.

From the environmental point of view, the carbonation strategy for recycling Mg $(OH)_2$ nanoadsorbent was eco-friendly, which did not require any other chemical reagents and generate any chemical byproducts. Therefore, CO_2 actually was the reusable medium in the cycle for developing this $Cr^{(VI)}$ concentration system with zero reagent consumption. The major advantages of this method are low-cost, high efficiency, regeneration of nanoadsorbent, zero reagent consumption, and possible heavy metal recovery.

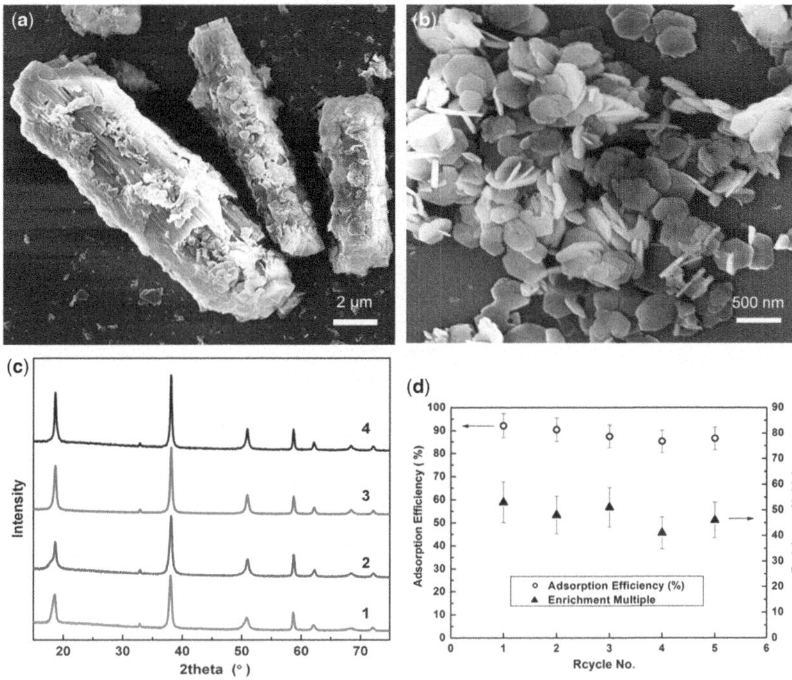

Figure 7.4 (a) SEM image of the calcined $MgCO_3{\cdot}3H_2O$, (b) SEM image of the recycled $Mg(OH)_2$ nanoadsorbent, (c) XRD patterns of $Mg(OH)_2$ samples in the repeated cycles and (d) $Cr^{(VI)}$ removal efficiency and enrichment multiple of $Cr^{(VI)}$ in each of the five continuous cycles. The initial concentration of $Cr^{(VI)}$ in wastewater was 5.0 mg/L, and the dosage of $Mg(OH)_2$ was 5.8 g/L.

The amount of consumed CO_2 during the desorption process (step II) was exactly equal to the amount of CO_2 released during the regeneration process (step III). As shown in Figure 7.5a, the $Cr^{(VI)}$ solution could be greatly enriched during the reversible cycle between Cr-adsorbed nano-$Mg(OH)_2$ and Cr-desorbed bulk-$MgCO_3{\cdot}3H_2O$ by using CO_2 as a phase transformation medium. As summarized in Figure 7.5b, if the nano-$Mg(OH)_2$ and medium CO_2 were combined together as a "black box", where the input substance was low concentration wastewater, the outputs were the purified water and enriched $Cr^{(VI)}$ solution (Figure 7.5). No extra chemical agents were required or released.

This carbonation strategy by using CO_2 is also developed for the treatment of $Cr^{(VI)}$-containing $Mg(OH)_2$ nanowaste from traditional industry. $NaHCO_3$ is an effective mineralizer for rapidly increasing the grain size of magnesium compounds. Therefore, the $Cr^{(VI)}$ can be successfully desorbed from the solid waste for further recovery (Liu & Huang *et al.*, 2008). This mineralizer may be not ideal for treating low concentrations of $Cr^{(VI)}$ in wastewater since it can leave chemical byproducts in the system. However, it has the advantage of dealing with the $Cr^{(VI)}$-adsorbed $Mg(OH)_2$ nanowaste, especially when the solid waste is treated at high temperature.

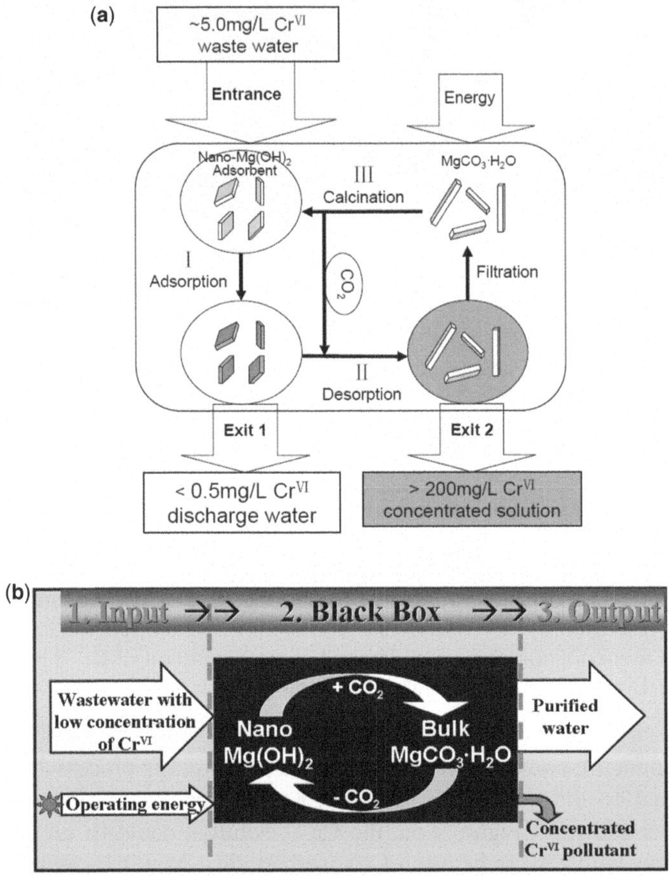

Figure 7.5 (a) Summary of $Cr^{(VI)}$ concentration system without any consumption of chemical reagent and (b) Schematic diagram of Figure 7.5a.

7.3 TREATMENT OF $Cr^{(VI)}$-CONTAINING $Mg(OH)_2$ NANOWASTE

7.3.1 $Cr^{(VI)}$-containing $Mg(OH)_2$ nanowaste generated by the chlorate industry

As mentioned above, a large amount of nano-$Mg(OH)_2$-containing sludge is generated from the sodium and potassium chlorate industries (Li *et al.*, 2007).

The $Mg(OH)_2$ nanowaste has such a high surface-adsorption affinity for the toxic $Cr^{(VI)}$. Therefore, it is difficult to remove the $Cr^{(VI)}$ ions from the solid waste through the traditional way, such as washing, filtration and centrifugation.

The typical $Cr^{(VI)}$-containing nanowaste from the chlorate industry is composed of about 50% water and 2,048 mg/kg of $Cr^{(VI)}$, and the main solid composition is 20-nm $Mg(OH)_2$ and 100-nm $CaCO_3$. The pH of the waste is around 9. If the $Cr^{(VI)}$-adsorbed nanowaste is stored directly, the adsorbed $Cr^{(VI)}$ ion gradually releases and penetrates into the soil and groundwater.

The conventional treatment for $Cr^{(VI)}$-containing sludge is landfilling after reduction (Cheremisinoff, 2003; Su & Ludwig, 2005) (Figure 7.6), leading to secondary pollution. When the $Cr^{(III)}$ is exposed to air, it can be oxidized into $Cr^{(VI)}$ and contaminate the environment again. Therefore, it is necessary to develop an effective and feasible disposal for this $Cr^{(VI)}$-adsorbed $Mg(OH)_2$ nanowaste.

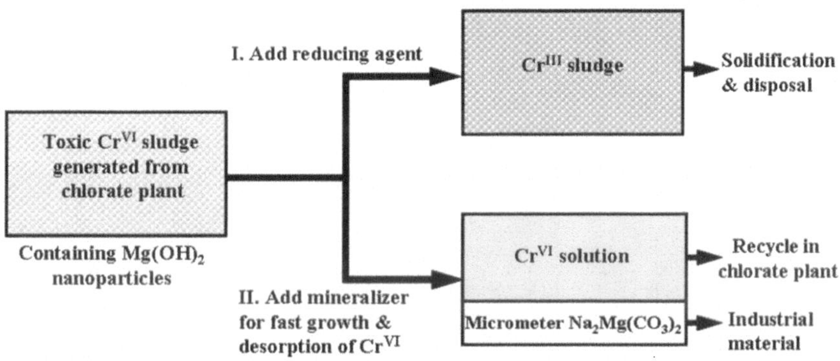

Figure 7.6 Disposal of $Cr^{(VI)}$-containing sludge (I: The conventional method, II: the new approach proposed).

A promising strategy for disposing the above-mentioned nanowaste was developed through transforming $Mg(OH)_2$ NPs into bulk materials by using $NaHCO_3$, which is a mineralizer and can release $Cr^{(VI)}$ into solution (Liu *et al.*, 2008). This $Cr^{(VI)}$-containing solution could subsequently be recycled in the chlorate process, and the detoxified solids could potentially be reutilized as additives in other applications, such as in ceramics, paint, metal coating or flame-retardant engineering plastics.

7.3.2 Treatment of $Mg(OH)_2$ nanowaste by mineralizer

Figure 7.7 shows that the original nanowaste contained a lot of $Mg(OH)_2$ nanoplatelets, which were difficult to settle in the solution (Fig. 7.7a, d). Hydrothermal coarsening in pure water did not appreciably alter the size and settling velocity of these $Mg(OH)_2$ nanoparticals (Figure 7.7b, e), indicating that the $Mg(OH)_2$ phase could remain stable on a nanoscale for a long time in the sludge under naturally occurring conditions. The most appropriate mineralizer was found to consist of 0.5 M Na_2CO_3 and 1.5 M $NaHCO_3$ (1:10, w/v; hereinafter referred to as "mineralizer A") (Liu *et al.*, 2008). Once mineralizer A ($NaHCO_3 + Na_2CO_3$) was introduced, the solids were settled to the bottom of the breaker in only 20 min. A subsequent SEM observation revealed that the NPs were already changed into micrometer-sized $Na_2Mg(CO_3)_2$ crystals (Figure 7.7c, f). Moreover, the leaching concentration of $Cr^{(VI)}$ in the third treated sample was only 0.052 mg/L, which was far below the standard of environmental protection technical specifications for treatment of chromium residues in China (HJ/T 301-2007, 0.5 mg/L).

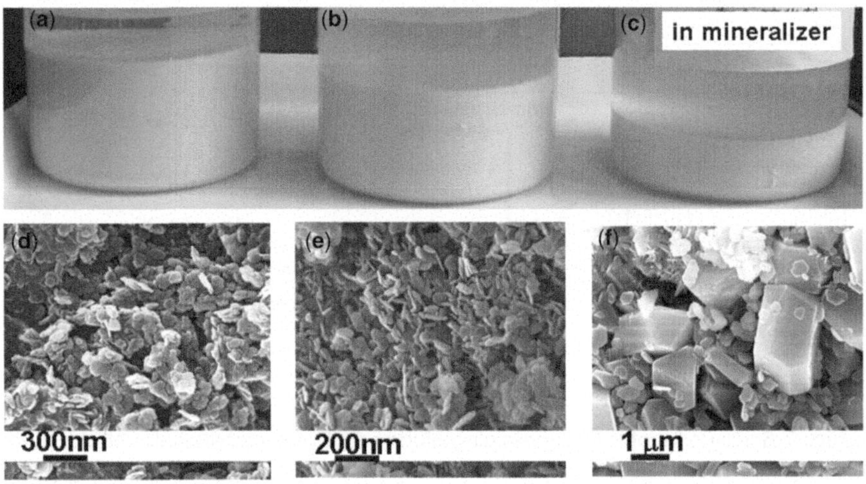

Figure 7.7 Settling test and SEM images of the samples: (a) and (d) original sludge; (b) and (e) sample coarsened in distilled water at 180°C for 24 hours; (c) and (f) sample coarsened in mineralizer A ($NaHCO_3 + Na_2CO_3$) at 180°C for 24 hours. The solid/liquid ratio used in the settle ability test was 1 g/10 mL, with a settling time of 3 hours.

The phase transformation of $Mg(OH)_2$ under the effect of mineralizer A at different temperatures was systematically investigated (Liu *et al.*, 2008). Figure 7.8 clearly shows that there was an intermediate phase $Mg_5(CO_3)_4(OH)_2{\cdot}3H_2O$ between the initial $Mg(OH)_2$ and final product $Na_2Mg(CO_3)_2$. The $Na_2Mg(CO_3)_2$ phase appeared within 1 hour during coarsening when the temperature exceeded 80°C. The $Mg(OH)_2$ phase was completely changed into $Na_2Mg(CO_3)_2$ when it was coarsened with mineralizer A at 180°C for 3 hours.

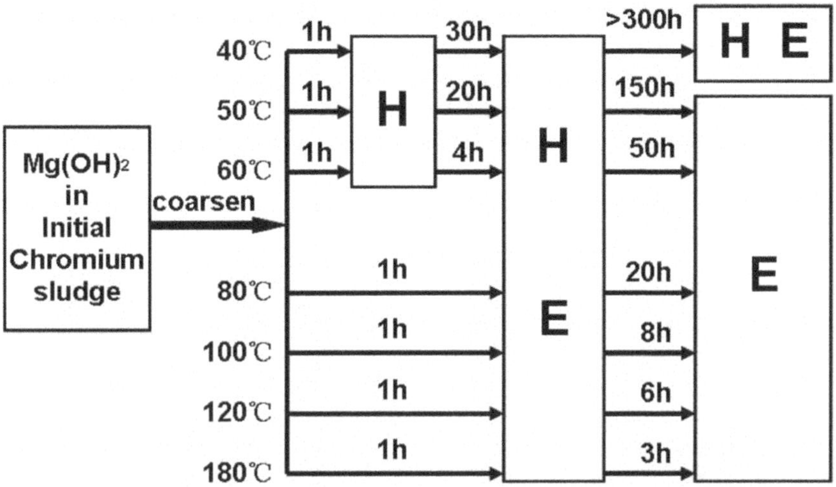

Figure 7.8 The main phase transformation of $Mg(OH)_2$ under the effect of mineralizer A at different temperatures. H: Hydromagnesite, $Mg_5(CO_3)_4(OH)_2{\cdot}3H_2O$; E: Eitelite, $Na_2Mg(CO_3)_2$.

SEM observations revealed that the $Na_2Mg(CO_3)_2$ was well crystallized with a crystal size about hundreds of times greater than that of $Mg(OH)_2$ (Fig. 7.9a), suggesting an extremely fast growth rate of $Na_2Mg(CO_3)_2$. Furthermore, the $Mg_5(CO_3)_4(OH)_2{\cdot}3H_2O$ microspheres were formed by the aggregation of nanolamellas. Further studies of the phase transformation and crystal growth from nano-$Mg(OH)_2$ to bulk-$Na_2Mg(CO_3)_2$ proposed a new mechanism of crystal growth, which was a "collective phase transformation mechanism" (Liu *et al.*, 2008).

Figure 7.9 SEM images of the sample coarsened in mineralizer A at 60°C for 6 hours: (a) an intermediate state containing both $Mg_5(CO_3)_4(OH)_2{\cdot}3H_2O$ microspheres and micrometer-sized $Na_2Mg(CO_3)_2$ crystals, (b) a typical $Na_2Mg(CO_3)_2$ microcrystal ("b" in Fig. 7.9a) and (c) a typical $Mg_5(CO_3)_4(OH)_2{\cdot}3H_2O$ microsphere ("c" in Figure 7.9a).

Figure 7.10 summarizes the mechanism of $Cr^{(VI)}$ removal during coarsening Mg $(OH)_2$ nanowaste by mineralizer A. The entire process included two steps: (i) desorption of $Cr^{(VI)}$ from the solid surface due to high degree of aggregation of $Mg_5(CO_3)_4(OH)_2{\cdot}3H_2O$ nanolamellas, and (ii) efficient separation of the $Cr^{(VI)}$-containing liquid from the solid due to fast crystal growth.

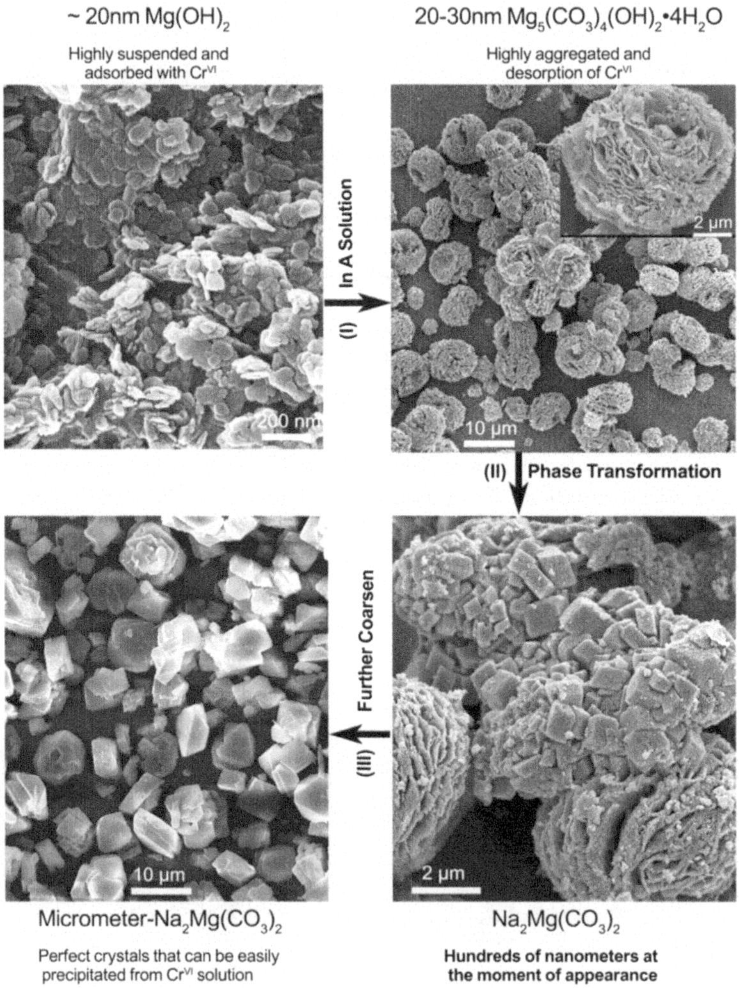

Figure 7.10 The mechanism of $Cr^{(VI)}$ removal during treatment of $Mg(OH)_2$ sludge by mineralizer A.

This waste-disposal strategy was proven to be successful in dealing with $Cr^{(VI)}$-containing $Mg(OH)_2$ nanowaste, because almost all of the $Cr^{(VI)}$ was removed from the solid waste after transforming nano-$Mg(OH)_2$ into bulk materials, which facilitated the reuse of $Cr^{(VI)}$ in the plant and reutilization of the detoxified solid product.

7.3.3 Pilotscale treatment of nanowaste in chlorate plant

Table 7.1 shows the pilot-scale treatment of $Cr^{(VI)}$-containing $Mg(OH)_2$ nanowaste of a chlorate plant. Briefly, 1 ton of initial nanowaste was mixed with 2 tons of $NaHCO_3$ solution (2.3 ~ 2.5 M). The mixture was stirred in a tank, and then transferred into a 5000-L autoclave and incubated at 120°C for 4 hours. Subsequently, after the autoclave was cooled down to room temperature, the supernatant was stored in a tank for recycling, the solid product was washed and filtered for 3 ~ 5 times, and then it was dried for later reutilization. The wash water could be treated for reuse in the plant.

Table 7.1 Ton-scale treatment of $Cr^{(VI)}$-adsorbed $Mg(OH)_2$ nanowaste in chlorate plant.

Batch No.	1	2	3
Water content of sludge (%)	48.5	51.3	47.3
$Cr^{(VI)}$ in initial sludge (mg/kg)	2,013	2,354	2,200
$Cr^{(VI)}$ released in solution (mg/L)	393.4	462.8	437.0
Residual Cr in solid product (mg/kg)	52.12	73.70	60.92
Removal efficiency (%)	97.41	96.87	97.23
Leaching $Cr^{(VI)}$ from solid product (mg/L)	0.26	0.48	0.39

More than 97% of $Cr^{(VI)}$ was effectively removed from the solid, and the $Cr^{(VI)}$ leaching from the solid products was less than 0.5 mg/L. At the end of the process, the supernatant liquid was mainly composed of NaCl, $KClO_3$, Na_2CO_3 and excess $NaHCO_3$. This $Cr^{(VI)}$-containing supernatant liquid could be reused in the chlorate production process, therefore, the $KClO_3$ could also be recovered.

The reutilization of the detoxified solid product has also been studied (Liu & Xu *et al.*, 2010), and the use of the detoxified solid product is developed in ceramics, metal coating and flame-retardant engineering plastics.

7.4 CONCLUSIONS

A reversible route between Cr-adsorbed nano-$Mg(OH)_2$ and Cr-desorbed bulk-$MgCO_3{\cdot}3H_2O$ by using CO_2 as a phase transformation medium was established. In the adsorption-desorption recycle, the low concentration of $Cr^{(VI)}$ in wastewater was enriched, which was beneficial to further recovery and reuse of $Cr^{(VI)}$. Furthermore, a carbonation strategy was used for the treatment of

$Cr^{(VI)}$-containing $Mg(OH)_2$ nanowaste. In addition, the $Cr^{(VI)}$-containing nanowaste was detoxified through altering NPs into bulk materials by mineralizer, whereupon the $Cr^{(VI)}$ was desorbed from the solid surface. This method facilitated the recovery of $Cr^{(VI)}$in the plant and reutilization of the detoxified solid product. The major advantages of this strategy were low-cost, easy operation, high efficiency, minimization of residue and possible heavy metal recovery.

$Mg(OH)_2$ can remove the anionic pollutants by adsorbing the toxic anions onto its positively charged surface. Therefore, besides $Cr^{(VI)}$, this recycling strategy could also be potentially applied for the enrichment and recovery of other highly toxic anionic pollutants, such as F^-, AsO_4^{3-}, AsO_3^{3-}, SeO_3^{2-}, $[UO_2(CO_3)_2]^{2-}$ and so on. On the other hand, $Mg(OH)_2$ is usually used to remove heavy metals by co-precipitation, such as Hg, Pb and Cd. Therefore, it was possible to recover these heavy metals by recycling $Mg(OH)_2$ according to the different solubility of these metal hydroxides or carbonates. Further work still needs to be done to test this new approach for the recycling of heavy metal cations.

REFERENCES

Aarthi T. and Madras G. (2008). Photocatalytic reduction of metals in presence of combustion synthesized nano-TiO_2. *Catal. Commun.*, **9**(5), 630–634.

Alvarez-Ayuso E. and Nugteren H. W. (2005). Purification of chromium(VI) finishing wastewaters using calcined and uncalcined Mg-Al-CO_3-hydrotalcite. *Water Res.*, **39** (12), 2535–2542.

Alvarez-Ayuso E., Garcia-Sanchez A. and Querol X. (2007). Adsorption of Cr(VI) from synthetic solutions and electroplating wastewaters on amorphous aluminium oxide. *J. Hazard. Mater.*, **142**(1–2), 191–198.

Babel S. and Kurniawan T. A. (2004). Cr(VI) removal from synthetic wastewater using coconut shell charcoal and commercial activated carbon modified with oxidizing agents and/or chitosan. *Chemosphere*, **54**(7), 951–967.

Banerjee S. S. and Chen D. H. (2007). Fast removal of copper ions by gum arabic modified magnetic nano-adsorbent. *J. Hazard. Mater.*, **147**(3), 792–799.

Baral S. S., Das S. N. and Rath P. (2006). Hexavalent chromium removal from aqueous solution by adsorption on treated sawdust. *Biochem. Eng. J.*, **31**(3), 216–222.

Basaldella E. I., Vazquez P. G., Iucolano F. and Caputo D. (2007). Chromium removal from water using LTA zeolites: effect of pH. *J. Colloid Interface Sci.*, **313**(2), 574–578.

Baysal A., Kahraman M. and Akman S. (2009). The solid phase extraction of lead using silver nanoparticles – attached to silica gel prior to its determination by FAAS. *Curr. Anal. Chem.*, **5**(4), 352–357.

Bhattacharya A. K., Naiya T. K., Mandal S. N. and Das S. K. (2008). Adsorption, kinetics and equilibrium studies on removal of Cr(VI) from aqueous solutions using different low-cost adsorbents. *Chem. Eng. J.*, **137**(3), 529–541.

Bhattacharyya K. G. and Gupta S. S. (2008). Adsorption of a few heavy metals on natural and modified kaolinite and montmorillonite: a review. *Adv. Colloid Interface Sci.*, **140**(2), 114–131.

Cheremisinoff N. P. (2003). *Handbook of Solid Waste Management and Waste Minimization Technologies*. Elsevier, Burlington.

Cieslak-Golonka M. (1996). Toxic and mutagenic effects of chromium(VI). A review. *Polyhedron*, **15**(21), 3667–3689.

Demirbas A. (2008). Heavy metal adsorption onto agro-based waste materials: a review. *J. Hazard. Mater.*, **157**(2–3), 220–229.

Deng Y. W., Stjernstrom M. and Banwart S. (1996). Accumulation and remobilization of aqueous chromium(VI) at iron oxide surfaces: application of a thin-film continuous flow-through reactor. *J. Contam. Hydrol.*, **21**(1–4), 141–151.

Dias J. M., Alvim-Ferraz M., Almeida M. F., Rivera-Utrilla J. and Sanchez-Polo M. (2007). Waste materials for activated carbon preparation and its use in aqueous-phase treatment: a review. *J. Environ. Manage.*, **85**(4), 833–846.

Dursun S., Guclu D., Berktay A. and Guner T. (2008). Removal of chromate from aqueous system by activated red-mud. *Asian J. Chem.*, **20**(8), 6473–6478.

Galan B., Castaneda D. and Ortiz I. (2005). Removal and recovery of Cr(VI) from polluted ground waters: a comparative study of ion-exchange technologies. *Water Res.*, **39**(18), 4317–4324.

Gilbert B., Ono R. K., Ching K. A. and Kim C. S. (2009). The effects of nanoparticle aggregation processes on aggregate structure and metal uptake. *J. Colloid Interface Sci.*, **339**(2), 285–295.

Golder A. K., Chanda A. K., Samanta A. N. and Ray S. (2011). Removal of hexavalent chromium by electrochemical reduction-precipitation: investigation of process performance and reaction stoichiometry. *Sep. Purif. Technol.*, **76**(3), 345–350.

Granados-Correa F. and Jimenez-Becerril J. (2009). Chromium (VI) adsorption on boehmite. *J. Hazard. Mater.*, **162**(2–3), 1178–1184.

Hu J. S., Zhong L. S., Song W. G. and Wan L. J. (2008). Synthesis of hierarchically structured metal oxides and their application in heavy metal ion removal. *Adv. Mater.*, **20**(15), 2977–2982.

Huang S. H. and Chen D. H. (2009). Rapid removal of heavy metal cations and anions from aqueous solutions by an amino-functionalized magnetic nano-adsorbent. *J. Hazard. Mater.*, **163**(1), 174–179.

Kanel S. R., Manning B., Charlet L. and Choi H. (2005). Removal of arsenic(III) from groundwater by nanoscale zero-valent iron. *Environ. Sci. Technol.*, **39**(5), 1291–1298.

Kaur A. and Gupta U. (2009). A review on applications of nanoparticles for the preconcentration of environmental pollutants. *J. Mater. Chem.*, **19**(44), 8279–8289.

Kent J. A. (2003). *Riegel's Handbook of Industrial Chemistry*. Springer, New York, US.

Kimbrough D. E., Cohen Y., Winer A. M., Creelman L. and Mabuni C. (1999). A critical assessment of chromium in the environment. *Crit. Rev. Environ. Sci. Technol.*, **29**(1), 1–46.

Kocaoba S., Orhan Y. and Akyuz T. (2007). Kinetics and equilibrium studies of heavy metal ions removal by use of natural zeolite. *Desalination*, **214**(1–3), 1–10.

Korzenowski C., Rodrigues M., Bresciani L., Bernardes A. M. and Ferreira J. Z. (2008). Purification of spent chromium bath by membrane electrolysis. *J. Hazard. Mater.*, **152**(3), 960–967.

Kugbe J., Matsue N. and Henmi T. (2009). Synthesis of Linde type A zeolite-goethite nanocomposite as an adsorbent for cationic and anionic pollutants. *J. Hazard. Mater.*, **164**(2–3), 929–935.

Lazaridis N. K. and Charalambous C. (2005). Sorptive removal of trivalent and hexavalent chromium from binary aqueous solutions by composite alginate-goethite beads. *Water Res.*, **39**(18), 4385–4396.

Lazaridis N. K., Karapantsios T. D. and Georgantas D. (2003). Kinetic analysis for the removal of a reactive dye from aqueous solution onto hydrotalcite by adsorption. *Water Res.*, **37**(12), 3023–3033.

Li M., Twardowski Z., Mok F. and Tam N. (2007). Sodium molybdate – a possible alternate additive for sodium dichromate in the electrolytic production of sodium chlorate. *J. Appl. Electrochem.*, **37**(4), 499–504.

Li Y. J., Gao B. Y., Wu T., Sun D. J., Li X., Wang B. and Lu F. J. (2009). Hexavalent chromium removal from aqueous solution by adsorption on aluminum magnesium mixed hydroxide. *Water Res.*, **43**(12), 3067–3075.

Liang P., Liu Y., Guo L., Zeng J. and Lu H. B. (2004). Multiwalled carbon nanotubes as solid-phase extraction adsorbent for the preconcentration of trace metal ions and their determination by inductively coupled plasma atomic emission spectrometry. *J. Anal. At. Spectrom.*, **19**(11), 1489–1492.

Liu W. Z., Huang F., Liao Y. Q., Zhang J., Ren G. Q., Zhuang Z. Y., Zhen J. S., Lin Z. and Wang C. (2008). Treatment of Cr(VI)-containing $Mg(OH)_2$ nanowaste. *Angew. Chem., Int. Ed.*, **47**(30), 5619–5622.

Liu W. Z., Xu X. J., Wang Y. J., He Z., Zhuo N., Huang F. and Lin Z. (2010). Treatment of Cr (VI)-containing nanowastes via the growth of nanomaterial. *Chin. Sci. Bull.*, **55**(4–5), 373–377.

Liu W. Z., Huang F., Wang Y. J., Zou T., Zheng J. S. and Lin Z. (2011). Recycling $Mg(OH)_2$ nanoadsorbent during treating the low concentration of Cr(VI). *Environ. Sci. Technol.*, **45**(5), 1955–1961.

Mishra S. P., Tiwari D., Prasad S. K., Dubey R. S. and Mishra M. (2007). Inorganic particulates in removal of toxic heavy metal ions – Part X: Removal behavior of aluminum hydroxide for Hg(II): a radiotracer study. *J. Radioanal. Nucl. Chem.*, **274**(2), 257–263.

Mohan D. and Pittman C. U. (2006). Activated carbons and low cost adsorbents for remediation of tri- and hexavalent chromium from water. *J. Hazard. Mater.*, **137**(2), 762–811.

Mondal M. K. (2007). Absorption equilibria of dilute sulphur dioxide in $Mg(OH)_2$ slurry. *Fluid Phase Equilib.*, **262**(1–2), 111–120.

Moore R. C. and Anderson D. R. (2007). Arsenic removal from water. US7247242-B1.

Owlad M., Aroua M. K., Daud W. and Baroutian S. (2009). Removal of hexavalent chromium-contaminated water and wastewater: a review. *Water Air Soil Pollut.*, **200** (1–4), 59–77.

Pagana A. E., Sklari S. D., Kikkinides E. S. and Zaspalis V. T. (2008). Microporous ceramic membrane technology for the removal of arsenic and chromium ions from contaminated water. *Microporous Mesoporous Mater.*, **110**(1), 150–156.

Pitcher S. K., Slade R. and Ward N. I. (2004). Heavy metal removal from motorway stormwater using zeolites. *Sci. Total Environ.*, **334**, 161–166.

Qin G., Mcguire M. J., Blute N. K., Seidel C. and Fong L. (2005). Hexavalent chromium removal by reduction with ferrous sulfate, coagulation, and filtration: a pilot-scale study. *Environ. Sci. Technol.*, **39**(16), 6321–6327.

Qu J. H. (2008). Research progress of novel adsorption processes in water purification: a review. *J. Environ. Sci.*, **20**(1), 1–13.

Rengaraj S., Yeon K. H. and Moon S. H. (2001). Removal of chromium from water and wastewater by ion exchange resins. *J. Hazard. Mater.*, **87**(1–3), 273–287.

Ruparelia J. P., Duttagupta S. P., Chatterjee A. K. and Mukherji S. (2008). Potential of carbon nanomaterials for removal of heavy metals from water. *Desalination*, **232**(1–3), 145–156.

Sada E., Kumazawa H. and Nishimura H. (1983). Absorption of sulfur dioxide into aqueous double slurries containing limestone and magnesium hydroxide. *AIChE J.*, **29**(1), 60–65.

Schlag S., Fujita K. and Glauser J. (2007). *Magnesium Oxide and Other Magnesium Chemicals. Chemical Economics Handbook.* Stanford Research Institute, California, USA, June.

Sedman R. M., Beaumont J., McDonald T. A., Reynolds S., Krowech G. and Howd R. (2006). Review of the evidence regarding the carcinogenicity of hexavalent chromium in drinking water. *J. Environ. Sci. Health C*, **24**(1), 155–182.

Sharma Y. C., Srivastava V., Singh V. K., Kaul S. N. and Weng C. H. (2009). Nano-adsorbents for the removal of metallic pollutants from water and wastewater. *Environ. Technol.*, **30**(6), 583–609.

Shin H. S. and Lee S. M. (1998). Removal of nutrients in wastewater by using magnesium salts. *Environ. Technol.*, **19**(3), 283–290.

Shin S. and Jang J. (2007). Thiol containing polymer encapsulated magnetic nanoparticles as reusable and efficiently separable adsorbent for heavy metal ions. *Chem. Commun.*, **7**(41), 4230–4232.

Su C. M. and Ludwig R. D. (2005). Treatment of hexavalent chromium in chromite ore processing solid waste using a mixed reductant solution of ferrous sulfate and sodium dithionite. *Environ. Sci. Technol.*, **39**(16), 6208–6216.

Vaghetti J., Lima E. C., Royer B., Brasil J. L., da Cunha B. M., Simon N. M., Cardoso N. F. and Norena C. (2008). Application of Brazilian-pine fruit coat as a biosorbent to removal of Cr(VI) from aqueous solution – kinetics and equilibrium study. *Biochem. Eng. J.*, **42** (1), 67–76.

Valix M., Cheung W. H. and Zhang K. (2006). Role of heteroatoms in activated carbon for removal of hexavalent chromium from wastewaters. *J. Hazard. Mater.*, **135**(1–3), 395–405.

Viswanathan K. and Tilak B. V. (1984). Chemical, electrochemical, and technologicak aspects of sodium chlorate manufacture. *J. Electrochem. Soc.*, **131**(7), 1551–1559.

Vlyssides A. G. and Israilides C. J. (1997). Detoxification of tannery waste liquors with an electrolysis system. *Environ. Pollut.*, **97**(1–2), 147–152.

Wang L., Wang Y. P., Xu C. X., An Z. Y. and Wang S. M. (2011). Analysis and evaluation of the source of heavy metals in water of the River Changjiang. *Environ. Monit. Assess.*, **173**(1–4), 301–313.

Wang Y. H., Lin S. H. and Juang R. S. (2003). Removal of heavy metal ions from aqueous solutions using various low-cost adsorbents. *J. Hazard. Mater.*, **102**(2–3), 291–302.

Weng C. H., Sharma Y. C. and Chu S. H. (2008). Adsorption of Cr(VI) from aqueous solutions by spent activated clay. *J. Hazard. Mater.*, **155**(1–2), 65–75.

Wu Q. Z. (2002). Potential Applications of Magnesium Hydroxide for Municipal Wastewater Treatment – Sludge Digestion Enhancement and Nutrient Removal. Ph.D. thesis, Dept. Civil Environ. Eng., University of Cincinnati, Ohio45221, USA.

Yang J. E., Kim J. S., Ok Y. S. and Yoo K. R. (2007). Mechanistic evidence and efficiency of the Cr(VI) reduction in water by different sources of zerovalent irons. *Water Sci. Technol.*, **55**(1–2), 197–202.

Zarazua G., Avila-Perez P., Tejeda S., Barcelo-Quintal I. and Martinez T. (2006). Analysis of total and dissolved heavy metals in surface water of a Mexican polluted river by total reflection x-ray fluorescence spectrometry. *Spectrochim. Acta Part B*, **61**(10–11SI), 1180–1184.

Zhang Q. R., Pan B. C., Pan B. J., Zhang W. M., Jia K. and Zhang Q. X. (2008). Selective sorption of lead, cadmium and zinc ions by a polymeric cation exchanger containing nano-$Zr(HPO_3S)_2$. *Environ. Sci. Technol.*, **42**(11), 4140–4145.

Chapter 8

Visible-light active doped titania for water purification: nitrogen and silver doping

I. El Saliby, H. K. Shon and A. McDonagh

8.1 INTRODUCTION

Photocatalysis over semiconductors, mainly TiO_2, has been employed and extensively studied over the past few decades (Malato *et al.*, 2009). The so-called "*first generation photocatalysts*" are characterised by their photoexcitation limited to the UV region of the solar spectrum. Nevertheless, many photocatalysts showed outstanding performance in decomposing a large number of pollutants of water and air. A number of water pollutants belong to the family of biorecalcitrant pollutants (non-biodegradable) and these include pharmaceutical and personal care products, industrial wastes, and a large number of pesticides/herbicides. Many of the emerging pollutants are resistant to conventional water treatment processes, noxious to the environment and some are classified as endocrine disrupting agents. The removal of non-biodegradable pollutants from wastewater poses a serious challenge for the water industry. Generally, advanced oxidation processes (AOP) are employed to achieve this goal. Heterogeneous photocatalysis, as part of AOP, is sometimes adopted as a tool to eliminate these contaminants. However, the high cost associated with the use of photocatalysis to treat such compounds has always been a disadvantage for its large-scale adoption in treatment processes. The high cost is mainly associated with:

(1) the energy consumed for the UV light irradiation of the photocatalyst,
(2) the energy required for the separation of the photocatalyst particles from water.

A total of 40% of the solar light reaching the lithosphere is visible light, whereas UV light accounts only for 5% of the total solar radiation. The development of new photocatalysts that can effectively harvest visible light is crucial for making this technology viable in the water treatment industry. The use of "*second-generation TiO_2 photocatalysts*" to harvest visible light from solar energy is gaining popularity (Anpo, 1997). Most of the research involving the use of second generation TiO_2 photocatalysts has focused on energy conversion and solar cell applications (Yuan *et al.*, 2006). However, interest is increasing in their potential for treating water pollutants (Chatterjee & Dasgupta, 2005; Malato *et al.*, 2009; Rehman *et al.*, 2009). In particular, the large surface areas of nanoparticulate TiO_2 materials make them particularly interesting in this regard. In addition, the ability to introduce dopants into the TiO_2 structure allows the optical properties of the nanomaterials to be manipulated.

In this chapter, doped second generation TiO_2 photocatalysts are reviewed in the context of water pollutant treatment.

8.2 OVERVIEW OF VISIBLE LIGHT ACTIVITY BY DOPING TiO_2

The role of doping, the accumulation of defects (Martyanov *et al.*, 2010) or oxygen vacancies (Ihara *et al.*, 2003; Wang & Doren, 2005; Graciani *et al.*, 2008) are important factors for the initiation of visible light sensitivity. In section 8.2.1, the role of nitrogen doping in triggering visible light activity is discussed while the use of silver dopants is highlighted in section 8.2.2.

8.2.1 Nitrogen doping

Sato (1986) reported the first visible-light-sensitive nitrogen-doped TiO_2 photocatalyst. N-doped TiO_2 showed better photocatalytic activity than undoped TiO_2 for the decomposition of CO and C_2H_6 under visible-light irradiation. Later, Asahi *et al.* (2001) prepared N-doped TiO_2 photocatalysts by two different methods. N-TiO_2 films were made by sputtering TiO_2 in N_2/Ar gas mixtures. Powders were prepared by treating anatase particles in an NH_3/Ar atmosphere at 600°C. Since then, many articles have proposed new methods for preparing N-doped TiO_2 (Table 8.1) and TiO_2 doped with various non-metal species such as C, S, and others have been discussed and developed (Chen & Burda, 2008; Zaleska, 2008).

Photocatalysis is initiated by the adsorption of photons with energy equal or greater than the band gap of TiO_2 (3.0 and 3.2 eV for rutile and anatase, respectively). Band gap narrowing enables the adsorption of photons of lower energy present in the visible light range. The electronic structures of anatase with substitutional N dopant have been investigated by the first-principles density functional theory calculations (Wang & Doren, 2005). The local density approximation calculations demonstrated that N doping led to a narrow TiO_2 band gap by introducing some states at the valence band maximum. These findings have been further validated by the study of Chen and Burda (2008) on the electronic origin of the visible light absorption properties of N-doped TiO_2. Additional electronic states above the valence band edge of pure TiO_2 were identified by x-ray photoelectron spectroscopy (XPS) analysis. The red-shift adsorption of the rutile N-doped TiO_2 was attributed to this additional electronic density of states. On the other hand, Graciani *et al.* (2008) conducted density-functional calculations on the structural and electronic properties of N-doped rutile TiO_2. They concluded that N doping of rutile TiO_2 did not improve its photocatalytic activity without the presence of oxygen vacancies. They also reported that the presence of oxygen vacancies lowered the implantation energy of nitrogen. The on-going debate on the electronic and structural origins of the visible-light sensitivity of N-TiO_2 is becoming more interesting day by day. New findings using advanced characterisation technologies are expected to clarify this dilemma in the near future.

Table 8.1 Selected methods for the preparation of N doped TiO_2.

Nitrogen precursors	Method of preparation	Target pollutant	References
Urea, guanidine and ammonium fluoride	Spray pyrolysis	Degradation of acetaldehyde (air pollutant) and trichloroethylene (water pollutant)	Li *et al.* (2005)
Ammonia solution (28%)	Hydrolysis of $Ti(OCH(CH_3)_2)_4$, TTIP and $TiCl_4$	Photooxidation of CO	Sato *et al.* (2005)
Urea	Hydrolysis of $Ti(OH)_4$	Photooxidation of KI	Kobayakawa *et al.* (2005)
Nitrogen gas	RF magnetron sputtering	Not tested	Mardare *et al.* (2007)
N_2O	Infrared laser pyrolysis technique	Not tested	Alexandrescu *et al.* (2009)
NH_4Cl	Sol-gel method using TTIP	Not tested	Livraghi *et al.* (2009)
Ammonia solution	Sol-gel method using Degussa P25	Decolourisation of methylene blue (MB)	El Saliby *et al.* (2011)

Selected methods for the preparation of nitrogen doped titanium dioxide are shown in Table 8.1. Generally, nitrogen doped titania nanopowders exhibit a yellow colour. Nitrogen doping can be detected by XPS analysis and diffuse reflectance spectroscopy (DRS) measurements (Sato, 1986; Asahi *et al.*, 2001). Nevertheless, XPS peak assignment is still a controversial issue, and the shift in the absorbance spectra in DRS cannot guarantee that the powder is photoactive. Therefore, a photocatalytic activity test is always required. Chen and Burda (2004) reported an extended XPS analysis of nitrogen incorporation in TiO_2 nanostructured materials. The binding energy at 401.3 eV for N 1s was attributed to the energy of the N atom in the environment of O-Ti-N. This substitutional doping was reported to induce the significant increase in photocatalytic activity in N doped TiO_2 nanoparticles. In a recent study, Asahi and Morikawa (2007) observed that nitrogen doping resulted in such species as substitutional N_S or interstitial N_i, substitutional $(NO)_s$, interstitial $(NO)_i$ and substitutional $(NO_2)_s$. The role of a band gap modification and the assignment of XPS peaks to N species are discussed in detail by Emeline *et al.* (2008) and Rehman *et al.* (2009).

8.2.2 Silver doping

Metal implantation into TiO_2 was initiated in the late 1970's. At that time, an oil supply shortage triggered research interest in the production of alternative or synthetic fuels such as H_2. A popular semiconductor material was TiO_2, mainly due to its chemical stability. In 1982, the first visible light active TiO_2 particles were reported by Borgarello *et al.* (1982). Later, Anpo (1997) presented a new metal-doping approach by ion-implantation. TiO_2 was bombarded by high energy metal ions to dramatically modify its electronic states and induce visible light response. The same method was successfully employed to produce V, Cr, Ni, Mn and Fe-doped TiO_2 (Yamashita *et al.*, 2002). Recent patents on doped nano-TiO_2 highlighted a number of new techniques to produce visible light active semiconductors by metal doping through chemical routes as well as ion implantation (Zaleska *et al.*, 2008).

Silver metal was used in ancient times by the Phoenicians to preserve water, wine and vinegar from spoiling. Silver nanoparticles are used nowadays as antibacterial agents in many products (see Chapter 11). Because of the low cost and ease of preparation, silver doping can be employed to enhance photocatalytic activity under UV light as well as to promote the visible light response of TiO_2 (Sung-Suh *et al.*, 2004; Chen *et al.*, 2007). The metallic silver particles on TiO_2 surfaces acted as electron scavengers and contributed dramatically to the decrease of electron-hole recombination in TiO_2. Thus, Ag-TiO_2 exhibited an overall higher photocatalytic activity in the UV range than undoped TiO_2 (Liu *et al.*, 2003; Rao *et al.*, 2003; Gupta *et al.*, 2006; Chen *et al.*, 2007; Park & Kang, 2008; Behnajady *et al.*, 2008; Vinu & Mardas, 2009). On the other hand, Ag ions caused a blue shift in the adsorption spectrum of synthesised Ag-TiO_2 films suggesting a change in the film nanostructure instead of the introduction of impurity states (Stathatos *et al.*, 2001). Several methods have been adopted for the preparation of Ag-doped

TiO_2 (Table 8.2). Sol-gel methods are the most popular because they are straight forward, relatively cheap, and do not require sophisticated equipment.

Ag deposition can enhance the photooxidation power of TiO_2 by the following mechanisms:

(1) Ag nanoparticles enhance the electron-hole separation in TiO_2 by trapping the excited electrons after illumination and transferring them to O_2 as an electron acceptor (Herrmann *et al.*, 1986).
(2) The excitation of the surface electrons and interfacial electron transfer are enabled after the surface plasmon resonance of Ag particles is excited by visible light ($\lambda = 400$ nm) (Herrmann *et al.*, 1997; Sung-Suh *et al.*, 2004; Rehman *et al.*, 2009).
(3) In the case of dye degradation, Ag-doping also enhances the dye adsorption on TiO_2 and promotes the dye photosensitisation process (Rehman *et al.*, 2009).

Ag doping onto TiO_2 can be detected by XPS analysis, DRS measurements and can be further verified by the photocatalytic experiments (Falaras *et al.*, 2003; Vinu & Mardas, 2009). Binding energy peaks around 367–368 eV and 373–374 eV indicate the presence of silver in different ionic forms in powders (Lützenkirchen-Hecht & Strehblow, 2009). Generally, the doped Ag ion has an ionic radius greater than that of Ti^{4+}, therefore it is mostly deposited on the surface of TiO_2 and is not readily incorporated at Ti sites in the TiO_2 lattice (Rehman *et al.*, 2009).

8.3 DOPED TiO_2 AND WATER RECLAMATION

The photocatalytic activity of doped titania is a valuable characteristic of the synthesised doped material. Here, we examine the photocatalytic activity for the decomposition and mineralisation of water pollutants. A number of N-TiO_2 materials have been assessed for the photodegradation of specific water pollutants, for example, dyes, phenols and other organic contaminants although the use of N-doped TiO_2 in large scale water treatment plants is currently limited.

8.3.1 Nitrogen-doped TiO_2

8.3.1.1 Photodegradation of dyes

Dyeing industries often generate contaminated and coloured wastewater. Eutrophication or algal blooms from dye wastewater discharge into rivers and water bodies may result if no treatment measures are adopted. Moreover, azo dyes are designed to resist degradation making them difficult to remove using conventional wastewater treatment processes. Sometimes, the degradation by-products are of great concern because they are more toxic, mutagenic or carcinogenic than the original dyes themselves (Oh *et al.*, 1997).

Table 8.2 Selected methods for the preparation of Ag doped TiO_2 using $AgNO_3$ as a silver precursor.

Method of preparation	Light source	Target pollutant	References
Sol-gel method using titanium tetraisopropoxide (TTIP)	Simulated solar radiation (250 W tungsten halogen lamp)	Photodegradation of Basic Blue 41	Stathatos *et al.* (2001)
Photodeposition method	Visible light (450 W Xe-arc lamp with a UV cut filter)	Photodegradation of trichloroacetate and carbon tetrachloride	Bae & Choi (2003)
Hydrothermal method using $TiCl_4$	None	Not tested	Liu *et al.* (2003)
Sol-gel method using TTIP	Visible light (200 W halogen lamp)	Photodegradation of Rhodamine B	Sung-Suh *et al.* (2004)
Ultrasonic-assisted sol-gel method using $Ti(O\text{-}Bu)_4$	UVA (8 W medium pressure mercury lamp)	Photodegradation of 2,4,6-trichlorophenol	Rengaraj & Li (2006)
Sol-gel method using Degussa P25	Visible light	Photooxidation of o-Cresol	Chen *et al.* (2007)
Liquid impregnation and photodeposition methods	UVC (15 W lamp)	Photodegradation of Acid Red 88	Behnajady *et al.* (2008)
Sol-gel method using Ti $(OC_4H_9)_4$	UVA lamps	Degradation of acetaldehyde (air pollutant)	Yang *et al.* (2007)
Sol-gel method	UVA (6 x 6W lamps)	Photooxidation of methanol and H_2 production	Park & Kang (2008)
Sol-gel method using metatitanic acid	Direct sunlight	Photodegradation of methyl orange	Binitha *et al.* (2009)

Although some dyes are used as TiO_2 photosensitisers, model dye pollutants are widely used for testing the photocatalytic activity of N-doped TiO_2 (Asahi *et al.*, 2001; Liu *et al.*, 2009a). The photocatalytic activity of $TiO_{2-x}N_x$ was evaluated by measuring the decomposition of Methylene Blue under visible light. Significantly more degradation was observed at an irradiation wavelength of 500 nm compared to the reference TiO_2 (Asahi *et al.*, 2001). Also, no loss in the UV photocatalytic activity was recorded for the N doped photocatalyst over 2 h. Qiu *et al.* (2007) reported the photodegradation of Methylene Blue under visible light irradiation ($\lambda > 400$ nm) for an N-doped TiO_2 prepared by reacting Ti $[OCH_2(CH_3)_2]_4$ (4 mL) with ethylenediamine (20 mL) in 1-hexanol (150 mL). The resultant nanoparticles were washed and sintered at 200°C for 1 h. The photocatalytic activity of the N-TiO_2 increased with N doping levels (reported as up to 8% N). At 8% N doping, around 85% of Methylene Blue was removed after 340 min of visible light irradiation. Although the reaction time might appear lengthy, the photocatalytic activity of N-doped TiO_2 varies with many factors. The degradation of pollutants with relatively short irradiation times have been reported (Asahi *et al.*, 2001; Liu *et al.*, 2006) but also other studies suggest that the visible light degradation of pollutants in water is slow (Yang & Gao, 2008).

The photocatalytic activity of N-doped crystalline rutile nanostructures was reported for experiments using a Pyrex photoreactor (400 mL) containing 0.16 g TiO_2 and a 20 mg/L Methylene Blue solution (Yang & Gao, 2008). The reactor was irradiated with a 300 W medium pressure lamp equipped with a filter to exclude UV light. Dye degradation was correlated to the shape of the nanomaterial as well as to the percentage of N doping. Complete visible-light photocatalytic degradation of Methylene Blue was achieved after 7.5 h and 5.5 h using floc-like aggregates and rod-like aggregates of N-doped nanostructures, respectively. Although N-doping improved the visible light catalysis, the N-doped photocatalysts also retained good activity in the UV.

Nitrogen-doped titanate nanotubes (N-TNT) were prepared by using guanidine carbonate to post-treat nanotubes produced by the conventional hydrothermal method (Geng *et al.*, 2009). A simulated solar light photoreactor equipped with a 150 W xenon arc lamp was used to decompose 100 mL of 30 mg/L MB solution over 30 mg of N-TNT. The photocatalytic degradation rate of N-TNT was higher than their TNT precursors under solar light irradiation. Calcination at temperatures above 350°C had a negative effect on the overall performance of N-TNT, mainly because the amount of N decreased with the increase in calcination temperatures. Liu *et al.* (2010) compared several doped and co-doped photocatalysts for cleaning Methylene Blue wastewater under visible light (electron saving lamp, 14 W, emission at $\lambda = 545$ nm). A Methylene Blue solution (10 mg/L) was mixed with 1 g/L of photocatalyst and irradiated for 6 h to check the photodegradation rate. The apparent kinetic constant fitted from the $\ln(C_0/C_t) \sim t$ curve of N-TiO_2 was 0.277 h^{-1}. This was much higher than the Fe doped and Fe-N co-doped TiO_2 rate constants of 0.1975 h^{-1} and 0.1694 h^{-1}, respectively. The decrease of photoactivity in the co-doped sample is apparently due to the capture of charge carriers.

Recently, Liu *et al.* (2009a) photocatalytically degraded Rhodamine B (RhB) under visible light (fluorescent lamp, 18 W, glass filter for UV cut off, $\lambda > 420$ nm). Nitrogen-doped TiO_2 with different ratios of anatase to rutile were prepared by employing air and gaseous NH_3 milling atmospheres. The photocatalytic activity of the photocatalysts (100 mg) was assessed in 100 mL of 2×10^{-5} M of RhB solution under magnetic stirring, visible light irradiation and after 30 min of dark adsorption. It was concluded that these mixed-phase N-doped TiO_2 materials, with higher rutile content, had greater photocatalytic activity because of more abundant surface water and hydroxyl groups (which may facilitate the formation of hydroxyl radicals). Similarly, Liu *et al.* (2009b) reported the photodecomposition of RhB (4×10^{-5} M) using the experimental reactor described above over nitrogen-doped mesoporous TiO_2. Substantially higher photocatalytic activity was observed for the N-doped mesoporous TiO_2 compared to the non-doped mesoporous TiO_2 and P25. N-TiO_2 completely decomposed RhB in solution after 4 h of visible light irradiation. The high photocatalytic activity was attributed to the abundant surface states in nitrogen doped mesoporous TiO_2.

Acid orange 7 can be degraded under solar light irradiation (12 mW/cm^2) using N-doped TiO_2 prepared by Liu *et al.* (2006). Experiments were conducted in Petri dishes (15 ml of Acid orange 7 solution covered with a plastic film to avoid evaporation) containing a solution of 20 mg/L of Acid orange 7 and 10 mg/L of N-TiO_2 at pH of 4.5–5. The degradation followed first order reaction kinetics and full decolourisation was achieved after 4 h of solar irradiation, while only 5% of colour was removed with P25. Smaller nanoparticles size (10 nm) and high surface hydroxyl group density in N-TiO_2 were reported to cause the increase in photocatalytic activity. On the other hand, Sun *et al.* (2008) reported the photocatalytic degradation of Orange Gunder visible and sunlight irradiation over N-TiO_2. A visible light photoreactor equipped with a 500 W xenon-arc lamp and a UV cut-off filter ($\lambda > 400$ nm) was used to test several photocatalysts under different experimental conditions. Outdoor experiments under direct sunlight were also carried out. The results showed that N-TiO_2 degraded 96.3% of Orange G and P25 degraded 42.6% of the dye under visible light in 150 min. The opposite trend was recorded under solar light irradiation where P25 and N-TiO_2 decolourised 90% and 75% of Orange G after 20 min of solar irradiation, respectively. Gao *et al.* (2009) synthesised N-doped anatase TiO_2 by a hydrothermal method and photodecomposed Methyl Orange using a quartz cylinder irradiated with a 300 W high pressure mercury lamp. A total of 99% of Methyl Orange (20 mg/L) was degraded after 20 minutes of irradiation. The photocatalytic degradation efficiency was affected by the calcination temperature of the TiO_2 powders with powders calcined at 300°C being more efficient than those calcined at 500°C and to a lesser extent by the pH of the solution.

Liu *et al.* (2008) prepared N-doped TiO_2 films on heated quartz glass substrate. Aqueous Methyl Orange was photodegraded using the prepared film (1 cm^2) under a 15 W daylight lamp equipped with a UV cut-off filter ($\lambda > 400$ nm). The visible light photoactivity of the N-doped films was superior to the undoped films. It was found that N-doping and the high adsorption of Methyl Orange on the surface of the films (causing photosensitisation) contributed to the visible light photocatalytic activity. High adsorption capacities of azo dyes onto N-TiO_2 were reported by Janus *et al.* (2009), where the surface area of their N-modified material was 183 m^2/g, significantly greater than that of P25 (~55 m^2/g). Unfortunately, no photocatalytic activity tests have been reported to date investigating any synergetic effects of N-doping and dye photosensitisation.

8.3.1.2 *Phenolic compounds*

Phenolic compounds are produced by several industrial activities. They can be discharged with wastewater and are often highly soluble in industrial effluents. Wastewater from industries such as resin manufacturing, iron smelting and paper making contains high concentration of phenol and their derivatives (Wang *et al.*, 2005). Phenolic compounds can be toxic, carcinogenic, difficult to degrade and are ubiquitous in natural environments. Many experiments have been conducted to decompose phenolic compounds over N-doped TiO_2 under visible or solar light irradiation (Sun *et al.*, 2008; Huo *et al.*, 2008; Xing *et al.*, 2009; Lee *et al.*, 2010; Baruwati & Varma, 2011; Triantis *et al.*, 2011).

Different irradiation sources were used to evaluate the photocatalytic activity of N-doped TiO_2 produced through a sol-gel method followed by calcination of the powder (Wang *et al.*, 2005). A 1000 W xenon lamp (with a UV cut-off filter $\lambda > 400$ nm) irradiated 1 L of 20 ppm phenol solution mixed with 0.5 g of N-TiO_2 to conduct the visible light photocatalytic test. Under relatively similar conditions, outdoor experiments were carried out to evaluate the photocatalytic activity under solar irradiation. N-doped TiO_2 calcined at 400°C decomposed 35.6% of phenol after 2 h of photoreaction under visible light. The photocatalytic activity of catalysts calcined at 350, 400 or 500°C was measured and with the material calcined at 400°C displaying the highest photocatalytic activity. All three photocatalysts were anatase phase but the material calcined at 500°C contained less nitrogen although had higher crystallinity compared to the 400°C material while the reverse was the case for the 350°C material. Increasing the calcination temperature to 600°C resulted in no nitrogen being incorporated. P25 performed better than N-doped TiO_2 under solar light conditions and degraded 90% of phenol in 3 h while the N-doped TiO_2removed 40% after 7 h.

Although N-doping may increase the visible light response of photocatalysts, it sometimes reduces the photocatalytic activity in the UV excitation range. The gain of activity in the visible range does not always compensate for the loss of activity in the UV range and thus decreases the overall activity under solar light irradiation. Lee *et al.* (2010) studied the influence of nitrogen chemical states on the photocatalytic activities of N-doped TiO_2 nanoparticles under visible light. A 10 ppm phenol solution was used to investigate the photocatalytic activity of the nanoparticles (1 g/L) irradiated with a 200 W halogen lamp equipped with a UV cut-off filter ($\lambda > 420$ nm). They reported different photocatalytic activities with regard to the chemical states of N in the prepared samples. The complete degradation of phenol was not reached before 12, 17 and 24 h for the binding energies peaks recorded at 398.3 eV, both 398.3 eV and 401.6 eV and 401.5 eV alone. The authors attributed the peaks to substitutional (398.3 eV) and interstitial (401.5 eV) N and concluded that the former led to better photocatalytic activity than the latter.

The degradation of 4-chlorophenol over TiO_2 doped with a series of lower-concentration nitrogen species was reported by Sun *et al.* (2008). The 6 h photoreaction under visible light (250 W xenon lamp, UV cut-off filter $\lambda > 400$ nm) revealed that the photocatalyst removed 51% of the solution TOC, while undoped TiO_2 failed to decrease the TOC of the phenolic solution. Sun *et al.* (2009) carried out a similar experiment with different light sources (visible light: as described above, UV: 250 W high-pressure mercury lamp and natural solar light). N-doped TiO_2 showed higher photoactivity under UV light than P25 (apparent-reaction-rate constants equal 0.163 min^{-1} and 0.143 min^{-1} for N-doped TiO_2 and P25, respectively). Huo *et al.* (2008) studied the photodegradation of aqueous *p*-chlorophenol using nitrogen-doped TiO_2 prepared under supercritical conditions (the advantage of such a method was to obtain a high surface area, highly crystalline anatase). Up to 76% of the 10 mg/L *p*-chlorophenol solution was photodegraded after 4 h of irradiation using 50 mg/L of photocatalyst (with the highest N content, N/Ti = 3.4%). The irradiation source was a 500 W xenon lamp topped with a UV cut-off filter ($\lambda > 420$ nm). Here again, the degradation percentage was shown to be affected by the percentage of N-doped onto TiO_2. The photocatalysis of other phenolic compounds was studied over several N-doped TiO_2 powders by Xing *et al.* (2009) and Ananpattarachai *et al.* (2009). They both reported very good visible light activity of the N-doped material. However, the materials were not tested for photoactivity under solar light or compared to the standard P25.

Recently, Baruwati and Varma (2011) photodegraded 2,4-dichlorophenol (DCP) over N-doped TiO_2 prepared using guanidine nitrate as a nitrogen source. The photocatalyst showed good performance under indoor light bulb irradiation and very good photocatalytic activity under natural solar light. It was also found that the N-TiO_2 was stable for five degradation cycles. It is widely recognised that a superior N-doped TiO_2 should not only be able to degrade phenolic compounds under visible light for prolonged periods of time, but also it has to show better overall photocatalytic activity under solar light (40% visible light).

8.3.1.3 Other organic water contaminants

The photodegradation of pollutants such as benzoic acid and herbicides have been investigated using N-doped TiO_2 (Shifu *et al.*, 2005; Wang *et al.*, 2007; Abramovic *et al.*, 2009). Herbicides are widely used in agriculture, horticulture and landscape activities to kill weeds. They are transported by precipitation and irrigation to the groundwater and runoff water and they contaminate rivers and waterways. The photocatalytic degradation of herbicides in water is an efficient treatment to mineralise them, and cost-effective methods may use solar light.

The photocatalytic degradation of benzoic acid under simulated solar light (125 W xenon lamp) removed 65% after 2 h. With an initial benzoic acid concentration of 25 mg/L and a photocatalyst loading of 0.1 g/L, the photoreaction was performed in a borosilicate photoreactor at 25°C (Wang *et al.*, 2007). Moreover, total organic carbon (TOC) decreased by 53.4% after 2 h. Single wavelength experiments were conducted to check the performance of the synthesised N-TiO_2 and a non-doped TiO_2 produced by a similar procedure. N-doped TiO_2 showed photocatalytic activity at $\lambda = 334$ nm, 365 nm, 500 nm and 600 nm, while the undoped TiO_2 showed only activity in the UV range. N-doped TiO_2 prepared by ball milling titania in aqueous ammonia solution followed by drying in air at 110°C (Shifu *et al.*, 2005) decomposed monocrotophos. Experiments utilised a photochemical quartz reactor equipped with a 375 W medium pressure mercury lamp. The photocatalytic activity of N-doped TiO_2 was assessed using a 10^{-4} M monocrotophos solution at a photocatalyst loading of 2 g/L and a pH of 5 for 4 h. With monochromatic light irradiation ($\lambda = 365$ nm), TiO_2 (ball milled in water) and N-doped TiO_2 (ball milled in ammonia) showed similar activity with 17.6% and 19.3% degradation after 4 h, respectively. At $\lambda = 405$ nm, N-doped TiO_2 (6.1% degradation efficiency) showed better photoactivity than TiO_2 (0.97% degradation efficiency). N-doping was confirmed by a red-shift in the DRS spectra and two N 1s peaks at 396.5 eV and 402 eV binding energies as revealed by XPS. Abramovic *et al.* (2009) studied the use of N-doped TiO_2 (prepared by a sol-gel method) for the degradation of *RS*-2-(-4-chloro-*o*-tolyloxy) propionic acid (mecoprop) and (4-chloro-2-methylphenoxy) acetic acid (MCPA) herbicides. The photocatalytic degradation was performed in a Pyrex glass cell using 2 g/L of photocatalysts in herbicides' solution (concentration of herbicides ranged from 1 to 8 g/L) under different light sources. N-doped TiO_2 showed superior photocatalytic activity under the visible light irradiation and the degradation rates are shown in Table 8.3.

Recently, a nanostructured N-doped TiO_2 was synthesised by Triantis *et al.* (2011) through the hydrolysis condensation of tetrabutyltitanate reaction with urea. The photocatalyst was tested for the photocatalytic destruction of microcystin-LR (MC-LR) under UVA (four F15W/T8 black light tubes, 71.7 mW·cm^{-2}), visible (nine LED, 3 mW·cm^{-2}) and solar light (Oriel 150 W Xe arc lamp, 115 mW·cm^{-2}). The photocatalytic degradation was carried out using

5 mL of MC-LR (10 mg/L) in a cylindrical Pyrex cell (N-TiO_2 = 200 mg/L, initial pH = 5.7, T = 25°C). Results revealed that N-TiO_2 was active under UVA, visible and solar light. The full degradation of MC-LR was achieved after 20 min under UVA and 4 h under solar. Moreover, 14% of MC-LR was deactivated after 6 h of visible light irradiation.

Table 8.3 Mecoprop and MCPA degradation rates without (direct photolysis) and over N-doped TiO_2 and P25 (Abramovic *et al.*, 2009).

	Halogen Lamp (50 W)	Mercury Lamp (125 W)	Sunlight
	Mecoprop degradation rate (μmol dm^3 min^{-1})		
N-doped TiO_2	1.36*	7.27^	0.05^n
P25	0.86*	9.66^	0.08^n
Direct photolysis	0.22*	0.79^	0.00^n
	MCPA degradation rate (μmol dm^3 min^{-1})		
N-doped TiO_2	1.02*	9.31^	0.05^n
P25	0.74*	9.97^	0.05^n
Direct photolysis	0.18*	1.08^	0.00^n

Degradation rate calculated for time period of *21 days; ^19 h; ^{n}210 min.

8.3.2 Silver-doped TiO_2

8.3.2.1 Photodegradation of dyes

An increase in the photocatalytic activity of TiO_2 under visible light can be attained by doping with silver (Sung-Suh *et al.*, 2004; Gupta *et al.*, 2006). Also, the antibacterial properties of silver particles can be incorporated into antibacterial inorganic Ag-TiO_2 complexes (Wu *et al.*, 2010).

Mesoporous titania films can be doped with Ag^+ ions to enhance their photocatalytic activity under visible light. Additionally, the photoreduction of Ag^+ ions (UV light irradiation with a mercury 400 W lamp) in the Ag-doped films increased the rate of photodegradation of Basic Blue 41 solution (illuminated using a 250 W tungsten-halogen spot (0.6 W/m^2)) compared to the untreated films. The UV pre-treated films showed better photocatalytic activity (full degradation in 20 min) than the untreated films (full degradation in 60 min) due to the creation of metal centres in the film (Stathatos *et al.*, 2001). Recently, Seery *et al.* (2007) tested the effect of pre-irradiating Ag-TiO_2 (irradiation was carried out by using a 250 W bulb to photoreduce Ag ions) on the photo-oxidation activity under simulated solar and natural sunlight for degrading rhodamine 6G (R6G, reaction volume = 50 mL, concentration 5×10^{-5} M, catalyst loading = 1.2 g/L). The intensity of the natural solar light (0.4–0.68 W/m^2) was generally less than that of the simulated source (0.68 W/m^2) which lowered the rate of degradation in the outdoor experiments. Pre-irradiation reduced the degradation rate of the Ag-TiO_2 from 0.34 min^{-1} for the un-irradiated material to 0.24 min^{-1} for the pre-irradiated Ag-TiO_2. The change in rate was assigned to several factors: (i) low Ag dispersion; (ii) Ag acting as a light adsorption barrier; (iii) Ag preventing the adsorption of the organic substrate onto the TiO_2 and (iv) Ag acting as an electron-hole recombination centre (Xin *et al.*, 2005; Seery *et al.*, 2007). A 2 atomic% Ag deposition onto TiO_2 nanoparticles was found to be the optimum for a high photocatalytic degradation of Rhodamine B (RhB) under visible light (Sung-Suh *et al.*, 2004). The Ag-TiO_2 was prepared by a peptization method (titanium isopropoxide as a titanium precursor) followed by a photocatalytic deposition process (AgNO$_3$ as a silver precursor). A 100 mL photoreactor was used to degrade 50 mL of RhB solution (10^{-5} M) irradiated with a 200 W halogen lamp as a visible light source. Ag-TiO_2 decomposed 30% more RhB than TiO_2 under visible light, moreover, it showed better photocatalytic activity than P25 and Ag-P25 (prepared by depositing silver onto P25). The high RhB adsorption on Ag-TiO_2 apparently boosted the photocatalytic degradation due to the photosensitisation mechanism. On the other hand, Binitha *et al.* (2009) reported the degradation of Methyl Orange using nano silver-doped mesoporous TiO_2 under sunlight (between 11:00 to 14:00). The photoactivity was studied at a Methyl Orange concentration of 10^{-5} M. After 1 h of solar irradiation, the Ag-doped TiO_2 degraded 94% of the Methyl Orange.

8.3.2.2 Other prominent water contaminants

Cresols are used in the manufacturing of pesticides, dyes and PCs and are found in drinking and irrigation water. Chen *et al.* (2007) studied the photodegradation of o-Cresol under visible light over Ag-TiO_2. The maximum photocatalytic activity was recorded for the 0.5 wt.% Ag-doping. The deactivation of bacteria (*E. coli*) present in wastewater is also possible under solar light using Ag-TiO_2 (Ashkarran, 2011). The Ag-TiO_2/bacteria suspension was illuminated with a 90 W halogen lamp to conduct the photocatalytic inactivation of bacteria. Later on, the bacteria were isolated and cultured on the sterilised glass. A significant decrease in the survival rate for the samples treated with Ag-TiO_2 (from 400 to almost 0 viable bacteria) was observed compared with non-doped TiO_2 (from 400 to 300 viable bacteria). Thus, the antimicrobial properties of Ag in addition to inducing visible light response in TiO_2 had photooxidised the bacteria.

8.4 CO-DOPING WITH METAL/NON-METAL

Co-doping with metal and non-metal elements can further increase the visible light photocatalytic activity of TiO_2. As discussed above, N is a cost-effective non-metal dopant capable of narrowing the band gap of TiO_2. When coupled with other metal compounds, the co-doped material is characterised by a higher photocatalytic activity and faster degradation compared to the single element doped photocatalyst (Wu *et al.*, 2010; Yuan *et al.*, 2010). In recent years, metals like V, W, Ag, Eu, Ta, Fe and Cd have been used in conjunction with N to boost the photocatalytic activity of TiO_2 (Table 8.4).

Yuan *et al.* (2010) have co-doped N-TiO_2 with silver nanoparticles to test the photocatalytic activity under fluorescent light for the inactivation of model bacterial cells (*E. coli* and *B. subtilis*). The DRS spectra of Ag-N co-doped TiO_2 revealed that the adsorption band shifted towards longer wavelengths (into the visible range). The photocatalyst was synthesised by the hydrolysis precipitation method using ammonia and silver nitrate as N and Ag precursors, respectively. Two bacteria were inhibited by irradiating the culture with a fluorescent light in the presence of N-doped TiO_2 and Ag-N-doped TiO_2. The antibacterial property was investigated by the agar hole diffusion-inhibition zone method. Photocatalytic activity was compared by measuring the bacteriostasis circle around the deposition hole (inhibition zone). The 1 at.% Ag co-doping had the highest antibacterial activity showing the widest bacteriostasis circles. On the other hand, Wu *et al.* (2010) prepared Ag-N-doped TiO_2 by a sol-gel method. They studied the effect of co-doping on the inactivation of *E. coli* AN 387 cells irradiated by a halogen lamp (1.6 mW/cm_2) equipped with a UV cut-off filter. Similarly to the results reported by Yuan *et al.* (2010), Ag-N-TiO_2 showed higher photocatalytic activity under visible light. The inactivation was more photocatalytic than bactericidal by the production of hydroxyl radicals. The photocatalytic killing mechanism involved oxidative damage on the cell wall and cell membrane as well as alterations of the internal DNA molecules.

Recently, a sol-gel method for producing Ag-N-doped TiO_2 co-doped photocatalyst was reported (El Saliby *et al.*, 2011). The material was tested for the discolouration of Methylene Blue solutions under visible light (Table 8.4). The co-doped material showed superior photocatalytic activity under visible light as well as very high adsorption of Methylene Blue. Research is continuing to test the new material for the degradation of several water pollutants.

8.5 SUMMARY

Heterogeneous TiO_2 photocatalysis processes that decompose aqueous contaminants are highly dependent on the energy of excitation of the titania, which in its pure form is UV light active only. The doping and co-doping of titania is a successful method to prepare visible light active TiO_2 nanomaterials. Second generation photocatalysts that can harvest visible light have been studied and advances in nanotechnologies for fabricating interesting photocatalysts have impacted on the methods of synthesis. Straightforward methods are now being

Table 8.4 Recent methods for the preparation of N/metal co-doped TiO_2.

Method of preparation and co-doped metal	Light source	Photoreaction conditions	References
Sol-gel method using tetrabutyltitanate, **Iron (III)**	Visible light (200 W tungsten halogen lamp with a UV cut-off filter)	Photodegradation of RhB (10^{-5} M), 0.1 g TiO_2 in 60 mL solution	Pingxiao *et al.* (2007)
Precipitation hydrothermal method, **Cadmium**	Visible light (1000 W halogen lamp with a UV cut-off filter)	Photodegradation of RhB (20 mg/L), 1 g/L of photocatalyst	Cong *et al.* (2007)
Radio frequency magnetron sputtering (film), **Tantalum**	Xenon lamp (Hayashi Tokei, Luminar Ace 251 Xe) with a glass filter	Decomposition of the Oleic acid measured by the contact angle	Obata *et al.* (2007)
Sol-gel method using tetrabutyltitanate followed by ball milling in urea, **Tungsten**	Visible light (13 W fluorescent lamp with UV cut-off filter)	Photodegradation of sulfosalicylic acid (40 mg/L), 1 g/L TiO_2 Photodegradation of MB (20 mg/L), 0.5 g/L TiO_2	Shen *et al.* (2008)
Sol-gel method using $Ti(O\text{-}Bu)_4$, **Europium**	Visible light (250 W halogen lamp with UV cut-off filter)	Photodegradation of reactive red 2 (50 mg/L), 70 mg/L of TiO_2 in 200 mL solution	Xu *et al.* (2008)
Wet chemical method using $Ti(O\text{-}Bu)_4$, **Vanadium**	Visible light (400 W xenon lamp with a UV cut-off filter)	Photooxidation of MB (0.005 mM), 40 mg of TiO_2 in 100 mL solution	Gu *et al.* (2008)
Sol-gel method using TTIP, **Silver**	Visible light (Metal halogen lamp with a UV cut-off filter, 1.6 mW/cm^2)	Photocatalytic inactivation of *E. coli* AN 387 cells	Wu *et al.* (2010)
Hydrolysis precipitation method using $Ti(OBu)_4$, **Silver**	30 W fluorescent lamp	Photocatalytic inactivation of *E. coli* and *B. subtilis*	Yuan *et al.* (2010)
Sol-gel method using P25, **Silver**	Visible light (2×10 W lamps with a UV cut-off filter)	Photooxidation of MB (23 mg/L), 0.5 g/L of TiO_2	El Saliby *et al.* (2011)

used and interesting photocatalysts are being generated, which have great potential uses in water treatment.

In addition to the general benefits of using heterogeneous photocatalysis for water treatment, other advantages of using the second generation photocatalysts for water reclamation include: (i) water treatment can be achieved using indoor lamps or small photoreactors equipped with immobilised photocatalysts (e.g. household photoreactors); and (ii) a reduction of water treatment costs can be achieved by high efficiency water treatment using a greater proportion of sunlight energy in water treatment plants.

Water reclamation suffers from the high energy costs of the processes used to meet the discharged or treated water guidelines. The use of renewable energy to produce power for water treatment plants is still in the early developmental stages and the production is far too low compared to conventional methods. However, future research could focus on the use of alternative water treatment processes that employs second generation photocatalysts characterised by their high photocatalytic activity under solar light, stability and reusability.

ACKNOWLEDGEMENTS

This book chapter was funded by an ARC LP grant (0991544).

REFERENCES

Abramovic B. F., Sojic D. V., Anderluh V. B., Abazovic N. D. and Comor M. I. (2009). Nitrogen-doped TiO_2 suspensions in photocatalytic degradation of mecoprop and (4-chloro-2-methylphenoxy) acetic acid herbicides using various light sources. *Desalination*, **244**(1–3), 293–302.

Alexandrescu R., Scarisoreanu M., Morjan I., Birjega R., Fleaca C., Luculescu C., Soare I., Cretu O., Negrila C. C., Lazarescu N. and Ciupina V. (2009). Preparation and characterization of nitrogen-doped TiO_2 nanoparticles by the laser pyrolysis of N_2O-containing gas mixtures. *Appl. Surf. Sci.*, **255**(10), 5373–5377.

Ananpattarachai J., Kajitvichyanukul P. and Seraphin S. (2009). Visible light absorption ability and photocatalytic oxidation activity of various interstitial N-doped TiO_2 prepared from different nitrogen dopants. *J. Hazard. Mater.*, **168**(1), 253–261.

Anpo M. (1997). Photocatalysis on titanium oxide catalysts approaches in achieving highly efficient reactions and realizing the use of visible light. *Catal. Surv. Jpn.*, **1**, 169–179.

Asahi R. and Morikawa T. (2007). Nitrogen complex species and its chemical nature in TiO_2 for visible-light sensitized photocatalysis. *Chem. Phys.*, **339**(1–3), 57–63.

Asahi R., Morikawa T., Ohwaki T., Aoki K. and Taga Y. (2001). Visible-light photocatalysis in nitrogen-doped titanium oxides. *Science*, **293**(5528), 269–271.

Ashkarran A. A. (2011). Antibacterial properties of silver-doped TiO_2 nanoparticles under solar simulated light. *J. Theor. Appl. Phys.*, **4**(4), 1–8.

Bae E. and Choi W. (2003). Highly enhanced photoreductive degradation of perchlorinated compounds on dye-sensitized metal/TiO_2 under visible light. *Environ. Sci. Technol.*, **37**(1), 147–152.

Baruwati B. and Varma R. S. (2011). Synthesis of N-doped nano TiO_2 using guanidine nitrate: an excellent visible light photocatalyst. *J. Nanosci. Nanotechnol.*, **11**(3), 2036–2041.

Behnajady M. A., Modirshahla N., Shokri M. and Rad B. (2008). Enhancement of photocatalytic activity of TiO_2 nanoparticles by silver doping: photodeposition versus liquid impregnation methods. *Glob. NEST J.*, **10**(1), 1–7.

Binitha N. N., Yaakob Z., Reshmi M. R., Sugunan S., Ambili V. K. and Zetty A. A. (2009). Preparation and characterization of nano silver-doped mesoporous titania photocatalysts for dye degradation. *Catal. Today*, **147**(1), S76–S80.

Borgarello E., Kiwi J., Graetzel M., Pelizzetti E. and Visca M. (1982). Visible light induced water cleavage in colloidal solutions of chromium-doped titanium dioxide particles. *J. Am. Chem. Soc.*, **104**(11), 2996–3002.

Chatterjee D. and Dasgupta S. (2005). Visible light induced photocatalytic degradation of organic pollutants. *J. Photochem. Photobiol.,C*, **6**(2–3), 186–205.

Chen H. W., Ku Y. and Kuo Y. L. (2007). Photodegradation of *o*-cresol with Ag deposited on TiO_2 under visible and UV light irradiation. *Chem. Eng. Technol.*, **30**(9), 1242–1247.

Chen X. and Burda C. (2004). Photoelectron spectroscopic investigation of nitrogen-doped titania nanoparticles. *J. Phys. Chem. B*, **108**(40), 15446–15449.

Chen X. and Burda C. (2008). The electronic origin of the visible-light absorption properties of C-, N- and S-doped TiO_2 nanomaterials. *J. Am. Chem. Soc.*, **130**(15), 5018–5019.

Cong Y., Zhang J., Chen F., Anpo M. and He D. (2007). Preparation, photocatalytic activity, and mechanism of nano-TiO_2 co-doped with nitrogen and iron (III). *J. Phys. Chem. C*, **111**(28), 10618–10623.

El Saliby I., Erdei L., Shon H. K. and Kim J.-H. (2011). Development of visible light sensitive titania photocatalysts by combined nitrogen and silver doping. *J. Ind. Eng. Chem.*, **17**(2), 358–363.

Emeline A. V., Kuznetsov V. N., Rybchuk V. K. and Serpone N. (2008). Visible-light-active titania photocatalysts: the case of N-doped TiO_2s – properties and some fundamental issues. *Int. J. Photoenergy*, 2008, Article ID 258394, 19.

Falaras P., Arabatzis I. M., Stergiopoulos T. and Bernard M. C. (2003). Enhanced activity of silver modified thin-film TiO_2 photocatalysts. *Int. J. Photoenergy*, **5**(3), 123–130.

Gao H., Zhou J., Dai D. and Qu Y. (2009). Photocatalytic activity and electronic structure analysis of N-doped anatase TiO_2: a combined experimental and theoretical study. *Chem. Eng. Technol.* **32**(6), 867–872.

Geng J., Yang D., Zhu J., Chen D. and Jiang Z. (2009). Nitrogen-doped TiO_2 nanotubes with enhanced photocatalytic activity synthesized by a facile wet chemistry method. *Mater. Res. Bull.*, **44**(1), 146–150.

Graciani J., Alvarez L. J., Rodriguez J. A. and Sanz J. F. (2008). N doping of rutile TiO_2 (110) surface. A theoretical DFT study. *J. Phys. Chem. C*, **112**(7), 2624–2631.

Gu D.-E., Yang B.-C. and Hu Y.-D. (2008). V and N co-doped nanocrystalanatase TiO_2 photocatalysts with enhanced photocatalytic activity under visible light irradiation. *Catal. Commun.*, **9**(6), 1472–1476.

Gupta A. K., Pal A. and Sahoo C. (2006). Photocatalytic degradation of a mixture of crystal violet (Basic Violet 3) and methyl red dye in aqueous suspensions using Ag^+ doped TiO_2. *Dyes Pigments*, **69**(3), 224–232.

Herrmann J. M., Disdier J. and Pichat P. (1986). Photoassisted platinum deposition on TiO_2 powder using various platinum complexes. *J. Phys. Chem.* **90**(22), 6028–6034.

Herrmann J. M., Tahiri H., Ait-Ichou Y., Lassaletta G., Gonzalez-Elipe A. R. and Fernandez A. (1997). Characterization and photocatalytic activity in aqueous medium of TiO_2 and Ag-TiO_2 coatings on quartz. *Appl. Catal. B Environ.*, **13**(3–4), 219–228.

Huo Y., Bian Z., Zhang X., Jin Y., Zhu J. and Li H. (2008). Highly active $TiO_{2-x}N_x$ visible photocatalyst prepared by N-doping in Et_3N/EtOH fluid under supercritical conditions. *J. Phys. Chem. C*, **112**(16), 6546–6550.

Ihara T., Miyoshi M., Iriyama Y., Matsumoto O. and Sugihara S. (2003). Visible-light-active titanium oxide photocatalyst realized by an oxygen-deficient structure and by nitrogen doping. *Appl. Catal. B Environ.*, **42**(4), 403–409.

Janus M., Choina J. and Morawski A. W. (2009). Azo dyes decomposition on new nitrogen-modified anatase TiO_2 with high adsorptivity. *J. Hazard. Mater.*, **166**(1), 1–5.

Kobayakawa K., Murakami Y. and Sato Y. (2005). Visible-light active N-doped TiO_2 prepared by heating of titanium hydroxide and urea. *J. Photochem. Photobiol. A*, **170** (2), 177–179.

Lee S., Cho I.-S., Lee D. K., Kim D. W., Noh T. H., Kwak C. H., Park S., Hong K. S., Lee J.-K. and Jung H. S. (2010). Influence of nitrogen chemical states on photocatalytic activities of nitrogen-doped TiO_2 nanoparticles under visible light. *J. Photochem. Photobiol. A*, **213**(2–3), 129–135.

Li D., Haneda H., Hishita S. and Ohashi N. (2005). Visible-light-driven nitrogen-doped TiO_2 photocatalysts: effect of nitrogen precursors on their photocatalysis for decomposition of gas-phase organic pollutants. *Mater. Sci. Eng. B*, **117**(1), 67–75.

Liu C., Tang X., Mo C. and Qiang Z. (2008). Characterization and activity of visible-light-driven TiO_2 photocatalyst codoped with nitrogen and cerium. *J. Solid. State. Chem.*, **181**(4), 913–919.

Liu G., Wang X., Chen Z., Cheng H.-M. and Lu G. Q. (2009a). The role of crystal phase in determining photocatalytic activity of nitrogen doped TiO_2. *J. Colloid Interface Sci.*, **329**(2), 331–338.

Liu G., Wang X., Wang L., Chen Z., Li F., Lu G. Q. and Cheng H.-M. (2009b). Drastically enhanced photocatalytic activity in nitrogen doped mesoporous TiO_2 with abundant surface states. *J. Colloid Interface Sci.*, **334**(2), 171–175.

Liu Y., Li J., Qiu X. and Burda C. (2006). Novel TiO_2 nanocatalysts for wastewater purification: tapping energy from the sun. *Water. Sci. Technol.*, **54**(8), 47–54.

Liu Y., Liu C.-y., Rong Q.-h. and Zhang Z. (2003). Characteristics of the silver-doped TiO_2 nanoparticles. *Appl. Surf. Sci.*, **220**(1–4), 7–11.

Liu Z., Wang Y., Chu W., Li Z. and Ge C. (2010). Characteristics of doped TiO_2photocatalysts for the degradation of methylene blue waste water under visible light. *J. Alloy. Compd.*, **501**(1), 54–59.

Livraghi S., Czoska A. M., Paganini M. C. and Giamello E. (2009). Preparation and spectroscopic characterization of visible light sensitized N doped TiO_2 (rutile). *J. Solid. State Chem.*, **182**(1), 160–164.

Lützenkirchen-Hecht D. and Strehblow H.-H. (2009). Anodic silver (II) oxides investigated by combined electrochemistry, *ex situ* XPS and *in situ* x-ray absorption spectroscopy. *Surf. Interf. Sci.*, **41**, 820–829.

Malato S., Fernández-Ibáñez P., Maldonado M. I., Blanco J. and Gernjak W. (2009). Decontamination and disinfection of water by solar photocatalysis: recent overview and trends. *Catal. Today*, **147**(1), 1–59.

Mardare D., Luca D., Teodorescu C. M. and Macovei D. (2007). On the hydrophilicity of nitrogen-doped TiO_2 thin films. *Surf. Sci.*, **601**(18), 4515–4520.

Martyanov I. N., Berger T., Diwald O., Rodrigues S. and Klabunde K. J. (2010). Enhancement of TiO_2 visible light photoactivity through accumulation of defects during reduction–oxidation treatment. *J. Photochem. Photobiol. A*, **212**(2–3), 135–141.

Obata K., Irie H. and Hashimoto K. (2007). Enhanced photocatalytic activities of Ta N co-doped TiO_2 thin films under visible light. *Chem. Phys.*, **339**(1–3), 124–132.

Oh S. W., Kang M. N., Cho C. W. and Lee M. W. (1997). Detection of carcinogenic amines from dyestuffs or dyed substrates. *Dyes Pigments*, **33**(2), 119–135.

Park M.-S. and Kang M. (2008). The preparation of the anatase and rutile forms of Ag-TiO_2 and hydrogen production from methanol/water decomposition. *Mater. Lett.*, **62**(2), 183–187.

Pingxiao W., Jianwen T. and Zhi D. (2007). Preparation and photocatalysis of TiO_2 nanoparticles doped with nitrogen and cadmium. *Mater. Chem. Phys.*, **103**(2–3), 264–269.

Qiu X., Zhao Y. and Burda C. (2007). Synthesis and characterization of nitrogen-doped group IVB visible-light-photoactive metal oxide nanoparticles. *Adv. Mater.*, **19**(22), 3995–3999.

Rao K. V. S., Lavédrine B. and Boule P. (2003). Influence of metallic species on TiO_2 for the photocatalytic degradation of dyes and dye intermediates. *J. Photochem. Photobiol. A*, **154**(2–3), 189–193.

Rehman S., Ullah R., Butt A. M. and Gohar N. D. (2009). Strategies of making TiO_2 and ZnO visible light active. *J. Hazard. Mater.*, **170**(2–3), 560–569.

Rengaraj S. and Li X. Z. (2006). Enhanced photocatalytic activity of TiO_2 by doping with Ag for degradation of 2,4,6-trichlorophenol in aqueous suspension. *J. Mol. Catal. A Chem.*, **243**(1), 60–67.

Sato S. (1986). Photocatalytic activity of NO_x-doped TiO_2 in the visible light region. *Chem. Phys. Lett.*, **123**(1–2), 126–128.

Sato S., Nakamura R. and Abe S. (2005). Visible-light sensitization of TiO_2 photocatalysts by wet-method N doping. *Appl. Catal. A Gen.*, **284**(1–2), 131–137.

Seery M. K., George R., Floris P. and Pillai S. C. (2007). Silver doped titanium dioxide nanomaterials for enhanced visible light photocatalysis. *J. Photochem. Photobiol. A*, **189**(2–3), 258–263.

Shen Y., Xiong T., Li T. and Yang K. (2008). Tungsten and nitrogen co-doped TiO_2nano-powders with strong visible light response. *Appl. Catal. B Environ.*, **83**(3–4), 177–185.

Shifu C., Lei C., Shen G. and Gengyu C. (2005). The preparation of nitrogen-doped photocatalyst $TiO_{2-x}N_x$ by ball milling. *Chem. Phys. Lett.*, **413**(4–6), 404–409.

Stathatos E., Petrova T. and Lianos P. (2001). Study of the efficiency of visible-light photocatalytic degradation of Basic Blue adsorbed on pure and doped mesoporoustitania films. *Langmuir*, **17**(16), 5025–5030.

Sun H., Bai Y., Liu H., Jin W. and Xu N. (2009). Photocatalytic decomposition of 4-chlorophenol over an efficient N-doped TiO_2 under sunlight irradiation. *J. Photochem. Photobiol. A*, **201**(1), 15–22.

Sun J., Qiao L., Sun S. and Wang G. (2008). Photocatalytic degradation of Orange G on nitrogen-doped TiO_2 catalysts under visible light and sunlight irradiation. *J. Hazard. Mater.*, **155**(1–2), 312–319.

Sung-Suh H. M., Choi J. R., Hah H. J., Koo S. M. and Bae Y. C. (2004). Comparison of Ag deposition effects on the photocatalytic activity of nanoparticulate TiO_2 under visible and UV light irradiation. *J. Photochem. Photobiol. A*, **163**(1–2), 37–44.

Triantis T. M., Fotiou T., Kaloudis T., Kontos A. G., Falaras P., Dionysiou D. D., Pelaez M. and Hiskia A. (2011). Photocatalytic degradation and mineralization of microcystin-LR under UV-A, solar and visible light using nanostructured nitrogen doped TiO_2. *J. Hazard. Mater.* **211–212**, 196–202.

Vinu R. and Madras G. (2009). Photocatalytic activity of Ag-substituted and impregnated nano-TiO_2. *Appl. Catal. A Gen.*, **366**(1), 130–140.

Wang Y. and Doren D. J. (2005). First-principles calculations on TiO_2 doped by N Nd, and vacancy. *Solid State Commun.*, **136**(3), 186–189.

Wang Y. Q., Yu X. J. and Sun D. Z. (2007). Synthesis, characterization, and photocatalytic activity of $TiO_{2-x}N_x$ nanocatalyst. *J. Hazard. Mater.*, **144**(1–2), 328–333.

Wang Z., Cai W., Hong X., Zhao X., Xu F. and Cai C. (2005). Photocatalytic degradation of phenol in aqueous nitrogen-doped TiO_2 suspensions with various light sources. *Appl. Catal. B Environ.*, **57**(3), 223–231.

Wu P., Xie R., Imlay K. and Shang J. K. (2010). Visible-light-induced bactericidal activity of titanium dioxide codoped with nitrogen and silver. *Environ. Sci. Technol.*, **44**(18), 6992–6997.

Xin B., Jing L., Ren Z., Wang B. and Fu H. (2005). Effects of simultaneously doped and deposited Ag on the photocatalyticactivity and surface states of TiO_2. *J. Phys. Chem. B*, **109**(7), 2805–2809.

Xing M., Zhang J. and Chen F. (2009). New approaches to prepare nitrogen-doped TiO_2 photocatalysts and study on their photocatalytic activities in visible light. *Appl. Catal. B Environ.*, **89**(3–4), 563–569.

Xu J., Ao Y., Fu D. and Yuan C. (2008). A simple route for the preparation of Eu N-codoped TiO_2 nanoparticles with enhanced visible light-induced photocatalytic activity. *J. Colloid Interface Sci.*, **328**(2), 447–451.

Yamashita H., Harada M., Misaka J., Takeuchi M., Ikeue K. and Anpo M. (2002). Degradation of propanol diluted in water under visible light irradiation using metal ion-implanted titanium dioxide photocatalysts. *J. Photochem. Photobiol. A*, **148**(1–3), 257–261.

Yang G.-J., Li C.-J., Huang X.-C., Li C.-X. and Wang Y.-Y. (2007). Influence of silver doping on photocatalytic activity of liquid-flame-sprayed-nanostructured TiO_2 Coating. *J. Therm. Spray. Technol.*, **16**(5), 881–885.

Yang S. and Gao L. (2008). Photocatalytic activity of nitrogen doped rutile TiO_2 nanoparticles under visible light irradiation. *Mater. Res. Bull.*, **43**(7), 1872–1876.

Yuan J., Chen M., Shi J. and Shangguan W. (2006). Preparations and photocatalytic hydrogen evolution of N-doped TiO_2 from urea and titanium tetrachloride. *Int. J. Hydrogen Energy*, **31**(10), 1326–1331.

Yuan Y., Ding J., Xu J., Deng J. and Guo A. J. (2010). TiO_2 nanoparticles co-doped with silver and nitrogen for antibacterial application. *J. Nanosci. Nanotechnol.*, **10**, 1–7.

Zaleska A. (2008). Doped-TiO_2: a review. *Recent Patents Eng.*, **2**, 157–164.

Chapter 9

Pd nanocatalysts for PCB removal

T. Hennebel, J. P. Fitts, S. Van Nevel and N. Boon

9.1 INTRODUCTION

Polychlorinated biphenyls (PCBs) are a class of compounds in which the aromatic biphenyl core is substituted with one to ten chlorine atoms (Figure 9.1). Theoretically, 209 different PCB congeners exist, varying in extent and location of the chlorine substitutions. Polychlorinated biphenyls were synthesized and sold as complex mixtures containing between 60–90 congeners. These mixtures were produced at large scales and due to their superior thermal and chemical stability widely deployed for more than five decades beginning in the 1930s. PCBs are viscous insulating liquids that are essentially non-corrosive and non-flammable. These properties led to their widespread use in a diverse range of industrial applications and consumer products including capacitors, heat transfer systems and hydraulic equipment. The characteristics that made PCBs industrially attractive also resulted in their persistence following their inevitable dispersal in the environment. Many studies reported bio-accumulation in the food chain and demonstrated how PCBs negatively affect the immune, reproductive, nervous and endocrine systems. This resulted in their withdrawal from commercial production and use in the mid-1970s.

Due to their high hydrophobicity and K_{ow} values, a large fraction of the PCBs released into the environment were deposited on the bottom of rivers, lakes and oceans either as sludge or adsorbed onto sediments. The aqueous transport of PCBs occurs partly via dissolved PCBs, however, the movement of PCB-laden particles is one of the most important dispersion mechanisms. The sediment can also act as a source of PCBs through desorption and resuspension. Major sources of PCB release into the environment include emission by open burning, incomplete waste incineration, leakage from landfill sites and leakage from industrial or household devices. Despite international efforts to control and regulate persistent halogenated organic pollutants, past and ongoing releases have resulted in widespread contamination of soils and sediments. The U.S. Environmental Protection Agency (EPA) Toxics Release Inventory Database indicates that 1.13 million kg of PCBs were released from monitored industries in 2001 alone. It was estimated that between 1929 and 1980, 370 million kg of PCBs has been released to the environment (Tanabe, 1988). This contamination with PCBs is of great public concern due to bioaccumulation and potential toxicity to humans and wildlife (Kobayashi *et al.*, 2011, Zhang *et al.*, 2011).

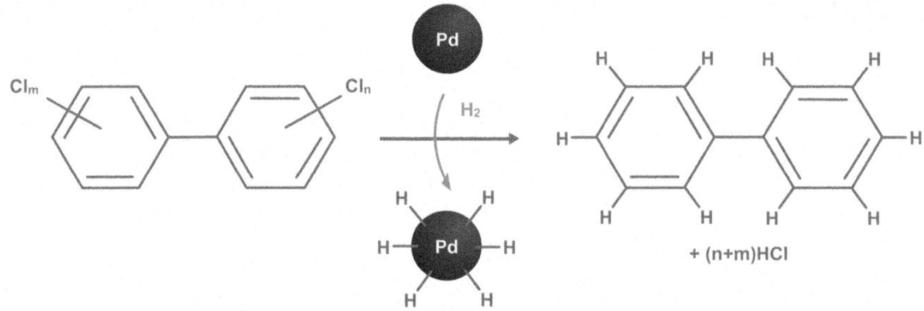

Figure 9.1 Dechlorination mechanism of PCBs using Pd nanoparticles as catalyst and hydrogen gas as hydrogen donor.

9.2 REMEDIATION OF PCBs FROM THE ENVIRONMENT

In this chapter, the potential use of nanoparticles for PCB remediation will be evaluated. To get a better insight in the possibilities of introducing nanoparticles into the market of halogenated pollutant remediation, the currently applied technologies and their specific shortcomings are discussed. Indeed, due to their recalcitrance and hydrophobicity, PCBs have required specially developed remediation technologies. In general, all available technologies can be divided in three classes according to the remediation goal (Nollet, 2005). (I) Stabilization technologies, for example, landfilling and capping enclose the PCBs to limit their further spread in the environment. (II) Separation technologies, for example, soil washing and thermal desorption transfer the PCBs from soil and sediments to reduce the waste volume that needs to be landfilled or destroyed. (III) Destructive technologies, for example, thermal, chemical and microbial remediation strategies aim for the total abatement of PCBs by breaking them down to non-toxic chemical constituents. The choice of the preferred technology strongly depends on the characteristics of the contaminated site and the PCB mixture.

One of the most applied technologies is soil washing in which sludge or sediment is excavated and mixed ex-situ with surfactants or solvents to transfer PCBs from the soil to the liquid phase (Nollet 2005). Subsequently, this liquid stream undergoes additional treatment using destructive technologies such as incineration or dehalogenation (Lajoie *et al.*, 1997). An alternative to soil washing is thermal desorption, whereby the contaminated soil is heated to evaporate the PCB. Temperatures near 600°C are advised for this process, and therefore, the large energy input means that these technologies are typically not economically viable.

These technologies require large amounts of solvents and energy and can therefore be considered as unsustainable. The last decade, researchers have explored the possibilities of introducing nanotechnology in the field of environmental remediation. Two metal based nanocatalysts, namely zerovalent iron and palladium, have shown to be very promising to be applied in the removal of halogenated contaminants from environmental compartments. Both nanometals are effective catalysts in dehalogenation reactions and can be used to transform a variety of halogenated contaminants such as chlorinated solvents or PCBs into less or non-persistent molecules. This chapter will give an overview of the working mechanism of palladium nanocatalyst based dehalogenation with a focus on PCB removal.

9.3 DEHALOGENATION REACTIONS BY PALLADIUM (NANO-)CATALYSIS

Palladium (Pd), along with platinum, rhodium, ruthenium, iridium and osmium, form the platina group metals (PGM). Pd is electrically stable, chemically and thermally inert but its most remarkable characteristic is the interaction with hydrogen. Hydrogen is absorbed by Pd and stored in the interstitial spaces of the crystal grid. During this chemisorption process, hydrogen molecules are homolytically dissociated and taken up by the Pd as a sponge (De Windt, 2006). Pd has a broad range of applications and it is arguably the most versatile and the most widely applied catalytic metal (Tavakoli *et al.*, 2007). Its catalytical application has been described in hydrogenation, dehydrogenation, oxidation, hydrogenolysis, and C-C or C-N coupling reactions. Moreover, Pd is by far the most active metal for the hydrogenolysis of aromatic halogen atoms (Klabunde, 2001). During the last decades, especially nanoscale Pd materials have gained much interest because they show new or altered properties and characteristics compared to the bulk material. The main reason for this is their extremely small size and consequently large surface to volume ratio, that is, the specific surface area. Moreover, given the high cost of Pd (15–20 € g^{-1} Pd), it is essential to minimize the Pd amounts in catalytic reactions. Over the last years, the use of Pd as dehalogenation catalyst for environmental applications gained an enormous interest. Especially applications of heterogeneous Pd catalysts, in which metallic Pd is supported on a carrier, are intensively studied for the treatment of aqueous matrices or air flows containing environmental pollutants such as chlorinated solvents, PCBs, PBDEs, CFCs and so on PCBs are used in this chapter as model compound to elucidate the dehalogenation reaction based on Pd nanoparticle catalysis given the extreme recalcitrance of PCBs. The production and type of Pd nanoparticles and the reaction mechanism will be described in detail.

9.3.1 Production and supports for Pd nanoparticles

The preparation of nanomaterials can be classified into physical and chemical methods. The physical methods are based on reducing the size of bulk metals, while chemical methods are based on the reduction of metal ions or decomposition of precursors to form atoms, followed by aggregation of the atoms. Nanoparticles prepared by chemical methods usually have a more narrow size distribution than physical methods (Liz-Marzán & Kamat, 2003). In the case of Pd nanoparticles, especially chemical methods are applied starting from a Pd (II) salt, a reductant and a stabilizer.

Pd nanoparticles on varying support materials have been developed and their activity towards PCB dechlorination has been tested. The selection of the support is of utmost importance as it strongly affects the hydrodechlorination selectivity, the catalytic activity and the catalyst stability. An ideal palladium support is stable under corrosive reaction conditions such as high HCl concentrations and high temperatures. Secondary reactions like chloride-exchange or coupling reactions should be avoided since they would lead to selectivity loss (van de Sandt *et al.*, 1998).

Supports include both organic and inorganic structures. The mostly applied inorganic support is zerovalent iron (ZVI). Grittini *et al.* (1995) were the first to use palladium on ZVI and observed that palladization of ZVI increased the dehalogenation rates for PCB mixtures. Technically, ZVI functioned not only as support material but also as hydrogen donor (Korte, 2000). Furthermore, palladium was not applied under the form of nanoparticles but as a coating on the microscale zerovalent iron (Grittini *et al.*, 1995; West *et al.*, 1996). Nowadays, the production of nanoscale ZVI with hot-spots of Pd is widely described and the performance of these nanoparticles towards PCB dechlorination has been largely improved.

Other examples of inorganic Pd-supports used for PCB dehalogenation are alumina and zinc (Schuth & Reinhard, 1998; Wafo *et al.*, 1997). Nanocomposites consisting of Pd on Mg, Al, Mn, or Sn are recently considered by some authors (Yang *et al.*, 2011). The main disadvantage of inorganic supports is their sensibility towards acids formed in the dehalogenation reaction. To avoid this, one could use organic supports such as the frequently used activated carbon (Choi *et al.*, 2009). Other innovative organic carriers are carbon nanotubes and polyethylene beads (Chen *et al.*, 2010; Liao *et al.*, 2007). These carriers are hydrophobic, favouring the sorption of the hydrophobic PCBs to the carrier.

Consequently, the hydrogen transfer from the catalyst to the PCB is enhanced (Anwer *et al.*, 1989). The acid-base properties of the carrier can also have an influence on the catalyst. Alkaline supports like alumina are able to capture the chlorine atoms released after hydrodechlorination and enhance therefore the reaction rate (Schoonenboom *et al.*, 1995). Furthermore, the formed HCl gets neutralized, allowing the reaction to carry on until a further conversion without the acidification of the medium (Aramendia *et al.*, 1999).

An innovative support for Pd nanoparticles, which showed to be very effective in PCB degradation, is bacterial biomass. When a hydrogen donor is provided (such as formate), bacteria can reduce Pd(II) and subsequently precipitate it as Pd nanocrystals on their cell wall and in their periplasmatic space, that is, bio-Pd. De Windt *et al.* (2005) and Mertens *et al.* (2007) used the metal-respiring bacterium *Shewanella oneidensis* to produce bio-Pd. They showed the superior characteristics of this catalyst towards the dehalogenation of respectively PCB's and lindane. Baxter-Plant *et al.* (2003) used the bacterium *Desulfovibrio desulfuricans* to produce bio-Pd and used this as a catalyst for the dehalogenation of PCBs and chlorophenols. The production process of bio-Pd is a simple, cost effective and environmentally friendly technology. The presence of a bacterial cell offers also advantages in the dehalogenation process: (i) the Pd is attached to a biological carrier which stabilizes the nanoparticles and keeps them in suspension without sedimentation or conglomeration and (ii) the bacterial carrier can interact with the pollutant to offer a more efficient removal (Hennebel *et al.*, 2009). However, the production and the use of bio-Pd also have some disadvantages. Indeed, the formation of Pd(0) nanoparticles is a delicate biological process and depends very much on the growth parameters of the bacteria. Furthermore, the stability of its cell wall has never been studied over longer periods.

9.3.2 Hydrogen donors for dehalogenation reactions

The main mechanism of dehalogenation reactions with Pd is based on the formation of reactive hydrogen species that are produced by the contact between a hydrogen donor and the catalyst surface. Subsequently, these hydrogen radicals attack the R-X bound to result in R-H and HX molecules. Hydrogen gas is commonly used due to its efficiency and low poisoning properties (Figure 9.1).

Not only hydrogen gas but also formic acid, formate, hydrazine and isopropanol are described as practicable alternatives (Kopinke *et al.*, 2004; Mertens *et al.*, 2007, Urbano & Marinas, 2001). The preferred donor in the Pd-catalyzed hydrodechlorination is hydrogen as only in the case of formic acid, similar dehalogenation rates were obtained (Hennebel *et al.*, 2009; Kopinke *et al.*, 2004). However, due to the low water solubility of hydrogen gas (0.4 mg L^{-1} at atmospheric conditions), more hydrogen than stoichiometrically required have to be dosed to obtain complete dehalogenation (Hennebel *et al.*, 2011b). Moreover, also practical implications are encountered by hydrogen handling. The repeated dosage or even in situ providing of gaseous hydrogen into contaminated aquifers is a difficult procedure. Hence, it is clear that the study of alternative reductants is a worthwhile task, although they can show a decreased dechlorinating reactivity.

An alternative for using hydrogen donors could be the application of zerovalent metal based nanoparticles. ZVI is known for many years to enable dechlorination reactions (Tratnyek *et al.*, 1997, Vogan *et al.*, 1999). The hydrogen-generating reaction in this system is the corrosion of iron according to:

$$Fe^0 + 2H_2O \longrightarrow Fe^{2+} + H_2 + 2OH^- \quad \text{(Eq. 9.1)}$$

This concept is schematically shown in Figure 9.2. ZVI impregnated with Pd systems are more and more investigated in this context (Choi *et al.*, 2007; Li *et al.*, 2011). In these bimetallic particles, galvanic cells are formed by the coupled iron and a noble metal such as palladium. Iron serves as the anode in the galvanic cells and becomes preferably oxidized whereas the noble metal (cathode) is protected. The oxidation of Fe(0) generates the indispensable hydrogen donor for Pd-catalyzed dechlorination (see Eq. 9.2). The galvanic coupling between the iron and the noble metal accelerates contaminant reduction by electron transfer, and it also favours water reduction and formation of activated H-species, which in turn can contribute to increased rates of contaminant reduction.

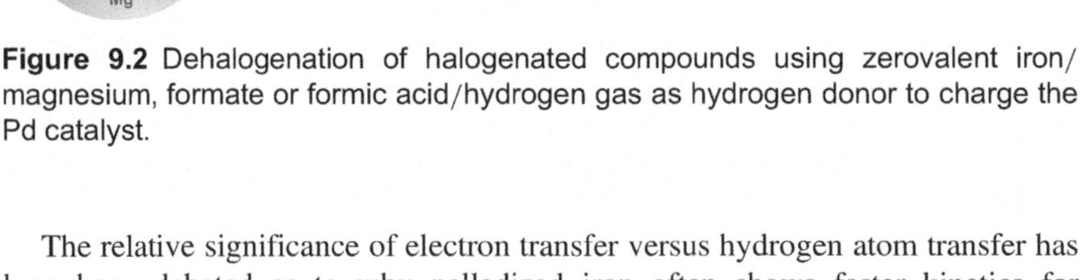

Figure 9.2 Dehalogenation of halogenated compounds using zerovalent iron/ magnesium, formate or formic acid/hydrogen gas as hydrogen donor to charge the Pd catalyst.

The relative significance of electron transfer versus hydrogen atom transfer has long been debated as to why palladized iron often shows faster kinetics for contaminant reduction. For example, the half-life of PCB 21 by nZVI alone was found to be 1 year even under conditions with excess nZVI, indicating the necessity of introducing Pd to enhance the corrosion of the Fe(0) for reaction with the PCB. By comparing the nZVI content normalized rate, palladization was found to increase the rate of dechlorination for PCB 21 by 500 times, which was greater than the increase of the H_2 production rate (Zhuang *et al.*, 2011).

Therefore, galvanic couple formation could not account for the whole portion of the rate enhancements for PCB 21. Alternatively, the iron can be replaced by magnesium to form Mg/Pd bimetallic systems. Magnesium shows several advantages over the Fe/Pd system including the ability of the Mg/Pd to dechlorinate in the presence of oxygen which causes the complete corrosion of micro- and nano-scale iron. This is possible for magnesium because it forms thin, oxygen-impermeable but slightly soluble water layers (Coutts *et al.*, 2011).

The nZVI in combination with a catalyst is already applied on large scale and was for example successfully tested at a chlorinated solvent contaminated site in New Jersey, USA (USEPA, 2008). This technology has proven to be extremely effective in the dehalogenation of chlorinated solvents and pesticides. Moreover, barriers using Pd on Fe for treating sites at full-scale are already commercially available (cfr. KEOpS® from Sita remediation). The fact that iron corrosion is the rate-limiting step in the reaction represents the main disadvantage. Moreover, corrosion rates usually decreases over time and a passivation layer of Fe-oxides/hydroxides is formed (Van Nooten *et al.*, 2008). Nevertheless, iron is an environmentally friendly hydrogen source with a relatively long lifetime, depending on the water chemistry (pH and E_H values, salinity) in the order of magnitude of 1000 days. This may be an advantage for long-term in situ application in contaminated aquifers, where continuous hydrogen supply is desirable. It is not, however, an appropriate solution for fast reactions in relatively short time intervals.

9.3.3 Reaction kinetics and selectivity

A variety of sources is available describing dechlorination of PCBs with Pd based catalysts. Both the differences in type and amount of PCBs congeners studied and the applied reaction conditions make it very hard to compare catalysts. Indeed, the dehalogenation speed is influenced by the degree of chlorination and the substitution position of the chlorine. Moreover, the reaction largely depends on the catalyst concentration, the hydrogen donor, the solvent, additives, temperature and pressure. When describing catalyst performance, different values are reported. Most authors present percentages or amounts of PCB removal in a certain time frame. Other authors give the chloride release in function of time or when (apparent) first order kinetics are detected, first order rate constants (k). In Table 9.1, some examples of different Pd catalysts and their recalculated activity expressed in μg PCB mg^{-1} Pd^{-1} h^{-1} are shown. The complexity of catalyst comparison becomes clear from this table.

When addressing the activity of palladium against several different PCB congeners, a difference is made between ortho-, meta-, and para-substituted PCBs. Most authors conclude that para-elimination is preferred over meta-elimination (Choi *et al.*, 2009; De Windt *et al.*, 2005; Lassova *et al.*, 1998; West *et al.*, 1996) although this can be the other way around (Rodriguez &

Table 9.1 Overview of different supports, reaction conditions and reaction rates. (*): Aroclor 1240 = commercial PCB mixture with 54% Cl.

Support	Reaction condition	Rate (μg PCB.mg Pd $^{-1}\cdot h^{-1}$)	Reference
0.5% Pd on carbon	Decahydronaphtalene with $Ca(OH)_2$, 53 * 10^3 mg · L^{-1} Aroclor 1240$^{(*)}$, 6 atm H_2, 180°C, 30 seconds	8.48 * 10^6	Forni *et al.* (1997)
1% Pd on alumina	Deionized water without gasphase, 2 mg · L^{-1} PCB 3, initial H_2 saturation, 21°C, 10 min	2400	Schuth & Reinhard (1998)
10% Pd on carbon	Hexane, 206 mg/L PCB 29, room temperature, 15 min, 487 kHz ultrasonication	4994	Rodriguez & Lafuente (2002)
Biomass: *Shewanella oneidensis*	Mineral medium, 1 mg · L^{-1} PCB 21, H_2 atmosphere, 28°C, 5 hours	0.18	De Windt *et al.* (2006)
3.5 mg Pd.cm^{-1} on Ni-foam electrode	Electrochemical dechlorination in 50% methanol/H_2O, 100 mg · L^{-1} PCB 29, 30°C, 90 min	9257	Yang *et al.* (2007)
10% Pd on activated carbon	Methanol with triethylamine, 600–4600 mg · L^{-1} PCB 77, H_2 atmosphere, 20°C, 24 hours	4167	Kume *et al.* (2008)
0.61% Pd in HDPE beads	Acetonitrile, 31 mg · L^{-1} PCB 101, 10 atm H_2, 150 atm CO_2, 80°C (supercritical conditions), 120 min	0.106	Wang & Chiu (2009)
0.087% Pd on iron	Soil in deionized water, 50 mg · kg^{-1} PCB 101, 25°C, 4 hours	0.166	He *et al.* (2009)
0.1257% Pd on Mg	Methanol, 10 mg · L^{-1} PCB 151, air, room temperature, 2 min	9547	Coutts *et al.* (2011)

Lafuente, 2002). Ortho-elimination is consequently seen to be the slowest step. According to Choi *et al.* (2009), this can be explained mainly by steric effects. The first step of the dechlorination reactions is adsorption of the PCB to the active metallic surface (Bucheli & Gustafsson, 2003). This can be affected by steric hindrances; a parallel configuration of the phenyl rings of PCBs to the catalyst surface is most favorable. However, the inhibition of the intra molecular rotation around the 1,1′ bond, exerted by chlorines close to the bond (inhibition ortho > meta > para), hinders the parallel configuration and hereby restricts surface interaction of PCBs with the catalyst surface. This effect is the most pronounced for ortho-chlorines, as reflected in their lowest dehalogenation rates. Consequently, ortho-substituted congeners are generally the only endproducts if the dehalogenation is not completed to biphenyl (Lowry & Johnson, 2004). These ortho-substituted congeners are non-coplanar, which is regarded to be less toxic to microbiota than para- or meta-substituted coplanar congeners, making them more susceptible for further anaerobic biodegradation.

9.3.4 Solvents and additives in Pd catalysis

The dehalogenation of PCBs by Pd nanoparticles has been studied in several solvents or solvent mixtures such as water, methanol, ethanol, hexane, and so on Some examples coupled to the reaction rates are listed in Table 9.1. Urbano *et al.* (2001) stated that the choice of solvent for liquid-phase hydrodehalogenation is of utmost importance. The following aspects are to be considered in the selection of a solvent: (i) dissolvability of the reactants, (ii) influence on catalytic reaction mechanism, (iii) recovery, high availability and fair price and (iv) toxic effects and environmental aspects. However towards large scale applications, it is clear that the medium in which the PCBs are present or retrieved, is the most realistic reaction solvent. Alcohols (mainly methanol and ethanol) are the most usual choice as solvent for PCB hydrodehalogenation. The fact that alcohols such as ethanol and isopropanol may also play a role as a hydrogen donor can represent an additional advantage (Zhang *et al.*, 1994). The solvent can largely affect the solubility of the pollutant, which improves the mass transfer to and away from the catalyst. Moreover, a good selection of the solvent can also minimize poisoning effects of either reactants or products. For example, aging experiments showed that a 50/50 mixture of ethanol/water could improve the Pd/C catalysts dramatically compared to reactions in plain ethanol (Hoke *et al.*, 1992).

Some of the most applied additives are bases and the rate and performance of catalytic hydrodechlorination often depends on the type of base utilised. Although the exact function of the base in the catalytic hydrodechlorination is not completely understood, it serves, at least, as a proton scavenger for hydrogen halide liberated in the reaction (Hoke *et al.*, 1992; Urbano & Marinas, 2001). It was for example observed that sodium hydroxide (NaOH) dissolved palladium when it was used as proton acceptor (Balko *et al.*, 1993) but on the other hand, dosing NaOH resulted in the highest reaction rates and catalyst stability. This difference in effects of NaOH supply could be attributed to differences in reaction conditions. In case of PCB dechlorination, one of the most applied bases is triethylamine (Monguchi *et al.*, 2011) and quaterny ammonium salts are often used as phase transfer agent (Marques *et al.*, 1994). The latter may help to transfer the alkaline ions from the aqueous phase forward to the catalysts by a phase-transfer process.

9.4 PCB DEGRADATION BY BIMETALLIC Pd CATALYSTS

Research efforts have only just begun to investigate bimetallic formulations to improve catalyst activity and selectivity for PCB degradation (Choi *et al.*, 2008; Coutts *et al.*, 2011; Yang *et al.*, 2011). The potential benefits of developing bimetallic catalysts are numerous, and therefore, the design goals extend beyond improved activity, selectivity and stability to include the need for cheaper and more abundant metals, more environmentally benign synthesis and disposal as well as lower cost and lower energy input. The pairing of metals and options for how they are arranged, synthesized, stabilized and supported are also numerous, and as a result systematic selection criteria have yet to be proposed. What follows is a review of potential selection criteria for developing bimetallic catalysts, placing the few examples of bimetallics developed to catalyze PCB degradation in the proper context by borrowing from the wider catalyst literature. The systematic development of bimetallic catalysts places unique challenges on characterizing nanometer-scale morphology and atomic-level structures, and therefore, applications of the current state-of-the-art characterization methods to study bimetallics are reviewed.

This part of the chapter divides bimetallics into alloyed and composite materials. Alloyed bimetallics are defined as catalysts where two metals occur in the same phase; sharing a common lattice or asymmetric unit. In contrast, composite bimetallics employ two metals in separate phases that share an interface or boundary. Core-shell bimetallics and nanoparticles on active supports are examples of composite catalyst designs that have been applied for dechlorination reactions. Both phases play an active role in composite bimetallics, and therefore, are distinct from the previously reviewed supported Pd catalysts. The distinctions between the two classifications are not absolute. For example, nanocatalysts can have a core-shell composite structure where an alloyed shell is the catalytically active phase.

Formal selection criteria have been proposed to pair metals in bimetallic catalysts for fuel cell applications (Demirci, 2007). Fundamental properties of individual metals such as the center of the d-band relative to the Fermi level (Hammer & Norskov, 2000) and operational experimental parameters such as standard electrode potential have been applied to select metals for bimetallic catalysts

(Yang *et al.*, 2011). Different metal pairings along with synthesis and conditioning methods can significantly change the adsorbate bond-energy and the geometry of the active site (De Corte *et al.*, 2011; Shao *et al.*, 2006). Therefore, bimetallic alloys have the potential to improve selectivity for the most recalcitrant PCB congeners.

Core–shell composite structures with a precious metal shell and a base metal core have been most successfully used to lower the cost of the catalyst by reducing the amount of precious metal needed to achieve comparable catalytic activity (Gong *et al.*, 2011). The synthesis requires matching the structures of the core and shell materials so that reactivity is maintained, the particles are stable, and diffusion from either phase does not poison the catalyst. Alternatively, different degrees and types of structural mismatches between the core and shell phases can be used to promote strain in the shell phase that in turn modifies the activity and/or selectivity of the catalyst (Wang *et al.*, 2011). Core-shell bimetallic catalyst designs have not yet been applied to PCBs to date, but Wong and co-workers catalyzed chlorinated solvent degradation using Pd-shell and Au-core nanoparticles (Nutt *et al.*, 2005).

Catalysts supported by a metal phase that actively participates in the catalytic process by facilitating the diffusion of reactants and/or products offer another type of composite catalyst design that may prove useful for PCB degradation. Active support materials have been used to modify sorption strength and diffusion of reactants to improve catalyst activity (Shao *et al.*, 2006). Improved diffusion and desorption of reaction products away from active sites was shown to prevent catalyst poisoning (Yang *et al.*, 2011).

Researchers have demonstrated the need to construct a fundamental understanding of how a catalyst functions by characterizing the coordination environment, bonding and electronic structure of active sites (Adzic *et al.*, 1982). The complex composite structures and surface coordination environments of the different types of bimetallic catalysts, however, require a nuanced approach to characterizing active sites and their distribution on the catalyst surface. Scanning and transmission electron microscopy have become standard methods for characterizing nanometer-scale dimensions and morphology of catalysts. Technical advances in electron microscopy, especially with regard to electron and x-ray detection, have enabled high resolution imaging capable of revealing domains within single nanoparticles and, in idealized model catalysts, the geometry of individual active sites. In contrast, synchrotron-based x-ray absorption spectroscopy and scattering methods provide an ensemble average over all of the structural domains and active sites within a catalyst sample. These X-ray methods are also unique in that *in operando* studies are possible, where electronic and/or physical structural characteristics of the catalysts are observed as they work in small scale reactors (Foettinger *et al.*, 2011). Finally, the next major advance toward developing *ab initio* models of catalyst functioning is to take advantage of the complementary nature of these methods and experimental approaches by combining results in a single data analysis framework. This framework will then generate structural input parameters and constraints for *ab initio* models of catalyst function (Billinge & Levin, 2007).

9.5 APPLICATIONS OF Pd NANOPARTICLES FOR PCB TREATMENT

9.5.1 Treatment of PCBs spread in the environment

Thus far, only a limited number of publications have reported on the pilot- or full-scale applications of Pd for catalytic PCB dehalogenation. This is probably due to their extreme recalcitrance and hydrophobicity, which makes them inaccessible for reaction. Indeed, PCBs are often strongly adsorbed to soil and sediment particles and are not transported along with the groundwater. This is in contrast with other, more hydrophilic organohalogens such as chlorinated solvents or pesticides, which can be treated by installing reactive barriers containing Pd particles. At present, several remediation companies construct these reactive walls consisting of cartridges containing Pd(0) on nanoZVI. While the groundwater migrates through the barriers, the chlorinated solvents are dehalogenated and only non-toxic dechlorinated products leave the barriers. A similar system cannot be used as the largest fraction of PCBs is related to soil particles and to the organic fraction in the sediment. For this specific reason, a pump and treat technology consisting of pumping groundwater containing organohalogens and catalytic Pd treatment, will also not be effective for PCBs. Several pilot plants based on this principle were previously described in the case of trichloroethylene and lindane contamination (Davie *et al.*, 2008; Hennebel *et al.*, 2011b). Therefore, alternative technologies should be developed.

In this regard, researchers applying biogenic Pd nanoparticles as catalyst have done considerable efforts. De Windt *et al.* (2005) treated sediments containing 7 PCB congeners at a Pd concentration of 100 mg Pd kg^{-1} sediment dry weight. The sum of the 7 indicator PCBs was decreased to 27% of their initial concentration, within 48 h at 28°C. There was no selective effect towards dechlorination of specific PCB congeners, since the catalytic effect was observed over a wide range of highly and lightly chlorinated PCBs. Moreover, the catalytic activity obtained by biogenic Pd was equal to a 10-fold higher concentration of Pd black (a black precipitate of zerovalent Pd often used in catalysis). This was attributed to the hydrophobic interactions between the cell biomass and the PCBs. Alternatively, Choi *et al.* (2008) developed the concept of Pd/Fe on activated carbon pellets. The activated carbon improved the adsorption and hence the availability to react on the catalytic surface. The authors suggested to implement the catalyst between thin geo-textile membranes and install these horizontally at the bottom of estuaries and riverbanks containing PCB-contaminated sediments (Choi *et al.*, 2008). Alternatively, the reactive activated carbon (RAC) could be directly mixed with the PCB-contaminated matrix for more prompt capture and degradation of PCBs, while a RAC barrier can also be installed vertically for flow-through treatment of PCBs-contaminated groundwater.

An alternative in-situ remediation technology would consist of the supply of a Pd (II) salt and a carbon source (e.g. glucose) to a soil. It was shown that Pd nanoparticles were formed by biofilms on sand grains in aquifer mesocosms and that this Pd catalyzed the reduction of the toxic Cr(VI) into the less toxic Cr(III) (Chidambaram *et al.*, 2010). The authors inoculated an aquifer column with *Clostridium pasteurianum* BC1. However, the inoculation is very likely not required in a remediation program, as many bacteria that reduce Pd and form Pd nanoparticles belong to the endogenous population. The advantage of such a technology would be the good contact between catalyst, PCB and soil matrix. A drawback of this technology might be the low efficiency of Pd reduction and hence the environmental implications and costs related to the unreduced Pd(II). Alternatively, one could produce the nanoparticles off site and inject them in the soil. This technology has been previously applied using Fe(0) nanoparticles. One could mix the Pd nanoparticles and a slow degradable C-source with these sludges at extremely low concentrations approaching the background values of metals. Autochthonous bacteria would degrade the C-source and form hydrogen gas in these anaerobic environments. The latter could be used as hydrogen donor to charge the Pd nanoparticles. The applications of Pd based technologies that we foresee, are schematically shown in Figure 9.3.

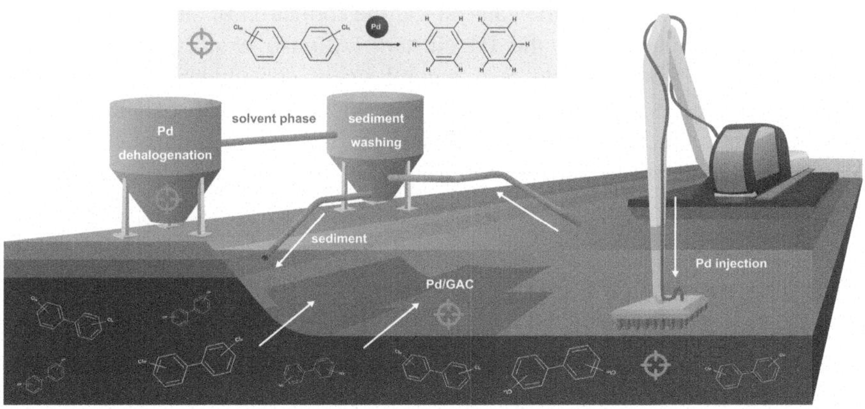

Figure 9.3 Schematic presentation of possible applications for PCB remediation using Pd catalysts. Pd nanoparticles could be injected in contaminated sediments or encapsulated in a Pd on activated carbon (GAC) cloth to cover the sediment. Alternatively, sediments could be excavated and washed with solvents (e.g. ethanol). These PCB charged washing liquids can subsequently be treated in a Pd-based reactor.

This bioremediation strategy, however, has not yet been developed to the extent of pump-and-treat strategies, which are commonly used to treat hydrophilic compounds. Many aspects such as the durability of the dehalogenation process, the sensitivity for poisoning and the *in situ* hydrogen production need to be demonstrated in batch and pilot scale experiments. Moreover, technological aspects such as a homogeneous dosing of the catalyst and the electron donor to soils or sludges have to be studied.

Also, ecological aspects such as environmental effects of Pd supply, the fate and effect of nanoparticles in the soil need to be considered. Finally, in order to be economically competitive, the costs will need to be lower relative to incineration. Therefore, one should work at maximal Pd concentrations between 10 and 40 g Pd ton^{-1} (assuming a Pd cost of 20 € g^{-1}). *Ex situ* treatment of PCBs extracted from the sediment can be a more costly alternative, which can only be applied when the soil or sludges can be excavated. As mentioned above, one of the current remediation technologies consists of contaminant extraction using solvents or surfactants or thermal desorption, both followed by incineration. Both liquid and gaseous streams can subsequently be treated in reactors containing Pd as catalyst. All kinds of reactors have been developed to treat aqueous streams contaminated with organohalogens. Among them, fixed bed reactors are the most popular on lab-, pilot- and full-scale but also fluidized bed reactors, membrane reactors and (bio-)electrochemical cells have been suggested (Hennebel *et al.*, 2009; Hennebel *et al.*, 2011a; Hennebel *et al.*, 2011b; Hennebel *et al.*, 2011c). A possible limitation for the use of Pd might imply the poisoning of the catalyst. For example sulfides are known to have a strong affinity for the Pd metal and may block the active sites of the catalyst via the formation of strong Pd–S bonds and layers of sulfide around the Pd clusters (Alfonso *et al.*, 2003; Gravil & Toulhoat, 1999). Since sulfide is a natural water constituent under reducing conditions produced by microbial sulfate reduction, sulfide induced catalyst deactivation is a crucial issue which hinders the full exploitation of the catalyst potential as a treatment technology for groundwater remediation (Angeles-Wedler *et al.*, 2009).

9.5.2 Treatment of PCB containing sources

Not only sites contaminated with PCBs after leakage into the environment but also electric devices containing PCBs need to be disposed properly. Indeed, PCBs were used as electrical insulating oil in capacitors and transformers. Therefore, some researchers focused on the development of a safe and reliable degradation method of spent PCBs that have been used in these devices. The group of Sajiki explored hydrodechlorination of PCBs practically used as high-tension capacitor oil for over a quarter-century (Monguchi *et al.*, 2011). A Pd on carbon catalyst was able to fully dechlorinate the PCBs and produce biphenyl using hydrogen gas, methanol and trimethylamine. Recently, the same researchers expanded the system to pilot scale, which is to our knowledge the first demonstration that PCBs can be treated effectively in these difficult matrices. These results open up many perspectives to enhance the dechlorination by using more novel Pd or bimetallic Pd catalysts.

9.6 CONCLUSIONS AND PERSPECTIVES

Pd nanoparticle based dehalogenation of PCBs is an innovative and promising technology that is well-studied in literature. However, thus far a limited amount of studies reported on full-scale PCB remediation using this catalyst. This is very likely due to two reasons: (i) the production and application of Pd nanoparticles is costly. Indeed, the price of Pd is extremely variable but can add up to 20 € g^{-1}. Moreover, one can assume an additional cost of around 0.1 € g^{-1} when using a carrier material. Hence, one kg of 0.1 wt% Pd catalysts typically costs 120 € and (ii) Pd is considered as a heavy metal from which exact (eco-)toxicological effects are not known. The leaching of ions is commonly reported for catalysts based on transition metals. Little is known about the fate of Pd in the environment and potential adverse health effects, but due to the precautionary principle, release needs to be prevented. Although minimal amounts of Pd doping result in a spectacular activity increase of zerovalent iron, companies refuse to mention the active compound of the catalyst. Regarding the release of metallic nanoparticles into the environment, it is of paramount importance for technology providers to clearly communicate the composition of their catalysts. Another main reason for the absence of large-scale remediation is the complexity of PCB contaminations. PCBs are often present as mixtures in the environment and associated with oils. Due to their high K_{ow} values, PCBs have also a large affinity for soil particles. Therefore, remediation companies often opt to excavate sediments and use incineration as well established technology. Besides the rather unexplored economical and ecological costs and unaddressed practical issues of Pd nanoparticles, more attention should be directed to the fundamentals of nanochemistry in the environment, that is, contaminant transformation at the nanoparticle–water-solid interface. More field-scale studies to investigate the transport and reaction mechanisms are clearly needed to further determine the potentials and the limits of the technology. Through improved understanding of the geochemistry of nanoparticles in both basic and field demonstrations in well-characterized environments, the prospect for successful exploiting of the nanoscale technology for environmental applications appears very good (Zhang 2003).

Given the costs of the Pd nanoparticles, the activity should be increased to apply minimal catalyst concentrations. Several approaches on combining metals have been suggested. Indeed, composites of Pd and Mg, Al and Fe have been extensively reported with regard to PCB remediation (Coutts *et al.*, 2011). Although bimetallic Pd/Au nanoparticles have shown to increase selectivity and activity with a factor 10 to 100 towards trichloroethylene (De Corte *et al.*, 2011; Nutt *et al.*, 2006), their efficiency towards PCB dehalogenation has never been tested. The goal would be to produce a catalyst with maximal activity at reasonable costs by combining Pd with other metals such as Ni, Ag and Au. Recent nanoresearch has focused on functionalization of nanoparticles with biomolecules, such as DNA or proteins. For instance, Nykypanchuk *et al.* (2008) described the DNA-guided crystallization of colloidal gold nanoparticles that have been mediated by interactions between complementary DNA molecules that were attached to the surface of the nanoparticles. These bio-gold crystals could for example be used in medical imaging. Likewise, there is a great potential in functionalizing Pd nanoparticles to optimize activity in for example dehalogenation of PCBs. One could design the biomolecules in a way to maximize the interaction between the nanoparticle and pollutant.

For the dehalogenation reactions using palladium, reducing power needs to be supplemented. This can be with formate, however the most efficient activator remains hydrogen gas (Hennebel *et al.*, 2009) although there are safety issues with handling and dosing hydrogen gas. Also in soils and sediments, the supplementation of hydrogen gas is practically impossible. An attractive and green alternative is the in situ production of hydrogen gas. This can be achieved by using for example electricity to hydrolyse water (De Gusseme *et al.*, 2011; Hennebel *et al.*, 2011a), although again practically difficult for soils and sediments and economically not feasible. Another alternative is the *in situ* biological hydrogen production using bacteria, cultivated under fermentative conditions. In this way, costs and problems coupled to hydrogen supply are avoided. In an MBR system, this microbially produced hydrogen (bio-H_2) was used to catalyze the reductive removal of diatrizoate from a waste stream with glucose as hydrogen donor and fermentative bacteria, such as *Citrobacter braakii* (Hennebel *et al.*, 2011c). Similar technologies used bio-hydrogen, produced by fermentation of sugar waste streams in a separate reactor, as hydrogen donor for Pd nanoparticle catalyzed chromate reduction (Humphries *et al.*, 2007). This approach could also be used for soils and sediments, contaminated with PCBs.

Finally, Pd nanoparticles could be combined with other remediation technologies to speed these up or to complete the degradation in order to obtain full mineralization. He *et al.* (2009) demonstrated that PCB 101 can be degraded by a sequential treatment system involving a Pd/Fe reducing process followed by an aerobic biodegradation process. PCB 101 was partly dehalogenated in the presence of Pd/Fe to PCB 17, which was almost completely degraded in the presence of a H1 bacterium after 28 d incubation. Complete dehalogenation of PCBs and further hydrogenation of the aromatic structures can be obtained by Pd catalysis (Schuth & Reinhard, 1998). However, both biphenyl and bicyclohexane are toxic compounds of which release in the environment should be prevented. Hence, a combination of catalytic dechlorination followed by bacterial mineralization could represent an integrated treatment technology.

REFERENCES

Adzic R. R., Tripkovic A. V. and Ogrady W. E. (1982). Structural effects in electrocatalysis. *Nature*, **296**(5853), 137–138.

Alfonso D. R., Cugini A. V. and Sholl D. S. (2003). Density functional theory studies of sulfur binding on Pd, Cu and Ag and their alloys. *Surface Science*, **546**(1), 12–26.

Angeles-Wedler D., Mackenzie K. and Kopinke F. D. (2009). Sulphide-induced deactivation of Pd/Al_2O_3 as hydrodechlorination catalyst and its oxidative regeneration with permanganate. *Applied Catalysis B-Environmental*, **90**(3–4), 613–617.

Anwer M. K., Sherman D. B., Roney J. G. and Spatola A. F. (1989). Applications of ammonium formate catalytic transfer hydrogenation: analysis of catalyst, donor quantity and solvent effects upon the efficacy of dechlorination. *Journal of Organic Chemistry*, **54**(6), 1284–1289.

Aramendia M. A., Borau V., Garcia I. M., Jimenez C., Lafont F., Marinas A., Marinas J. M. and Urbano F. J. (1999). Influence of the reaction conditions and catalytic properties on the liquid-phase hydrodechlorination of chlorobenzene over palladium-supported catalysts: activity and deactivation. *Journal of Catalysis*, **187**(2), 392–399.

Balko E. N., Przybylski E. and Vontrentini F. (1993). Exhaustive liquid-phase catalytic hydrodehalogenation of chlorobenzenes. *Applied Catalysis B-Environmental*, **2**(1), 1–8.

Baxter-Plant V., Mikheenko I. P. and Macaskie L. E. (2003). Sulphate-reducing bacteria, palladium and the reductive dehalogenation of chlorinated aromatic compounds. *Biodegradation*, **14**(2), 83–90.

Billinge S. J. L. and Levin I. (2007). The problem with determining atomic structure at the nanoscale. *Science*, **316**(5832), 1698–1698.

Bucheli T. D. and Gustafsson R. (2003). Soot sorption of non-ortho and ortho substituted PCBs. *Chemosphere*, **53**(5), 515–522.

Chen S., Qin Z., Quan X., Zhang Y. and Zhao H. (2010). Electrocatalytic dechlorination of 2,4,5-trichlorobiphenyl using an aligned carbon nanotubes electrode deposited with palladium nanoparticles. *Chinese Science Bulletin*, **55**(4–5), 358–364.

Chidambaram D., Hennebel T., Taghavi S., Mast J., Boon N., Verstraete W., van der Lelie D. and Fitts J. P. (2010). Concomitant microbial generation of palladium nanoparticles and hydrogen to immobilize chromate. *Environmental Science and Technology*, **44**(19), 7635–7640.

Choi H., Agarwal S. and Al-Abed S. R. (2008). Adsorption and simultaneous dechlorination of PCBs on GAC/Fe/Pd: mechanistic aspects and reactive capping barrier concept. *Environmental Science and Technology*, **43**(2), 488–493.

Choi H., Al-Abed S. R. and Agarwal S. (2009). Catalytic role of palladium and relative reactivity of substituted chlorines during adsorption and treatment of PCBs on reactive activated carbon. *Environmental Science and Technology*, **43**(19), 7510–7515.

Choi J. H., Kim Y. H. and Choi S. J. (2007). Reductive dechlorination and biodegradation of 2,4,6-trichlorophenol using sequential permeable reactive barriers: laboratory studies. *Chemosphere*, **67**(8), 1551–1557.

Coutts J. L., Devor R. W., Aitken B., Hampton M. D., Quinn J. W., Clausen C. A. and Geiger C. L. (2011). The use of mechanical alloying for the preparation of palladized magnesium bimetallic particles for the remediation of PCBs. *Journal of Hazardous Materials*, **192**(3), 1380–1387.

Davie M. G., Cheng H., Hopkins G. D., LeBron C. A. and Reinhard M. (2008). Implementing heterogeneous catalytic dechlorination technology for remediating TCE-contaminated groundwater. *Environmental Science and Technology*, **42**(23), 8908–8915.

De Corte S., Hennebel T., Fitts J. P., Sabbe T., Biznuk V., Verschuere S., van der Lelie D., Verstraete W. and Boon N. (2011). Biosupported bimetallic Pd-Au nanocatalysts for dechlorination of environmental contaminants. *Environmental Science and Technology*, **45**(19), 8506–8513.

De Gusseme B., Hennebel T., Vanhaecke L., Soetaert M., Desloover J., Wille K., Verbeken K., Verstraete W. and Boon N. (2011). Biogenic palladium enhances diatrizoate removal from hospital wastewater in a microbial electrolysis cell. *Environmental Science and Technology*, **45**(13), 5737–5745.

De Windt W., Aelterman P. and Verstraete W. (2005). Bioreductive deposition of palladium (0) nanoparticles on Shewanella oneidensis with catalytic activity towards reductive dechlorination of polychlorinated biphenyls. *Environmental Microbiology*, **7**(3), 314–325.

De Windt W. (2006). Environmental Applications on Precipitations of Nanoparticles by *Shewanella oneidensis*. PhD thesis, Ghent University, Gent, Belgium.

De Windt W., Boon N., Van den Bulcke J., Rubberecht L., Prata F., Mast J., Hennebel T. and Verstraete W. (2006). Biological control of the size and reactivity of catalytic Pd(0) produced by *Shewanella oneidensis*. *Antonie Van Leeuwenhoek International Journal of General and Molecular Microbiology*, **90**(4), 377–389.

Demirci U. B. (2007). Theoretical means for searching bimetallic alloys as anode electrocatalysts for direct liquid-feed fuel cells. *Journal of Power Sources*, **173**(1), 11–18.

Foettinger K., van Bokhoven J. A., Nachtegaal M. and Rupprechter G. (2011). Dynamic structure of a working methanol steam reforming catalyst: *in situ* quick-EXAFS on Pd/ZnO nanoparticles. *Journal of Physical Chemistry Letters*, **2**(5), 428–433.

Forni P., Prati L. and Rossi M. (1997). Catalytic dehydrohalogenation of polychlorinated biphenyls Part II: Studies on a continuous process. *Applied Catalysis B: Environmental*, **14**(1–2), 49–53.

Gong K., Su D. and Adzic R. R. (2011). Platinum monolayer shell on AuNi(0.5)Fe nanoparticle core electrocatalyst with high activity and stability for the oxygen reduction reaction. *Journal of the American Chemical Society*, **132**(41), 14364–14366.

Gravil P. A. and Toulhoat H. (1999). Hydrogen, sulphur and chlorine coadsorption on Pd (111): a theoretical study of poisoning and promotion. *Surface Science*, **430**(1–3), 176–191.

Grittini C., Malcomson M., Fernando Q. and Korte N. (1995). Rapid dechlorination of polychlorinated-biphenyls on the surface of a Pd/Fe bimetallic system. *Environmental Science and Technology*, **29**(11), 2898–2900.

Hammer B. and Norskov J. K. (2000). *Advances in Catalysis, Vol 45: Impact of Surface Science on Catalysis*, B.C. Gates, H. Knozinger, Elsevier Academic Press Inc, San Diego, CA, USA, pp. 71–129.

He N., Li P., Zhou Y., Fan S. and Ren W. (2009). Degradation of pentachlorobiphenyl by a sequential treatment using Pd coated iron and an aerobic bacterium (H1). *Chemosphere*, **76**(11), 1491–1497.

Hennebel T., Simoen H., De Windt W., Verloo M., Boon N. and Verstraete W. (2009). Biocatalytic dechlorination of trichloroethylene with bio-Pd in a pilot-scale membrane reactor. *Biotechnology and Bioengineering*, **102**(4), 995–1002.

Hennebel T., Benner J., Clauwaert P., Vanhaecke L., Aelterman P., Boon N. and Verstraete W. (2011a). Dehalogenation of environmental pollutants in microbial electrolysis cells with biogenic palladium nanoparticles. *Biotechnology Letters*, **33**, 89–95.

Hennebel T., Simoen H., Verhagen P., De Windt W., Dick J., Weise C., Pietschner F., Boon N. and Verstraete W. (2011b). Biocatalytic dechlorination of hexachlorocyclohexane by immobilized bio-Pd in a pilot scale fluidized bed reactor. *Environmental Chemistry Letters*, **9**(3), 417–422.

Hennebel T., Van Nevel S., Verschuere S., De Corte S., De Gusseme B., Cuvelier C., Fitts J. P., Van der Lelie D., Boon N. and Verstraete W. (2011c). Palladium nanoparticles produced by fermentatively cultivated bacteria as catalyst for diatrizoate removal with biogenic hydrogen. *Applied Microbiology and Biotechnology*, **91**(5), 1435–1445.

Hoke J. B., Gramiccioni G. A. and Balko E. N. (1992). Catalytic hydrodechlorination of chlorophenols. *Applied Catalysis B-Environmental*, **1**(4), 285–296.

Humphries A. C., Penfold D. W. and Macaskie L. E. (2007). Cr(VI) reduction by bio and bioinorganic catalysis via use of bio-H-2: a sustainable approach for remediation of wastes. *Journal of Chemical Technology and Biotechnology*, **82**(2), 182–189.

Klabunde K. J. (2001). *Nanoscale Materials in Chemistry*. Wiley Interscience, New York (NY).

Kobayashi J., Kinoshita K., Mizukawa K., Sakurai T., Imaizumi Y., Takada H. and Suzuki N. (2011). Dietary uptake kinetics of polychlorinated biphenyls from sediment-contaminated sandworms in a marine benthic fish (*Pseudopleuronectes yokohamae*). *Chemosphere*, **82**(5), 745–750.

Kopinke F. D., Mackenzie K., Koehler R. and Georgi A. (2004). Alternative sources of hydrogen for hydrodechlorination of chlorinated organic compounds in water on Pd catalysts. *Applied Catalysis A-General*, **271**(1–2), 119–128.

Korte N. (2000). Dechlorination of polychlorinated biphenyls (PCBs) in soil.

Kume A., Monguchi Y., Hattori K., Nagase H. and Sajiki H. (2008). Pd/C-catalyzed practical degradation of PCBs at room temperature. *Applied Catalysis B-Environmental*, **81**(3–4), 274–282.

Lajoie C. A., Layton A. C., Easter J. P., Menn F. M. and Sayler C. S. (1997). Degradation of nonionic surfactants and polychlorinated biphenyls by recombinant field application vectors. *Journal of Industrial Microbiology & Biotechnology*, **19**(4), 252–262.

Lassova L., Lee H. K. and Hor T. S. A. (1998). Catalytic dehalogenation of highly chlorinated benzenes and aroclors using $PdCl_2$(dppf) and $NaBH_4$: efficiency, selectivity, and base support. *Journal of Organic Chemistry*, **63**(11), 3538–3543.

Li Z. R., Yuan S. H., Wan J. Z., Long H. Y. and Tong M. (2011). A combination of electrokinetics and Pd/Fe PRB for the remediation of pentachlorophenol-contaminated soil. *Journal of Contaminant Hydrology*, **124**(1–4), 99–107.

Liao W. H., Chen Y. C., Wang J. S., Yak H. K. and Wai C. M. (2007). Hydrodechlorination of chlorinated biphenyls in supercritical CO_2 catalyzed by polymer-stabilized palladium nanoparticles. *Industrial & Engineering Chemistry Research*, **46**(15), 5089–5093.

Liz-Marzán L. M. and Kamat P. V. (2003). *Nanoscale Materials*. Kluwer Academic Publishers, Boston (MA).

Lowry G. V. and Johnson K. M. (2004). Congener-specific dechlorination of dissolved PCBs by microscale and nanoscale zerovalent iron in a water/methanol solution. *Environmental Science & Technology*, **38**(19), 5208–5216.

Marques C. A., Selva M. and Tundo P. (1994). Facile hydrodehalogenation with H_2 and Pd/C catalyst under multiphase conditions. Part 2. Selectivity and kinetics. *The Journal of Organic Chemistry*, **59**(14), 3830–3837.

Mertens B., Blothe C., Windey K., De Windt W. and Verstraete W. (2007). Biocatalytic dechlorination of lindane by nano-scale particles of Pd(0) deposited on *Shewanella oneidensis*. *Chemosphere*, **66**(1), 99–105.

Monguchi Y., Ishihara S., Ido A., Niikawa M., Kamiya K., Sawama Y., Nagase H. and Sajiki H. (2011). Pilot-plant study of the PCB degradation at ambient temperature and pressure. *Organic Process Research & Development*, **14**(5), 1140–1146.

Nollet H. (2005). *Physico-chemical and Biocatalytical PCB Remediation Strategies*. Ghent University, Gent.

Nutt M. O., Hughes J. B. and Wong M. S. (2005). Designing Pd-on-Au bimetallic nanoparticle catalysts for trichloroethene hydrodechlorination. *Environmental Science and Technology*, **39**(5), 1346–1353.

Nutt M. O., Heck K. N., Alvarez P. and Wong M. S. (2006). Improved Pd-on-Au bimetallic nanoparticle catalysts for aqueous-phase trichloroethene hydrodechlorination. *Applied Catalysis B-Environmental*, **69**(1–2), 115–125.

Nykypanchuk D., Maye M. M., van der Lelie D. and Gang O. (2008). DNA-guided crystallization of colloidal nanoparticles. *Nature*, **451**(7178), 549–552.

Rodriguez J. G. and Lafuente A. (2002). A new advanced method for heterogeneous catalysed dechlorination of polychlorinated biphenyls (PCBs) in hydrocarbon solvent. *Tetrahedron Letters*, **43**(52), 9581–9583.

Schoonenboom M. H., Zoetemeijer H. E. and Olie K. (1995). Dechlorination of octachlorodibenzo-*p*-dioxin and octachlorodibenzofuran on an alumina support. *Applied Catalysis B-Environmental*, **6**(1), 11–20.

Schuth C. and Reinhard M. (1998). Hydrodechlorination and hydrogenation of aromatic compounds over palladium on alumina in hydrogen-saturated water. *Applied Catalysis B-Environmental*, **18**(3–4), 215–221.

Shao M. H., Huang T., Liu P., Zhang J., Sasaki K., Vukmirovic M. B. and Adzic R. R. (2006). Palladium monolayer and palladium alloy electrocatalysts for oxygen reduction. *Langmuir*, **22**(25), 10409–10415.

Tanabe S. (1988). PCB problems in the future – foresight from current knowledge. *Environmental Pollution*, **50**(1–2), 5–28.

Tavakoli A., Sohrabi M. and Kargari A. (2007). A review of methods for synthesis of nanostructured metals with emphasis on iron compounds. *Chemical Papers*, **61**(3), 151–170.

Tratnyek P. G., Johnson T. L., Scherer M. M. and Eykholt G. R. (1997). Remediating ground water with zero-valent metals: chemical considerations in barrier design. *Ground Water Monitoring and Remediation*, **17**(4), 108–114.

Urbano F. J. and Marinas J. M. (2001). Hydrogenolysis of organohalogen compounds over palladium supported catalysts. *Journal of Molecular Catalysis A-Chemical*, **173**(1–2), 329–345.

USEPA. (2008). *Nanotechnology for Site Remediation Fact Sheet*. United States Environmental Protection Agency, Washington D.C.

van de Sandt E., Wiersma A., Makkee M., van Bekkum H. and Moulijn J. A. (1998). Selection of activated carbon for the selective hydrogenolysis of CCl_2F_2 (CFC-12) into CH_2F_2 (HFC-32) over palladium-supported catalysts. *Applied Catalysis A-General*, **173**(2), 161–173.

Van Nooten T., Springael D. and Bastiaens L. (2008). Positive impact of microorganisms on the performance of laboratory-scale permeable reactive iron barriers. *Environmental Science and Technology*, **42**(5), 1680–1686.

Vogan J. L., Focht R. M., Clark D. K. and Graham S. L. (1999). Performance evaluation of a permeable reactive barrier for remediation of dissolved chlorinated solvents in groundwater. *Journal of Hazardous Materials*, **68**(1–2), 97–108.

Wafo W., Coen S., BruschiniChircop C., Perichaud A. and Rossi C. (1997). Palladium-catalyzed dechlorination of polychlorobiphenyls: a study of the kinetic mechanism. *Analusis*, **25**(7), 230–236.

Wang J. S. and Chiu K. (2009). Destruction of pentachlorobiphenyl in soil by supercritical CO_2 extraction coupled with polymer-stabilized palladium nanoparticles. *Chemosphere*, **75**(5), 629–633.

Wang J. X., Ma C., Choi Y., Su D., Zhu Y., Liu P., Si R., Vukmirovic M. B., Zhang Y. and Adzic R. R. (2011). Kirkendall effect and lattice contraction in nanocatalysts: a new strategy to enhance sustainable activity. *Journal of the American Chemical Society*, **133**(34), 13551–13557.

West O. R., Liang L., Holden W. L., Korte N. E., Fernando Q. and Clausen J. L. (1996). *Degradation of Polychlorinated Biphenyls (PCBs) using Palladized Iron, R*, ORNL/TM-13217, Oak Ridge, TN, USA.

Yang B., Yu G. and Huang J. (2007). Electrocatalytic hydrodechlorination of 2,4,5-trichlorobiphenyl on a palladium-modified nickel foam cathode. *Environmental Science & Technology*, **41**(21), 7503–7508.

Yang B., Deng S., Yu G., Zhang H., Wu J. and Zhuo Q. (2011). Bimetallic Pd/Al particles for highly efficient hydrodechlorination of 2-chlorobiphenyl in acidic aqueous solution. *Journal of Hazardous Materials*, **189**(1–2), 76–83.

Zhang W. X. (2003). Nanoscale iron particles for environmental remediation: an overview. *Journal of Nanoparticle Research*, **5**(3–4), 323–332.

Zhang X.-L., Luo X.-J., Liu J., Luo Y., Chen S.-J. and Mai B.-X. (2011). Polychlorinated biphenyls and organochlorinated pesticides in birds from a contaminated region in South China: association with trophic level, tissue distribution and risk assessment. *Environmental Science and Pollution Research*, **18**(4), 556–565.

Zhang Y. P., Liao S. J. and Xu Y. (1994). Highly active polymer anchored palladium catalyst for the hydrodehalogenation of organic halides under mild conditions. *Tetrahedron Letters*, **35**(26), 4599–4602.

Zhuang Y., Ahn S., Seyfferth A. L., Masue-Slowey Y., Fendorf S. and Luthy R. G. (2011). Dehalogenation of polybrominated diphenyl ethers and polychlorinated biphenyl by bimetallic, impregnated, and nanoscale zerovalent iron. *Environmental Science and Technology*, **45**(11), 4896–4903.

Chapter 10

Activated carbon-supported palladized iron nanoparticles: applications to contaminated site remediation

H. Choi, P. Nfodzo, S. R. Al-Abed, S. Agarwal and D. D. Dionysiou

This chapter describes the potential of nanotechnology to provide new solutions to managing and cleaning our contaminated water and soil and improving the performance of conventional technologies used in cleanup efforts. Our initial efforts have been focused on key pollutants of concern to the US Environmental Protection Agency (EPA), which have been historically difficult to manage such as sources that release low-level but toxic chemicals in complex cleanup sites. In particular, development of a strategy using nanomaterials for in situ cleanup of the Superfund sites contaminated with polychlorinated biphenyls (PCBs) has been one of the highest priorities of the EPA and other associated agencies.

A series of our studies using novel nanoparticles to improve efforts to clean up PCBs-contaminated sites have shown promising results. Traditional adsorptive activated carbon was modified with iron/palladium (Fe/Pd) bimetallic nanoparticles, introducing the new concept of reactive activated carbon (RAC), which is more effective not only in physically trapping and isolating PCBs in the environment but also in chemically degrading the contaminants to less harmful and biodegradable compounds. The findings lead to the development of innovative practices to advance caps or barriers to contain and degrade PCBs and they have implications for nanotechnological treatment of other contaminants in the environment. In this chapter, we overview: (i) tailor-designing of the properties of RAC at nanoscale; (ii) working mechanisms of RAC for the adsorption and dechlorination of PCBs; (iii) reactivity of various PCB congeners with RAC; and (iv) technical challenges, on-going progress, and potential applications.

10.1 INTRODUCTION

The remediation of the environment contaminated with hydrophobic organic compounds such as polychlorinated biphenyls (PCBs) remains a scientific and technical challenge. PCBs are a class of organic compounds with 2 to 10 chlorine atoms attached to biphenyl (BP), which is a molecule composed of two benzene rings (Figure 10.1). PCBs were widely used mainly as insulators, coolants, and dielectric fluids in electrical equipment because of their inertness, stability, and flame-retarding properties. The use of PCBs was, however, banned in 1979 due to their toxicity and persistence in the environment. In spite of the discontinued use of PCBs, past disposal practices and accidental releases coupled with their highly hydrophobic nature have resulted in significant amounts of PCBs being still present predominantly in soils and sediments. The solid matrix serves as a source of the slow and long term release of PCBs into the aqueous environment. Along with their high chemical stability (i.e. strong C-Cl bond), the low aqueous solubility and high organic affinity of PCBs make them difficult to treat (Borja *et al.*, 2005).

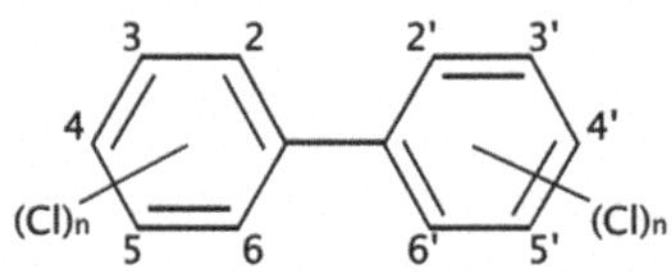

Figure 10.1 Molecular structure of PCBs. Depending on the number of chlorines (up to 10) and their position, PCBs are a group of 209 congeners with unique properties.

10.1.1 Physical capping approaches

Sites contaminated with PCBs have been dredged and treated by solvent extraction, incineration, and chemical processes. However, such ex situ processes are expensive and ineffective. In some cases, the processes might disrupt ecosystems and move the contaminants elsewhere. In an attempt to find more effective in situ technologies, studies have suggested that capping using adsorptive and/or reactive materials is the most practical strategy for contaminated site remediation (Wiegel & Wu, 2000; Murphy *et al.*, 2006; Agarwal *et al.*, 2007a). As shown in Figure 10.2, capping entails the isolation of a contaminated site by the placement of a clean layer (called a cap) over the site, inducing a physical separation of the site from the surrounding environment. Its effectiveness is greatly improved by adding a layer composed of adsorptive materials such as activated carbon (AC) or mixing AC directly with contaminated sediment (McDonough *et al.*, 2007; Werner *et al.*, 2005). Due to its high affinity and adsorption capacity for PCBs, AC is also effective to remove aqueous PCBs desorbed from sediment and reduce their bioavailability (Morales *et al.*, 2002).

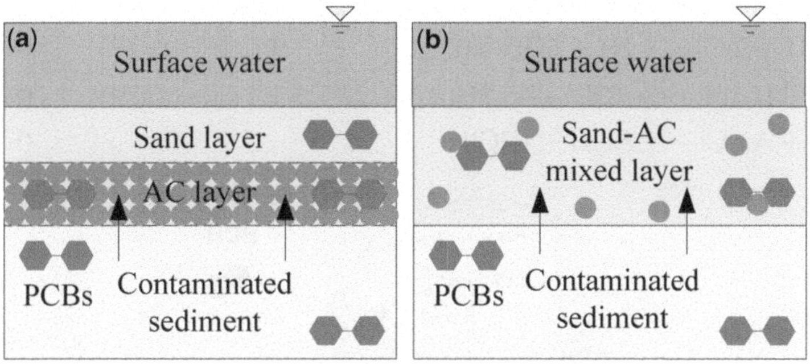

Figure 10.2 Capping of PCBs-contaminated sites with a clean sand layer to sequester the sediment from the surrounding environment. In order to further improve its performance, the sand layer is: (a) modified with an AC layer and (b) mixed with AC granules. AC removes the PCBs available in the aqueous phase through the adsorption process.

10.1.2 Chemical dechlorination approaches

Many chemical approaches have been proposed to decompose PCBs in particular by employing reactive metal particles. The synthesis and use of reactive zerovalent metals such as iron (Fe) and magnesium (Mg) have been documented to work efficiently for the electrochemical dechlorination of PCBs found in groundwater and sediment (Wang & Zhang, 1997; Zhang, 2003; Liu *et al.*, 2005). The dechlorination process to lower congeners and eventually to biphenyl (BP) is important because less chlorinated PCBs are generally less toxic and more bio-available.

The working mechanism of reactive metal particles for the dechlorination of PCBs is shown in Figure 10.3 (Liu *et al.*, 2005). As summarized in Equations 10.1 and 10.2, zerovalent Fe (ZVI) in the aqueous phase oxidizes, releasing electrons. PCBs (denoted as Ar_2Cl) accept the electrons (i.e. electrochemical reduction), replacing chlorine atoms with hydrogen atoms. This direct reduction of PCBs is known to be not so effective. The metallic system can be advanced for enhanced reactivity through deposition of a discontinuous layer of noble metals onto the substrate metal surface. Incorporation of less active second metals, for example Pd, into ZVI surface can greatly affect the overall reaction due to the facilitated oxidation of ZVI and Pd-meditated hydrogenation (Grittini *et al.*, 1995; Elliott & Zhang, 2001; Agarwal *et al.*, 2007b). Hydrogen, which is formed via electrolysis of water by the released electrons, can be used to dechlorinate PCBs *via* Pd-catalyzed hydrodechlorination (Equations 10.3–10.5) (Lowry & Reinhard, 1999; Mitsui *et al.*, 2003).

$$\mathrm{Fe^0} \longrightarrow \mathrm{Fe^{2+}(aq)} + 2\mathrm{e^-} \quad \mathrm{E_H^0} = +0.477V \qquad \text{(Eq. 10.1)}$$

$$\mathrm{Ar_2Cl} + \mathrm{xH^+ye^-} \longrightarrow \mathrm{Ar_2}(y-x)\mathrm{H} + (y-x)\mathrm{Cl^-} \qquad \text{(Eq. 10.2)}$$

$$2H_2O + 2e^- \longrightarrow H_2 + 2OH(aq)^- \qquad E_H^0 = -0.379V \qquad \text{(Eq. 10.3)}$$

$$H_2(g) \longleftrightarrow 2H_{ad} \longleftrightarrow 2H(aq) + 2e^- \qquad E_H^0 = +0.379V \qquad \text{(Eq. 10.4)}$$

$$PdH_2 + Ar_2Cl \longrightarrow Ar_2H + PdClH \qquad \text{(Eq. 10.5)}$$

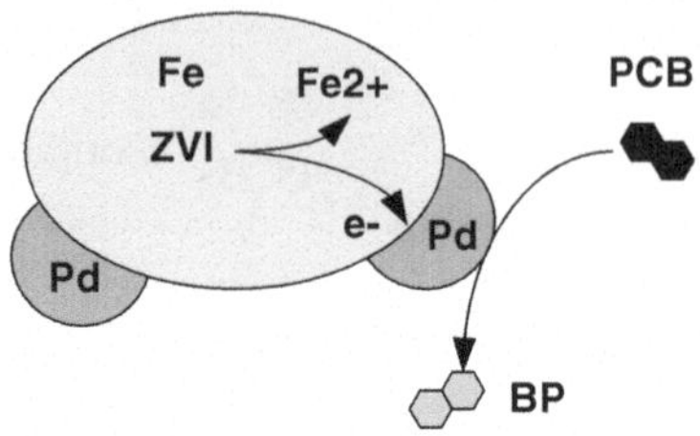

Figure 10.3 Surface reaction between PCBs with Fe/Pd bimetallic particles. Fe in this case is zerovalent iron (ZVI) (i.e. elemental Fe (Fe^0)). For their dechlorination, PCBs should physically contact the Fe/Pd particles.

More recently, the properties of metallic particles have been precisely manipulated in the nanoscale regime. The reactivity of ZVI is mainly determined by the composition and availability of its near surface because the dechlorination reaction is surface-mediated (Gaspar *et al.*, 2002). The small size and high specific surface area of nanoscale Fe particles make them attractive for the reaction, compared to typical micron size Fe particles. The application of metallic nanoparticles in groundwater treatment and contaminated site remediation has received considerable attention (Agarwal *et al.*, 2007a).

10.1.3 Concept of reactive activated carbon (RAC)

In spite of the strong sequestration capability of the AC-modified capping layer, the high capillary effect of AC can facilitate desorption of PCBs in a stable sediment matrix to the aqueous phase. The AC approach does not degrade PCBs but only sequesters them. Slight changes in the capping environment can disturb the capping layer and re-suspend the sequestered PCBs at high concentration. Meanwhile, nanoscale metallic systems (in particular Fe/Pd) seem promising for the treatment of aqueous phase PCBs (see Chapter 9). However, they might not work properly for PCBs adsorbed to sediment matrix (Ponder *et al.*, 2001; Schrick *et al.*, 2004; Saleh *et al.*, 2007). Hydrophobic chemicals, such as PCBs, exhibit extremely high affinity for organic substances and have low water solubility (Luthy *et al.*, 1997; Jonker & Koelmans, 2002; Choi & Al-Abed, 2009). PCBs adsorbed onto a solid matrix in the environment are much less available for the surface reaction on Fe/Pd particles added. The fast dechlorination by the metallic system with non-catalytic reactivity has to be controlled to respond to the slow release of PCBs to the aqueous phase. The injection of colloidal metal particles into sites also poses a critical challenge (Schrick *et al.*, 2004; Saleh *et al.*, 2007).

In a series of studies, we have developed a new strategy employing granular activated carbon (GAC) composites impregnated with reactive ZVI and Pd bimetallic nanoparticles (ZVI/Pd) (Choi *et al.*, 2008, 2009a, 2009b, 2009c; Choi & Al-Abed, 2010). Fe/Pd bimetallic nanoparticles were embedded into the pores of GAC. The composite was named "reactive activated carbon (RAC)" (Figure 10.4). The RAC system can synergistically combine two strategies: i) the use of GAC for the physical adsorption of PCBs and ii) the use of ZVI/Pd for the chemical dechlorination of PCBs (Choi *et al.*, 2009a). The RAC is a "smart" composite for the dechlorination of PCBs, because RAC with high adsorption capacity actively attracts hydrophobic PCBs in the sediment matrix towards itself. The ability of GAC to place PCBs into its pores impregnated with Fe/Pd nanoparticles with high reactivity to dechlorinate PCBs makes the RAC system attractive.

Figure 10.4 Granular activated carbon (GAC) impregnated with palladium and zerovalent iron nanoparticles, denoted reactive activated carbon (RAC). The RAC could be used in a reactive barrier or cap to sequester and dechlorinate PCBs in contaminated environmental sites.

10.2 SYNTHESIS OF REACTIVE ACTIVATED CARBON

Through a novel synthesis procedure that includes the consecutive steps of the incipient wetness impregnation method, heat-treatment, borohydride reduction, and Pd doping, the physicochemical properties of GAC/ZVI/Pd can be tailor-designed for the enhanced adsorption and dechlorination of PCBs (Choi *et al.*, 2008).

10.2.1 Impregnation of GAC with Fe/Pd particles

The specific synthesis approaches include: (i) introduction of mesoporous GAC as a support material for ZVI particles and as an adsorptive material for PCBs; (ii) in situ formation and incorporation of Fe particles into the GAC pores to increase their mechanical stability; (iii) modification of the Fe surface with a discontinuous layer of noble metal Pd to facilitate the dechlorination of PCBs, and (iv) nanoscaling of ZVI and Pd particles through the restriction of Fe oxide growth in the GAC pores and reductive deposition of Pd to the ZVI surface.

Figure 10.5 shows the graphical procedure of fabricating a series of modified AC composites. In order to control the physicochemical properties of GAC composites at the nano level, Fe was imbedded in situ into the pores of mesoporous GAC via an incipient wetness impregnation method employing iron nitrate. For total incorporation of Fe to GAC, the slurry was completely dried. GAC/Fe_{salt} was further heat-treated in a furnace at 300°C to remove nitrate ions and transform Fe ions to Fe oxides. It is notable that the crystal growth of Fe oxides was restricted within the GAC pores in mainly the meso (1.7–50 nm) and partially micro (less than 1.7 nm) ranges. The heat treatment procedure was critical to increase the mechanical stability of Fe through Fe-C metal-support interaction and thus keep Fe firmly attached to GAC in the subsequent steps. To reduce Fe^{3+} to Fe^0, $NaBH_4$ in water was mixed with GAC/Fe in methanol/water (Equation 10.6).

$$4Fe^{3+} + 3BH_4^- + 9H_2O \longrightarrow 4Fe^0 + 3H_2BO_3^- + 12H^+ + 6H_2 \qquad \text{(Eq. 10.6)}$$

$$Fe^0 + Pd^{2+} \longrightarrow Fe^{2+} + Pd^0\downarrow \qquad \text{(Eq. 10.7)}$$

Figure 10.5 Synthesis step of GAC composites: (a) embedding of mesoporous GAC with Fe salt via incipient wetness impregnation method, (b) formation of nanosize iron oxides, Fe_2O_3 in the pores of GAC during calcination, (c) reduction of Fe_2O_3 to elemental Fe using $NaBH_4$ solution, and (d) reductive deposition of Pd to Fe surface from $Pd(CH_3CO_2)_2$. "Reprinted with permission from *Chem. Mater.* 2008, **20**(11), 3649–3655. Copyright 2011 American Chemical Society."

For Pd doping, palladium acetate dissolved in methanol was mixed with GAC/ZVI. Pd was reductively deposited as elemental Pd islands onto the ZVI surface (Equation 10.7). The Fe/Pd system is known to create numerous isolated galvanic cells where bulk Fe particles as an anode are preferably oxidized while Pd islands as a cathode are protected and remain uncharged (Agarwal *et al.*, 2007b; Grittini *et al.*, 1995).

10.2.2 Physical and chemical properties of RAC

The visual images of a series of RAC materials are shown in Figure 10.6. Depending on the species and content of Fe, the reactivity and longevity of RAC, and adsorption and decomposition pathways and kinetics of PCBs vary, which is helpful for responding to the characteristics of PCBs-contaminated sites. Immobilization of ZVI/Pd particles in the porous structure of GAC improves their mechanical stability and dechlorination capability. Microporous AC can provide a large number of sorption sites, but its small size pores limit Fe immobilization to the grain boundary of AC. Selection of appropriate AC with intrinsic microporous as well as well-defined mesoporous structures is critical.

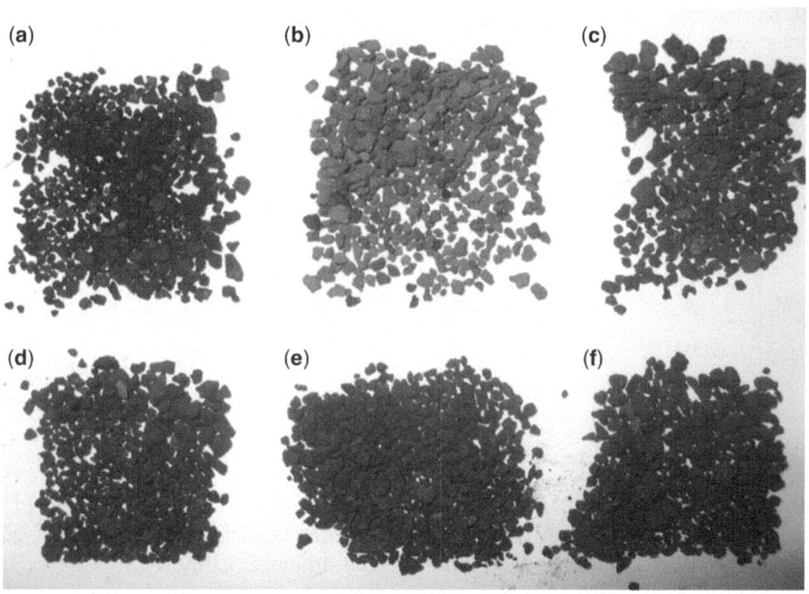

Figure 10.6 RAC composites: (a) pristine GAC in size of 1-3 mm, (b) GAC/$Fe(NO_3)_3$, (c) GAC/Fe_2O_3, (d) GAC/ZVI, and (e) and (f) GAC/ZVI/Pd with different amounts of Pd (0.68% and 0.01%, respectively).

Table 10.1 and Figure 10.7 show the structural properties of RAC and its metal contents. In addition to the microporous structure which is useful for the adsorption of PCBs, GAC showed type IV N_2 adsorption/desorption isotherms typical for well-developed mesoporous structures with pore size of 10–40 nm. The mesoporous structure was well-occupied by Fe/Pd particles. The small hysteresis loop between the adsorption/desorption isotherms of GAC/ZVI/Pd supports partial pore blocking by ZVI/Pd particles. The Fe and Pd contents were 14.4% and 0.68%, respectively. About 38% of the surface area and 45% of the pore volume of GAC were occupied by Fe/Pd particles with dechlorination function. The RAC still had high surface area of 358 m^2/g and pore volume of 0.352 cm^3/g, which provides enough space for PCBs to adsorb.

Table 10.1 Physical and chemical properties of GAC composites.

Composition	SA[a] (m^2/g)	PV[b] (cm^3/g)	PS[c] (nm)	Fe (%)	Fe size (nm)	Pd (%)	Pd size (nm)
GAC	574	0.639	4.45	–	–	–	–
GAC/ZVI/Pd	358	0.352	3.94	14.4	9–12	0.68	2–3

[a] Brunauer, Emmett, and Teller surface area. [b] Single point adsorption total pore volume.
[c] Adsorption average pore size.

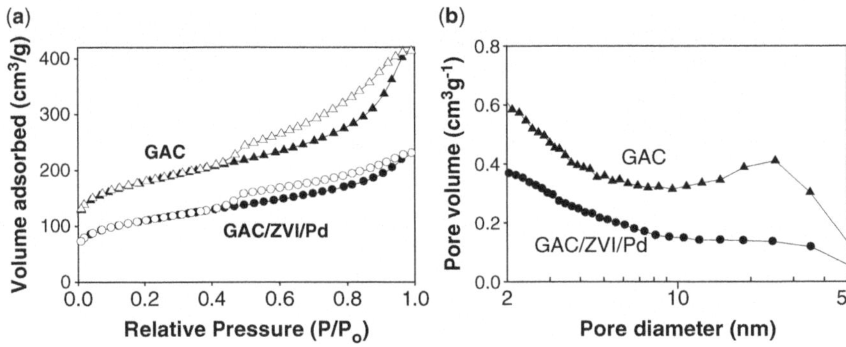

Figure 10.7 (a) N_2 adsorption/desorption isotherms of GAC and GAC/ZVI/Pd and (b) their pore size distribution. "Reprinted with permission from *Chem. Mater.* 2008, **20**(11), 3649-3655. Copyright 2011 American Chemical Society."

The visual structure of GAC composites at nanoscale is shown in Figure 10.8. A RAC grain in 1–3 mm size is clearly seen in Figure 10.8a. The cross-section of GAC/ZVI/Pd in Figure 10.8b shows agglomerated ZVI/Pd particles in the macro and mesopores of GAC. In Figure 10.8c, the low contrast area (or dark spots) indicates a ZVI/Pd nanoparticle-rich phase. As shown in Figure 10.8d, 7–8 nm well-defined Fe nanoparticles were impregnated into GAC. Nice conjunction of the lattice planes of Fe particles with the graphitic sheets in the surrounding carbon shells implies complete incorporation of Fe particles into the GAC matrix. D spaces of 3.65 Å observed in Figure 10.8e matched with the theoretical values of Fe_2O_3 (2000 JCPDS 87-1164) at Miller indices of (012). Figure 10.8f shows an image of Pd nanoislands on the surface of GAC/Fe. Distinct 2–3 nm Pd particles were well-distributed. Figure 10.9 shows a two-dimensional X-ray mapping of a 2 μm × 2 μm area of the RAC surface to investigate the elemental composition in the GAC matrix. Most of GAC surface was covered with Fe but some regions were still C-rich phases. The weak O signal in the Fe-rich phase implied that most of the Fe species was ZVI. Some O-rich areas (e.g. highlighted with an arrow) were also found, implying the presence of Fe_2O_3, which was not transformed to ZVI during its borohydride reduction. Pd seemed uniformly deposited.

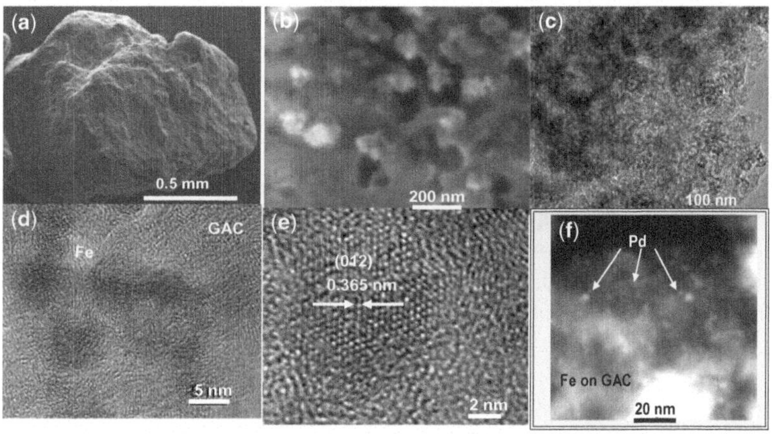

Figure 10.8 Visual morphology of a GAC/ZVI/Pd (RAC) composite: (a) RAC grain, (b) cross-section of RAC, (c) Fe and Pd-embedded GAC matrix, (d) and (e) Fe nanoparticles impregnated into GAC, and (f) Pd nanoislands on the surface of Fe, based on Z-contrast. "Reprinted with permission from *Chem. Mater.* 2008, **20**(11), 3649–3655. Copyright 2011 American Chemical Society."

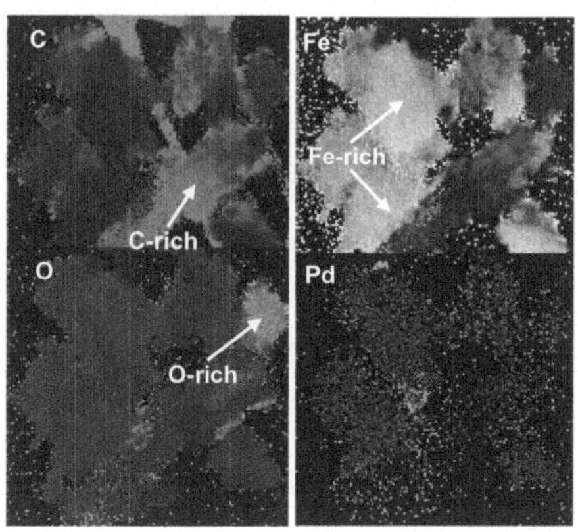

Figure 10.9 X-ray mapping of Fe-rich phase (2 μm × 2 μm area) in GAC/ZVI/Pd for C, Fe, O, and Pd elements. "Reprinted with permission from *Chem. Mater.* 2008, **20** (11), 3649–3655. Copyright 2011 American Chemical Society."

10.2.3 Control of Fe and Pd contents

In RAC, a small amount of Pd as a catalyst governs the dechlorination kinetics, while ZVI as an electron donor determines the performance longevity. Considerable amount of Pd might be needed for the prompt treatment of PCBs available in the liquid phase at high concentration, while a high content of Fe coupled with a trace amount of Pd might be preferred to respond to long term and low level release of PCBs from a solid matrix such as aquatic sediments. Control of the metal contents in RAC was attempted during the RAC synthesis, as shown in Figure 10.10 (Choi *et al.*, 2009c). All Fe added at low dosage was completely incorporated into GAC, while high Fe loading showed a non-linear increase in the amount of Fe embedded. A level-off was observed at around 30%. This proved that Fe was primarily incorporated into the GAC pores rather than the GAC grain boundary, which is good for the mechanical stability of Fe and protection of its reactivity from prompt oxidation. The driving force of Pd deposition onto Fe is completely different from that of physical placement of Fe into GAC pores. As shown in Figure 10.10b, Pd added to GAC/ZVI was completely deposited (i.e., linear response), as expected from the reductive deposition of Pd onto the abundant Fe surface.

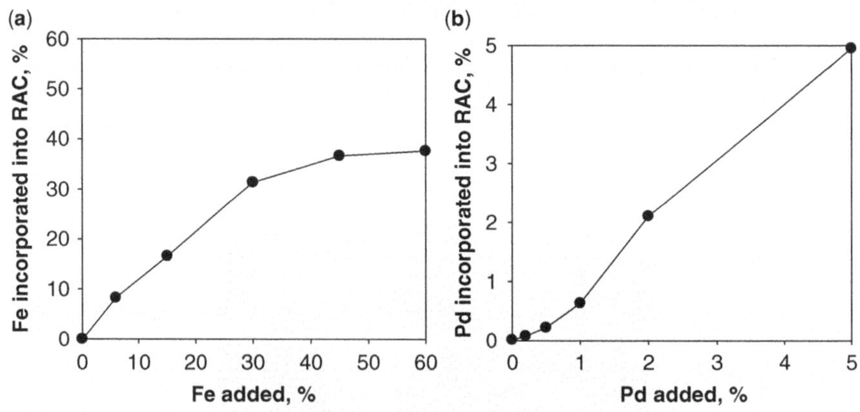

Figure 10.10 Relation between the amount of metal added to GAC and that incorporated into RAC: (a) Fe and (b) Pd. "Reprinted with permission from *Environ. Sci. Technol.* 2009, **43**(19), 7510–7515. Copyright 2011 American Chemical Society."

10.3 WORKING MECHANISMS OF RAC

The uniqueness of the RAC system is that the dechlorination of PCBs is mediated by adsorption. The RAC system is expected to be highly complex even for the treatment of liquid phase PCBs since it consists of three materials (AC, Fe,

Pd), at least two chemicals (PCB, BP), and two phases (liquid, AC solid), and several reactions occur in parallel and/or sequentially. Setting a carbon mass balance helps to configure the overall reaction in the RAC system. Meanwhile, treatment of PCBs on RAC largely depends on the reaction pH, mainly because of the pH dependency of Fe corrosion and the dechlorination reaction. In most cases, RAC summarized in Table 10.1 (14.4% Fe and 0.68% Pd) was used at 100 g/L while PCB concentration was at its solubility (e.g. 4 mg/L for 2-chlorobiphenyl).

10.3.1 Adsorption and dechlorination of PCBs

In order to understand the physical and chemical reactions in the RAC system, the simplest aqueous PCB congener, 2-chlorobiphenyl (2-ClBP), was introduced to the system, as shown in Figure 10.11. (Choi *et al.*, 2009a). Use of 2-ClBP, as a probe compound, made system interpretation simple and clear because 2-ClBP and BP (as a reaction product) were the only chemicals present in the system. Most of 2-ClBP added disappeared from the liquid phase while no significant amount of BP was found in the liquid phase, suggesting a strong adsorption ability of AC for the chemicals. For GAC, disappearance of 2-ClBP from the liquid phase was due to its adsorption onto GAC. Since there was no dechlorination reaction, accumulation of 2-ClBP was observed in the GAC solid phase.

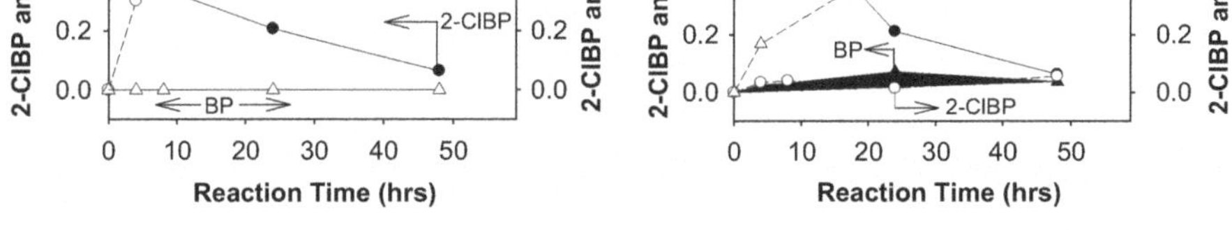

Figure 10.11 Adsorption of 2-ClBP to the AC phase, transformation of the adsorbed 2-ClBP to BP, and partitioning of the formed BP between the liquid and AC phases over time in: (a) GAC and (b) RAC. The concentrations of 2-ClBP and BP are expressed in terms of their mass fraction in the overall system. The concentrations of 2-ClBP and BP in the liquid phase are their actual concentrations while those in the solid phase are different from (lower than) their actual concentrations due to their incomplete extraction from AC. The adsorption and dechlorination kinetics above are subject to change upon the amount of 2-ClBP and RAC (or GAC) added. "Reprinted with permission from *Environ. Sci. Technol.* 2009, **43**(2), 488–493. Copyright 2011 American Chemical Society."

Meanwhile, the RAC system was more complex thanks to its dechlorination function. 2-ClBP in the liquid phase disappeared initially by its adsorption to the RAC phase. Time evolution of the concentrations of 2-ClBP and BP in the RAC solid phase was interesting. Assuming that the dechlorination reaction is slow compared to adsorption and thus limits the overall reaction in the RAC system, some 2-ClBP would accumulate in the RAC solid phase. However, 2-ClBP was negligible in the solid phase while BP was predominantly observed. This suggests that the dechlorination reaction was fast enough to immediately treat 2-ClBP once adsorbed. Importantly, BP was found only in the RAC phase, implying BP, when formed, was strongly adsorbed.

Partitioning and carbon mass balance in the RAC system are demonstrated in Figure 10.12. In this simple case with 2-ClBP added and BP formed, they are partitioned to the liquid and RAC phases. Unfortunately, incomplete extraction of 2-ClBP and BP from the AC phase, due to their high affinity for the AC material, makes it impossible to determine the actual amounts of 2-ClBP and BP in the solid phase. Based on the mass balance, however, the sum of 2-ClBP and BP in the un-extracted portion can be calculated. Since an adsorption and extraction experiment performed with individual 2-ClBP and BP showed no preferential affinity of GAC for them, it was assumed that the ratio of 2-ClBP to BP in the unextracted portion is the same as that observed from the extracted portion. This assumption was useful to interpret the experimental data with uncertainty and thus applied through this study.

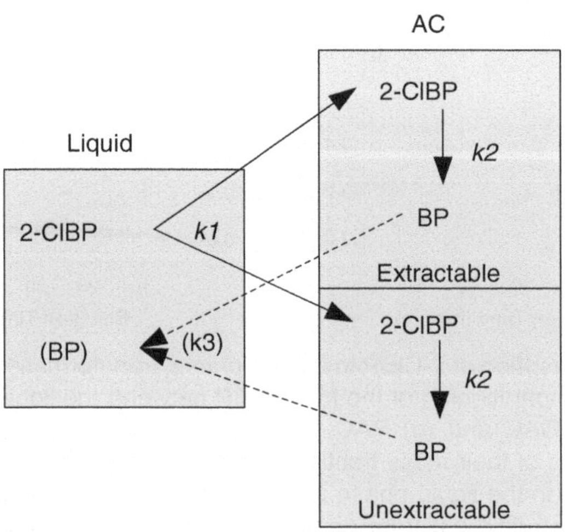

Figure 10.12 Partitioning of 2-ClBP remained and BP formed during adsorption and dechlorination of 2-ClBP. Desorption of BP from the AC phase to the liquid phase is thermodynamically unfavourable. "Reprinted with permission from *Environ. Sci. Technol.* 2009, **43**(2), 488–493. Copyright 2011 American Chemical Society."

10.3.2 Catalytic role of palladium

The modification of the Fe surface with Pd nanoparticles drastically increased the dechlorination kinetics, as shown in Figure 10.13 (Choi *et al.*, 2009c). The dechlorination occurred on ZVI surface even without Pd, but it was significantly slow. Lien and Zhang (2007) reported that for trichloroethylene dechlorination, Fe alone exhibited around 80 times lower first order reaction rate than Fe with 1–5% Pd. In case of only ZVI, direct electrochemical reduction of 2-ClBP occurred via formation of an organic complex adsorbed to the surface of ZVI which served as an electron donor (note Equations 10.1–10.2). Meanwhile, indirect but faster reduction of 2-ClBP at the Pd surface also occurred in the ZVI/Pd system, involving atomic hydrogen as a powerful secondary reducing agent (note Equations 10.3–10.5). In the ZVI/Pd system, the surface of Pd with the low cathodic hydrogen overpotential was covered by atomic hydrogen due to the low activation barrier for H_2 dissociation, while such an indirect reduction was not considerable on the Fe surface with high hydrogen overpotential (Grittini *et al.*, 1995). The significant increase in the reactivity over Pd doping strongly supported the critical role of Pd.

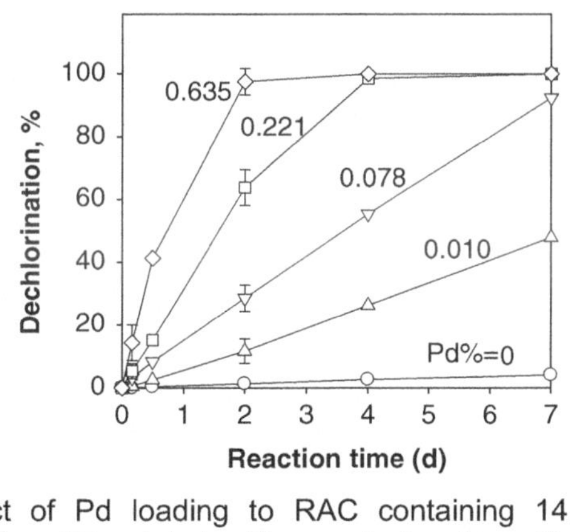

Figure 10.13 Effect of Pd loading to RAC containing 14.4% Fe on 2-ClBP dechlorination kinetics. "Reprinted with permission from *Environ. Sci. Technol.* 2009, **43**(19), 7510–7515. Copyright 2011 American Chemical Society."

10.3.3 Dependency of RAC performance on pH

Fe corrosion chemistry changes over pH and consequent Fe oxidation products exhibit a different affinity for PCBs. Decreasing the pH in metallic systems generally enhances the dechlorination rate due to prompt electron release as a result of facilitated Fe corrosion. However, in this RAC system, the degree of PCBs adsorption over pH might be more important for their dechlorination rather than the pH-dechlorination chemistry because the dechlorination is mediated by adsorption. As shown in Figure 10.14, neutral pH showed much more 2-ClBP adsorption than acidic and basic conditions (Choi & Al-Abed, 2010). According to the Fe corrosion chemistry, the basic pH leads to the prompt precipitation of Fe oxide species (initially from unstable $Fe(OH)_2$, FeOOH, and FeO eventually to more stable Fe_2O_3 and Fe_3O_4). Meanwhile, Fe is soluble at pH 4.0 and thus the formation of an oxide layer on the core Fe surface is negligible. Fe corrosion products somehow affect PCBs adsorption event. Noubactep (2007) reported that in situ generated Fe corrosion products have a strong ability to sequester PCBs. In addition, Fe rust might reduce the permeability of RAC by occupying available porous structure (Furukawa *et al.*, 2002).

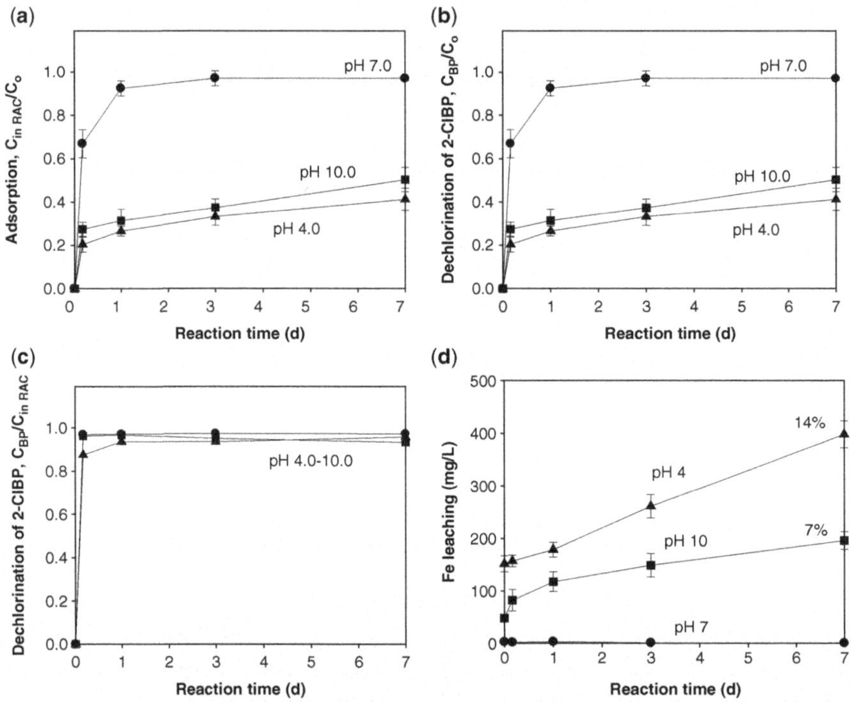

Figure 10.14 Treatment of 2-ClBP on RAC at different pH conditions: (a) adsorption, (b) overall dechlorination, (c) solid phase dechlorination, and (d) Fe leaching. Overall dechlorination is based on 2-ClBP and BP in all phases while solid phase dechlorination is based on 2-ClBP and BP in the RAC solid phase. The % number in Figure 10.14(d) is the leaching fraction of Fe. "Reprinted with permission from *J. Hazard. Mater.* 2010, **179**(1–3), 869–874. Copyright 2011 Elsevier."

The Fe corrosion behaviour was consistent with Fe leaching results demonstrated in Figure 10.14d (Pd leaching was not so significant because Pd, once released, is believed to immediately re-deposit to ZVI surface). Apparently, Fe leaching was more significant at pH 4.0 > 10.0 while it was negligible at pH 7.0. Fe at pH 4.0 was soluble while unstable ferrous hydroxides and hydrated oxides at pH 10.0 were present as disintegrates in the solution. Fe leaching at pH 4.0 kept increasing over time since Fe surface at pH 4.0 was in direct contact with the acid solution and thus the leaching proceeded fast. Meanwhile, Fe leaching at pH 10.0 was relatively slow because the Fe oxide passivating layer prevented the core Fe from oxidizing (Choi *et al.*, 2009b).

Many metallic systems have shown higher reaction rates in acidic pH due to the enhancement of Fe oxidation and thus electron release (Liu & Lowry, 2006). As clearly seen in Figures 10.14a and 10.14b, the adsorption degree of 2-ClBP under different pH conditions was almost identical to its dechlorination degree in the whole system. The solid phase dechlorination at around 100% shown in Figure 10.14c reflected the simultaneous dechlorination of adsorbed 2-ClBP. The results implied that adsorption step controlled the overall reaction. As a result, the common pH-dechlorination chemistry, which has been applied to neat metallic systems, was less meaningful for this RAC system. Since the dechlorination in the RAC system was uniquely mediated by adsorption, the pH-adsorption chemistry was more significant.

10.4 TREATMENT CAPACITY AND LONGEVITY

The PCB adsorption capacity of RAC is closely related to its structural properties, such as surface area and pore volume. The content of Fe^0 (zerovalent iron, ZVI) as a primary electron donor determines dechlorination capacity and longevity. The adsorption ability, dechlorination capacity, and oxidation and ageing of RAC are considered the most critical factors in determining the practical applications of RAC system.

10.4.1 Dechlorination capacity

Comparing an empirical dechlorination capacity with an estimated dechlorination capacity from the properties of the metal particles and nature of the associated reactions makes it possible to evaluate the practical effectiveness of the metallic systems. Based on current understanding of the reactions, oxidation of one mole of Fe^0 releases two moles of electrons via the most often assumed route to Fe^{2+} (Equation 10.1). In the direct dechlorination (Equation 10.2, pathway I), one mole of electrons dechlorinate half mole of mono PCBs (2-ClBP). However, this pathway is known to be 2–3 fold slower than indirect Pd-mediated hydrodechlorination (Equations 10.3–10.5, pathway II) (Choi *et al.*, 2008). One mole of electrons produces a half mole of H_2, which is able to dechlorinate half mole of 2-ClBP. In either pathway I or II, one mole of Fe^0 can dechlorinate one mole of 2-ClBP. If all Fe is zerovalent and completely oxidized and all electrons generated are used for the dechlorination reaction, 1 g of RAC with 14.4% Fe here can theoretically dechlorinate 488.1 mg of 2-ClBP (RAC:2-ClBP = 2:0.98). Figure 10.15 shows the adsorption and dechlorination of 2-ClBP in the presence of excessive amount of 2-ClBP (RAC:2-ClBP ratio of 2:1.06) (Choi *et al.*, 2009b). Approximately, 23.2% of 2-ClBP added disappeared from the liquid phase by adsorption (i.e. 122.6 mg of 2-ClBP/g RAC). After 8 days, 2-ClBP dechlorination reached a maximum point at 10.7%, corresponding to 56.5 mg 2-ClBP/g RAC.

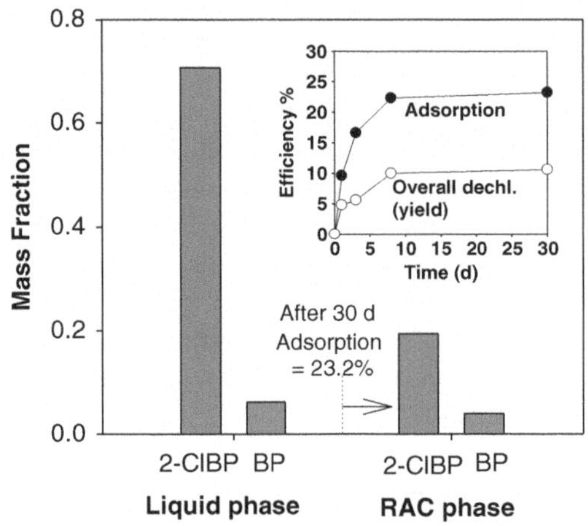

Figure 10.15 Adsorption and dechlorination of 2-ClBP on RAC for 30 days in the presence of excessive amount of 2-ClBP for investigating the 2-ClBP treatment capacity of RAC (insert), and fraction of 2-ClBP and BP partitioned to the liquid and solid phases after 30 days. "Reprinted with permission from *Environ. Sci. Technol.* 2009, **43**(11), 4137–4142. Copyright 2011 American Chemical Society."

It is useful to introduce a term "dechlorination yield": the ratio of the amount of actually dechlorinated PCBs to that of PCBs which can be theoretically dechlorinated by the system. The yield in this case was calculated as 0.12. The low yield was mainly due to the overestimation of the capacity. Numerous undesired reactions occurred, competing with the dechlorination reaction for electrons generated. As indicated in Figure 10.9, in fact, the Fe species was a mixture of Fe^0 and Fe^{3+} since not all Fe^{3+} was transformed to Fe^0 during borohydride reduction. Partial oxidation of Fe^0 also occurred in advance during reductive deposition of Pd on ZVI surface (Equation 10.7). Surface corrosion precipitates also masked the active redox sites and reduced the permeability of the RAC. Loss of some reactivity of RAC from the passivation of Pd surface might partly explain the low yield (Lowry & Reinhard, 2000). However, there was also a possibility of underestimating the capacity. It was proposed that Fe^0/Fe_3O_4 coupling might be more thermodynamically favourable than Fe^0/Fe^{2+} coupling at pH above 6.0 (Equation 10.8) (Kohn *et al.*, 2005; Nurmi *et al.*,

2005). In that case, the Fe^0/Fe_3O_4 coupling can generate 0.67 mole more electrons per one mole of Fe^0.

$$3Fe^0 + 4H_2O \longrightarrow Fe_3O_4 + 8H^+ + 8e^- \qquad \text{(Eq. 10.8)}$$

10.4.2 Short term treatability

Since an immediate decrease in 2-ClBP adsorption and dechlorination kinetics was expected, the response of GAC and RAC systems to short term successive 2-ClBP spiking was tested, as shown in Figure 10.16 (Choi *et al.*, 2009b). All 2-ClBP provided in each cycle immediately disappeared from the liquid phase, resulting in its accumulation in the GAC phase. RAC exhibited similar 2-ClBP adsorption behavior. As expected, BP instead of 2-ClBP accumulated in the RAC solid phase. Actual dechlorination in each cycle was at around 80% while apparent dechlorination kept increasing because non-reacted 2-ClBP in the RAC solid during a cycle was dechlorinated in the following cycle.

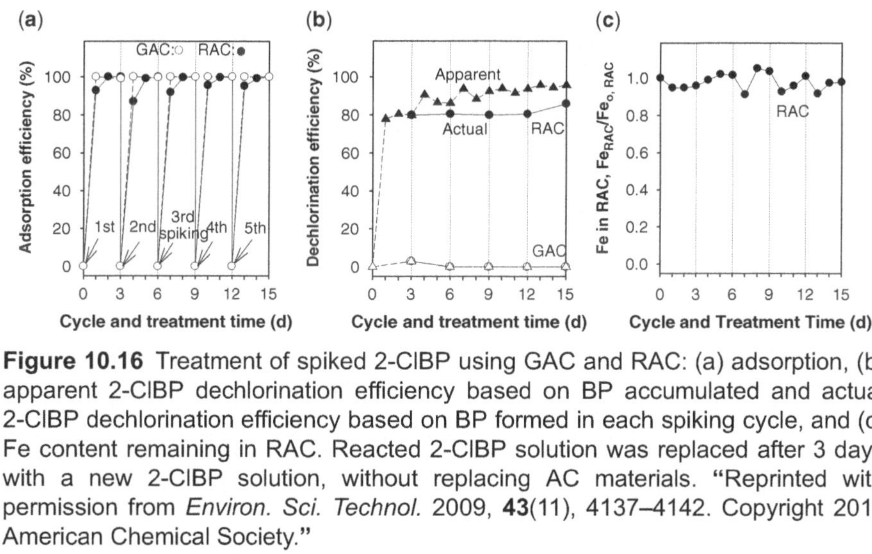

Figure 10.16 Treatment of spiked 2-ClBP using GAC and RAC: (a) adsorption, (b) apparent 2-ClBP dechlorination efficiency based on BP accumulated and actual 2-ClBP dechlorination efficiency based on BP formed in each spiking cycle, and (c) Fe content remaining in RAC. Reacted 2-ClBP solution was replaced after 3 days with a new 2-ClBP solution, without replacing AC materials. "Reprinted with permission from *Environ. Sci. Technol.* 2009, **43**(11), 4137–4142. Copyright 2011 American Chemical Society."

10.4.3 Long term ageing and oxidation

Klausen *et al.* (2003) reported a slow reduction in the reactivity of micron size Fe particles over years. In case of nanosize Fe particles with nano Pd islands, more remarkable changes were expected due to their high specific surface area (Liu & Lowry 2006). As shown in Figure 10.17, the reactivity of RAC aged for one year under N_2, $H_2O + N_2$, and $H_2O + O_2$ (O_2 level at 0.1 mg/L) environments was compared with that of fresh RAC (Choi *et al.*, 2009b). Ageing of RAC under N_2 did not affect its 2-ClBP adsorption ability while the presence of H_2O and/or O_2

during RAC ageing affected the adsorption ability of RAC. Exposure of RAC to water under O_2 or N_2 for a short time decreased its adsorption capability while that for a long time eventually enhanced the adsorption. The reactivity of RAC decreased over ageing under more oxidizing environments: $N_2 < H_2O + N_2 < H_2O + O_2$. The reactivity reduction was 5.6%, 19.5%, and 32.5% over one year, respectively. The decline was considered to be much slower than what is expected from the nanoscale Fe/Pd corrosion chemistry. Fe^0 fraction at 57.2% decreased to 46.2% for N_2, 33.4% for $H_2O + N_2$, 26.5% for $H_2O + N_2$, which was consistent with the decrease in RAC reactivity.

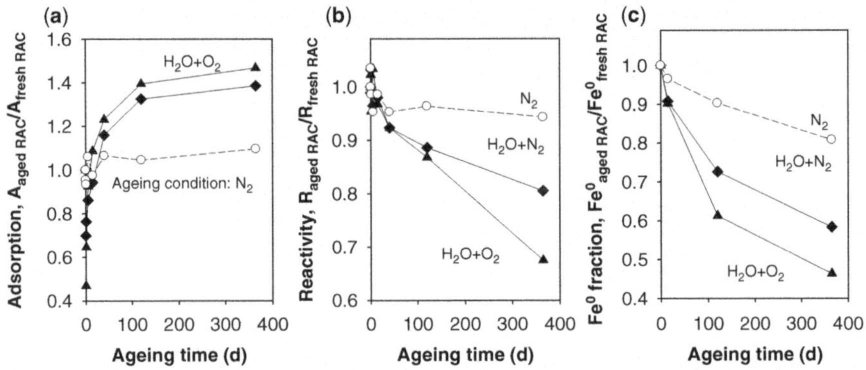

Figure 10.17 Changes in (a) adsorption ability, (b) dechlorination reactivity, and (c) ZVI content over one year ageing under the conditions of anaerobic storage (N_2), exposure to water in N_2 environment ($H_2O + N_2$), and exposure to water open to air ($H_2O + O_2$). The results of aged RAC were normalized with those of fresh RAC. "Reprinted with permission from *Environ. Sci. Technol.* 2009, **43**(11), 4137–4142. Copyright 2011 American Chemical Society."

The Fe oxidation and corrosion is complex (Choi *et al.*, 2009b). The formation of iron oxyhydroxides and other iron oxides may be good for the immobilization of certain contaminants (Lackovic *et al.*, 2000). Formation of the oxide passivating films might explain the slow decreases in the Fe^0 fraction and reactivity of RAC. The decrease, however, was significantly slower than that of suspended nanosize ZVI particles reported by Liu and Lowry (2006). In one Fe particle, the ZVI core/Fe oxide shell structure is formed and thus the oxidation of the core ZVI can be protected by the oxide layer. Meanwhile, in one RAC grain, Fe particles closely located in the RAC grain boundary oxidize faster than those in the RAC grain core (i.e. inner pores of RAC). The oxide layer in the RAC grain boundary grows and protects ZVI in the RAC grain core. This protecting effect can be a characteristic of this supported ZVI type, which is different from highly dispersed metal particles susceptible to oxidation. The formation of Fe oxide layer apparently decreases dechlorination kinetics on RAC, whereas it eventually prolongs the dechlorination longevity of RAC. This feature is good for the RAC system to respond to low level of PCBs slowly released from sites.

Fe leaching from RAC during the one year oxidation is shown in Figure 10.18 (Choi *et al.*, 2009b). The presence of O_2 facilitated more Fe dissolution due to the faster Fe oxidation. Fe leaching was negligible at below 0.2 mg/L because the pH of the reaction solution at 6.5 immediately increased to pH 8.9 where the Fe species was Fe oxides. The amount of dissolved Fe corresponded to only 0.3×10^{-5} in terms of its relative fraction to that in a fresh RAC.

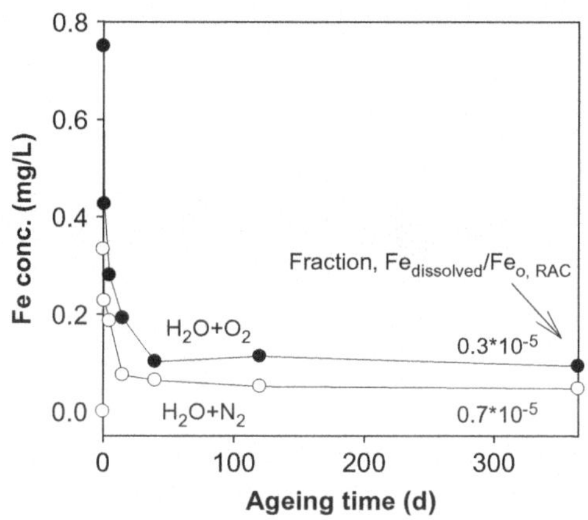

Figure 10.18 Fe leaching due to the oxidation of RAC over one year ageing under $H_2O + N_2$ and $H_2O + O_2$ shown in Figure 10.17. "Reprinted with permission from *Environ. Sci. Technol.* 2009, **43**(11), 4137–4142. Copyright 2011 American Chemical Society."

10.5 REACTVITIY OF RAC WITH PCBs

Treatability of RAC towards highly chlorinated and positional PCBs is another important issue. In the previous sections, 2-ClBP was used as a probe PCB due to the simplicity of the system analysis. However, tri-, tetra-, penta-, and hexa-chlorinated biphenyls are the most problematic in real world treatment. In general, *ortho* dechlorination may not be the same as *meta* or *para* dechlorination, and dechlorination of a single chlorine may not be similar to that of multiple chlorines. Treatability of RAC towards various substituted chlorines in PCBs is important for elucidating PCB structure-dechlorination relation and subsequent dechlorination pathways of PCBs.

10.5.1 Mono PCB congeners

Chlorines in different positions within the PCB structure had their unique reactivity with RAC, as shown in Figure 10.19 (Choi *et al.*, 2009c). The dechlorination efficiency increased in order of 4(*para*)-ClBP > 3(*meta*)-ClBP > 2(*ortho*)-ClBP, implying the presence of the relative resistance of the positional isomers (Yak *et al.*, 2000; Lowry & Johnson, 2004). The same tendency was also observed even in case of the dechlorination of a mixture of the three positional mono PCBs. Comparing the results under the independent (left side) and competitive (right side) conditions, the dechlorination of *meta* or *para* mono PCB was not affected by the competition while that of *ortho* PCB significantly decreased from 77.0% to 62.1% in the competing environment.

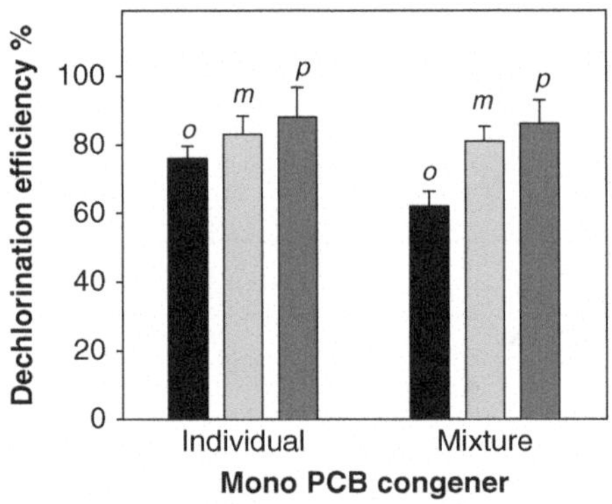

Figure 10.19 Dechlorination of positional mono PCB congeners on RAC: (a) reaction with individual PCB congener (left) and (b) reaction with a mixture of the mono PCB congeners (right). "Reprinted with permission from *Environ. Sci. Technol.* 2009, **43**(19), 7510–7515. Copyright 2011 American Chemical Society."

The chlorine in *ortho* position causes a significant increase in the electron density at the biphenyl bond, twisting the two phenyl rings in order to minimize the steric repulsion. According to the calculation by Greaves *et al.* (1994), the twist angle between the two phenyl rings increases from 41° for BP to 56° for 2-ClBP. Meanwhile, the distance between the biphenyl bond and *meta* or *para* chlorine is too far to induce this effect (but not negligible). As a result, the presence of *ortho* chlorines results in the nonplanarity of the phenyl rings. The first step of the dechlorination reaction is surface interaction of chlorines in PCBs with ZVI/Pd in RAC. The parallel configuration of the phenyl ring of PCBs to the active metallic surface was more favourable for their adsorption since more adsorption occurred between more planar PCBs (4-ClBP > 3-ClBP > 2-ClBP) and the GAC. As a result, more significant inhibition of the intra molecular rotation around the phenyl bond, by chlorines closer to the bond (*ortho* > *meta* > *para*), restricts surface interactions between PCBs and GAC (Bucheli & Gustafsson, 2003). It was also reported that the electron cloud of the *ortho* chorine on one phenyl ring hangs on the top of the other phenyl ring, causing partial deactivation to electrophilic substitution (Hagh & Allen, 1990).

10.5.2 Tri-chlorinated biphenyl

The highest dechlorination susceptibility of the *para* positional chlorine was also confirmed during the dechlorination of 2,3,4-trichlorinated biphenyl (2,3,4-triClBP), as shown in Figure 10.20 (Choi *et al.*, 2009c). Looking at the first round dechlorination, 2,3-diClBP was a dominant intermediate than any other diClBPs found. In the second round dechlorination, 2-ClBP was present much more than 3-ClBP and 4-ClBP. Noma *et al.* (2003) found that the reactivity of a chlorine in aromatics decreases upon the presence of adjacent chlorines. Based on the conformation of 2,3,4-triClBP, the *para* chlorine (Cl–'Cl'–H) > *meta* chlorine (Cl–'Cl'–Cl) > *ortho* chlorine (phenyl–'Cl'–Cl) was the preferred target to remove from 2,3,4-triClBP. Promisingly, this indicated that more toxic coplanar PCB congeners were preferentially reduced.

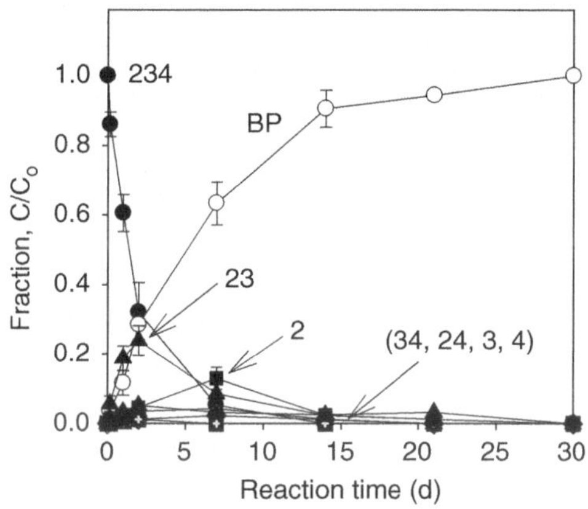

Figure 10.20 Stepwise dechlorination of 2,3,4-trichlorinated biphenyl (2,3,4 -ClBP) on RAC. Relative fraction of 2,3,4-ClBP, reaction intermediates, and BP is reported. "Reprinted with permission from *Environ. Sci. Technol.* 2009, **43**(19), 7510–7515. Copyright 2011 American Chemical Society."

10.5.3 Higher PCB congeners

Figure 10.21 shows treatment of higher *ortho* PCB congeners, up to tetra chlorinated biphenyls (Choi *et al.*, 2009c). Instead of tracing their dechlorination pathway, which occurs stepwise with a chlorine atom being replaced by a hydrogen atom in each single reduction step, the carbon mass balance (mole fraction) was monitored after 24 h reaction. The adsorption efficiency of RAC was consistently decreased from 77.5% for the smallest size monoClBP to 49.0% for the largest size tetraClBP. In the RAC solid phase, lesser chlorinated PCBs showed more dominant BP fraction than their mother fraction while higher chlorinated PCBs exhibited more dominant mother PCB fraction than BP fraction. BP was detected in all cases, indicating that it is possible to remove all chlorines from the higher PCBs.

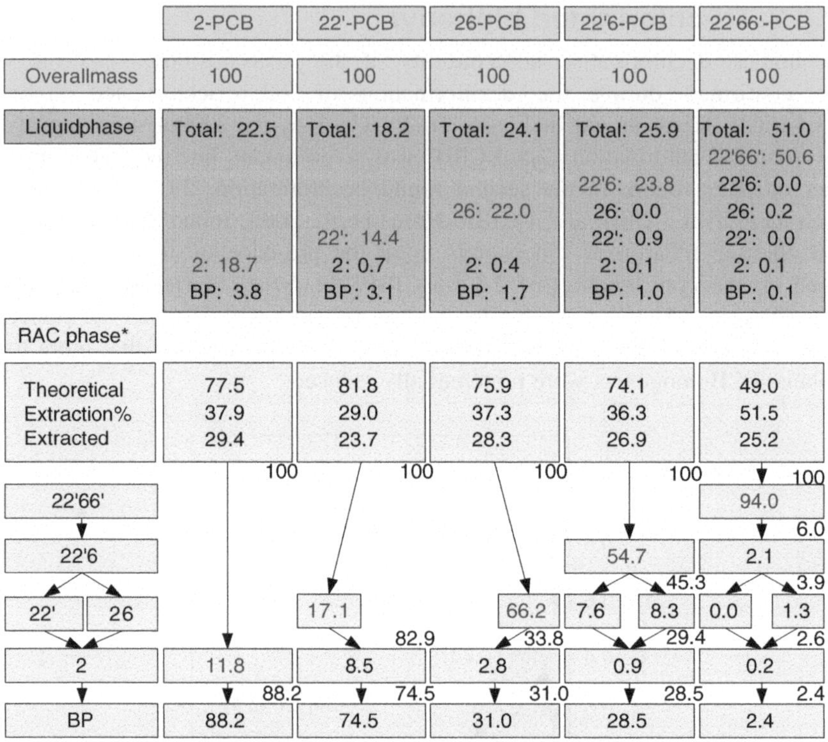

Figure 10.21 PCBs mass balance (mole fraction) during adsorption and dechlorination of *ortho* PCB congeners. Take 22′66′-PCB for an example of data interpretation. After 24 h, 51.0% of PCBs were partitioned to the liquid phase, consisting of 22′66′: 50.6%, 22′6: 0.0%, 26: 0.2%, 22: 0.0%, 2: 0.1% and BP: 0.1%. As a result, 49.0% of PCBs should be present in the RAC solid phase (i.e. PCBs adsorption efficiency). However, due to incomplete extraction of PCBs from the solid phase, only 25.2% out of 49.0% was extracted, yielding a solid phase extraction efficiency 51.5%. Since complete extraction of PCBs is difficult to achieve, the relative ratio of PCB congeners actually recovered from the system is important for indirectly estimating the dechlorination reaction of PCBs, no matter how much of PCBs are extracted. To reflect this, the total mass fraction of PCB congeners in the RAC solid at 25.2% was reset as 100%, giving ratios of PCB congeners in the RAC solid phase. "Reprinted with permission from *Environ. Sci. Technol.* 2009, **43**(19), 7510–7515. Copyright 2011 American Chemical Society."

Comparing 22′-ClBP with 26-ClBP as mother PCBs, 22′-ClBP in RAC survived much less than 26-ClBP. If no selective dechlorination among the chlorines within the *ortho* position exists in the first round dechlorination of 22′6-ClBP and the second round dechlorination of 22′66′-ClBP, the ratio of 26-ClBP to 22′-ClBP should be 0.5 on the basis of the conformation of 22′6-ClBP. However, the ratio of 26-ClBP to 22′-ClBP was rather somewhat reversed. This means that the 2′ positional chlorine was preferentially removed compared to the 2 or 6 positional chlorine in 22′6-ClBP. In addition, 26-ClBP, once formed, was more resistant to dechlorination than 22′-ClBP. This suggests the isolated chlorine in a phenyl ring was more susceptible to dechlorination than one of the co-existing chlorines in the other phenyl ring, consistent with the finding by Noma *et al.* (2003) that the less chlorinated phenyl ring should be prone to electrophilic attack. The same was found for *meta* PCBs (Choi *et al.*, 2009c).

Even though not all 209 PCB congeners have been tested, many meaningful conclusions were drawn from the detailed interpretation of Figure 10.21, including: (i) favourable dechlorination of more toxic coplanar PCB congeners; (ii) preferential electrophilic attack to chlorines in a less substituted phenyl ring; (iii) prominent dechlorination inhibition for higher PCB congeners; and (iv) eventual complete dechlorination of higher PCB congeners to BP.

10.6 APPLICATION PROSPECTS

10.6.1 Reactive capping barrier concept

The RAC can be directly mixed with the PCBs-contaminated matrix for prompt sequestration of PCBs, as demonstrated in Figure 10.2. A "reactive" cap or barrier composed of RAC pellets is also proposed. The concept of a permeable barrier has been commercially established to resolve some technical issues associated with materials installation. The cap can be installed horizontally to contain PCBs-contaminated sites. Alternatively, the RAC barrier can be installed vertically for flow-through treatment of contaminated surface water and groundwater, as shown in Figure 10.22.

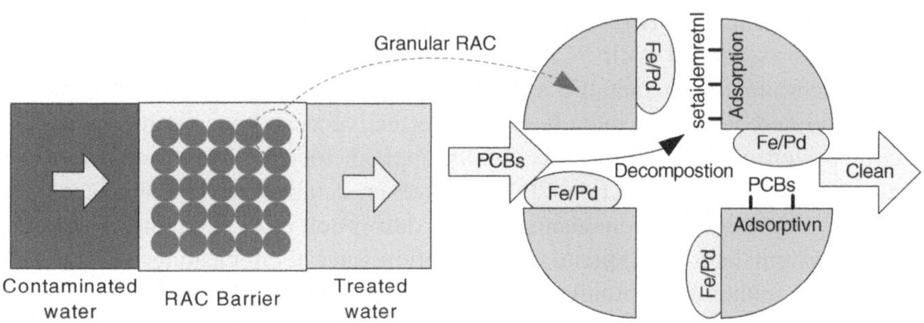

Figure 10.22 Application example of a RAC barrier to treat PCBs (or similar chemicals)-contaminated surface water and groundwater. The scheme shows a flow-through treatment strategy.

The reactions in the barrier/cap are complex. As time goes on, most PCBs might still remain in the stable solid matrix, especially organic fraction. Upon desorption of PCBs to the aqueous phase and movement to the surrounding environment, PCBs are sequestrated in the RAC. Some of the PCBs are strongly adsorbed and some are dechlorinated to form reaction intermediates and eventually BP, which are also adsorbed to RAC or diffused in negligible amounts to the bulk liquid phase.

10.6.2 Challenges

The results described previously have not addressed the complexities of treating actual contaminated sites, and rather neglected the potentially adverse impacts of many co-existing species in the treatment environment (Choi & Al-Abed, 2010). Some practical reaction conditions relevant in natural systems should be explored for the full scale application of RAC system to contaminated sites. The co-existing chemicals serve as proton donors or electron shuttles, compete with PCBs for electron utilization and active sites, form deactivation layers on the metal surface, and/or form and attack metal oxide films (Drissi *et al.*, 1994; Lowry & Reinhard, 2000; Choi & Al-Abed, 2010). Anions increase the ionic strength of an aquatic system. Some ionic species are subject to reduction or oxidation and they also adsorb to active sites in RAC. These generally decrease the RAC reactivity. Some species are known to be aggressive, readily attack metal surfaces, and thus increase the Fe corrosion rate. Interestingly, it was reported that natural organic matter can also serve as an electron shuttle and thus PCBs can be oxidatively decomposed by ZVI system (Kang & Choi, 2009).

Treatment of aqueous phase PCBs has been proven through many studies. However, as mentioned earlier, the significant and practical problem with PCBs treatment is associated with their strong adsorption to solid matrix. As a result, the dechlorination kinetics and efficiency of RAC might be slow and depreciated when it treats actual PCBs-contaminated sites. Most likely because of real world complexities associated with PCBs desorption from a solid matrix, it is hard to find in literature that a metallic system alone was tested to treat contaminated aquatic soil and sediment, regardless of its effectiveness. Aqueous mixing of the metallic system with sediment was simply applied, focusing on the feasibility of the treatment (Gavaskar & Chen, 2002). However, detailed mechanistic aspects of the underlying removal mechanisms, such as desorption of PCBs, their transport to metallic system and subsequent dechlorination, and their ensuing partitioning, have not been either conceptually considered or experimentally elucidated.

In addition, before its scale up and field applications, the high cost of the RAC material due to the expensive Pd doping as well as time and material-consuming synthesis procedures needs to be addressed. In general, GAC as an adsorbent determines PCBs adsorption, ZVI as electron donor initiates the dechlorination reaction, and Pd as a catalyst rules the dechlorination kinetics. The three materials, all together, affect the overall reaction. Fe, when coupled with a less active metal such as Cu, Co, and Ni, oxidizes faster and thus shows better reactivity, but its longevity to dechlorinate PCBs is shortened (vice versa in coupling with more active metal). Although potential leaching of toxic Pd might be a huge debate, a small amount of Pd incorporated onto Fe is helpful for increasing the dechlorination kinetics (Figure 10.13).

10.6.3 Applications

AC, traditionally used to absorb contaminants, was modified with Fe/Pd bimetallic nanoparticles. This made the AC more effective not only in trapping and isolating PCBs but also degrading the contaminants. The findings here have significant implications for nanotechnological treatment of other contaminants in the environment, although more detailed research should be conducted. This RAC research focuses on a really "reactive" (not just active) material with "multiple" of physical adsorption and chemical decomposition, and aims at truly "in situ" application, which has been overlooked in previous studies. A series of the studies represent an important first step in a new direction of permeable reactive barrier research to provide a practical solution to contaminated sites with a variety of hydro-geological settings. The RAC system immediately sequesters PCBs and "then" treats them. Reaction intermediates are also essentially sequestrated, awaiting further dechlorination. As a result, the RAC system is fail-safe, unlike other chemical strategies (i.e. at least sequestration of PCBs occurs). The RAC can also treat other co-existing toxic chemicals in the environment. The strategy with the concept of a reactive cap or barrier is more practical and easy to apply in situ.

This RAC strategy provides an effective tool to manage PCBs-contaminated sites in a manner that gains regulatory acceptance. The main goal of this study has a strong tie with the US Environmental Protection Agency (EPA)'s perspective direction to cutting-edge scientific research for developing new sustainable methods and technologies for removing toxic chemicals from the Superfund sites (Lubick, 2009). Developing effective strategies to respond to current regulations on PCBs contamination might be one of the highest priorities of industrial companies which manufactured PCBs and related products, U.S. Department of Defense managing numerable sites contaminated with PCBs released by their military activities, and EPA regulating and controlling contaminated sites. Since we believe that the strategy technically works for other historically notorious toxic compounds as well as emerging chemicals of concern (in particular, halogenated organic compounds), the approach and technology adopted in this study will initiate researchers into strategic research plans to understand the transport, reaction, and fate of such contaminants on RAC as a new environmental risk management option.

ACKNOWLEDGEMENT

This research was funded in part by the University of Texas at Arlington through new faculty start-up funds, the US Geological Survey/Texas Water Research

Institute through the Graduate Student Grant Program (G11AP20108), and the National Risk Management Research Laboratory of EPA. This paper has not been subjected to internal policy review of the EPA. Therefore, the research results do not necessarily reflect the views of the agency or its policy. Mention of trade names and commercial products does not constitute endorsement or recommendation for use.

REFERENCES

Agarwal S., Al-Abed S. R. and Dionysiou D. D. (2007a). *In situ* technologies for reclamation of PCB-contaminated sediments: current challenges and research thrust areas. *J. Environ. Eng.*, **133**, 1075.

Agarwal S., Al-Abed S. R. and Dionysiou D. D. (2007b). Enhanced corrosion-based Pd/Mg bimetallic systems for dechlorination of PCBs. *Environ. Sci. Technol.*, **41**(10), 3722–3727.

Borja J., Taleon D. M., Auresenia J. and Gallardo S. (2005). Polychlorinated biphenyls and their biodegradation. *Process Biochem.*, **40**(6), 1999–2013.

Bucheli T. D. and Gustafsson O. (2003). Soot sorption of non-*ortho* and *ortho* substituted PCBs. *Chemosphere*, **53**(5), 515–522.

Choi H., Al-Abed S. R., Agarwal S. and Dionysiou D. D. (2008). Synthesis of reactive nano-Fe/Pd bimetallic system-impregnated activated carbon for the simultaneous adsorption and dechlorination of PCBs. *Chem. Mater.*, **20**(11), 3649–3655.

Choi H., Agarwal S. and Al-Abed S. R. (2009a). Adsorption and simultaneous dechlorination of PCBs on GAC/Fe/Pd: Mechanistic aspects and reactive capping barrier concept. *Environ. Sci. Technol.*, **43**(2), 488–493.

Choi H., Al-Abed S. R. and Agarwal S. (2009b). Effects of aging and oxidation of palladized iron embedded in activated carbon on the dechlorination of 2-chlorobiphenyl. *Environ. Sci. Technol.*, **43**(11), 4137–4142.

Choi H., Al-Abed S. R. and Agarwal S. (2009c). Catalytic role of palladium and relative reactivity of substituted chlorines during adsorption and treatment of PCBs on reactive activated carbon. *Environ. Sci. Technol.*, **43**(19), 7510–7515.

Choi H. and Al-Abed S. R. (2009). PCB congener sorption to carbonaceous sediment components: macroscopic comparison and characterization of sorption kinetics and mechanism. *J. Hazard. Mater.*, **165**(1–3), 860–866.

Choi H. and Al-Abed S. R. (2010). Effect of reaction environments on PCB reactivity with activated carbon impregnated with palladized iron. *J. Hazard. Mater.*, **179**(1–3), 869–874.

Drissi H., Refait P. and Genin J. (1994). The oxidation of $Fe(OH)_2$ in the presence of carbonate ions: structure of carbonate green rust. *Hyperfine Interact.*, **90**(1), 395–400.

Elliott D. W. and Zhang W.-X. (2001). Field assessment of nanoscale bimetallic particles for groundwater treatment. *Environ. Sci. Technol.*, **35**(24), 4922–4926.

Furukawa Y., Kim J. W., Watkins J. and Wilkin R. T. (2002). Formation of ferrihydrite and associated iron corrosion products in permeable reactive barriers of zero-valent iron. *Environ. Sci. Technol.*, **36**(24), 5469–5475.

Gaspar D. J., Lea A. S., Engelhard M. H. and Baer D. R. (2002). Evidence for localization of reaction upon reduction of carbon tetrachloride by granular iron. *Langmuir*, **18**(20), 7688–7693.

Gavaskar A. R. and Chen A. S. C. Eds. (2002). Remediation of chlorinated and recalcitrant compounds. Proceedings of the Third International Conference, 20–23 May 2002, Monterey, CA, Battelle Memorial Institute, Columbus, OH.

Greaves J., Harvey E. and MacIntyre G. (1994). Correlation between electron capture negative chemical ionization mass spectrometric fragmentation and calculated internal energies for polychlorinated biphenyls. *J. Am. Soc. Mass. Spectrom.*, **5**(1), 44–52.

Grittini C., Malcomson M., Fernando Q. and Korte N. (1995). Rapid dechlorination of polychlorinated biphenyls on the surface of a Pd/Fe bimetallic system. *Environ. Sci. Technol.*, **29**(11), 2898–2900.

Hagh B. F. and Allen D. T. (1990). Catalytic hydroprocessing of chlorinated benzens. *Chem. Eng. Sci.*, **45**(8), 2695–2701.

Jonker M. T. O. and Koelmans A. A. (2002). Sorption of polycyclic aromatic hydrocarbons and polychlorinated biphenyls to soot and soot-like materials in the aqueous environment: mechanistic considerations. *Environ. Sci. Technol.*, **36**(17), 3725–3734.

Kang S. H. and Choi W. (2009). Oxidative degradation of organic compounds using zero-valent iron in the presence of natural organic matter serving as an electron shuttle. *Environ. Sci. Technol.*, **43**(3), 878–883.

Klausen J., Vikesland P. J., Kohn T., Burris D. R., Ball W. P. and Roberts A. L. (2003). Longevity of granular iron in groundwater treatment processes: solution composition effects on reduction of organohalides and nitroaromatic compounds. *Environ. Sci. Technol.*, **37**(6), 1208–1218.

Kohn T., Livi K. J. T., Robers A. L. and Vikesland P. J. (2005). Longevity of granular iron in groundwater treatment process: corrosion product development. *Environ. Sci. Technol.*, **39**(8), 2867–2879.

Lackovic J. A., Nikolaidis N. P. and Dobbs G. M. (2000). Inorganic arsenic removal by zero-valent iron. *Environ. Eng. Sci.*, **17**(1), 29–39.

Lien H. L. and Zhang W. X. (2007). Nanoscale Pd/Fe bimetallic particles: catalytic effects of palladium on hydrodechlorination. *Appl. Catal. B: Environ.*, **77**(1–2), 110–116.

Liu Y., Choi H., Dionysiou D. and Lowry G. V. (2005). Trichloroethene hydrodechlorination in water by highly disordered monometallic nanoiron. *Chem. Mater.*, **17**(21), 5315–5322.

Liu Y. and Lowry G. V. (2006). Effect of particle age (Fe0 content) and solution pH on NZVI reactivity: H_2 evolution and TCE dechlorination. *Environ. Sci. Technol.*, **40**(19), 6085–6090.

Lowry G. V. and Reinhard M. (1999). Hydrodehalogenation of 1- to 3-carbon halogenated organic compounds in water using palladium catalyst and hydrogen gas. *Environ. Sci. Technol.*, **33**(11), 1905–1910.

Lowry G. V. and Reinhard M. (2000). Pd-catalyzed TCE dechlorination in groundwater: solute effects, biological control, and oxidative catalyst regeneration. *Environ. Sci. Technol.*, **34**(15), 3217–3223.

Lowry G. V. and Johnson K. M. (2004). Congener-specific dechlorination of dissolved PCBs by microscale and nanoscale iron in a water/methanol solution. *Environ. Sci. Technol.*, **38**(19), 5208–5216.

Lubick N. (2009). Cap and degrade: A reactive nanomaterial barrier also serves as a cleanup tool. *Environ. Sci. Technol.*, **43**(2), 235–235.

Luthy R. G., Aiken G. R., Brusseau M. L., Cunningham S. D., Gschwend P. M., Pignatello J. J., Reinhard M., Traina S. J., Weber W. J. Jr. and Westall J. C. (1997). Sequestration of hydrophobic organic contaminants by geosorbents. *Environ. Sci. Technol.*, **31**(12), 3341–3347.

McDonough K. M., Murphy P., Olsta J., Zhu Y., Reible D. and Lowry G. V. (2007). Development and placement of a sorbent-amended thin layer sediment cap in the anacostia river. *Soil Sediment Contam.: Int. J.*, **16**(3), 313–322.

Mitsui T., Ross M. K., Formin E., Ogletree D. F. and Salmeron M. (2003). Dissociative hydrogen adsorption on palladium requires aggregates of three or more vacancies. *Nature*, **422**, 705–707.

Morales J., Hutcheson R. and Cheng I. F. (2002). Dechlorination of chlorinated phenols by catalyzed and uncatalyzed Fe (0) and mg (0) particles. *J. Hazard. Mater.*, **90**(1), 97–108.

Murphy P., Marquette A., Danny Reible P. and Lowry G. V. (2006). Predicting the performance of activated carbon-, coke-, and soil-amended thin layer sediment caps. *J. Environ. Eng.*, **132**, 787–794.

Noma Y., Ohno M. and Sakai S. (2003). Pathways for the degradation of PCBs by palladium-catalyzed dechlorination. *Fresenius Environ. Bull.*, **12**(3), 302–308.

Noubactep C. (2007). Processes of contaminant removal in Fe0-H_2O systems revisited: the importance of co-precipitation. *Open Environ. J.*, **1**(1), 9–13.

Nurmi J. T., Tratnyek P. G., Sarathy V., Baer D. R., Amonette J. E., Pecher K., Wang C., Linehan J. C., Matson D. W. and Penn R. L. (2005). Characterization and properties of metallic iron nanoparticles: spectroscopy, electrochemistry, and kinetics. *Environ. Sci. Technol.*, **39**(5), 1221–1230.

Ponder S. M., Darab J. G., Bucher J., Caulder D., Craig I., Davis L., Edelstein N., Lukens W., Nitsche H. and Rao L. (2001). Surface chemistry and electrochemistry of supported zerovalent iron nanoparticles in the remediation of aqueous metal contaminants. *Chem. Mater.*, **13**(2), 479–486.

Saleh N., Sirk K., Liu Y., Phenrat T., Dufour B., Matyjaszewski K., Tilton R. D. and Lowry G. V. (2007). Surface modifications enhance nanoiron transport and NAPL targeting in saturated porous media. *Environ. Eng. Sci.*, **24**(1), 45–57.

Schrick B., Hydutsky B. W., Blough J. L. and Mallouk T. E. (2004). Delivery vehicles for zerovalent metal nanoparticles in soil and groundwater. *Chem. Mater.*, **16**(11), 2187–2193.

Wang C. B. and Zhang W. (1997). Synthesizing nanoscale iron particles for rapid and complete dechlorination of TCE and PCBs. *Environ. Sci. Technol.*, **31**(7), 2154–2156.

Werner D., Higgins C. P. and Luthy R. G. (2005). The sequestration of PCBs in lake hartwell sediment with activated carbon. *Water Res.*, **39**(10), 2105–2113.

Wiegel J. and Wu Q. (2000). Microbial reductive dehalogenation of polychlorinated biphenyls. *FEMS Microbiol. Ecol.*, **32**(1), 1–15.

Yak H. K., Lang Q. and Wai C. M. (2000). Relative resistance of positional isomers of polychlorinated biphenyls toward reductive dechlorination by zerovalent iron in subcritical water. *Environ. Sci. Technol.*, **34**(13), 2792–2798.

Zhang W. (2003). Nanoscale iron particles for environmental remediation: an overview. *J. Nanopart. Res.*, **5**(3), 323–332.

Chapter 11

Microbial manufactured silver nanoparticles for water disinfection

L. Sintubin

11.1 INTRODUCTION

Despite the serious efforts taken by international organizations to improve water sanitation facilities, over 600 million people still have no access to safe drinking water (WHO, 2004). This results in the occurrence of a number of waterborne diseases with diarrhea being the predominant one. Worldwide 3.7% of all deaths are related to diarrhea, a number which can increase to 6.9% in low-income countries (WHO, 2010). For children aged under five years, diarrhea is responsible for 15% of children's mortality. More specific, this means that over 2.5 million children die every year due to diarrhea (WHO, 2011). The lack and/or malfunctioning of existing centralized and sophisticated water treatment systems in developing countries cause an urgent need for point of use (POU) water treatment systems. These POU water treatments need to be efficient, rapid and inexpensive. A couple of currently used POU treatments are solar disinfection (SODIS) using a combination of heat and UV irradiation, boiling, filtration and chlorination. Boiling is often unfeasible from an economical and practical point of view since large amounts of wood are needed (Burch & Thomas, 1998). Moreover, boiling does not provide a residual effect; hence, recontamination of the disinfected water upon storage is not unimaginable. Also SODIS and filtration deal with the trouble of recontamination (Berney *et al.*, 2006). Chlorination, in contrast, does provide residual disinfection. However, chlorination can cause the formation of toxic disinfection byproducts (Clasen *et al.*, 2007). Hence, there is a need for other POU treatments which provide a residual disinfecting effect without the potential danger of toxic byproduct formation. In the last years interest has arisen to use silver for POU water treatment.

Mankind has been using silver for thousands of years to decontaminate drinking water but also in the medical field to disinfect wounds and treat burns (Silver *et al.*, 2005). Not only is silver active towards a broad range of microorganisms, such as viruses, bacteria and fungi, it also has been shown to have a low toxicity toward humans (Edward-Jones, 2009; Elechiguerra *et al.*, 2005; Morones *et al.*, 2005; Lansdown, 2007). The use of bulk silver to disinfect water dates back to Alexander The Great and the Romans who stored water and wine in silver vessels or used silver coins to prevent spoilage. While back then mainly bulk silver metal (Ag^0) was used, many other silver formulations have been used during history. From the Middle Ages on until the beginning of the twentieth century, silver ions (Ag^+) were used by the application of inorganic silver salts such as silver nitrate for the treatment of wound infections. Also later in the twentieth century silver ions were used, but instead of inorganic silver salts, silver was applied in organic complexes with proteins (Klasen, 2000a; Klasen, 2000b). More recently, Ag^0 became popular again, not in its bulk formulation but as nanoparticles.

11.2 THE MICROBIAL PRODUCTION OF SILVER NANOPARTICLES

Silver nanoparticles can be synthesized using chemical and physical methods such as liquid-phase synthesis, high-temperature reduction, vapor-phase condensation, laser ablation, photoreduction and electrolysis (Abid *et al.*, 2002; Mafune *et al.*, 2000; Malynych & Chumanov, 2003; Wang *et al.*, 2002; Zhu *et al.*, 2000). Depending on the method being used, advantages and disadvantages occur during production with common problems being costs, scalability and wide size distribution (Evanoff & Chumanov, 2005). Also the instability of the nanoparticles upon application can be problematic since aggregation decreases the specific surface, lowering the antimicrobial activity (Mafune *et al.*, 2000). Moreover, there is a demand for eco-friendly production processes which avoid the use of solvents and toxic reagents. Even though there are some eco-friendly chemical methods reported (Huang & Yang, 2004; Vigneshwaran *et al.*, 2006b), the current green nanotechnology focuses more on the use of biological systems such as microorganisms and plants to produce silver nanoparticles. Although plants can be used for the green biological production of silver nanoparticles (cfr. Chapter 18), this chapter will focus mainly on the microbial production with bacteria, fungi and yeasts.

11.2.1 The basic principles

The chemical production of silver nanoparticles requires three components: a silver salt (usually $AgNO_3$), a reducing agent (e.g. sodium borohydride) and a stabilizer or capping agent (e.g. polivinylpyrrolidone) to control the growth of the nanoparticles and prevent aggregation by steric hindrance or electrostatic repulsion (Ledwith *et al.*, 2007). In the case of the microbial synthesis of silver nanoparticles, the reducing agent and the stabilizer are replaced by molecules produced by bacteria, fungi or yeasts (Figure 11.1). Hence, microbial processes do not require the separate addition of a reducing and a stabilizing agent, since the microbial cells function as both.

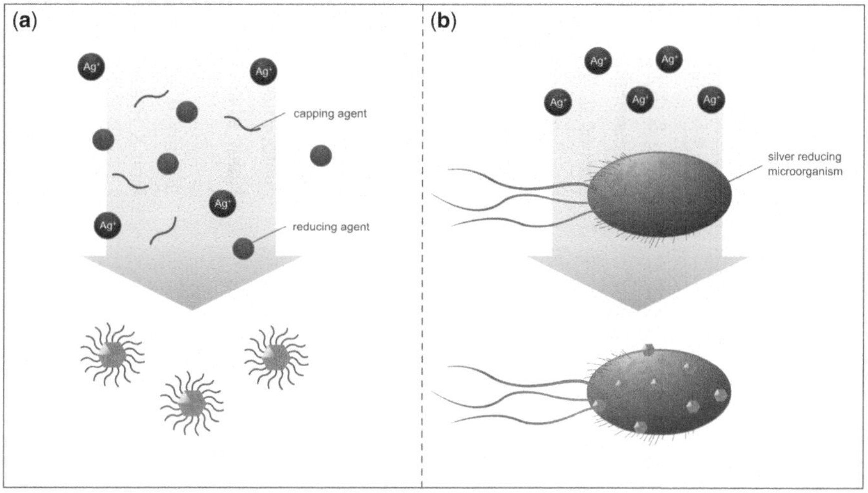

Figure 11.1 Chemical (a) versus microbial (b) synthesis of silver nanoparticles. In the case of chemical synthesis, silver ions, provided as a silver salt, are reduced by a chemical reducing agent forming a silver nucleus. The growth of the nucleus to a particle is controlled by a capping agent which also prevents aggregation. In the case of biological synthesis the reducing and capping agent are both provided by microorganisms. In this example (b) the ions are precipitated and stabilized by functional groups on the cell wall of the microbial cells.

The microbial manufacturing of silver nanoparticles can occur enzymatically or non-enzymatically with polysaccharides or proteins, and in many cases proteins function also as stabilizer or capping agent. A variety of nanoparticles with different sizes has been reported depending on the microorganism responsible for production (Table 11.1). Next to different sizes, microorganisms can manufacture nanoparticles with diverse shapes. Spherical particles, however, are the most occurring nanoparticle shape. The nanoparticles are situated intracellularly or extracellularly on the cell wall when entire cells are used for the nanoparticle production or they are produced extracellularly in the cell supernatant. The latter case allows the use of cell free extracts, containing the reducing and stabilizing agents, which minimizes the post-production extraction and purification of the nanoparticles.

Table 11.1 Overview of bacteria, fungi, and yeasts and plants which synthesize silver nanoparticles (NP) with their size distribution and shape.

Species	NP size (nm)	NP shape	Reference
Bacteria			
Bacillus cereus	5	Spherical	Babu & Gunasekaran (2009)
Bacillus licheniformis	40–50	–	Kalishwaralal *et al.* (2008)
Bacillus thuringiensis	10–20	Various	Jain *et al.* (2010)
Brevibacterium casei	10–50	Spherical	Kalishwaralal *et al.* (2010)
Corynebacterium	5–15	–	Zhang *et al.* (2005)
Escherichia coli	50 [42.2–89.6]	Spherical	Gurunathan *et al.* (2009)
Escherichia coli	40–60	–	Natarajan *et al.* (2010)
Klebsiella pneumoniae	52.5 [28.2–122]	–	Shahverdi *et al.* (2007
Lactobacillus farciminis	17.0 [14.3–19.7]	Spherical	Sintubin *et al.* (2009)
Lactobacillus fermentum	11.2 [10.3–12.1]	Spherical	Sintubin *et al.* (2009)
Lactobacillus plantarum	19.4 [16.8–21.9]	Spherical	Sintubin *et al.* (2009)
Lactobacillus rhamnosus	15.7 [13.6–17.8]	Spherical	Sintubin *et al.* (2009)
Pseudomonas stutzeri	$<$200	Triangular, hexagonal, spherical	Klaus *et al.* (1999)
Ureibacillus thermosphaericus	1–100	Spherical	Juibari *et al.* (2011)
Lactobacillus sp.	15–500	Polymorphus	Nair & Pradeep (2002)
Plectonema boryanum	1–15 1–40 5–200	Spherical Spherical Octahydral platelets	Lengke *et al.* (2007
Staphylococcus aureus	160–180	–	Nanda & Saravanan (2009)

Fungi			
Aspergillus niger	20	Spherical	Binupriya *et al.* (2009); Gade *et al.* (2008)
Aspergillus ochraceus	20	Spherical	Vijayakumar & Prasad (2009)
Aspergillus oryzae var. viridis	5–50	Spherical	Binupriya *et al.* (2010)
Cladosporium cladosproides	10–100	Spherical	Balaji *et al.* (2009)
Coriolus versicolor (proteins)	10 25–75	Spherical Spherical	Sanghi & Verma (2009)
Fusarium oxysporum	1.6	Spherical	Duran *et al.* (2007)
Fusarium oxysporum (enzyme)	25 ± 12	–	Kumar *et al.* (2007)
Fusarium solani	16.23 ± 0.78 [5–35]	Spherical	Ingle *et al.* (2009)
Neurospora crassa	11	Spherical, ellipsoidal	Castro-Longoria *et al.* (2011)
Penicillium brevicompactum	58.35 ± 17.88	–	Shaligram *et al.* (2009)
Penicillium fellutanum	5–25	Spherical	Kathiresan *et al.* (2009)
Phaenerochaete chrysosporium	50–200	Spherical	Vigneshwaran *et al.* (2006a)
Pleurotus sajor-caju (protein)	30.5 ± 4	Spherical	Vigneshwaran *et al.* (2007)
Plerurotus sajor-caju	5–50	Spherical	Nithya & Ragunathan (2009)
Phoma glomerata	60–80	Spherical	Birla *et al.* (2009)
Phoma sp. 3.2883	71.06 ± 3.48	Spherical	Chen *et al.* (2003)
Phyllanthus amarus (protein)	30	Spherical	
Trichoderma viride	2–4	Spherical	Fayaz *et al.* (2010)
Trichoderma viride	10–40	Spherical	Thakkar *et al.* (2010)
Verticillium sp.	25 ± 12	–	Sastry *et al.* (2003)
Volvariella volvacea	15	Spherical, ellipsoidal	Philip (2009)
Yeasts			
Silver-tolerant strain MKY3	2–5 [2–20]	Hexagonal, twinned cubic	Kowshik *et al.* (2003)

11.2.2 Reduction mechanism

11.2.2.1 Enzymatic reduction

The exact biological reduction mechanism of Ag^+ to Ag^0 is not always known, but generally enzymatic reduction can be distinguished from non-enzymatic reduction. The possible involvement of enzymes in the reduction of silver ions by microorganisms was pointed out for the first time by Mukherjee *et al.* (2001). They observed the formation of silver nanoparticles on the mycelia of *Verticillium* sp. cells but not into solution. Hence, they speculated that the negatively charged carboxylated groups of enzymes in the cell wall had trapped and subsequently reduced Ag^+ to Ag^0. The idea of enzymatic silver reduction was shared by Ahmad *et al.* (2003) who proved with a protein assay that an NADPH-dependent reductase was responsible for the production of silver nanoparticles by the fungus *Fusarium oxysporum*. Kumar *et al.* (2007) confirmed this by manufacturing silver nanoparticles with α-NADPH-dependent nitrate reductase extracted from *F. oxysporum*. Apart from fungi, bacteria have also shown the ability to reduce silver in an enzymatic way. Through the addition of piperitone, an inhibitor of nitroreductases, Shahverdi *et al.* (2007) prevented the production of nanoparticles, indicating that the oxygen-insensitive nitroreductase NfsA is a key player in the reduction of Ag^+. Also *Bacillus licheformis* uses a nitrate reductase to reduce silver ions (Kalishwaralal *et al.*, 2008).

One of the possible disadvantages of the enzymatic reduction of silver is the restricted pH and temperature conditions in which enzymes function, hence this limits the level in which the reduction speed and/or efficiency can be influenced by changing the environmental parameters. Moreover, silver ions inhibit enzymes (Section 11.3.1.3). Therefore, low Ag^+ concentrations need to be applied during enzymatic production of silver nanoparticles. Microorganisms that manufacture silver nanoparticles in a non-enzymatic way might therefore be a valuable alternative.

11.2.2.2 Non-enzymatic reduction

The non-enzymatic reduction of Ag^+ is comparable to the chemical reduction, but in this case the reducing and stabilizing compounds originate from microorganisms (Kumar & Yadav, 2009; Sharma *et al.*, 2009; Sintubin *et al.*, 2009). As pointed out earlier, the stabilizing compounds are quite often proteins which cap the nanoparticles and prevent them from aggregating (Binupriya *et al.*, 2010; Sathyavathi *et al.*, 2010; Zhang *et al.*, 2005). Reducing sugars such as glucose have been reported several times as the reducing compounds which are responsible for the reduction of Ag^+. This is, for example, the case for *Lactobacillus* sp. bacteria and for the fungus *Phaenerochaete chrysosporium* (Lin *et al.*, 2005; Sintubin *et al.*, 2009; Vigneshwaran *et al.*, 2006a). Proteins and polyols have also been mentioned as reducing compounds (Jha *et al.*, 2009).

As mentioned earlier, the non-enzymatic production of silver nanoparticles can handle more extreme parameters such as a high pH or temperature which can possibly accelerate the synthesis rate. For example, at a pH of 11.5, Sintubin *et al.* (2009) succeeded in producing nanoparticles with *Lactobacillus* sp. within one minute. Due to a high pH more negatively charged groups occur at the cell wall favouring the entrapment of Ag^+. Moreover, an alkaline pH stimulates the ring opening of reducing sugars liberating the reducing aldehyde group which is now free to react with Ag^+ and reduce it to Ag^0. Sanghi and Verma (2009) and Fu *et al.* (2006) also noticed a faster reaction rate and a higher recovery or yield of reduced silver when working in alkaline conditions. The recovery with *Lactobacillus fermentum* at pH 11.5 was 1.6 times higher compared to the recovery at pH 7, and 6.4 times higher compared to pH 2 (Sintubin *et al.*, 2009). Also *Corynebacterium* SH09 recovered more silver in alkaline conditions. The yield was 1.4 times higher at pH 8 compared to pH 2 (Zhang *et al.*, 2005). In the case of *Aeromonas* SH10, the recovery depended on the initially added Ag^+ concentration, with the silver adsorption and/or reduction being 2 to 4 times higher at pH 8 compared to pH 2 (Zhang *et al.*, 2007).

Also temperature can influence the reduction rate and the yield of reduced silver. Lengke *et al.* (2007) noticed a higher silver recovery when increasing the temperature: at 100°C all silver was taken up by the biomass from the environment. Also Zhang *et al.* (2005) reported that higher temperatures were favourable for bio-reduction by increasing the reduction rate. Since enzymes are inhibited by higher concentrations of Ag^+, the enzymatic reduction does not easily allow working with $AgNO_3$ concentrations above $0.1\ g \cdot L^{-1}$. The non-enzymatic reduction, in contrast, can easily handle $AgNO_3$ concentrations of 1 to $10\ g \cdot L^{-1}$ which is 10 to 100 times higher than for enzymatic reduction (Sintubin *et al.*, 2009; Zhang *et al.*, 2005; Zhang *et al.*, 2007). This can be a valuable argument when deciding which microorganism to use for industrial or upscaled production. On the other hand, working with low silver concentrations does not inhibit or eliminate microorganisms; hence a continuous production process could be possible (Mukherjee *et al.*, 2001).

This section described the difference between enzymatic and non-enzymatic reduction, but one must point out that it is not always easy to distinct which type of reduction takes place. In the case enzymes are the dominant factor for silver reduction, it is more than probable that a minor fraction of the silver nanoparticles is still formed by non-enzymatic reduction.

11.2.3 The use of entire cells vs. cell extracts

When entire microbial cells are used, silver precipitation can happen intracellularly, as in the case of *Pseudomonas stutzeri* and *Verticillium* sp. (Klaus *et al.*, 1999; Mukherjee *et al.*, 2001), or extracellularly with the formation of silver nanoparticles on the cell wall, as in the case of *Lactobacillus* sp. (Nair &

Pradeep, 2002; Sintubin *et al.*, 2009). Another case of extracellular synthesis is the formation of nanoparticles in the medium or supernatant by reducing and capping agents excreted by microorganisms (Duran *et al.*, 2007; Kowshik *et al.*, 2003) (Figure 11.2).

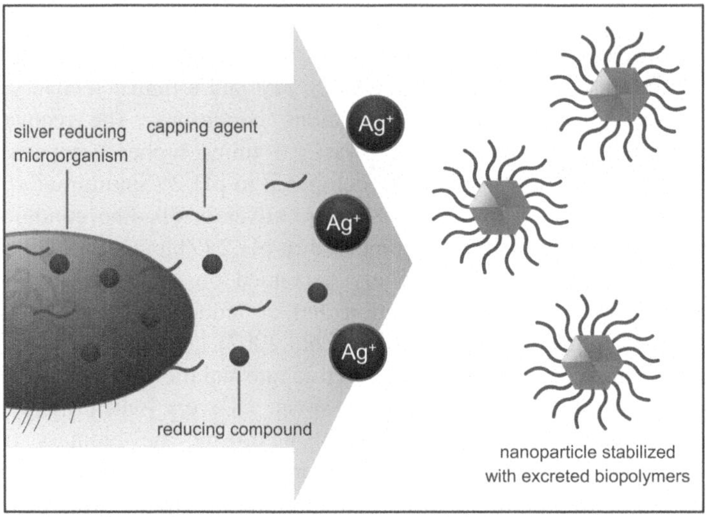

Figure 11.2 Reducing compounds and capping agents excreted by microorganisms in the supernatant result in the extracellular synthesis of silver nanoparticles.

Hence, silver nanoparticles can be produced using cell-free extracts or cell supernatant. Such cell extracts can be made in two ways: (1) by washing the biomass and dissolving the cells in water or a buffer for a certain amount of time to extract the reducing compounds and capping agents (Binupriya *et al.*, 2010; Duran *et al.*, 2007), or (2) by using the medium in which the biomass has grown (Nanda & Saravanan, 2009; Nithya & Ragunathan, 2009). The latter might contain next to excreted microbial components, such as enzymes or carbohydrates, also reducing compounds originating from the medium. Hence, one must take into mind that the production of silver nanoparticles might not be completely biological.

Microbial cells are constituted out of a complex structure of carbohydrates, proteins and fatty acids which form for example the cell wall and S-layers of the microbial cell. Hence, microbial cells usually contain more than one reducing and/or capping agent. Weak reducing agents generally produce larger nanoparticles whereas strong reducing agents produce smaller nanoparticles (Tolaymat *et al.*, 2010). Therefore, a combination of these reducing compounds with various capping agents will lead to the manufacturing of nanoparticles with a wide range of sizes and shapes or a high polydispersity. This is what generally happens when silver nanoparticles are produced microbially, which is in contrast to chemical production where only one reducing agent is used and more monodisperse nanoparticles are formed (Binupriya *et al.*, 2010).

A possibility to avoid this heterogeneity is to use only a limited number of biological molecules or cellular structures extracted and purified from the microbial biomass, which will give rise to more monodisperse silver nanoparticles. Such extraction and purification will, however, require one or more extra steps in the production process and, therefore, will be more time-consuming and less economic. An economic exception would be the use of simple molecules such as reducing sugars and polysaccharides like glucose which can be easily purchased at fairly low prices.

Since microorganisms consist of highly organized structures, the cells or biomass parts can be used as a template for nanoparticle precipitation with an external reducing agent. Saifuddin *et al.* (2009), for example, treated a *Bacillus subtilis* extract with microwave irradiation to create nanoparticles.

11.3 APPLICATIONS OF MICROBIAL MANUFACTURED SILVER NANOPARTICLES

11.3.1 Antimicrobial activity

11.3.1.1 General introduction on antimicrobial activity

Because of their strong antimicrobial properties, silver nanoparticles are applied in a wide variety of applications to control microbial growth. The antimicrobial activity, however, depends on the size of the nanoparticles. The smaller the nanoparticles, the higher the specific surface and therefore also their activity (Lok *et al.*, 2007; Morones *et al.*, 2005; Pal *et al.*, 2007). Also the shape of the nanoparticles is important for the antimicrobial activity. Pal *et al.* (2007) discovered that triangular nanoparticles display a higher antibacterial activity than spherical particles. Furthermore, the type of microbial molecule which stabilizes the nanoparticles can influence the activity by altering the interaction with the target microorganisms (Cho *et al.*, 2005; Goodman *et al.*, 2004). For example, positively charged polymers interact more easily with the negatively charged cell wall of bacteria. Because of their specific shape, size and the type of molecules which cover and functionalize the surface of the nanoparticles, microbial manufactured silver nanoparticles are unique. Hence, it is important to thoroughly characterize their antimicrobial activity in order to formulate the most effective concentration for every application.

11.3.1.2 Quantification of antimicrobial activity

The minimal inhibitory concentration (MIC) is often considered to be the 'golden standard' for the determination of microbial susceptibility towards an antimicrobial agent (Andrews, 2001). It is the minimal concentration of a disinfectant or biocide, in this case microbial manufactured silver nanoparticles, needed to inhibit microbial growth. The MIC can be determined in liquid medium (broth) or solid medium (agar). Depending on the user-friendliness of the method or the desired information needed for a certain application, also other techniques have been applied previously to measure the antimicrobial activity of microbial manufactured silver nanoparticles, such as the minimal bactericidal concentration (MBC), time-kill, well-diffusion method and disk-diffusion method (Table 11.2).

When the broth dilutions of the MIC are plated on silver-free agar plates, information can be gathered on the lethal concentrations (Andrews, 2001; Hobson & Bolsen, 2001). For bacteria, the lowest lethal concentration is called the MBC. In contrast to the MIC, which is a more suitable parameter for applications which prevent microbial growth, MBC can be used as a parameter for applications which also treat microbial contaminations by removing or eliminating the pathogens. For example, the MBC can be a useful parameter for drinking water disinfection. Time-kill curves also give information about the contact-time needed for biogenic silver to decrease the number of microbial colonies with a log-unit (Jorgensen & Ferraro, 2009). Well-diffusion and disk-diffusion methods examine the possible zone of growth inhibition (ZOI) which might occur around the well or disc impregnated with microbial manufactured silver nanoparticles. Just as the MIC, they are not able to differentiate between growth-inhibitory and lethal concentrations (Hobson & Bolsen, 2001). In contrast, these are not quantitative techniques expressing the inhibition in terms of concentrations, but rather qualitative techniques expressing the susceptibility of a microorganism in terms of "susceptible", "intermediate" and "resistant" according to the size of the ZOI in mm (Jorgensen & Ferraro, 2009).

Since each researcher uses the method which suites him most against the species and strain which interests him most, it is hard to compare the antimicrobial activity between the microbial manufactured silver nanoparticles produced with different microorganisms, or between microbial manufactured nanosilver and chemical nanosilver. According to Table 11.2, the MIC and MBC of different types of microbial manufactured nanosilver range from 10–200 $mg \cdot L^{-1}$ and the disk- and well-diffusion tests are characterized by 0 mm to 15 mm of ZOI. To complicate things even more, also the composition of the medium in which the test occurs, and the starting inocula have a major influence on the outcome of the antimicrobial tests. Sintubin *et al.* (2011b) showed previously that the minimal bactericidal concentration in plain water was up to 400 times lower compared to the MBC tested in a rich medium containing high concentrations of salts and proteins. Taking all these factors into mind, it is clear that a standardized protocol is needed to test the antimicrobial activity of nanosilver in general and microbial manufactured nanosilver in specific in order to be able to compare the activity of the nanoparticles and choosing the most suitable type of silver nanoparticle for a certain application.

Table 11.2 Overview of the antimicrobial activity of microbial manufactured silver produced by different microorganisms.
For each type of microbial silver, the used antimicrobial method and target pathogens are given. The antimicrobial activity is described as the concentration needed to inhibit pathogens (MIC), the necessary time (log kill) or the presence of a zone of inhibition (ZOI) (well-diffusion and disc-diffusion method). In the latter case, the applied biogenic silver concentration or volume is given where possible.

Microorganism	Antimicrobial method	Pathogen	Inhibition	Reference
Aspergillus oryzae	MIC[1] in liquid broth	*Staphylococcus aureus*	$50\ mg \cdot L^{-1}$	Binupriya *et al.* (2010)
Staphylococcus aureus	Well-diffusion method	MRSA	ZOI (2 μg)	Nanda & Saravanan (2009)
		MRSE	ZOI (2 μg)	
		Streptococcus pyogenes	ZOI (2 μg)	
		Salmonella Typhi	ZOI (2 μg)	
		Klebsiella pneumoniae	ZOI (2 μg)	
		Vibrio cholerae	No ZOI (2 μg)	
Pleurotus sajor caju	Well-diffusion method	*Pseudomonas aeruginosa*	ZOI	Nithya & Ragunathan (2009)
		Escherichia coli	ZOI	
		Staphylococcus aureus	ZOI	
Aspergillus clavatus	Well-diffusion method	MRSA	ZOI (10 μL)	Saravanana & Nanda (2010)
		MRSE	ZOI (10 μL)	
	Disk-diffusion method	MRSA	ZOI (10 μL)	
		MRSE	ZOI (10 μL)	
Phoma glomerata	Disk-diffusion method	*Escherichia coli*	ZOI (15 μL)	Birla *et al.* (2009)
		Staphylococcus aureus	ZOI (15 μL)	
		Pseudomonas aeruginosa	ZOI (15 μL)	
Aspergillus ochraceus	Time kill	*Escherichia coli*	$2\ h\ (7\ mg \cdot L^{-1})$	Vijayakumar & Prasad (2009)
		Bacillus subtilis	$2\ h\ (7\ mg \cdot L^{-1})$	
Pleurotus sajor-caju	Well-diffusion method	*Staphylococcus aureus*	ZOI	Vigneshwaran *et al.* (2007)
		Klebsiella pneumoniae	ZOI	

(*Continued*)

Table 11.2 Overview of the antimicrobial activity of microbial manufactured silver produced by different microorganisms. For each type of microbial silver, the used antimicrobial method and target pathogens are given. The antimicrobial activity is described as the concentration needed to inhibit pathogens (MIC), the necessary time (log kill) or the presence of a zone of inhibition (ZOI) (well-diffusion and disc-diffusion method). In the latter case, the applied biogenic silver concentration or volume is given where possible. (*Continued*).

Microorganism	Antimicrobial method	Pathogen	Inhibition	Reference
Lactobacillus fermentum	MIC[1]	*Escherichia coli*	12.5 mg · L^{-1}	Sintubin *et al.* (2011b)
		Pseudomonas aeruginosa	12.5 mg · L^{-1}	
		Staphylococcus aureus	50 mg · L^{-1}	
	MBC[2]	*Escherichia coli*	25 mg · L^{-1}	
		Pseudomonas aeruginosa	100 mg · L^{-1}	
		Staphylococcus aureus	200 mg · L^{-1}	
Bacillus thuringiensis	Disk-diffusion method	*Escherichia coli*	13 mm ZOI (50 μg · mL^{-1})	Jain *et al.* (2010)
		Pseudomonas aeruginosa	15 mm ZOI (50 μg · mL^{-1})	
		Staphylococcus aureus	9 mm ZOI (50 μg · mL^{-1})	

[1]MIC = Minimal Inhibitory Concentration.
[2]MBC = Minimal Bactericidal Concentration.

11.3.1.3 Antimicrobial mechanism of microbial manufactured nanoparticles

The antimicrobial activity of silver nanoparticles is still not fully understood. In the case of chemical nanoparticles it is believed that three different mechanisms determine the antimicrobial activity: (1) direct physical contact between the nanoparticles and the microbes causing structural damage (Kim *et al.*, 2007; Sondi & Salopek-Sondi, 2004); (2) the generation of reactive oxygen species (ROS) damaging the cell membrane (Kim *et al.*, 2007; Kumar *et al.*, 2003); and 3) the slow release of silver ions which interfere with the DNA replication and inhibit enzymes by reacting with thiol groups (Feng *et al.*, 2000; Liau *et al.*, 1997; Morones *et al.*, 2005). In the case of microbial manufactured silver nanoparticles, little research has been done to determine the mode of action. Sintubin *et al.* (2011b) did a first attempt elucidating the mechanism of microbial manufactured silver produced with *L. fermentum* and discovered the release of silver ions being the dominant factor, while the influence of ROS formation and direct contact seemed to be negligible. In this study, however, the silver nanoparticles were fixed on the cells of *L. fermentum*, which might have prevented true physical contact with the target bacteria, thus it cannot be excluded that direct contact might be an important antimicrobial mechanism for other types of microbial manufactured nanosilver (Sintubin *et al.*, 2011b). The exact mechanism still needs to be elucidated for other types of silver nanoparticles manufactured with different microorganisms.

11.3.1.4 Microbial vs. chemical manufactured nanosilver

As stated in the previous section (Section 11.3.1.2), it is almost impossible to compare the antimicrobial activity of microbial manufactured silver nanoparticles with chemical nanosilver, unless both types of nanosilver have been tested using the exact same set-up and the same microbial strain, inoculum and medium. At the moment, this is only the case when both types of nanosilver have been tested in the same study. This is exactly what the author of the current chapter has done. Silver nanoparticles produced with *L. fermentum* bacteria (diameter 11.2 nm) were compared with chemical nanosilver. Both the MIC and MBC were determined against three bacterial species: *S. aureus*, *P. aeruginosa* and *E. coli*. Chemical nanosilver needed 20 times higher concentrations compared to microbial nanosilver to inhibit or eliminate 7 log of bacteria. Further investigation revealed that the chemical nanoparticles were fairly unstable and aggregated in the medium forming clusters of over 250 nm. This resulted in a very low release of Ag^+ (0.04 mg/L). Microbial silver nanoparticles, in contrast, were stabilized by the bacterial matrix of *L. fermentum* on which they were formed during manufacturing. The fixation of the nanoparticles on the bacterial cells prevented them from aggregating, resulting in the maintenance of their high specific surface, hence in a high release of Ag^+ (0.56 mg/L) (Figure 11.3).

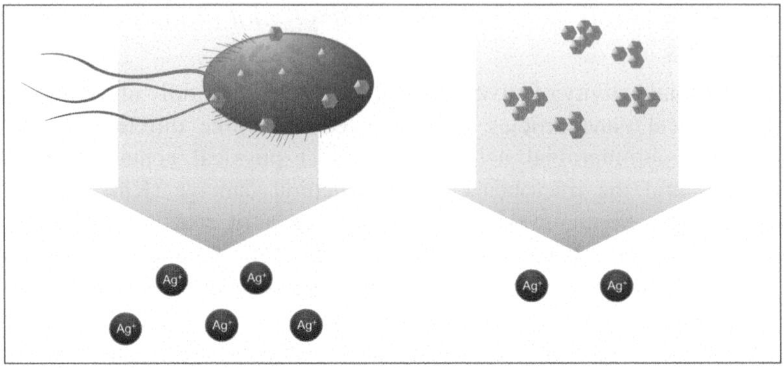

Figure 11.3 Nanoparticles which are stabilized on a microbial cell (left) result in a higher release of Ag^+ compared to chemical nanoparticles which aggregate upon application.

11.3.2 Microbial manufactured silver nanoparticles for water disinfection

More than 2000 years ago, Alexander the Great already consumed water disinfected with silver by storing it in bulk silver vessels (Silver *et al.*, 2005). Nowadays, silver is still popular to disinfect drinking water. Aquapure and QSI-Nano, for example, are commercially available home water purification systems (Maynard, 2007). These systems contain chemically produced nanoparticles. De Gusseme et al. (2010) was the first who investigated the use of microbial manufactured silver nanoparticles, produced with *L. fermentum*, for the removal of viruses from drinking water (De Gusseme *et al.*, 2010). A concentration of $5.4\ mg \cdot L^{-1}$ biogenic silver eliminated the two modelviruses–bacteriophage UZ1 and murine norovirus 1 (MNV-1)–within one to three hours. To disinfect drinking water in a continuous way, microbial nanosilver was coated on a cartridge filter which greatly enhanced the antiviral activity compared to an uncoated filter. The bacterial matrix, which served as a scaffold for the nanoparticles and stabilized them, had an additional advantage: the negatively charged cell walls secured a good attachment to the positively charged Nanoceram filters, resulting in less silver leaching into the drinking water and a longer life-time of the filter. In a second study, they used microbial nanosilver powder immobilized in PVDF membranes for the removal of viruses from drinking water (De Gusseme *et al.*, 2011). A virus decline of more than 3 log was observed at a flux rate of $72\ L\ m^{-2}\ h^{-1}$. The antiviral activity was the result of the slow release of Ag^+. This means that in time the antiviral activity will decrease making the membrane technology not applicable for long-term and large-scale continuous water treatment. However, for small scale water treatment and emergency drinking water units, these membranes show potential.

The same research group also investigated the influence of microbial nanosilver produced with *L. fermentum* on solar disinfection (SODIS). They found that low concentrations of biogenic silver enhanced SODIS during the first three hours of disinfection. Moreover, the presence of biogenic silver prevented regrowth of bacteria during storage. This was in contrast with SODIS alone or when TiO_2 was used in combination with SODIS (Sintubin *et al.*, 2011a).

11.3.3 Other applications of microbial manufactured silver nanoparticles

In theory, microbial nanoparticles can be implemented in every application which already uses chemically produced silver nanoparticles, for example, as catalysts and biosensors, in antimicrobial surfaces, biomedical applications, and so on. At the moment, studies have been performed to analyse the general antimicrobial activity of microbial nanosilver. However, except for water disinfection treatments, the implementation of microbial manufactured silver nanoparticles into practical applications has not yet been studied intensively, creating a major field of unexplored possibilities. The antimicrobial activity of silver nanoparticles in combination with their anti-inflammatory and wound healing features makes them interesting for biomedical applications such as bandages for the treatment of burns. However, one must take into mind that the microbial fabrication of biogenic silver could possibly cause a contamination with pathogens and subsequent infection (Chaloupka *et al.*, 2010) (Section 11.3.4). Hence, microbial nanosilver produced with pathogenic bacteria or fungi such as *Staphyloccocus aureus* and *Aspergillus niger* is not implementable in this case. Instead, microbial nanosilver manufactured with probiotic microorganisms are an appealing alternative.

The number of antimicrobial applications for microbial nanosilver does not limit itself to water treatment and the biomedical field but can also be extended to food packages, textiles and self-cleaning coatings of computer keyboards and mobile phones. Just as for biomedical applications, pathogens are not the preferred choice for the production of biogenic silver. Instead, probiotics or food-related microorganisms, such as eatable mushroom, are preferable production systems (Philip, 2009).

Silver nanoparticles are important catalysts for the oxidation of olefins, such as styrene, to epoxides and aldehydes (Chimentao *et al.*, 2004). Although it might be less important for a catalyst if microbial nanosilver is produced with a pathogenic microorganism or not, the choice of the microorganism is still crucial. The reason can be found in the shape- and size-dependent activity of the catalyst (Chimentao *et al.*, 2004; Xu *et al.*, 2006). Moreover, it is essential that the nanoparticles are synthesized extracellularly in the supernatant or on the cell wall. In the case of intracellular nanoparticles, the nanoparticle-surface is covered in layers of organic material and inaccessible for the substrates, preventing catalysis by the nanoparticles.

As noble metals, silver nanoparticles interact with visible light creating a plasmon band. Due to this optical feature, nanosilver can be used in biosensing and imaging. Since the plasmon resonance depends on the size and shape of the nanoparticles, the microorganism used for the production of biogenic silver will also be important in this case (Mock *et al.*, 2002; Mulvaney, 1996).

11.3.4 Silver nanoparticle toxicity

Toxicity of silver nanoparticles has been observed against several aquatic organisms such as zebrafish, and against fruitflies and mammalian cells such as germline stem cells and rat liver cells (Braydich-Stolle *et al.*, 2005; Griffitt *et al.*, 2009; Hussain *et al.*, 2005; Panacek *et al.*, 2011). What is remarkable, however, is that the observed toxic silver concentrations in these studies ranged between 5 mg/L and 100 mg/L. These concentrations are high and unlikely to occur in the environment. Mueller & Nowack (2008) calculated the release of nanosilver to the environment taking two scenarios into consideration: (1) a realistic scenario where the predicted environmental concentrations (PECs) of nanosilver in water, soil and air would amount to 0.03 μg/L, 0.02 μg/ton and 1.7×10^{-3} μg/m^3 respectively; and (2) a high exposure scenario where the PECs in water, soil and air would not go beyond 0.08 μg/L, 0.1 μg/ton and 4.4×10^{-3} μg/m^3. In both cases the PECs were below the predicted no-effect-concentration, therefore comprising no adverse effects (Mueller & Nowack, 2008).

In addition, Blaser *et al.* (2008) claimed that *"It is the toxicity of silver ions to aquatic organisms that causes concern about environmental effects"* (Blaser *et al.*, 2008). Indeed, Sintubin *et al.* (2011b) observed that the activity of silver nanoparticles manufactured with *L. fermentum* was comparable or less effective than ionic silver. Moreover, the antimicrobial activity of microbial nanosilver was mainly due to the release of silver ions, making it plausible that it also would be the main cause of its toxicity towards higher organisms. Also Griffith et al. (2008) remarked that based on mass added, the nanoparticles were less toxic than their soluble forms (Griffitt *et al.*, 2008). Nevertheless, free ions only occur in extremely low concentrations in natural waters. They bind strongly to sulfur containing compounds which exceed the predicted silver concentration with three magnitudes. These silver complexes tend to be far less toxic than the free ions. A similar effect was observed during time kill tests when organic compounds and salts were present, resulting in higher contact times and/or silver concentrations needed to maintain the antimicrobial properties of microbial nanosilver and silver ions (Sintubin *et al.*, 2011b).

Silver also sorbs easily to suspended particles (>0.45 μm) which enhances rapid incorporation in the sediments. Blaser *et al.* (2008) estimated that the total concentration of silver–nanosilver and silver originating from other sources–in river water would not exceed 140 ng/L and the concentration in the sediment never will never exceed 14 mg/ton (Blaser *et al.*, 2008). These estimations were made a couple of years ago based on the data available on silver production during that time. Meanwhile, the number of silver-containing biocides has increased. However, non-biocidal resources such as the photographic industry emits every year less silver, compensating for the increased concentration due to biocides. Hence, the previous numbers can still be considered a good estimation.

Next to the possible toxicity of silver nanoparticles and/or released silver ions, the biological system used to produce the nanoparticles could contain an additional risk. Chaloupka *et al.* (2010) warned against infections upon using products with microbial silver synthesized with pathogenic microbes. Also the biocompatibility of biogenic nanoparticles could pose troubles. *L. fermentum*, on the other hand, has a GRAS (generally regarded as safe) status (Zhang *et al.*, 2011) and might result in safe silver nanoparticles. Nevertheless, the absence of inflammatory or allergic reactions to present antigens of *L. fermentum* cannot be excluded without proper investigation. Hence, before using in biomedical applications, the biocompatibility and toxicity of each type of microbial nanosilver needs to be explored thoroughly.

11.3.5 Fate of microbial manufactured silver nanoparticles: legislation and regulation

According to Amara's law, we tend to overestimate the short-term effect of new technologies and underestimate their long-term effect (Bystrzejewska-Piotrowska *et al.*, 2009). Hence, despite the low predicted environmental concentrations, nanosilver still needs to be considered as possible hazardous, especially on the long term. Therefore, regulation is needed to control potential harmful exposures.

In the USA, the regulation of nanosilver falls under the Toxic Substances Control Act (TSCA) which regulates chemicals and the Federal Insecticide, Fungicide and Rodenticide Act (FIFRA) which regulates pesticides. The main problem with the TSCA is that it is not really attuned to nanoparticles. It considers nanosilver not as a new substance since its chemical composition is not unique from that of bulk material. Hence, nanosilver is regulated as bulk silver: its concentration cannot exceed 0.05 mg/L in drinking water or 100 $\mu g/m^3$ in air (Faunce & Watal, 2010; Seaton *et al.*, 2010).

In Europe, nanoparticles are regulated by the Registration Evaluation and Authorization of Chemicals (REACH). Similar to the TSCA, REACH has the limitation of recognizing nanomaterials only as new materials when they are different from bulk material. In 2010, however, the European Parliament voted to ban silver nanoparticles and carbon nanotubes out of electronic equipment because "there exists scientific uncertainty for human health and the environment" (NIA, 2010). Nevertheless, in the end a ban for nanosilver was not included in the Final Directive. The debate to amend the basis of nanomaterials regulation under REACH, however, is strong and a specific working group was created which will propose its preliminary conclusions in the near future. The commercial fate of nanosilver and biogenic silver will depend on these conclusions (NIA, 2011).

11.4 CONCLUSIONS

Microbial manufactured silver nanoparticles show potential as disinfectants, not only because of their eco-friendly production, but also because they have a higher antimicrobial activity than chemical nanoparticles due to the presence of organic structures or molecules that stabilize the nanoparticles. Moreover, the microbial matrix of such nanoparticles delivers additional advantages such as the increased adhesion to filters for water disinfection limiting the leaching of silver in drinking water.

Despite the potential of the microbial nanoparticles, further investigations are still needed regarding the suitability of the microbial nanoparticles for certain applications, the upscaling of the production process and the possible toxicity of the nanoparticles.

ACKNOWLEDGEMENTS

I would like to thank Tim Lacoere for the graphical work presented in this chapter.

REFERENCES

Abid J. P., Wark A. W., Brevet P. F. and Girault H. H. (2002). Preparation of silver nanoparticles in solution from a silver salt by laser irradiation. *Chem. Commun.*, (7), 792–793.

Ahmad A., Mukherjee P., Senapati S., Mandal D., Khan M. I., Kumar R. and Sastry M. (2003). Extracellular biosynthesis of silver nanoparticles using the fungus *Fusarium oxysporum*. *Colloid Surf. B-Biointerfaces*, **28**(4), 313–318.

Andrews J. M. (2001). Determination of minimum inhibitory concentrations. *J. Antimicrob. Chemother.*, **48**, 5–16.

Babu M. M. G. and Gunasekaran R. (2009). Production and structural characterization of crystalline silver nanoparticles from *Bacillus cereus* isolate. *Colloid Surf. B-Biointerfaces*, **74**(1), 191–195.

Balaji D. S., Basavaraja S., Deshpande R., Mahesh D. B., Prabhakar B. K. and Venkataraman A. (2009). Extracellular biosynthesis of functionalized silver nanoparticles by strains of *Cladosporium cladosporioides* fungus. *Colloid Surf. B-Biointerfaces*, **68**(1), 88–92.

Berney M., Weilenmann H. U. and Egli T. (2006). Flow-cytometric study of vital cellular functions in *Escherichia coli* during solar disinfection (SODIS). *Microbiology*, **152**, 1719–1729.

Binupriya A. R., Sathishkumar M., Vijayaraghavan K. and Yun S. I. (2009). Bioreduction of trivalent aurum to nano-crystalline gold particles by active and inactive cells and cell-free extract of *Aspergillus* oryzae var. viridis. *J. Hazard. Mater.*, **177**(1–3), 539–545.

Binupriya A. R., Sathishkumar M. and Yun S. I. (2010). Myco-crystallization of silver ions to nanosized particles by live and dead cell filtrates of *Aspergillus oryzae* var. viridis and its bactericidal activity toward *Staphylococcus aureus* KCCM 12256. *Ind. Eng. Chem. Res.*, **49**(2), 852–858.

Birla S. S., Tiwari V. V., Gade A. K., Ingle A. P., Yadav A. P. and Rai M. K. (2009). Fabrication of silver nanoparticles by *Phoma glomerata* and its combined effect against *Escherichia coli*, *Pseudomonas aeruginosa* and *Staphylococcus aureus*. *Lett. Appl. Microbiol.*, **48**(2), 173–179.

Blaser S. A., Scheringer M., MacLeod M. and Hungerbuhler K. (2008). Estimation of cumulative aquatic exposure and risk due to silver: contribution of nano-functionalized plastics and textiles. *Sci. Total Environ.*, **390**(2–3), 396–409.

Braydich-Stolle L., Hussain S., Schlager J. J. and Hofmann M. C. (2005). *In vitro* cytotoxicity of nanoparticles in mammalian germline stem cells. *Toxicol. Sci.*, **88**(2), 412–419.

Burch J. D. and Thomas K. E. (1998). Water disinfection for developing countries and potential for solar thermal pasteurization. *Sol. Energy*, **64**, 87–97.

Bystrzejewska-Piotrowska G., Golimowski J. and Urban P. L. (2009). Nanoparticles: their potential toxicity, waste and environmental management. *Waste Manage.*, **29**(9), 2587–2595.

Castro-Longoria E., Vilchis-Nestor A. R. and Avalos-Borja M. (2011). Biosynthesis of silver, gold and bimetallic nanoparticles using the filamentous fungus Neurospora crassa. *Colloid Surf. B-Biointerfaces*, **83**(1), 42–48.

Chaloupka K., Malam Y. and Seifalian A. M. (2010). Nanosilver as a new generation of nanoproduct in biomedical applications. *Trends Biotechnol.*, **28**(11), 580–588.

Chen J. C., Lin Z. H. and Ma X. X. (2003). Evidence of the production of silver nanoparticles via pretreatment of Phoma sp.3.2883 with silver nitrate. *Lett. Appl. Microbiol.*, **37**(2), 105–108.

Chimentao R. J., Kirm I., Medina F., Rodriguez X., Cesteros Y., Salagre P., Sueiras J. E. and Fierro J. L. G. (2004). *Sensitivity of Styrene Oxidation Reaction to the Catalyst Structure of Silver Nanoparticles*. Elsevier Science BV, Gdansk, Poland, pp. 793–800.

Cho K. H., Park J. E., Osaka T. and Park S. G. (2005). The study of antimicrobial activity and preservative effects of nanosilver ingredient. *Electrochim. Acta*, **51**(5), 956–960.

Clasen T., Saeed T. F., Boisson S., Edmondson P. and Shipin O. (2007). Household water treatment using dichloroisocyanurate (NADCC) tablets: a randomized, controlled trial to assess microbiological effectiveness in Bangladesh. *Am. J. Trop. Med. Hyg.*, **76**, 187–192.

De Gusseme B., Hennebel T., Christiaens E., Saveyn H., Verbeken K., Fitts J. P., Boon N. and Verstraete W. (2011). Virus disinfection in water by biogenic silver immobilized in polyvinylidene fluoride membranes. *Water Res.*, **45**(4), 1856–1864.

De Gusseme B., Sintubin L., Baert L., Thibo E., Hennebel T., Vermeulen G., Uyttendaele M., Verstraete W. and Boon N. (2010). Biogenic silver for disinfection of water contaminated with viruses. *Appl. Environ. Microbiol.*, **76**(4), 1082–1087.

Duran N., Marcato P. D., De Souza G. I. H., Alves O. L. and Esposito E. (2007). Antibacterial effect of silver nanoparticles produced by fungal process on textile fabrics and their effluent treatment. *J. Biomed. Nanotechnol.*, **3**(2), 203–208.

Edwards-Jones V. (2009). The benefits of silver in hygiene, personal care and healthcare. *Lett. Appl. Microbiol.*, **49**, 147–152.

Elechiguerra J. L., Burt J. L., Morones J. R., Camacho-Bragado A., Gao X., Lara H. H. and Yacaman M. J. (2005). Interaction of silver nanoparticles with HIV-1. *J Nanobiotechnol.*, **3**, 6–0.

Evanoff D. D. and Chumanov G. (2005). Synthesis and optical properties of silver nanoparticles and arrays. *Chemphyschem*, **6**(7), 1221–1231.

Faunce T. and Watal A. (2010). Nanosilver and global public health: international regulatory issues. *Nanomedicine*, **5**(4), 617–632.

Fayaz M., Tiwary C. S., Kalaichelvan P. T. and Venkatesan R. (2010). Blue orange light emission from biogenic synthesized silver nanoparticles using *Trichoderma viride*. *Colloid Surf. B-Biointerfaces*, **75**(1), 175–178.

Feng Q. L., Wu J., Chen G. Q., Cui F. Z., Kim T. N. and Kim J. O. (2000). A mechanistic study of the antibacterial effect of silver ions on *Escherichia coli* and *Staphylococcus aureus*. *J. Biomed. Mater. Res.*, **52**(4), 662–668.

Fu M. X., Li Q. B., Sun D. H., Lu Y. H., He N., Deng X., Wang H. X. and Huang J. L. (2006). Rapid preparation process of silver nanoparticles by bioreduction and their characterizations. *Chin. J. Chem. Eng.*, **14**(1), 114–117.

Gade A. K., Bonde P., Ingle A. P., Marcato P. D., Duran N. and Rai M. K. (2008). Exploitation of *Aspergillus niger* for synthesis of silver nanoparticles. *J. Biobased Mater. Bioenergy*, **2**(3), 243–247.

Goodman C. M., McCusker C. D., Yilmaz T. and Rotello V. M. (2004). Toxicity of gold nanoparticles functionalized with cationic and anionic side chains. *Bioconjug. Chem.*, **15**(4), 897–900.

Griffitt R. J., Hyndman K., Denslow N. D. and Barber D. S. (2009). Comparison of molecular and histological changes in Zebrafish gills exposed to metallic nanoparticles. *Toxicol. Sci.*, **107**(2), 404–415.

Griffitt R. J., Luo J., Gao J., Bonzongo J.-C. and Barber D. S. (2008). Effects of particle composition and species on toxicity of metallic nanomaterials in aquatic organisms. *Environ. Toxicol. Chem.*, **27**(9), 1972–1978.

Gurunathan S., Kalishwaralal K., Vaidyanathan R., Deepak V., Pandian S. R. K., Muniyandi J., Hariharan N. and Eom S. H. (2009). Biosynthesis, purification and characterization of silver nanoparticles using *Escherichia coli*. *Colloid Surf. B-Biointerfaces*, **74**(1), 328–335.

Hobson D. W. and Bolsen K. (2001). Methods of testing oral and topical antiseptics and antimicrobials. In: *Disinfection, Sterilization and Preservation*, S. S. Block (ed.), Lippincott Williams & Wilkins, Philadelphia.

Huang H. Z. and Yang X. R. (2004). Synthesis of polysaccharide-stabilized gold and silver nanoparticles: a green method. *Carbohydr. Res.*, **339**(15), 2627–2631.

Hussain S. M., Hess K. L., Gearhart J. M., Geiss K. T. and Schlager J. J. (2005). *In vitro* toxicity of nanoparticles in BRL 3A rat liver cells. *Toxicol. Vitro*, **19**(7), 975–983.

Ingle A., Rai M., Gade A. and Bawaskar M. (2009). *Fusarium solani*: a novel biological agent for the extracellular synthesis of silver nanoparticles. *J. Nanopart. Res.*, **11**(8), 2079–2085.

Jain D., Kachhwaha S., Jain R., Srivastava G. and Kothari S. L. (2010). Novel microbial route to synthesize silver nanoparticles using spore crystal mixture of *Bacillus thuringiensis*. *Indian J. Exp. Biol.*, **48**(11), 1152–1156.

Jha A. K., Prasad K. and Kulkarni A. R. (2009). Plant system: nature's nanofactory. *Colloid Surf. B-Biointerfaces*, **73**(2), 219–223.

Jorgensen J. H. and Ferraro M. J. (2009). Antimicrobial susceptibility testing: a review of general principles and contemporary practices. *Clin. Infect. Dis.*, **49**(11), 1749–1755.

Juibari M. M., Abbasalizadeh S., Jouzani G. S. and Noruzi M. (2011). Intensified biosynthesis of silver nanoparticles using a native extremophilic *Ureibacillus thermosphaericus* strain. *Mater. Lett.*, **65**(6), 1014–1017.

Kalishwaralal K., Deepak V., Pandian S. R. K., Kottaisamy M., BarathManiKanth S., Kartikeyan B. and Gurunathan S. (2010). Biosynthesis of silver and gold nanoparticles using *Brevibacterium casei*. *Colloid Surf. B-Biointerfaces*, **77**(2), 257–262.

Kalishwaralal K., Deepak V., Ramkumarpandian S., Nellaiah H. and Sangiliyandi G. (2008). Extracellular biosynthesis of silver nanoparticles by the culture supernatant of *Bacillus licheniformis*. *Mater. Lett.*, **62**(29), 4411–4413.

Kathiresan K., Manivannan S., Nabeel M. A. and Dhivya B. (2009). Studies on silver nanoparticles synthesized by a marine fungus, *Penicillium fellutanum* isolated from coastal mangrove sediment. *Colloid Surf. B-Biointerfaces*, **71**(1), 133–137.

Kim J. S., Kuk E., Yu K. N., Kim J. H., Park S. J., Lee H. J., Kim S. H., Park Y. K., Park Y. H., Hwang C. Y., Kim Y. K., Lee Y. S., Jeong D. H. and Cho M. H. (2007). Antimicrobial effects of silver nanoparticles. *Nanomed.-Nanotechnol. Biol. Med.*, **3**(1), 95–101.

Klasen H. J. (2000a). Historical review of the use of silver in the treatment of burns I. Early uses. *Burns*, **26**, 117–130.

Klasen H. J. (2000b). Historical review of the use of silver in the treatment of burns II. Renewed interest for silver. *Burns*, **26**, 131–138.

Klaus T., Joerger R., Olsson E. and Granqvist C. G. (1999). Silver-based crystalline nanoparticles, microbially fabricated. *Proc. Natl. Acad. Sci. U. S. A.*, **96**(24), 13611–13614.

Kowshik M., Ashtaputre S., Kharrazi S., Vogel W., Urban J., Kulkarni S. K. and Paknikar K. M. (2003). Extracellular synthesis of silver nanoparticles by a silver-tolerant yeast strain MKY3. *Nanotechnology*, **14**(1), 95–100.

Kumar S. A., Abyaneh M. K., Gosavi S. W., Kulkarni S. K., Pasricha R., Ahmad A. and Khan M. I. (2007). Nitrate reductase-mediated synthesis of silver nanoparticles from $AgNO_3$. *Biotechnol. Lett.*, **29**(3), 439–445.

Kumar V. and Yadav S. K. (2009). Plant-mediated synthesis of silver and gold nanoparticles and their applications. *J. Chem. Technol. Biotechnol.*, **84**(2), 151–157.

Kumar V. S., Nagaraja B. M., Shashikala V., Padmasri A. H., Madhavendra S. S., Raju B. D. and Rao K. S. R. (2003). *Highly Efficient Ag/C Catalyst Prepared by Electro-Chemical Deposition Method in Controlling Microorganisms in Water*. Elsevier Science BV, Hyderabad, India, pp. 313–319.

Lansdown A. B. (2007). Critical observations on the neurotoxicity of silver. *Crit. Rev. Toxicol.*, **37**, 237–250.

Ledwith D. M., Whelan A. M. and Kelly J. M. (2007). A rapid, straight-forward method for controlling the morphology of stable silver nanoparticles. *J. Mater. Chem.*, **17**(23), 2459–2464.

Lengke M. F., Fleet M. E. and Southam G. (2007). Biosynthesis of silver nanoparticles by filamentous cyanobacteria from a silver(I) nitrate complex. *Langmuir*, **23**(5), 2694–2699.

Liau S. Y., Read D. C., Pugh W. J., Furr J. R. and Russell A. D. (1997). Interaction of silver nitrate with readily identifiable groups: relationship to the antibacterial action of silver ions. *Lett. Appl. Microbiol.*, **25**(4), 279–283.

Lin Z. Y., Zhou C. H., Wu J. M., Zhou J. Z. and Wang L. (2005). A further insight into the mechanism of Ag + biosorption by Lactobacillus sp strain A09. *Spectroc. Acta Pt. A-Molec. Biomolec. Spectr.*, **61**(6), 1195–1200.

Lok C. N., Ho C. M., Chen R., He Q. Y., Yu W. Y., Sun H., Tam P. K. H., Chiu J. F. and Che C. M. (2007). Silver nanoparticles: partial oxidation and antibacterial activities. *J. Biol. Inorg. Chem.*, **12**(4), 527–534.

Mafune F., Kohno J., Takeda Y., Kondow T. and Sawabe H. (2000). Structure and stability of silver nanoparticles in aqueous solution produced by laser ablation. *J. Phys. Chem. B*, **104**(35), 8333–8337.

Malynych S. Z. and Chumanov G. (2003). Vacuum deposition of silver island films on chemically modified surfaces. *J. Vac. Sci. Technol. A*, **21**(3), 723–727.

Maynard A. D. (2007). Nanotechnologies: overview and issues. In: *Nanotechnology – Toxicological Issues and Environmental Safety*, P. P. Simeonova, N. Opopol and M. I. Luster (eds), Springer, Dordrecht, pp. 1–14.

Mock J. J., Barbic M., Smith D. R., Schultz D. A. and Schultz S. (2002). Shape effects in plasmon resonance of individual colloidal silver nanoparticles. *J. Chem. Phys.*, **116**(15), 6755–6759.

Morones J. R., Elechiguerra J. L., Camacho A., Holt K., Kouri J. B., Ramirez J. T. and Yacaman M. J. (2005). The bactericidal effect of silver nanoparticles. *Nanotechnology*, **16**(10), 2346–2353.

Mueller N. C. and Nowack B. (2008). Exposure modeling of engineered nanoparticles in the environment. *Environ. Sci. Technol.*, **42**(12), 4447–4453.

Mukherjee P., Ahmad A., Mandal D., Senapati S., Sainkar S. R., Khan M. I., Parishcha R., Ajaykumar P. V., Alam M., Kumar R. and Sastry M. (2001). Fungus-mediated synthesis of silver nanoparticles and their immobilization in the mycelial matrix: a novel biological approach to nanoparticle synthesis. *Nano Lett.*, **1**(10), 515–519.

Mulvaney P. (1996). Surface plasmon spectroscopy of nanosized metal particles. *Langmuir*, **12**(3), 788–800.

Nair B. and Pradeep T. (2002). Coalescence of nanoclusters and formation of submicron crystallites assisted by Lactobacillus strains. *Cryst. Growth Des.*, **2**(4), 293–298.

Nanda A. and Saravanan M. (2009). Biosynthesis of silver nanoparticles from *Staphylococcus aureus* and its antimicrobial activity against MRSA and MRSE. *Nanomed.-Nanotechnol. Biol. Med.*, **5**(4), 452–456.

Natarajan K., Selvaraj S. and Murty V. R. (2010). Microbial production of silver nanoparticles. *Dig. J. Nanomater. Biostruct.*, **5**(1), 135–140.

NIA. (2010). European Parliament approaches Nanomaterials in Electrical and Electronic Equipment with strong Language and a heavy Hand.

NIA. (2011). RoHS Recast: Nanomaterials Escape Explicit Regulation.

Nithya R. and Ragunathan R. (2009). Synthesis of silver nanoparticle using *Pleurotis sajor caju* and its antimicrobial study. *Dig. J. Nanomater. Biostruct.*, **4**(4), 623–629.

Pal S., Tak Y. K. and Song J. M. (2007). Does the antibacterial activity of silver nanoparticles depend on the shape of the nanoparticle? A study of the gram-negative bacterium *Escherichia coli*. *Appl. Environ. Microbiol.*, **73**(6), 1712–1720.

Panacek A., Prucek R., Safarova D., Dittrich M., Richtrova J., Benickova K., Zboril R. and Kvitek L. (2011). Acute and chronic toxicity effects of silver nanoparticles (NPs) on *Drosophila melanogaster*. *Environ. Sci. Technol.*, **45**(11), 4974–4979.

Philip D. (2009). Biosynthesis of Au, Ag and Au-Ag nanoparticles using edible mushroom extract. *Spectroc. Acta Pt. A-Molec. Biomolec. Spectr.*, **73**(2), 374–381.

Saifuddin N., Wong C. W. and Yasumira A. A. N. (2009). Rapid biosynthesis of silver nanoparticles using culture supernatant of bacteria with microwave irradiation. *E-J. Chem.*, **6**(1), 61–70.

Sanghi R. and Verma P. (2009). Biomimetic synthesis and characterisation of protein capped silver nanoparticles. *Bioresour. Technol.*, **100**(1), 501–504.

Saravanana M. and Nanda A. (2010). Extracellular synthesis of silver bionanoparticles from *Aspergillus clavatus* and its antimicrobial activity against MRSA and MRSE. *Colloid Surf. B-Biointerfaces*, **77**(2), 214–218.

Sastry M., Ahmad A., Khan M. I. and Kumar R. (2003). Biosynthesis of metal nanoparticles using fungi and actinomycete. *Curr. Sci.*, **85**(2), 162–170.

Sathyavathi R., Krishna M. B., Rao S. V., Saritha R. and Rao D. N. (2010). Biosynthesis of silver nanoparticles using *Coriandrum Sativum* leaf extract and their application in nonlinear optics. *Adv. Sci. Lett.*, **3**(2), 138–143.

Seaton A., Tran L., Aitken R. and Donaldson K. (2010). Nanoparticles, human health hazard and regulation. *J. R. Soc. Interface*, **7**, S119–S129.

Shahverdi A. R., Minaeian S., Shahverdi H. R., Jamalifar H. and Nohi A. A. (2007). Rapid synthesis of silver nanoparticles using culture supernatants of enterobacteria: a novel biological approach. *Process Biochem.*, **42**(5), 919–923.

Shaligram N. S., Bule M., Bhambure R., Singhal R. S., Singh S. K., Szakacs G. and Pandey A. (2009). Biosynthesis of silver nanoparticles using aqueous extract from the compactin producing fungal strain. *Process Biochem.*, **44**(8), 939–943.

Sharma V. K., Yngard R. A. and Lin Y. (2009). Silver nanoparticles: green synthesis and their antimicrobial activities. *Adv. Colloid Interface Sci.*, **145**(1–2), 83–96.

Silver S., Phung L. T. and Silver G. (2005). *Silver as Biocides in Burn and Wound Dressings and Bacterial Resistance to Silver Compounds*. Springer Heidelberg, Chicago, IL, pp. 627–634.

Sintubin L., Awoke A., Wang Y., Van der Ha D. and Verstraete W. (2011a). Enhanced disinfection efficiencies of solar irradiation by biogenic silver. *Ann. Microbiol.*, **1**, 1–5.

Sintubin L., De Gusseme B., Van der Meeren P., Pycke B. F. G., Verstraete W. and Boon N. (2011b). The antibacterial activity of biogenic silver and its mode of action. *Appl. Microbiol. Biotechnol.*, **91**(1), 153–162.

Sintubin L., De Windt W., Dick J., Mast J., van der Ha D., Verstraete W. and Boon N. (2009). Lactic acid bacteria as reducing and capping agent for the fast and efficient production of silver nanoparticles. *Appl. Microbiol. Biotechnol.*, **84**(4), 741–749.

Sondi I. and Salopek-Sondi B. (2004). Silver nanoparticles as antimicrobial agent: a case study on *E. coli* as a model for Gram-negative bacteria. *J. Colloid Interface Sci.*, **275**(1), 177–182.

Thakkar K. N., Mhatre S. S. and Parikh R. Y. (2010). Biological synthesis of metallic nanoparticles. *Nanomed.-Nanotechnol. Biol. Med.*, **6**(2), 257–262.

Tolaymat T. M., El Badawy A. M., Genaidy A., Scheckel K. G., Luxton T. P. and Suidan M. (2010). An evidence-based environmental perspective of manufactured silver nanoparticle in syntheses and applications: a systematic review and critical appraisal of peer-reviewed scientific papers. *Sci. Total Environ.*, **408**(5), 999–1006.

Vigneshwaran N., Kathe A. A., Varadarajan P. V., Nachane R. P. and Balasubramanya R. H. (2006a). Biomimetics of silver nanoparticles by white rot fungus, *Phaenerochaete chrysosporium. Colloid Surf. B-Biointerfaces*, **53**(1), 55–59.

Vigneshwaran N., Nachane R. P., Balasubramanya R. H. and Varadarajan P. V. (2006b). A novel one-pot 'green' synthesis of stable silver nanoparticles using soluble starch. *Carbohydr. Res.*, **341**(12), 2012–2018.

Vigneshwaran N., Kathe A. A., Varadarajan P. V., Nachane R. P. and Balasubramanya R. H. (2007). Silver-protein (core-shell) nanoparticle production using spent mushroom substrate. *Langmuir*, **23**(13), 7113–7117.

Vijayakumar P. S. and Prasad B. L. V. (2009). Intracellular biogenic silver nanoparticles for the generation of carbon supported antiviral and sustained bactericidal agents. *Langmuir*, **25**(19), 11741–11747.

Wang T. C., Rubner M. F. and Cohen R. E. (2002). Polyelectrolyte multilayer nanoreactors for preparing silver nanoparticle composites: controlling metal concentration and nanoparticle size. *Langmuir*, **18**(8), 3370–3375.

WHO (2004). Water, sanitation and hygiene links to health: facts and figures. http://www.who.int/water_sanitation_health/en/factsfigures04.pdf (accessed November 2011)

WHO (2010). Progress on sanitation and drinking-water. http://whqlibdoc.who.int/publications/2010/9789241563956_eng_full_text.pdf (accessed November 2011)

WHO (2011). World Health Statistics 2011. http://www.who.int/whosis/whostat/EN_WHS2011_Full.pdf (accessed November 2011)

Xu R., Wang D. S., Zhang J. T. and Li Y. D. (2006). Shape-dependent catalytic activity of silver nanoparticles for the oxidation of styrene. *Chem.-Asian J.*, **1**(6), 888–893.

Zhang H. R., Li Q. B., Lu Y. H., Sun D. H., Lin X. P., Deng X., He N. and Zheng S. Z. (2005). Biosorption and bioreduction of diamine silver complex by Corynebacterium. *J. Chem. Technol. Biotechnol.*, **80**(3), 285–290.

Zhang H. R., Li Q. B., Wang H. X., Sun D. H., Lu Y. H. and He N. (2007). Accumulation of silver(I) ion and diamine silver complex by Aeromonas SH10 biomass. *Appl. Biochem. Biotechnol.*, **143**(1), 54–62.

Zhang Y. C., Li S. Y., Zhang C. H., Luo Y. K., Zhang H. P. and Yang Z. A. (2011). Growth and exopolysaccharide production by *Lactobacillus fermentum* F6 in skim milk. *Afr. J. Biotechnol.*, **10**(11), 2080–2091.

Zhu J. J., Liu S. W., Palchik O., Koltypin Y. and Gedanken A. (2000). Shape-controlled synthesis of silver nanoparticles by pulse sonoelectrochemical methods. *Langmuir*, **16**(16), 6396–6399.

Chapter 12

Electrospun nanofibrous membranes for water treatment applications

T. E. Cloete, M. de Kwaadsteniet, N. P. Gule and B. Klumperman

12.1 INTRODUCTION

Recent times have seen an increase in the use of nanostructured materials, such as mesoporous media, nanoparticles, carbon nanotubes and nanofibers in filter media (Kim *et al.*, 2005). Among these materials, electrospun nanofibers offer a number of attractive features compared to the other nanostructures in filter media applications (Huang *et al.*, 2003). Electrospun nanofibers are produced during electrospinning (ES), which is a fiber producing technique that relies on electric forces to fabricate fibers in a nonwoven mat. Nanofibers have a diameter that is more than one order of magnitude smaller than conventional melt-drawn microfibers therefore having a higher surface-to-volume ratio, which could result in better filtration efficiencies (Yoon *et al.*, 2008). Electrospun nanofibrous membranes (ENMs) have an advantage over mesoporous media by relieving the mass transfer limitation of substrates/products due to their reduced thickness. Filtration efficiencies of ENMs are also enhanced by the interconnected pores that form due to the nonwoven morphology of the nanofibers, which in turn increase the bulk porosity (Chu & Hsiao, 2009). The above mentioned concepts are illustrated in Table 12.1, where a decrease in fiber diameter leads to a decrease in filter thickness and an increase in filter specific area (m^2/m^3). The basis weight also decreases with a decrease in fiber diameter indicating a decrease in raw materials required (Podgórski *et al.*, 2006).

The electrospinning process enables the surface of nanofibers to be functionalized to produce antimicrobial, antifouling and affinity filtration membranes. The high surface-to-volume ratio of the nanofibers also increases the exposure of the functionalized surface to the target compounds in the liquid media. Additionally, it is easier to recover and reuse nanofibers than nanoparticles or carbon nanotubes since ES can generate long nanofibers, which can also be further processed into various structures such as nonwoven mats, well aligned arrays, or membranes (Kim *et al.*, 2005). The length of the nanofibers also minimizes safety concerns (Yoon *et al.*, 2008). Furthermore, ENMs possess high acid/base and temperature resistance, light weight, longer membrane life spans as well as high flexibility which enable nanofiber membranes to be formed into various membrane modules for larger commercial applications (Huang *et al.*, 2003). The use of nanofibers in filter media has received a lot of attention from researchers worldwide (Barhate & Ramakrishna, 2007). In this chapter, the electrospinning techniques as well as the physical and chemical techniques to functionalize nanofibers are discussed. This is followed by the different possible applications of nanofibers and nanofiber composite materials in the water industry.

Table 12.1 Structural parameters of four fibrous filters.

Thickness (mm)	Fiber diameter (μm)	Solidity (dimensionless)	Basis weight (g/m^2)	Filter specific area (m^2/m^3)
3	10	0.1	273	0.40×10^5
1	2.0	0.05	45.5	1.00×10^5
1	0.7	0.02	18.2	1.14×10^5
0.2	0.1	0.005	0.91	2.00×10^5

Source: Podgórski *et al.* (2006).

12.2 ELECTROSPINNING TECHNIQUES

12.2.1 Electrospinning processes

The principles of the ES process are well known and described in this chapter in the basic ES set-up, which was first patented by Formhals in 1934 (Formhals 1934). The basic ES technique, needle ES, involves the application of a high voltage difference between a positively charged metallic needle, ejecting the polymeric solution, and a grounded collector. When the electrostatic force overcomes the cohesive force of the solution, an electrically charged jet emerges from the capillary. The fluid jet is accelerated and stretched by the external electric field and thins dramatically while travelling towards the collector, which leads to the formation of nanometer-sized fibers that are collected on the target as a non-woven fiber web (Voigt, 2009).

While the ES technique is very simple, the spinning mechanism is rather complicated due to the instabilities of the jet on its way to the collector. The theory behind these instabilities is known as jet modelling. The jet modelling mechanism generally takes place in three stages known as initiation, thinning and solidification of the jet (Voigt, 2009). Under the influence of a strong electrostatic field, charges are induced in the solution and the charged polymer is accelerated towards the grounded metal collector (Baji *et al.*, 2010). At low electrostatic field strength, the pendant drop emerging from the tip of the pipette becomes highly electrified and Coulombic forces are exerted from the external electric field, marking the beginning of the initiation state. Inside the droplet, the surface tension minimizes the surface area to hold the droplet in a spherical shape. As the intensity of the electric field is increased, the induced charges on the liquid surface repel each other and create shear stress. These repulsive forces act in a direction opposite to the surface tension, which results in the extension of the pendant drop into a conical shape (Taylor cone) and serves as an initiating surface (Taylor, 1969). When the critical voltage is reached, the equilibrium of the forces is disturbed and a charged jet emanates from the tip of the conical drop (Baji *et al.*, 2010). While moving, the jet undergoes instability due to repulsing forces between the electric charges inside the jet. This bending instability is mainly responsible for a decrease of the jet diameter into the nanometre range (Voigt, 2009). The last stage, which is jet solidification, depends on the solvent vapour pressure and on the evaporation rate of the solvent as well as on the available time for the solvent to evaporate and is directly affected by the distance between spinneret (needle) and target (collector).

Since 1934, numerous electrospinning techniques have been developed to increase production rates and to overcome technical problems such as needle clogging. Electrospinning production rates can be increased through two approaches. In the first approach, the number of needles in the electrospinning process is increased therefore producing numerous nanofibers simultaneously (Chu *et al.*, 2004; Ding *et al.*, 2004). In the second approach, a needleless electrospinning set-up is used. One of the most successful needleless electrospinning processes, the Nanospider^{TM} technology, has been developed and patented by Jirsak *et al.* (2005) and commercialized for large scale production through Elmarco s.r.o (Czech Republic). During needleless electrospinning (also known as roller electrospinning) a high voltage is applied to a polymer solution or melt that is submerged into a tub containing a steel cylinder. The high voltage charges the thin polymer layer on the surface of the rolling cylinder, which results in the formation of numerous Taylor cones simultaneously.

12.2.2 Factors affecting the electrospinning process

Parameters influencing the basic electrospinning process can be divided into the system, process and environmental parameters. System parameters include polymer properties, such as molecular weight, molecular weight distribution and architecture, and polymer solution properties such as viscosity, conductivity, dielectric constant and surface tension (Bhardwaj & Kundu 2010; Deitzel *et al.*, 2001; Frenot & Chronakis, 2003; Li & Xia, 2004). Process parameters include the electric potential, flow rate and distance between the electrodes. During roller electrospinning, the roller diameter, roller velocity and velocity of substrate material are additional parameters that influence the process (Cengiz & Jirsak, 2009). Environmental parameters that need to be controlled during any electrospinning process include the temperature, humidity and air velocity (Deitzel *et al.*, 2001; Frenot & Chronakis, 2003; Li & Xia, 2004; Cengiz & Jirsak 2009). A review on the effects of the various parameters on the electrospinning process was published by Bharwaj and Kundu (2010) and Chronakis (2010) and will not be discussed any further in this chapter.

12.3 FUNCTIONALIZED NANOFIBERS

Various degradable and non-degradable synthetic nanofibers have been surface-modified with bioactive molecules for advanced applications (Yoo *et al.*, 2009). Synthetic polymers offer easier processability for electrospinning and more controllable nanofibrous morphology than natural polymers. Due to the processing benefits of synthetic polymers, a wide variety of functional groups as well as other compounds can be immobilized onto the nanofibrous surface of synthetic polymers without compromising bulk properties. The modification of polymer surfaces is a research field that has received much attention, mainly because of the lack of some specific surface properties needed in a number of applications, such as adhesion, biomaterials and protective coatings, although the polymers might possess excellent bulk properties (Sun & Onebby, 2006). Various surface modification techniques have been explored for functionalizing nanofibers and these include physical coating, plasma treatment, and chemical modifications using wet treatment, surface grafting polymerization and co-electrospinning.

Surface modification to improve hydrophilicity and therefore the water permeability is of vital importance during membrane development. The synthetic polymers commonly used in ENMs such as polytetrafluoroethylene (PTFE), polypropylene (PP), polyethersulfone (PES) and polyvinylidene fluoride (PVDF) are hydrophobic. The surface of these hydrophobic polymers could be rendered hydrophilic through plasma treatment (Shabani *et al.*, 2009), chemical oxidation and surface grafting (Bamford & Al-Lamee, 1996; Yoon *et al.*, 2009a) and blending with hydrophilic polymers (Sun *et al.*, 2003; Hayama *et al.*, 2004). The different physical and chemical techniques to functionalize nanofibers are discussed in the following sections.

12.3.1 Physical coating

Buldum *et al.* (2005) demonstrated physical coating of nanofibers through sputtering and/or deposition. They successfully coated fibers with films of copper, aluminium, titanium, zirconium, and aluminium nitride by using a plasma-enhanced physical vapour deposition sputtering process (Buldum *et al.*, 2005). Kim *et al.* (2005) also covalently bound enzymes to poly(styrene-*co*-maleic anhydride) nanofibers through physical coating and these fibers were stable as well as highly active (Kim *et al.*, 2005).

12.3.2 Plasma treatment

Plasma treatment of polymer substrates has been commonly used to tailor surface adhesion and wetting properties by changing the surface chemical composition. Appropriate selection of the plasma source enables the introduction of diverse functional groups on the target surface to improve biocompatibility or to allow subsequent covalent immobilization of various bioactive molecules. Plasma treatments with oxygen, ammonia, or air can generate carboxyl groups or amine groups on the surface for covalent immobilization of a specific molecule (Yoo *et al.*, 2009). Höcker (2002) reported the use of this technique to achieve shrink resistant fibers, while Ladizesky and Ward (1995) plasma-treated polyethylene fibers to reinforce them for clinical applications (Ladizesky & Ward, 1995; Höcker, 2002).

12.3.3 Wet chemical method

A popular wet chemical method is based on the random chemical scission of ester linkages on the polyester backbones located on the surface, resulting in the surface generation of carboxylic and hydroxyl groups from degraded, yet water-insoluble polymer fragments (Yoo *et al.*, 2009). The plasma treatment for nanofibrous mesh cannot effectively modify the surface of buried nanofibers deeply located in the mesh due to the limited penetration depth of plasma in the nanopores. Wet chemical etching methods can offer the flexibility for surface modification of thick nanofibrous meshes. An example of this method was reported by Sun and Onebby (2006) who investigated surface modification of biodegradable polycaprolactone films *via* base-catalysed hydrolysis in sodium hydroxide solution (Sun & Onebby, 2006).

12.3.4 Surface graft polymerization

Virtually all types of synthetic biodegradable polymers retain their hydrophobic surface nature, often requiring hydrophilic surface modification for desired cellular responses. Surface graft polymerization has been introduced not only to confer surface hydrophilicity, but also to introduce multi-functional groups on the surface (Yoo *et al.*, 2009). The surface graft polymerization is often initiated with plasma and UV radiation treatment to generate free radicals for the polymerization. There are two methods for chemically attaching polymer chains to a solid substrate: the “grafting to” and “grafting from” techniques. In the “grafting to” method, pre-formed, end-functionalized polymer chains are reacted with a chemically activated substrate (Rhodes, 2007). One advantage of this method is that polymer chains can be characterized before being attached to the substrate (Ranjan, 2008). The drawback, however, is that only relatively low grafting densities are obtained due to steric crowding of already attached chains on the surface, which hinder diffusion of additional chains to reactive sites. The "grafting from" method involves formation of an initiator layer on the surface of the nanofibers followed by polymerization of monomer (Rhodes, 2007; Ranjan, 2008).

12.3.5 Co-electrospinning of surface active agents and polymers

While the aforementioned surface modification methods are intended to be used for prefabricated electrospun nanofibers, nanoparticles and functional polymer segments can be directly exposed on the surface of nanofibers by co-electrospinning with bulk polymers (Yoo *et al.*, 2009). An example of this work was performed by Voigt (2009) and du Plessis (2011) who fabricated water-stable antimicrobial nanofibers from poly(vinyl alcohol) (PVA) and poly (acrylonitrile) (PAN), respectively, by adding silver nitrate to the spinning solutions prior to the electrospinning process (Voigt, 2009; du Plessis, 2011).

12.4 POTENTIAL APPLICATIONS OF ELECTROSPUN POLYMERIC NANOFIBERS IN THE WATER INDUSTRY

12.4.1 Microfiltration membranes

Microfiltration membranes have a pore size between 0.1 and 10 μm retaining suspended solids and even bacteria in cases where the pore size is less than 0.2 μm. Advantages of ENMs over conventional filter media include controllable pore size, high porosity and high hydraulic permeability. Microfiltration membranes produced during the electrospinning process could be used as a pretreatment for water filtration (Barhate *et al.*, 2006). Electrospun nylon-6 membranes for example have been shown to separate all particles between 1–10 μm and had a 90% separation efficiency for 0.5 μm particles (Aussawasathein *et al.*, 2008).

A disadvantage of electrospun microfiltration membranes over conventional cast membranes is the support substrate required for additional strength. ENMs are therefore based on hybrid systems with the nanofibers either placed over a support substrate, imbedded between various layers ("sandwiched") or blended with micron fibers. Bazargan *et al.* (2011) tried to overcome this problem by developing self-supporting electrospun PAN nanofibrous membranes. The high level of fiber consistency produced membranes with high water permeability and mechanical strength properties. Furthermore, the membrane was able to remove more than 90% of microparticles in the 1–20 μm range when using an optical particle counter (Bazargan *et al.*, 2011). Polyvinylidene fluoride (PVDF) (Gopal *et al.*, 2006) and polysulfone (PSU) (Gopal *et al.*, 2007) were electrospun into self-supporting membranes and their removal efficiency tested using polystyrene particles. Electrospun PVDF membranes had a 96% separation factor for 10 μm particles and a 91% separation factor for 5 μm particles (Gopal *et al.*, 2006).

Electrospun PSU membranes had a 99% separation factor for particles between 7–10 μm (Gopal *et al.*, 2007). In both cases a drop in flux were observed during the filtration of 1–2 μm particles with the formation of a cake layer (Gopal *et al.*, 2006, 2007).

12.4.2 Ultrafiltration membranes

Ultrafiltration (UF) membranes have a pore size between 1 and 100 nm and can filter out a range of particulates including viruses, emulsions, proteins and colloids. Thin film composites (TFC) are one of the most preferred formats for ultrafiltration, nanofiltration and reverse osmosis membranes. Conventional thin film composites consist of a microfibrous substrate, an asymmetric porous mid-layer and a barrier layer. Membrane fouling which results in flux decline is a major obstacle in conventional TFC membranes (Petersen, 1993). Thin film nanocomposite (TFNC) membranes have been shown through numerous studies to have a higher flux than conventional TFC membranes (Homaeigohar *et al.*, 2010; Ma *et al.*, 2010, 2011a, b; Tang *et al.*, 2009a, b; Yoon *et al.*, 2006, 2009b; Wang *et al.*, 2010). TFNC membranes consist of a conventional non-woven microfibrous support, an electrospun nanofibrous mid-layer and a 'non-porous' hydrophilic electrospun barrier layer (Yoon *et al.*, 2006). The advantages of TFNC membranes over conventional TFC membranes include an increase in flux while maintaining the same rejection ratio and reduced fouling due to the hydrophilic barrier layer (Ma *et al.*, 2011a, b).

Yoon *et al.* (2006) developed a TFNC membrane that contains a non-woven (polyethylene terephthalate) PET substrate, an asymmetric electrospun PAN nanofibrous mid-layer support and a water resistant chitosan coating (Figure 12.1). This TFNC membrane outperformed a commercial nanofiltration filter (NF 270) with an order of magnitude higher flux rate over 24 h of operation while maintaining the same rejection efficiency (>99%) (Yoon *et al.*, 2006). Yoon *et al.*, (2009b) further developed the TFNC membrane by replacing the water resistant chitosan layer with a crosslinked PVA barrier layer. The membrane was optimized according to the pore size of the PAN scaffold and the thickness of the crosslinked PVA barrier. This membrane demonstrated a 12 times higher flux than conventional TFC membranes (PAN10 and PAN400 ultrafiltration membranes) while maintaining the same rejection rate (99.5%) for separation of an oil/water emulsion over 190 h of operation (Yoon *et al.*, 2009b). The same research group developed similar TFNC membranes once again replacing the barrier layer, this time with either cellulose or chitin. These TFNC membranes exhibited between a 10 and 12 times higher permeation flux over conventional TFC membranes while still maintaining high rejection ratios for separation of an oil/water emulsion. These polysaccharides not only had high flux properties but were also produced through an environmentally friendly process and the cost of the raw material was low (Ma *et al.*, 2011a, b). However, cellulose can be degraded by microorganisms present in the waste water and therefore chitin, which is an inert material, would be better suited for waste water treatment membranes (Ma *et al.*, 2011b).

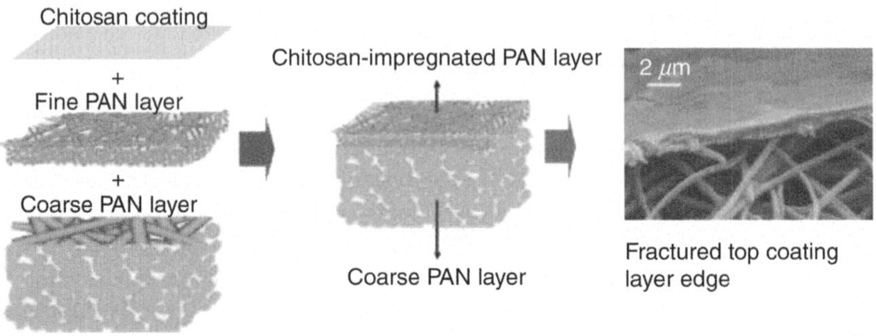

Figure 12.1 TFNC membrane developed by Yoon *et al.* (2006) that consists of a non-woven PAN substrate, an asymmetric electrospun PAN nanofibrous mid-layer support and a water resistant chitosan coating. (Reprinted from Polymer, 47 (7), Yoon K., Kim K., Wang X., Fang D., Hsiao B.S., Chu B, High flux ultrafiltration membranes based on electrospun PAN scaffolds and chitosan coating, 141–147, 2010, with permission from Elsevier.)

TFNC membranes for bacterial and virus removal for water purification have also been developed. Sato *et al.* (2011) developed a TFNC membrane that consisted out of a nonwoven PET substrate, an eletrospun PAN nanofibrous scaffold with either an ultrafine cellulose nanofibrous network barrier or microcrystalline cellulose nanofiber barrier. Both filters were able to simultaneously retain bacteria and adsorb viruses while maintaining high retention rates (Sato *et al.*, 2011).

Electrospun nanofibers used to fabricate micro- and ultrafiltration membranes can be further functionalized to have antimicrobial, antifouling and adsorption capabilities and will be discussed below.

12.4.3 Antimicrobial membranes

Antimicrobial nanofibers have been fabricated and characterized by numerous researchers to produce membranes with high filtration efficiencies and that are resistant to biofouling. Metal elements such as silver, copper, zinc, nickel, manganese, iron and lithium have been reported to exhibit antimicrobial properties (Botes & Cloete, 2010). Some of these metals have been blended with polymers and electrospun into fibers for use in filter media. These include silver nanoparticles, which have been extensively studied due to their robust antimicrobial nature (Daels *et al.*, 2011; Du Plessis, 2011). Other polymeric nanofibers for water treatment applications have been functionalized with silver nanoparticles after the electrospinning process (Bjorge *et al.*, 2009; Zhang *et al.*,

2011). In all of the above mentioned studies, the silver functionalized membranes were more efficient in the removal of bacteria than their non-functionalized counterparts. Care must be taken with the use of silver in water treatment applications, since microbial resistance against silver has been noted by different researchers (Silver, 2003). Antimicrobial nanofibers have also been fabricated by the incorporation of other metal elements including copper ions (Gule, 2012; Gule, *et al.*, 2012b, c) and copper nanoparticles (Sheikh, *et al.*, 2011).

The addition of antimicrobial additives in some cases also influences the morphology and diameter of nanofibers and hence the filtration capabilities of the filter. The diameter of nanofibers have been shown to decrease with the addition of a copper and zinc ion solution to PVA (Gule *et al.*, 2012b, c) and quaternary ammonium salt (benzyl triethylammonium chloride, BTEAC) to polycarbonate (Kim *et al.*, 2007). Researchers found that the change in conductivity and viscosity of the polymer solution due to the additives were important parameters affecting the outcome of the end product (Gule, 2012; Gule *et al.*, 2012b, c; Kim *et al.*, 2007). ENMs incorporated with quaternary ammonium salts showed good filtration efficiencies and antimicrobial activity against *Escherichia coli* and *Staphylococcus aureus* (Jeong *et al.*, 2007; Kim *et al.*, 2007; Yao *et al.*, 2008). Other antimicrobial additives that have been incorporated into antimicrobial ENMs include titanium dioxide (TiO_2) nanoparticles (Pant *et al.*, 2011), poly (4-vinyl-*N*-alkylpyridinium) bromide (Yao *et al.*, 2009) and *N*-halamine additives (Tan & Obendorf, 2007).

12.4.4 Antifouling membranes

Nanofibers can be modified to inhibit or decrease the attachment of microbial cells through the incorporation or grafting of functional groups. Antifouling nanofibers have been fabricated with functional groups that target extracellular polymer degradation (du Plessis *et al.*, 2013) and quorum sensing disruption (Gule *et al.*, 2012a). Extracellular polymers (EPS) produced during biofilm formation is an important component that contributes to the biofouling of surfaces, including membranes (Characklis, 1990). PAN nanofibers containing immobilized protease and amylase enzymes were resistant to biofilm formation when compared to pristine PAN nanofibers. The large specific surface area of the nanofibers increased the loading capability of the immobilized enzymes so that above 80% specific activity of the free enzymes were retained. The immobilized enzymes could also be re-used with 30% of initial activity retained after 10 cycles of reuse (du Plessis *et al.*, 2013). PAN nanofibrous membranes have also been immobilized with lipase, in this case not to produce antifouling membrane but for the use in oil hydrolysis (Li & Wu, 2009). In both studies, the PAN nanofibers were activated by an amination reaction for immobilizing the respective enzymes with covalent binding (du Plessis *et al.*, 2013; Li & Wu, 2009). Obstacles that need to be overcome for the large scale applications include the low catalytic efficiency and stability of enzymes, difficulties in industrial fabricating and the lack of appropriate analytical tools for nanofiber immobilized enzymes (Theron *et al.*, 2005).

Complex biofilm formation by environmental strains, which is the major cause of biofouling and biofilm formation, is in most cases controlled by *N*-acyl homoserine lactone (AHL)-mediated quorum sensing. Furanones are analogs of homoserine lactones that appear to interfere with the development of typical biofilm structure, leaving these organisms more susceptible to treatment with biocides. Furanones act by mimicking the AHL signal, presumably by occupying the binding site on the putative regulatory protein, rendering it highly unstable and accelerating its turnover rate and thus resulting in the rapid disruption of the quorum sensing-mediated gene regulation (Manefield *et al.*, 2002). Nanofibrous mats emanating from PVA and poly(styrene-*co*-maleic anhydride) (SMA) modified with furanone derivatives demonstrated promising results for water treatment applications. The nanofibrous membranes containing furanone derivatives not only demonstrated antimicrobial activity against *E. coli, S. aureus, Salmonella typhimurium, Klebsiella pneumoniae* and *Pseudomonas aeruginosa* but also the ability to inhibit attachment of these bacteria onto the surfaces of the nanofibrous membranes. These dual action nanofibers have great potential in water treatment applications (Gule, 2012; Gule *et al.*, 2012a).

12.4.5 Affinity membranes

Nanofibers can potentially be used to fabricate selective membranes for the removal of organic and inorganic compounds from water by the incorporation of functional groups into or onto the fibers. The large surface to volume ratio of nanofibers is an ideal characteristic for affinity membranes, since this is one of the most important requirements for selective membranes. The large surface area increases the availability of adsorption sites for binding of the compound of interest. Electrospinning of wool keratose/silk fibroin nanofibers produced membranes with similar flux as other microfiltration membranes, but with a high adsorption capacity for metal ions. The binding sites for metal adsorption were the carboxylic acid and cysteric acid side residues in the wool keratose resulting in an adsorption capacity Cu^{2+} of 2.88 μg/mg. The high density of the binding sites on the wool keratose/silk fibroin nanofibers resulted in 50 times higher Cu^{2+} adsorption capacity when compared to conventional filter media (Ki *et al.*, 2007). Silk fibroin/chitosan acetate nanofibrous membranes had high adsorption capacities for Cu^{2+} with a maximum adsorption value of 22.8 mg/g (Zhou *et al.*, 2011). Chitosan nanofibers have also shown high affinity for heavy metals with equilibrium adsorption capacities of 485.44 mg/g and 263.15 mg/g for Cu^{2+} and Pb^{2+} respectively (Haider & Park, 2009). Nano-boehmite impregnated into electrospun fiber membranes produced membranes with a high adsorption capacity for Cd^{2+} (0.2 mg/ml). Although higher adsorption capacities for boehmite electrospun membranes than the pure electrospun membranes were observed, boehmite nanofibrous materials had the highest adsorption value. This may be due to some of the active boehmite sites in the electrospun membranes being covered with polymers during the electrospinning process. On the other hand, the boehmite impregnated nanofibers can be more difficult to wash off during post water treatment processes (Hota *et al.*, 2008). Nanofibers modified with functional groups on their surfaces after the electrospinning process have in many cases showed higher adsorption capacities than nanofibers with impregnated functional groups. This is due to the exposure of the functional groups on the surfaces of the nanofibers to the metal ions of interest. Polymer nanofibers reinforced with multiwalled carbon nanotubes have been immobilized with zerovalent iron nanoparticles (ZVI NPs). The maximum adsorption capacities of these nanofibers for Cu^{2+} were 107.8 and 66.8 mg/g for the fibers with and without the functionalized ZVI NPs (Xiao *et al.*, 2011).

Affinity nanofibers for removing organic molecules from waste water have also been manufactured. For instance, poly (methyl methacrylate) nanofiber membranes were functionalized with a cyclic oligosaccharide, ß-cyclodextrin. The hydrophobic interior and hydrophilic exterior of ß-cyclodextrin captured hydrophobic organic molecules from water by forming an inclusion complex (Kaur *et al.*, 2006). Proteins can be present in waste water from food production factories and therefore researchers have investigated the use of nanofibers as protein affinity membranes. One example is the surface of PAN nanofibers that has been modified to capture bromelain (Zhang *et al.*, 2010). Disinfection byproducts, trihalomethanes and haloacetic acid present in water are known carcinogens and ENMs have been investigated as membrane adsorption media for their removal. In these studies, chloroform was used as the model disinfection byproduct. High adsorption capacities for chloroform were observed with carbonized ENMs fabricated from PAN (Singh *et al.*, 2010) and PVDF nanofibers (Feng *et al.*, 2011).

Positively charged electrospun nanofibrous membranes have the capacity to adsorb viruses. NanoCeram cartridge filters that consist of positively charged alumina nanofibers have been investigated as a tool to concentrate enteroviruses and noroviruses from respectively large volumes of freshwater (Karim *et al.*, 2009) and seawater samples (Gibbons *et al.*, 2010).

12.5 MULTIFUNCTIONAL COMPOSITE NANOFIBROUS MEMBRANES

The ES process could enable researchers to fabricate composite water filtration membranes with multiple functions, that is antimicrobial properties as well as adsorption capacities for a specific inorganic or organic material. Different approaches could be followed to produce multi-functional composite membranes. Different functionalized nanofibers might be added together to form a heterogeneous membrane. Multiple needles each filled with a specific functionalized polymer may possibly be used. A multi-tiered membrane with alternating functionalized nanofiber layers could also be fabricated. The proposed model will be to have the antimicrobial or antifouling layer on the outer surface to inhibit fouling of the membrane. The next layer could be an adsorbing nanofiber layer for removing inorganic or organic molecules from the water (Kidoaki *et al.*, 2005; Teo *et al.*, 2009). Other examples of multi-tiered nanofibrous membranes are TFNC ultrafiltration membranes (see Section 12.4.2).

12.6 CONCLUSIONS

Research on ENMs is increasing in laboratories due to the relatively low cost and low quantity of raw materials required, the simplicity of the fabrication technique as well as easy maintenance. Attractive features of electrospun nanofibers for water filtration membranes include their high surface to volume ratio, high porosity, interconnected pore structure, reduced thickness and the large length of the fibers, which reduces potential safety concerns and enhances their reusability. Nanofibers electrospun from synthetic polymers could also be further functionalized to produce antimicrobial, antifouling and affinity filtration membranes. Although the development of industrial scale electrospinning equipment enables the transformation of the technology to commercial applications, few ENMs are available on the market. Nonetheless, research on the use of nanofibrous membranes in filter media has given rise to a new era of high technology and efficient media for filtering various compounds from water.

REFERENCES

Aussawasathien D., Teerawattananon C. and Vongachariya A. (2008). Separation of micron to sub-micron particles from water: electrospun nylon-6 nanofibrous membranes as pre-filters. *J. Membr. Sci.*, **315**(1/2), 11–19.

Baji A., Mai Y.-W., Wong S.-C., Abtahi M. and Chen P. (2010). Electrospinning of polymer nanofibers: effects on orientated morphology, structures and tensile properties. *Compos. Sci. Technol.*, **70**(5), 703–718.

Bamford C. H. and Al-Lamee K. G. (1996). Studies in polymer surface modification and grafting for biomedical uses: 2. Application to arterial blood filters and oxygenators. *Polymer*, **37**(22), 4885–4889.

Barhate R. S. and Ramakrishna S. (2007). Nanofibrous filtering media: filtration problems and solutions from tiny materials. *J. Membr. Sci.*, **296**(1/2), 1–8.

Barhate R. S., Loong C. K. and Ramakrishna S. (2006). Preparation and characterization of nanofibrous filtering media. *J. Membr. Sci.*, **283**(1/2), 209–218.

Bazargan A. M., Keyanpour-rad M., Hesari F. A. and Ganji M. E. (2011). A study on the microfiltration behaviour of self-supporting electrospun nanofibrous membrane in water using an optical particle counter. *Desalination*, **265**(1/3), 148–152.

Bhardwaj N. and Kundu S. C. (2010). Electrospinning: a fiber fabrication technique. *Biotechnol. Adv.*, **28**(3), 325–347.

Bjorge D., Daels N., De Vrieze S., Dejans P., Van Camp T., Audenaert W., Hogie J., Westbroek P., De Clerk K. and Van Hulle S. W. H. (2009). Performance assessment of electrospun nanofibers for filter applications. *Desalination*, **249**(3), 942–948.

Botes M. and Cloete T. E. (2010). The potential of nanofibers and nanobiocides in water purification. *Crit. Rev. Microbiol.*, **36**(1), 68–81.

Buldum A., Clemons C. B., Dill L. H., Kreider K. L., Young G. W., Zheng X., Evans E. A., Zhang G. and Hariharan S. I. (2005). Multiscale modelling, simulations, and experiments of coating growth on nanofibers. Part II. Deposition. *J. Appl. Phys.*, **98**(4), 044304.

Cengiz F. and Jirsak O. (2009). The effect of salt on the roller electrospinning of polyurethane nanofibers. *Fibers Polym.*, **10**(2), 177–184.

Characklis W. G. (1990). Microbial fouling. In: *Biofilms*. W. G. Characklis and K. C. Marshall (eds), John Wiley & Sons, New York, NY. pp. 523–584.

Chronakis I. S. (2010). Micro-/Nano-Fibers by electrospinning technology: processing, properties and applications. In: *Micro-Manufacturing Engineering and Technology*, L. Qin, (ed.), William Andrew, Oxford, UK, pp. 264–286.

Chu B. and Hsiao B. S. (2009). The role of polymers in breakthrough technologies for water purification. *J. Polym. Sci. Polym. Phys.*, **47**(24), 2431–2435.

Chu B., Hsiao B. S. and Fang D. (2004). Apparatus and methods for electrospinning polymeric fibers and membranes. US006730011B2.

Daels N., De Vrieze S., Sampers I., Decostere B., Westbroek P., Dumoulin A., Dejans P., De Clerck K. and Van Hulle S. W. H. (2011). Potential of a functionalized nanofibre microfiltration membrane as an antimicrobial water filter. *Desalination*, **275**(1/3), 285–290.

Deitzel J. M., Kleinmeyer J., Harris D. and Tan N. C. B. (2001). The effect of processing variables on the morphology of electrospun nanofibers and textiles. *Polymer*, **42**(1), 261–272.

Ding B., Kimura E., Sato T., Fujita S. and Shiratori S. (2004). Fabrication of blend biodegradable nanofibrous nonwoven mats via multi-jet electrospinning. *Polymer*, **45**(6), 1895–1902.

Du Plessis D. M. (2011). Fabrication and Characterization of Anti-microbial and Biofouling Resistant Nanofibers with Silver Nanoparticles and Immobilized Enzymes for Application in Water Filtration. MSc thesis, Stellenbosch University, South Africa.

Du Plessis D. M., Botes M., Dicks L. M. T. and Cloete T. E. (2013). Immobilization of commercial hydrolytic enzymes on poly (acrylonitrile) nanofibers for anti-biofilm activity. *J. Chem. Technol. Biot.*, **88**(4), 585–593.

Feng C., Khulbe K. C. and Tabe S. (2011). Volatile organic compound removal by membrane gas stripping using electro-spun nanofiber membrane. *Desalination*, **287**, 98–102.

Formhals A. (1934). Process and apparatus for preparing artificial threads. US Patent 1975504.

Frenot A. and Chronakis I. S. (2003). Polymer nanofibers assembled by electrospinning. *Curr. Opin. Colloid Interface Sci.*, **8**(1/2), 64–75.

Gibbons C. D., Rodríguez R. A., Tallon L. and Sobsey M. D. (2010). Evaluation of positively charged alumina nanofibre cartridge filters for the primary concentration of noroviruses, adenoviruses and male-spesific coliphages from seawater. *J. Appl. Microbiol.*, **109**(2), 635–641.

Gopal R., Kaur S., Feng C. Y., Chan C., Ramakrishna S., Tabe S. and Matsuura T. (2007). Electrospun nanofibrous polysulfone membranes as pre-filters: particulate removal. *J. Membr. Sci.*, **289**(1/2), 210–219.

Gopal R., Kaur S., Ma Z., Chan C., Ramakrishna S. and Matsuura T. (2006). Electrospun nanofibrous filtration membrane. *J. Membr. Sci.*, **281**(1/2), 581–586.

Gule N. P. (2012). Anti-biofouling Nanofibrous Membranes. PhD thesis, Stellenbosch University, South Africa.

Gule N. P., Bshena O., De Kwaadsteniet M., Cloete T. E. and Klumperman B. (2012a). Immobilized furanone derivatives as inhibitors for adhesion of bacteria on modified poly (styrene-co-maleic anhydride). *Biomacromolecules*, **13**(10), 3138–3150.

Gule N. P., De Kwaadsteniet M., Cloete T. E. and Klumperman B. (2012b). Electrospun poly(vinyl alcohol) nanofibres with biocidal additives for application in filter media, 1 – properties affecting fibre morphology and characterisation. *Macromol. Mater. Eng.*, **297**(7), 609–617.

Gule N. P., De Kwaadsteniet M., Cloete T. E. and Klumperman B. (2012c). Electrospun poly(vinyl alcohol) nanofibres with biocidal additives for application in filter media, 2 – antimicrobial activity, regeneration, leaching and water stability. *Macromol. Mater. Eng.*, **297**(7), 609–617.

Haider S. and Park S. Y. (2009). Preparation of the electrospun chitosan nanofibers and their applications to the adsorption of Cu(II) and Pb(II) ions from an aqueous solution. *J. Membr. Sci.*, **328**(1/2), 90–96.

Hayama M., Yamamoto K. I., Kohori F., Uesaka T., Ueno Y., Sugaya H., Itagaki I. and Sakai K. (2004). Nanoscopic behavior of polyvinylpyrrolidone particles on polysulfone/polyvinylpyrrolidone film. *Biomaterials*, **25**(6), 1019–1028.

Höcker H. (2002). Plasma treatment of textile fibres. *Pure Appl. Chem.*, **74**(3), 423–427.

Homaeigohar S. Sh., Buhr K. and Ebert K. (2010). Polyethersulfone electrospun nanofibrous composite membrane for liquid filtration. *J. Membr. Sci.*, **365**(1/2), 68–77.

Hota G., Kumar B. R., Ng W. J. and Ramakrishna S. (2008). Fabrication and characterization of a boehmite nanoparticle impregnated electrospun fiber membrane for removal of metal ions. *J. Mater. Sci.*, **43**(1), 212–217.

Huang Z.-M., Zhang Y.-Z., Kotaki M. and Ramakrishna S. (2003). A review on polymer nanofibers by electrospinning and their applications in nanocomposites. *Compos. Sci. Technol.*, **63**(15), 2223–2253.

Jeong E. H., Yang J. and Youk J. H. (2007). Preparation of polyurethane cationomer nanofiber mats for use in antimicrobial nanofilter applications. *Mater. Lett.*, **61**(18), 3991–3994.

Jirsak O., Sanetrnik F., Lukas D., Kotek V., Martinova L. and Chaloupek J. (2005). A method of nanofibres production from a polymer solution using electrostatic spinning and a device for carrying out the method. WO2005024101.

Karim M. R., Rhodes E. R., Brinkman N., Wymer L. and Fout G. S. (2009). New electropositive filter for concentrating enteroviruses and noroviruses from large volumes of water. *Appl. Environ. Microbiol.*, **75**(8), 2393–2399.

Kaur S., Kotaki M., Ma Z., Gopal R. and Ramakrishna S. (2006). Oligosaccharide functionalized nanofibrous membrane. *Int. J. Nanosci.*, **5**(1), 1–11.

Ki C. S., Gang E. H., Um I. C. and Park Y. H. (2007). Nanofibrous membrane of wool keratose/silk fibroin blend for heavy metal ion adsorption. *J. Membr. Sci.*, **302**(1), 20–26.

Kidoaki S., Kwon I. and Matsuda T. (2005). Mesoscopic spatial designs of nano- and micro-fiber meshes for tissue-engineering matrix and scaffold based on newly devised multi-layering and mixing electrospinning techniques. *Biomaterials*, **26**(1), 37–46.

Kim B. C., Nair S., Kim J., Kwak J. H., Grate J. W., Kim S. H. and Gu M. B. (2005). Preparation of biocatalytic nanofibers with high activity and stability via enzyme aggregate coating on polymer nanofibers. *Nanotechnology*, **16**(7), S382–S388.

Kim S. J., Nam Y. S., Rhee D. M., Park H.-S. and Park W. H. (2007). Macromolecular Nanotechnology: preparation and characterization of antimicrobial polycarbonate nanofibrous membrane. *Eur. Polym. J.*, **43**(8), 3146–3152.

Ladizesky N. H. and Ward I. M. (1995). A review of plasma treatment and the clinical application of polyethylene fibres to reinforcements of acrylic resins. *J. Mater. Sci.: Mater. M.*, **6**(9), 497–504.

Li D. and Xia Y. (2004). Electrospinning of nanofibers: reinventing the wheel? *Adv. Mater.*, **16**(14), 1151–1170.

Li S. F. and Wu W. T. (2009). Lipase-immobilized electrospun PAN nanofibrous membranes for soybean oil hydrolysis. *Biochem. Eng. J.*, **45**(1), 48–53.

Ma H., Burger C., Hsiao B. S. and Chu B. (2011a). Ultrafine polysaccharide nanofibrous membranes for water purification. *Biomacromolecules*, **12**(4), 970–976.

Ma H., Hsiao B. S. and Chu B. (2011b). Thin-film nanofibrous composite membranes containing cellulose or chitin barrier layers fabricated by ionic liquids. *Polymer*, **52**(12), 2594–2599.

Ma H., Yoon K., Rong L., Shokralla M., Kopot A., Wang X., Fang D., Hsiao B. S. and Chu B. (2010). Thin-film nanofibrous composite ultrafiltration membranes based on polyvinyl alcohol barrier layer containing directional water channels. *Ind. Eng. Chem. Res.*, **49**(23), 11978–11984.

Manefield M., Rasmussen T. B., Henzter M., Anderson J. B., Steinberg P., Kjelleberg S. and Givskov M. (2002). Halogenated furanones inhibit quorum sensing through accelerated LuxR turnover. *Microbiology*, **148**(4), 1119–1127.

Pant H. R., Bajgai M. P., Nam K. T., Seo Y. A., Pandeya D. R., Hong S. T. and Kim H. Y. (2011). Electrospun nylon-6 spider-net like nanofiber mat containing TiO_2 nanoparticles: a multifunctional nanocomposite textile material. *J. Hazard. Mater.*, **185**(1), 124–130.

Petersen R. J. (1993). Composite reverse osmosis and nanofiltration membranes. *J. Membr. Sci.*, **83**(1), 81–150.

Podgórski A., Balazy A. and Gradoń L. (2006). Application of nanofibers to improve the filtration efficiency of the most penetrating aerosol particles in fibrous filters. *Chem. Eng. Sci.*, **61**(20), 6804–6815.

Ranjan R. (2008). Surface Modification of Silica Nanoparticles. University of Akron, Akron, United States of America.

Rhodes S. M. (2007). Electrically Conductive Polymer Composites. University of Akron, Akron, United States of America.

Sato A., Wang R., Ma H., Hsiao B. S. and Chu B. (2011). Novel nanofibrous scaffolds for water filtration with bacteria and virus removal capability. *J. Electron Microsc.*, **60**(3), 201–209.

Shabani I., Haddadi-Asl V., Seyedjafari E., Babaeijandaghi F. and Soleimani M. (2009). Improved infiltration of stem cells on electrospun nanofibers. *Biochem. Biophys. Res. Commun.*, **382**(1), 129–133.

Sheikh F. A., Kanjwal M. A., Saran S., Chung W. J. and Kim H. (2011). Polyurethane nanofibers containing copper nanoparticles as future materials. *Appl. Surf. Sci.*, **257** (7), 3020–3026.

Silver S. (2003). Bacterial silver resistance: molecular biology and uses and misuses of silver compounds. *FEMS Microbiol. Rev.*, **27**(2/3), 341–353.

Singh G., Rana D., Matsuura T., Ramakrishna S., Narbaitz R. M. and Tabe S. (2010). Removal of disinfection byproducts from water by carbonized electrospun nanofibrous membranes. *Sep. Purif. Technol.*, **74**(2), 202–212.

Sun H. and Onebby S. (2006). Facile polyester surface functionalization via hydrolysis and cell-recognizing peptide attachment. *Polym. Int.*, **55**(11), 1336–1340.

Sun S., Yue Y., Huang X. and Meng D. (2003). Protein adsorption on blood-contact membranes. *J. Membr. Sci.*, **222**(1/2), 3–18.

Tan K. and Obendorf S. K. (2007). Fabrication and evaluation of electrospun nanofibrous antimicrobial nylon 6 membranes. *J. Membr. Sci.*, **305**(1/2), 287–298.

Tang Z., Wei J., Yung L., Ji B., Ma H., Qiu C., Yoon K., Wan F., Fang D., Hsiao B. S. and Chu B. (2009a). UV-cured poly(vinyl alcohol) ultrafiltration nanofibrous membrane based on electrospun nanofiber scaffolds. *J. Membr. Sci.*, **328**(1/2), 1–5.

Tang Z. H., Qiu C. Q., McCutcheon J. R., Yoon K., Ma H. Y., Fang D. F., Lee E., Kopp C., Hsiao B. S. and Chu B. (2009b). Design and fabrication of electrospun polyethersulfone nanofibrous scaffold for high flux nanofiltration membranes. *J. Polym. Sci. Pol. Phys.*, **47**(22), 2288–2300.

Taylor G. I. (1969). Disintegration of water drops in an electric field. *Proc. Roy. Soc. Lond. B. Bio.*, **280**(1382), 383–453.

Teo W. E. and Ramakrishna S. (2009). Electrospun nanofibers as a platform for multifunctional, hierarchically organized nanocomposite. *Compos. Sci. Technol.*, **69**(11/12), 1804–1817.

Theron S. A., Yarin A. L., Zussman E. and Kroll E. (2005). Multiple jets in electrospinning: experiment and modeling. *Polymer*, **46**(9), 2889–2899.

Voigt W. V. (2009). Water Stable, Antimicrobial Active Nanofibres Generated by Electrospinning from Aqueous Spinning Solutions. PhD thesis, RWTH Aachen University, Munich, Germany.

Wang X., Zhang K., Yang Y., Wang L., Zhou Z., Zhu M., Hsiao B. and Chu B. (2010). Development of hydrophilic barrier layer on nanofibrous substrate as composite membrane via a facile route. *J. Membr. Sci.*, **356**(1/2), 110–116.

Xiao S. L., Ma H., Shen M. W., Wang S. Y., Huang Q. G. and Shi X. Y. (2011). Excellent copper(II) removal using zero-valent nanoparticle-immobilized hybrid electrospun polymer nanofibrous mats. *Colloid, Surf. A*, **381**(1/3), 48–54.

Yao C., Li X., Neoh K. G., Shi Z. and Kang E. T. (2008). Surface modification and antibacterial activity of electrospun polyurethane fibrous membranes with quaternary moieties. *J. Membr. Sci.*, **320**(1/2), 259–267.

Yao C., Li X., Neoh K. G., Shi Z. and Kang E. T. (2009). Antimicrobial activities of surface modified electrospun poly(vinylidene fluoride-co-hexafluoropropylene) (PVDF-HFP) fibrous membranes. *Appl. Surf. Sci.*, **255**(6), 3854–3858.

Yoo H. S., Kim T. G. and Park T. G. (2009). Surface-functionalized electrospun nanofibers for tissue engineering and drug delivery. *Adv. Drug Deliv. Rev.*, **61**(12), 1033–1042.

Yoon K., Hsiao B. S. and Chu B. (2008). Functional nanofibers for environmental applications. *J. Mater. Chem.*, **18**(44), 5326–5334.

Yoon K., Hsiao B. S. and Chu B. (2009a). Formation of functional polyethersulfone electrospun membrane for water purification by mixed solvent and oxidation processes. *Polymer*, **50**(13), 2893–2899.

Yoon K., Hsiao B. S. and Chu B. (2009b). High flux ultrafiltration membranes based on polyacrylonitrile electrospun scaffolds and crosslinked polyvinyl alcohol coating. *J. Membr. Sci.*, **338**(1/2), 145–152.

Yoon K., Kim K., Wang X., Fang D., Hsiao B. S. and Chu B. (2006). High flux ultrafiltration membranes based on electrospun PAN scaffolds and chitosan coating. *Polymer*, **47**(7), 2434–2441.

Zhang H., Nie H., Yu D., Wu C., Zhang Y., White C. J. B. and Zhu L. (2010). Surface modification of electrospun polyacrylonitrile nanofiber towards developing an affinity membrane for bromelain adsorption. *Desalination*, **256**(1/3), 141–147.

Zhang L. F., Luo J. E., Menkhaus T. J., Varadaraju H., Sun Y. Y. and Fong H. (2011). Antimicrobial nano-fibrous membranes developed from electrospun polyacrylonitrile nanofibers. *J. Membr. Sci.*, **369**(1/2), 499–505.

Zhou W. T., He J. X., Cui S. Z. and Gao W. D. (2011). Preparation of electrospun silk fibroin/cellulose acetate blend nanofibers and their applications to heavy metal ions adsorption. *Fibers Polym.*, **12**(4), 431–437.

Chapter 13

Porous ceramic and metallic microreactors

H. C. Aran and R. G. H. Lammertink

13.1 INTRODUCTION

Micro- and mesoreactors are devices with reduced characteristic dimensions (classically in sub-millimeter range) for carrying out chemical reactions. They occupy less space and allow much more controlled processes than conventional large-scale reactors. Because of their small characteristic length (hydraulic diameter), the flow regime inside these reactors is typically laminar, which makes the hydrodynamic characteristics well-defined and controllable (Aran *et al.*, 2011a; Aran *et al.*, 2011b; Aran *et al.*, 2011c; Gavriilidis *et al.*, 2002; Jähnisch *et al.*, 2004; Kiwi-Minsker & Renken, 2005).

Using these reactors, accurate experimental information can be gathered very quickly within a small volume. Due to their small volume, reactant costs and waste streams are reduced and safe operation can be performed in these reactors. These properties provide sustainability, energy efficiency, high productivity rates and operational safety for a given process (Gavriilidis *et al.*, 2002; Hessel *et al.*, 2005a; Hessel *et al.*, 2005b; Hessel *et al.*, 2005c; Jähnisch *et al.*, 2004; Kashid & Kiwi-Minsker, 2009; Kiwi-Minsker & Renken, 2005).

The channel surface to volume ratio (typically 10000–50000 m^2/m^3) is much higher in these reactors compared to conventional chemical reactors (100–1000 m^2/m^3). This large ratio provides enhanced heat and mass transfer properties, which make these reactors very suitable for exploring and performing fast and exothermic reactions. Micro- and mesofluidic devices open new routes in chemical process technology, which previously could not be applied using conventional reactors (Gavriilidis *et al.*, 2002; Hessel *et al.*, 2005a; Hessel *et al.*, 2005b; Hessel *et al.*, 2005c; Jähnisch *et al.*, 2004; Kashid & Kiwi-Minsker, 2009; Kiwi-Minsker & Renken, 2005).

The work in our research group is focused on multiphase processes, such as heterogeneously catalyzed gas-liquid-solid (G-L-S) reactions, within micro- and mesoreactors. Membrane functionality was integrated within microreactors and applied in multiphase reactions (Aran *et al.*, 2011a; Aran *et al.*, 2011b; Aran *et al.*, 2011c).

In this reactor concept, the contact between gas and liquid for reaction purposes is enabled by membrane technology and selective wetting of porous ceramic and metallic membranes (see Figures 13.1, 13.7 and 13.13). The gas (G) and the liquid (L) phases are supplied from opposite sides of the membrane and the contact of the phases takes place precisely on the membrane surface, where also the solid (S) catalyst is immobilized. For the porous ceramic reactors, the G-L interface is stabilized by modifying the wetting properties of the reactor wall. For porous metallic reactors, a gas-permeable polymeric dense layer is integrated on the outer surface to stabilize the interface on the membrane surface.

These reactors are utilized for reduction and oxidation reactions in water, which are of great importance for wastewater treatment: catalytic hydrogenation of nitrite ions and photocatalytic degradation of phenol. Section 13.2 (based on Aran *et al.*, 2011a) and 13.3 (based on Aran *et al.*, 2011b) focus on the first reduction reaction by the use of tubular porous ceramic and metallic reactors. Section 13.4 (based on Aran *et al.*, 2011c) describes the latter oxidation reaction carried out by planar porous ceramic reactors.

13.2 POROUS CERAMIC MESOREACTORS

13.2.1 Reactor concept

In this section, the development and optimization of *porous ceramic mesoreactors* for gas-liquid-solid (G-L-S) reactions is described. This concept ensures that the gaseous (G) and liquid (L) reactants meet at the solid (S) catalyst surface with a simple contacting approach (Figure 13.1).

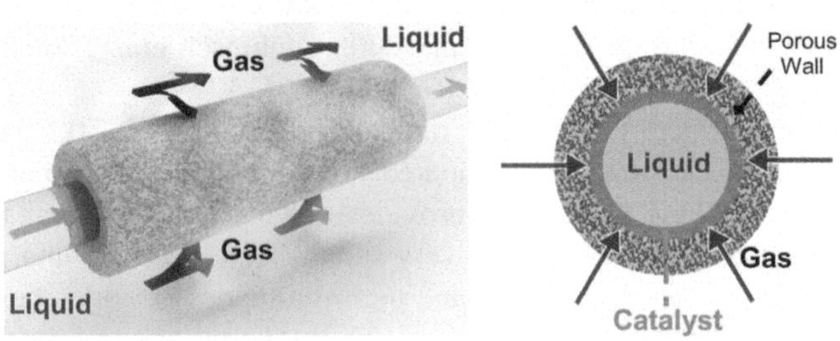

Figure 13.1 Contacting concept of porous ceramic mesoreactors for gas-liquid-solid reactions (Reprinted from Aran *et al.*, 2011a, Copyright (2011), with permission from Elsevier).

Using this continuous process concept, the G- and L-reactant can be independently controlled and no separation of reactants is needed at the reactor outlet. Furthermore, the G-phase composition can be kept constant along the full length of the reactor. In our studies, the G-L-S interface and the positioning of the reaction area are stabilized using surface modification (hydrophobization) techniques, as opposed to controlling trans-membrane pressure in conventional G-L-S membrane reactors.

The developed reactors are applied to heterogeneously catalyzed hydrogenation of harmful nitrite (NO_2^-) ions in water over a palladium (Pd) catalyst.

$$2NO_2^- + 3H_2 \xrightarrow{Pd} N_2 + 2OH^- + 2H_2O \quad \text{(Eq. 13.1)}$$

$$2NO_2^- + 6H_2 \xrightarrow{Pd} 2NH_4^+ + 4OH^- \quad \text{(Eq. 13.2)}$$

The conversion of the nitrite ions (NO_2^-) was the main performance criterion. In addition, production of the reaction side-product ammonia (NH_4^+) was measured, in order to confirm the NO_2^- reduction and to calculate the selectivity of the reaction.

The removal of nitrite (NO_2^-) and nitrate (NO_3^-) ions, commonly found in groundwater, is a relevant reaction from an environmental point of view. It can be carried out via biologic, physicochemical and catalytic processes. Because of the slow nature of the biologic processes and the formation of concentrated secondary streams by physicochemical methods, the catalytic hydrogenation process is referred to be the most promising for the nitrite removal (Aran *et al.*, 2011a; Chinthaginjala *et al.*, 2010; Chinthaginjala & Lefferts, 2010; Ebbesen *et al.*, 2008; Pintar *et al.*, 1998; Vorlop & Tacke, 1989).

13.2.2 Reactor preparation

Commercial porous hollow fibers (Alumina-α-Al_2O_3) with an inner diameter of 2.8 mm (HyFlux CEPAration) and a length of 13.5 cm were used as membrane support. The reactor preparation consists of three main stages as illustrated in Figure 13.2.

Below, a brief summary of the preparation steps are given, the details of the preparation steps can be found elsewhere (Aran *et al.*, 2011a):

First, the inner surface of the commercial α-Al_2O_3 membrane was coated with a γ-Al_2O_3 layer as catalyst support by a wash-coating method in order to increase the active surface area (thickness ≈ 80 μm, BET area ≈ 73 m^2/g). Subsequently, the palladium (Pd) catalyst was deposited on the reactor via a dry impregnation method using a Pd-precursor solution, followed by a calcination and reduction step. The resulting reactor had a Pd content of 0.073 wt%. The cross-section images of prepared reactors are displayed in Figure 13.3.

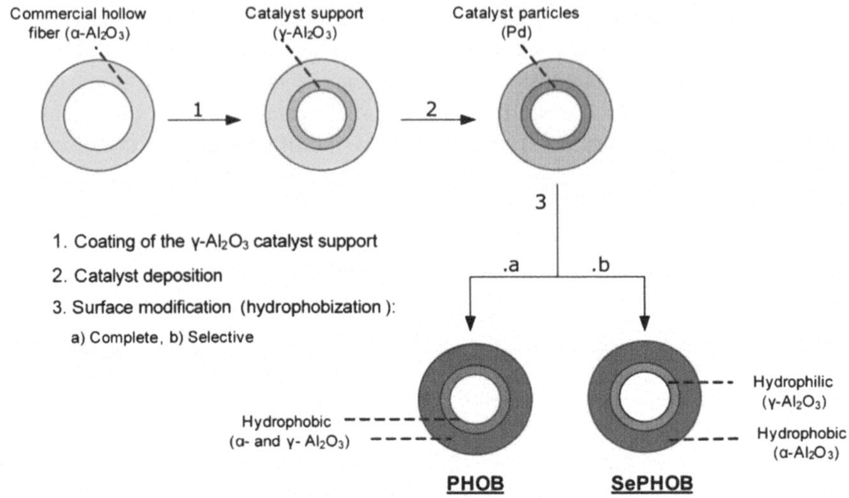

Figure 13.2 Preparation steps of the porous ceramic mesoreactors: drawings represent the cross-sections of the samples after each step (Reprinted from Aran *et al.*, 2011a, Copyright (2011), with permission from Elsevier).

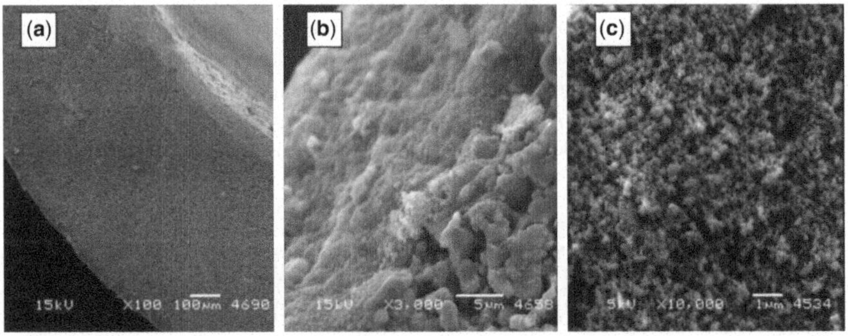

Figure 13.3 Cross-section images of porous ceramic mesoreactors: (a), (b) Intersection of α-Al_2O_3/γ-Al_2O_3 layers (c) Porous structure of the γ-Al_2O_3 catalyst support layer (Reprinted from Aran *et al.*, 2011a, Copyright (2011), with permission from Elsevier).

In the final stage of the reactor preparation, two different surface modification (hydrophobization) strategies were implemented (see Figure 13.2). This step was carried out in order to allow liquid flow inside the intrinsically hydrophilic porous reactor channel and to obtain a stabilized gas-liquid-solid interface. The reactor walls were hydrophobized by coating the surface with a fluorinated alkyl trichlorosilane (FOTS).

Two different routes were carried out for the surface modification (Figure 13.2): (1) Complete hydrophobization or (2) Selective hydrophobization. With the complete hydrophobization method, the whole sample was hydrophobized, both the membrane support α-Al_2O_3 layer and the catalyst support γ-Al_2O_3 layer. This reactor was labelled as PHOB. With the selective hydrophobization method, only the α-Al_2O_3 layer was hydrophobized, while the γ-Al_2O_3 layer remained hydrophilic. This reactor was labelled as SEPHOB.

The wetting behavior of the PHOB and SEPHOB samples was visualized by flowing an aqueous Phenol Red solution (as wetting indicator) through the fiber. In the PHOB (completely hydrophobized) sample, almost no coloring in the cross sections was observed along the length of the fiber, indicating that the entire sample was successfully hydrophobized. For the SEPHOB (selectively hydrophobized) samples, wetting was observed only in the γ-Al_2O_3 layer, showing that only the α-Al_2O_3 layer was selectively hydrophobized, as illustrated in Figure 13.4. In this case, the γ-Al_2O_3 catalyst support, where most of the Pd catalyst is located, is entirely accessible for the liquid reactant.

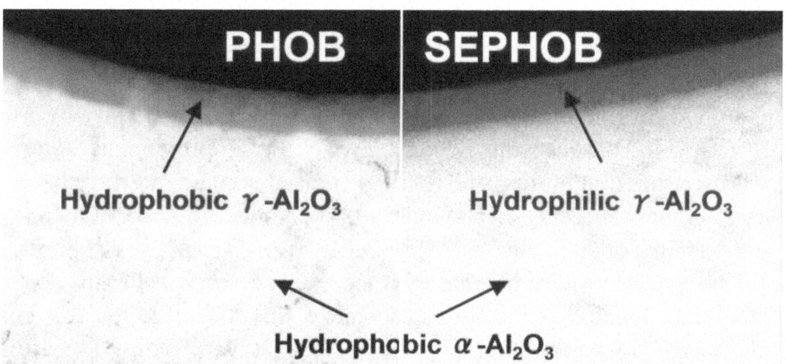

Figure 13.4 Wetting behavior for the completely (PHOB) and selectively hydrophobized (SEPHOB) reactors (adapted from Aran *et al.*, 2011a).

13.2.3 Reactor operation

The prepared reactors were utilized for the heterogeneously Pd catalyzed hydrogenation of nitrite (NO_2^-) ions in water. The main performance criterion was the nitrite conversion, which was determined by measuring the initial and final concentrations of the NO_2^- ions before and after the reactor.

Firstly, in order to determine the influence of the surface properties and the wetting behavior on the reactor performance, both the PHOB and SEPHOB reactors (Pd loading = 0.073 wt%) were tested. Figure 13.5 shows the overall conversions of nitrite ions for two liquid flow rates for both reactors.

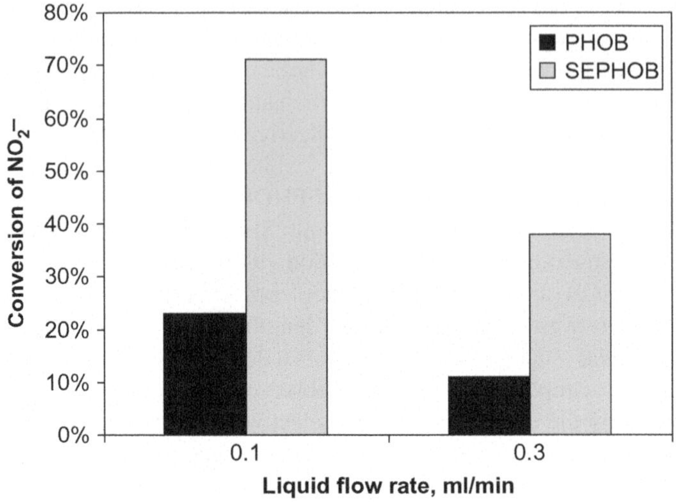

Figure 13.5 Effect of the wetting behavior on the reactor performance: NO_2^- conversions for PHOB and SEPHOB reactors (Initial nitrite concentration = 11 mg/l, H_2 concentration = 100%, Pd loading = 0.073 wt%).

Figure 13.5 shows that the NO_2^- conversion increases drastically by using the SEPHOB reactor. These results clearly prove the improvement achieved by altering wall wetting conditions, due to the increased contact interface between the liquid phase (containing NO_2^- ions) and the Pd catalyst. In the PHOB reactor the liquid phase is in contact with a hydrophobic γ-Al_2O_3 catalyst support, which reduces the NO_2^- solution in contact with the catalyst surface. On the other hand, in the SEPHOB reactor the liquid reactant is in contact with the complete hydrophilic γ-Al_2O_3 layer and the immobilized Pd catalyst leading to significantly higher conversion efficiencies.

In addition, the effect of the H_2 concentration in the gas phase of the reactor was investigated for both reactors. For these experiments, argon (Ar) was used as diluting agent and the volumetric concentration of the gaseous reactant H_2 in the gas flow was varied between 1% and 100% (H_2 partial pressures), while keeping the total gas flow rate constant (100 ml/min).

Figure 13.6 shows that the reactor performance remained constant even at dilute concentrations of H_2, for all the samples and flow conditions. Only below 5% H_2, a slight drop in the conversion efficiency was observed. Furthermore, there was no NO_2^- conversion observed when H_2 was absent in the gas phase, as expected.

These results prove that even at low values of the H_2 concentration, H_2 could sufficiently reach the catalytic reaction area (γ-Al_2O_3 layer) through the non-wetted pores of the hydrophobic α-Al_2O_3 membrane support. Even at low concentrations, enough H_2 is provided to maintain the reaction with dissolved nitrite (~0.24 mmol/l). This concept ensures negligible mass transport limitations for the gas reactant, enabling to work at dilute H_2 concentrations, which is of great advantage for operational safety.

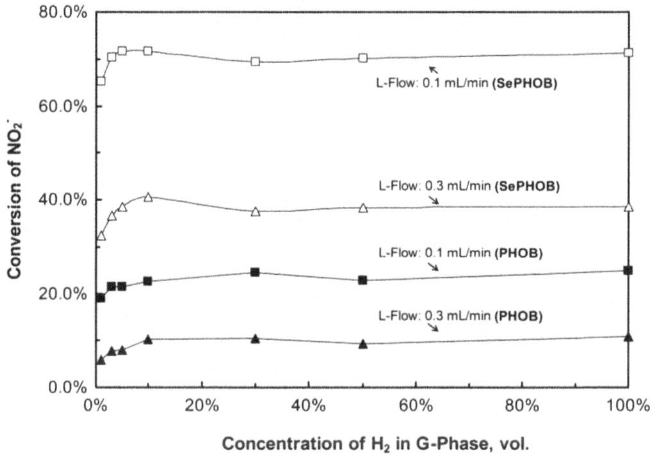

Figure 13.6 Effect of the H_2 concentration on the reactor performance: NO_2^- conversions for PHOB and SEPHOB reactors (Initial nitrite concentration = 11 mg/l, Pd loading = 0.073 wt%) (Reprinted from Aran *et al.*, 2011a, Copyright (2011), with permission from Elsevier).

These results prove that porous ceramic mesoreactors, with controllable wetting properties and catalytic activity, are very suitable for the catalytic hydrogenation of harmful nitrite ions in water. With this simple reactor concept, multiphase gas-liquid-solid reactions can be carried out even at low concentrations of the gaseous reactant.

13.3 POROUS METALLIC MICROREACTORS

13.3.1 Reactor concept

In this section, the preparation and operation of porous metallic microreactors is desctibed. The presented reactors consist of porous stainless steel tubes onto which carbon nanofibers (CNFs) are grown as catalyst support, all encapsulated by a gas permeable polymeric coating.

CNFs are very suitable materials as catalyst support thanks to their high surface area (>100 m^2/g) and large pore volumes (>0.5 cm^3/g). They exhibit low tortuosity and therefore efficient internal mass transfer. Furthermore, CNFs are mechanically strong and chemically inert (Chinthaginjala & Lefferts, 2009, 2010;

Chinthaginjala *et al.*, 2007; Chinthaginjala *et al.*, 2008; Jarrah *et al.*, 2006; Pacheco Benito & Lefferts, 2010; Rodriguez *et al.*, 1994).

In this continuous tubular reactor concept (Figure 13.7), the liquid phase flows at the tube side and the gas phase at the shell side of the hollow fiber. The contact between the liquid and gas phases takes place on the porous hollow fiber (in microscale) where the CNF catalyst support is located. A dense gas permeable polymeric layer is coated on the outer surface of the hollow fiber in order to stabilize the gas-liquid-solid (G-L-S) interface. This reactor concept was applied for the catalytic reduction of nitrite ions in water as in Section 13.2 (see Equations 13.1 and 13.2).

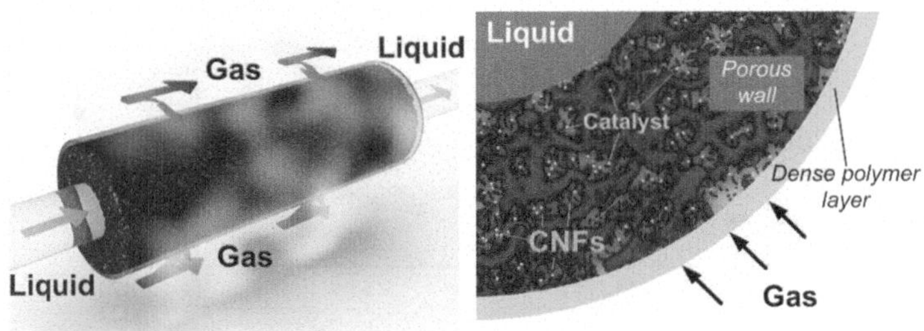

Figure 13.7 Gas-liquid contacting in porous metallic membrane microreactors for multiphase reactions: Porous stainless steel membrane decorated with carbon nanofibers (CNFs) as catalyst support. Reactor encapsulated with a selective, gas permeable PDMS layer. (Reprinted from Aran *et al.*, 2011b, Copyright (2011), with permission from Elsevier).

13.3.2 Reactor preparation

In order to demonstrate the influence of the reactor material in the reaction, three different types of samples were prepared: SS (Porous SS hollow fibers), SS + CNF (Porous SS hollow fibers with carbon nanofibers (CNF)), SS + CNF + Pd (Porous SS hollow fibers with CNF and palladium (Pd)).

The preparation steps of the three samples are displayed in Figure 13.8. A brief summary of the reactor preparation steps is given below. The details of the preparation steps can be found elsewhere (Aran *et al.*, 2011b).

Fabrication of the porous stainless steel hollow fibers was performed via spinning of a stainless steel containing suspension, followed by coagulation, drying and sintering steps (modified from (Luiten-Olieman *et al.*, 2011)). Growth of CNFs was carried out by catalytic chemical vapor deposition (CCVD) using ethylene as hydrocarbon source (modified from (Pacheco Benito & Lefferts, 2010)). By the dry impregnation method described in Section 13.2, the palladium (Pd) catalyst was deposited on the reactor. The prepared reactors are displayed in Figures 13.9 and 13.10.

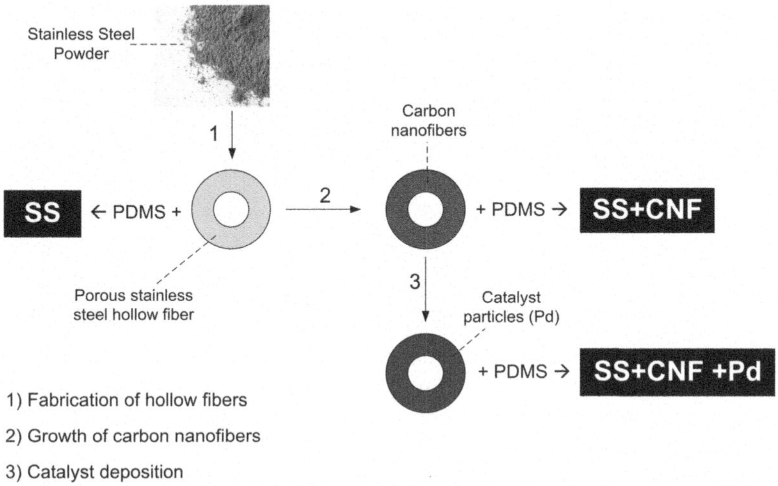

Figure 13.8 Preparation steps of porous metallic membrane microreactors: four samples with various properties (SS, SS + CNF and SS + CNF + Pd) represented in cross-sectional view (adapted from Aran *et al.*, 2011b).

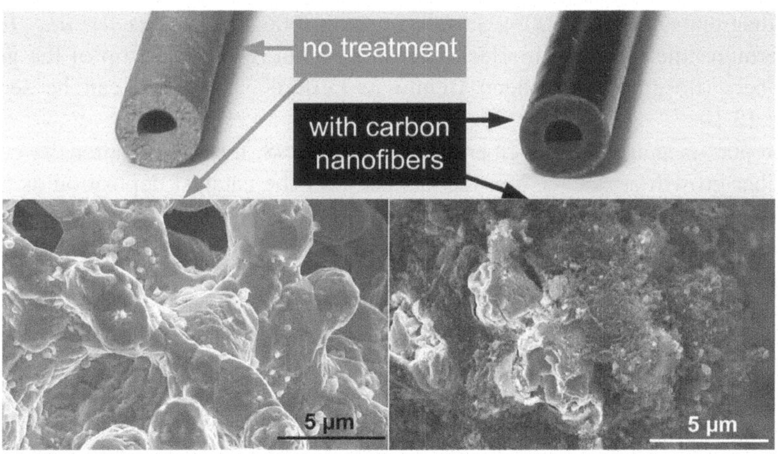

Figure 13.9 Porous stainless steel hollow fibers before and after the growth of the CNFs. (Reprinted from Aran *et al.*, 2011b, Copyright (2011), with permission from Elsevier).

The CNFs were grown on the porous stainless steel hollow fiber surface along the full length. The CNF growth on the surface can be observed both visually and by SEM (Scanning Electron Microscopy) analysis (Figure 13.9). The difference in morphology between samples with and without CNFs can clearly be noticed.

Figure 13.10 CNF growth on the porous stainless steel surface: (a) Metal-nano-particles (iron, nickel) on the tip of the CNFs, (b) Pd particles immobilized on CNFs (Reprinted from Aran *et al.*, 2011b, Copyright (2011), with permission from Elsevier).

With the growth of the CNFs, the weight of the sample increased ~16%. The BET surface area of the whole sample increased from ~0.4 m^2/g (SS) to ~21.3 m^2/g (SS + CNF). This corresponds to a carbon surface area of ~151 m^2/g (calculated), in similar range as previously reported values for CNFs (Chinthaginjala *et al.*, 2008; Jarrah *et al.*, 2006; Rodriguez *et al.*, 1994). Furthermore, metal-nano-particles (iron (Fe), nickel (Ni)) on the tip of the grown nanofibers (tip-growth (Pacheco Benito & Lefferts, 2010)), as can be seen in Figure 13.10.

The porous stainless steel fiber was labeled as SS, the sample after the carbon nanofiber growth as SS + CNF, the sample after the catalyst deposition as SS + CNF + Pd. All the samples were coated with PDMS prior to operation and tested for the catalytic reduction of nitrite ions in water. The coating procedure was modified from previous work (Dutczak *et al.*, 2011).

13.3.3 Reactor operation

The main performance criterion in this model reaction was the NO_2^- conversion. The inlet and outlet concentration of NO_2^- was measured to calculate the conversion efficiencies. In addition, the reaction side product NH_4^+ was determined to confirm the NO_2^- reduction process.

The reactors were tested under two different conditions: under the supply of hydrogen (H_2) as gaseous reactant or argon (Ar) as inert gas to the shell side of the membrane reactor. Figure 13.11 gives an overview of the NO_2^- conversion results of the porous metallic microreactors.

The conversion results show that the reactor SS + CNF + Pd displays the highest nitrite reduction values. On the other hand, the SS reactor shows no considerable NO_2^- reduction activity. An extraordinary result is that in the SS + CNF reactor also NO_2^- ions were reduced, even though no Pd catalyst was deposited in the reactor. With the SS + CNF and SS + CNF + Pd reactors, the reaction side-product ammonia (NH_4^+) was formed.

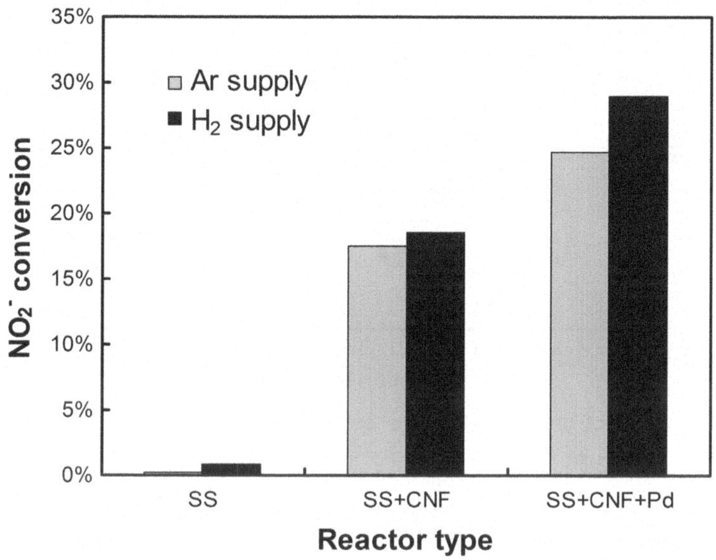

Figure 13.11 NO_2^- reduction performance (liquid flow rate = 0.1 ml/min, initial NO_2^- concentration = 50 mg/l) of the SS, SS + CNF, SS + CNF + Pd microreactors (Reprinted from Aran *et al.*, 2011b, Copyright (2011), with permission from Elsevier).

Another surprising result is that the reaction even proceeds without the supply of the H_2 to the reactor (Figure 13.11 – Ar supply). Under the supply of the inert Ar gas, both SS + CNF and SS + CNF + Pd reactors reduced significant amounts of NO_2^- and NH_4^+ was formed. For the SS + CNF reactor the NO_2^- conversion and NH_4^+ formation were comparable as for the experiments with H_2 and Ar supply, indicating no contribution of the gaseous reactant.

The only considerable contribution of the reactant H_2 to the reaction is observed for the reactor SS + CNF + Pd due to the presence of Pd as hydrogenation catalyst (Figure 13.12). The concentration changes of NO_2^- and NH_4^+ at the reactor outlet with H_2 and Ar supply during the reactor operation are displayed in Figure 13.12. At the first stage of the experiment (1), under inert Ar flow, the reaction proceeds with a NO_2^- conversion efficiency of ~25%. When the gas phase flow is switched to H_2 (2), the NO_2^- concentration decreases and shows a higher NO_2^- conversion of ~30%. At the same time, the NH_4^+ concentration increases. At the third stage of the experiment (3), under inert Ar flow, the outlet NO_2^- and NH_4^+ concentrations reach the same values as the first stage of the experiment.

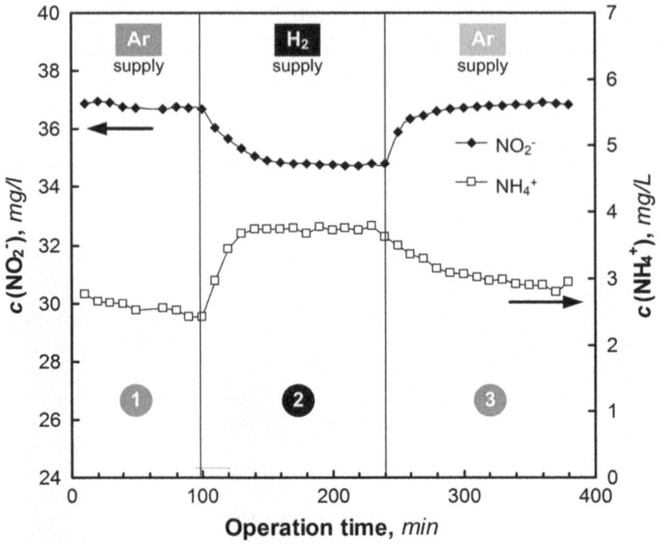

Figure 13.12 Effect of the gas phase composition on the performance of SS + CNF + Pd reactor: $c(NO_2^-)$, $c(NH_4^+)$ = Concentrations of NO_2^- (reactant) and NH_4^+ (product) at the reactor outlet. (Liquid flow rate = 0.1 ml/min, initial NO_2^- concentration = 50 mg/l). (Reprinted from Aran *et al.*, 2011b, Copyright (2011), with permission from Elsevier).

Interestingly, the NO_2^- reduction process proceeds in the absence of external hydrogen supply and Pd catalyst, as shown in Figures 13.11 and 13.12. This extraordinary performance of our reactors can be explained by the reactive properties of the reactor material.

The reductive properties of metal particles on the tip of the CNFs with high surface area may be responsible for the NO_2^- reduction process. Typically, CNFs grow catalytically on metal surfaces, such as iron (Fe) and nickel (Ni) (Baker *et al.*, 1972; Chinthaginjala & Lefferts, 2010; Chinthaginjala *et al.*, 2007; Chinthaginjala *et al.*, 2008). The growth catalysts in this study are presumably Fe and Ni, since the porous stainless steel hollow fibers were based on 316 L powder (~67% Fe, ~13% Ni; Epson Atmix Corporation). The CNFs growth takes place by deposition of carbon under these metal growth-catalysts (Fe, Ni) and these metal particles are squeezed out of the surface and stay at the tip of the fibers (tip growth) (Baker *et al.*, 1972; Pacheco Benito & Lefferts, 2010). Therefore, the CNFs are not only acting as catalyst support for Pd, but they also contain other metal-nano-particles (Fe, Ni) on their surfaces, as can be seen in Figure 13.10a.

Previous literature studies show that metallic iron (Fe^0) nanoparticles can act as a reductant for nitrite (NO_2^-) and nitrate (NO_3^-) ions without the need of an additional supply of H_2. Likewise, iron in its oxidized state (Fe^{2+}) was reported to be a reducing agent (Cheng *et al.*, 1997; Hu *et al.*, 2001; Huang & Zhang,

2006; Liang *et al.*, 2008; Shukla *et al.*, 2009). Liang *et al.* (2008) proposed the following reduction mechanism for the nitrite ions in water by iron:

$$3Fe^0 + NO_2^- + 8H^+ \longrightarrow 3Fe^{2+} + NH_4^+ + 2H_2O \quad \text{(Eq. 13.3)}$$

$$3Fe^0 + 2NO_2^- + 8H^+ \longrightarrow 3Fe^{2+} + N_2 + 4H_2O \quad \text{(Eq. 13.4)}$$

$$3Fe^0 + NO_2^- + 6H_2O \longrightarrow 3Fe(OH)_2 + NH_4^+ + 2OH^- \quad \text{(Eq. 13.5)}$$

The results showed higher NO_2^- conversions for the SS + CNF microreactor than that of our SS microreactor (Figure 13.11). This proves the importance of the CNFs for the reduction process. The CNFs do not only increase the surface area as a catalyst support, but makes the Fe particles more available on the surface.

Porous metallic microreactors decorated with carbon nanofibers show a great potential for the reduction of NO_2^- ions in water. The NO_2^- reduction proceeds by both catalytic reduction (with Pd and H_2) and reduction by the reactor material itself (by Fe on the CNFs). The latter effect will eventually exhaust in time due to corrosion of the reactor material, however, the reaction will still occur with the deposited Pd-catalyst and the membrane-assisted supply of H_2.

13.4 POROUS CERAMIC PHOTOCATALYTIC MICROREACTOR

13.4.1 Reactor concept

This section describes a porous ceramic (alumina, α-Al_2O_3) membrane microreactor concept for multiphase gas-liquid-solid (G-L-S) photochemical reactions. Microreactors are attractive devices for photochemistry and photocatalytic processes. The miniaturized dimensions and high surface to volume ratio does not only reduce mass transfer limitations, but also provides a reduced path for the light, thereby decreasing photon transfer limitations and leading to high illumination homogeneity along the reactor (Coyle & Oelgemoller, 2008; Gorges *et al.*, 2004; Matsushita *et al.*, 2008; Van Gerven *et al.*, 2007).

In the presented reactor (Figure 13.13), the liquid flows inside microchannels (fabricated on flat porous α-Al_2O_3 substrate), where the photocatalyst (Titanium dioxide: TiO_2) is immobilized (Figure 13.13). The gas permeates through the porous membrane wall and reaches the liquid that is flowing in the microchannel. In order to stimulate the photocatalytic reaction, UV-light is introduced from the top of the channel. As a model reaction, photocatalytic degradation of phenol on a TiO_2 catalyst and the influence of oxygen (O_2) in this process was studied.

Photocatalytic degradation of resistant organic pollutants in aqueous systems by oxidation in organic systems is an emerging technology. Hence, degradation of phenol in photocatalytic systems has been studied intensively (Lin *et al.*, 2011; Pardeshi & Patil, 2008; Roig *et al.*, 2003; Subramanian & Kannan, 2008). It is known, that in the presence of oxygen as an electron (e^-) acceptor enhances the conversion rate. This makes porous membrane reactor designs very suitable for this process (Lindstrom *et al.*, 2007; Matsushita *et al.*, 2007; Matsushita *et al.*, 2009; Subramanian & Kannan, 2008).

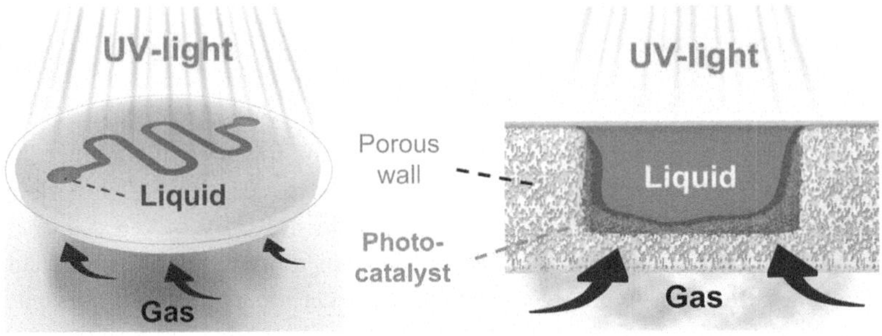

Figure 13.13 Contacting concept of Porous Ceramic Photocatalytic Microreactor (Reprinted from Aran *et al.*, 2011c, Copyright (2011), with permission from Elsevier).

13.4.2 Reactor preparation

The preparation of the Porous Ceramic Photocatalytic Microreactor consists of 4 main steps (Figure 13.14). The reactor preparation method is briefly summarized below. The details of the preparation steps can be found elsewhere (Aran *et al.*, 2011c).

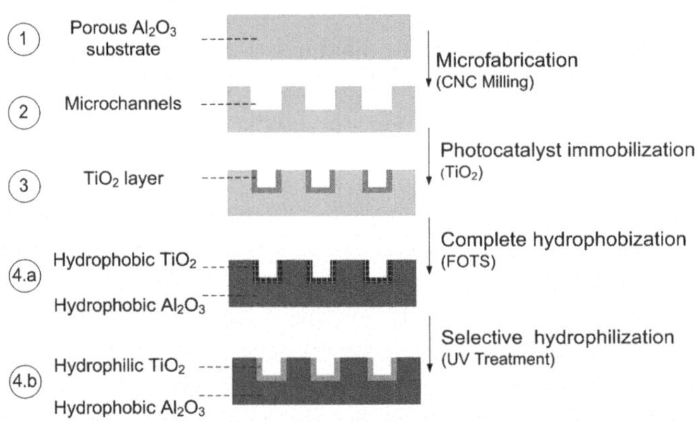

Figure 13.14 Summary of reactor preparation steps: The illustrations represent the cross-section after each preparation step. (Reprinted from Aran *et al.*, 2011c, Copyright (2011), with permission from Elsevier).

Porous α-Al_2O_3 substrates were prepared by using a (water-based) suspension, colloidal filtration technique, followed by a sintering step. The resulting circular flat substrates had a diameter of 39 mm, thickness of 2 mm, average pore diameter of 80 nm and a porosity of 35%. Next, the microchannel (depth = 500 μm, width = 1 mm, length = 65 mm, volume = 32.5 μl) on the porous substrate was fabricated using an automated CNC-mill. For the photocatalyst immobilization, a TiO_2 suspension was applied in the microchannels, followed by drying and a calcination step. The resulting TiO_2 layer had a BET surface area of 45 m^2/g.

In the final stage of the reactor preparation, the surface modification steps (hydrophobization and hydrophilization) were carried out in order to obtain a stabilized gas-liquid-solid (G-L-S) interface in the intrinsically hydrophilic Al_2O_3 and TiO_2 surfaces. The surface modification was performed in two steps: complete hydrophobization, followed by selective hydrophilization.

For complete hydrophobization, the sample was coated with FOTS by a liquid phase immobilization, so that the complete sample (both Al_2O_3 and TiO_2 surfaces) became hydrophobic (water contact angle $\approx 150°$). In the next step, the selective hydrophilization of the TiO_2 photocatalyst layer was performed. The completely hydrophobized sample was exposed to UV-light for 1 h, resulting in a hydrophilic TiO_2 layer (water contact angle $\approx 0°$), while Al_2O_3 remained hydrophobic.

Figure 13.15 demonstrates the resulting reactor after the preparation steps.

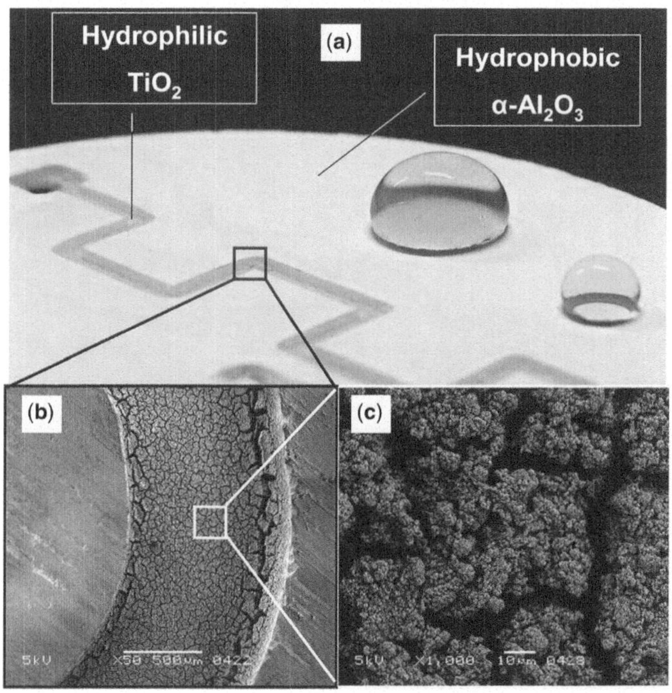

Figure 13.15 Porous ceramic photocatalytic microreactor with TiO_2 photocatalyst and modified surface properties. (a) Visualization of the surface properties after the selective hydrophilization step, (b) and (c) Ceramic membrane after photocatalyst immobilization: immobilized TiO_2 in the porous Al_2O_3 microchannel. (Reprinted from Aran *et al.*, 2011c, Copyright (2011), with permission from Elsevier).

13.4.3 Reactor operation

The prepared reactor was applied for the photocatalytic degradation of phenol, an important reaction for wastewater treatment. For the analysis of the phenol degradation, the UV-absorbance (at wavelengths of $250 < \lambda < 370$ nm) of the in- and outlet water were monitored during operation.

The aqueous phenol solution (1 mM) was flown through the reactor with a flow rate of 10 μl/min (residence time = 3.25 min). The channel was illuminated by UV-light from the top. The reactor was operated under various gas phase compositions. Nitrogen (N_2), air or oxygen (O_2) was supplied through the porous membrane to the reaction area.

The porous photocatalytic microreactor showed significant photocatalytic activity, as shown in Figure 13.16. The degradation of phenol increased drastically with the additional O_2 supply. As seen in Figure 13.16, under O_2 (air or pure O_2), the phenol absorbance peak ($\lambda = 270$ nm) is significantly lower, compared to the experiments with N_2 supply. A great advantage is that the degradation performance is very similar for pure O_2 compared to air (21% O_2), showing that O_2 is very efficiently transported through the membrane.

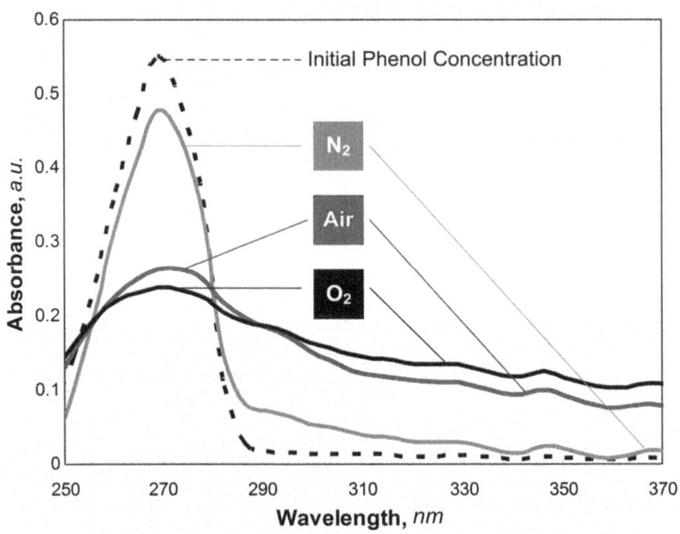

Figure 13.16 Effect of gas phase composition on the photocatalytic phenol degradation (initial phenol concentration = 1 mM, liquid flow rate = 10 μl/min) in the porous photocatalytic ceramic microreactor. UV-absorbance spectrum of the reactor in- and outlet streams is presented. (Reprinted from Aran *et al.*, 2011c, Copyright (2011), with permission from Elsevier).

Literature studies on photocatalytic phenol degradation showed that several aromatic intermediates are formed (e.g. hydroquinone, benzoquinone and catechol) in this process. The formation of such organic intermediates can also be observed in the reactor effluent, indicated by the increased UV-absorbance at $\lambda > 290$ nm (Figure 13.16).

The results of this study show that the phenol degradation is enhanced with additional membrane-assisted O_2 supply to the reaction zone. The porous photocatalytic ceramic microreactor is an attractive and promising reactor concept for photocatalytic degradation processes.

13.5 CONCLUSIONS

This chapter describes three different microreactor concepts utilizing porous ceramic and metallic membranes for multiphase gas-liquid-solid (G-L-S) reactions. Fabrication, surface modification (preparation) and application (operation) of porous ceramic (Alumina: Al_2O_3) and metallic (stainless steel) reactors in planar and in tubular geometry are described.

Porous ceramic and metallic microreactors were applied for aqueous reduction and oxidation reactions, which are very relevant for water and wastewater treatment. Catalytic hydrogenation of nitrite ions and photocatalytic oxidation of phenol in water were successfully carried out in these reactors.

With these reactors, the gaseous reactant (in this case H_2 and O_2) can be efficiently supplied to the reaction zone, gas and liquid phases can be independently controlled during the operation and they meet directly at the solid catalyst surface. Porous ceramic and metallic microreactors provide a stable multiphase interface with a simple reactor design.

ACKNOWLEDGEMENTS

The research described in this chapter is based on publications (Aran *et al.*, 2011a; Aran *et al.*, 2011b; Aran *et al.*, 2011c). It was a collaborative work of several groups in University of Twente: Soft matter, Fluidics and Interfaces (SFI), Catalytic Processes and Materials (CPM), Photocatalytic Syntheses (PCS), Inorganic Membranes (IM) and the Membrane Technology Group (MTG). The authors greatly acknowledge the efforts and scientific input of L. Lefferts, M. Wessling, G. Mul, N. E. Benes, J. K. Chinthaginjala, D. Salamon, S. Pacheco Benito, M. Luiten-Olieman, S. Er, T. Roelofs, R. Groote and T. Rijnaarts.

REFERENCES

Aran H. C., Chinthaginjala J. K., Groote R., Roelofs T., Lefferts L., Wessling M. and Lammertink R. G. H. (2011a). Porous ceramic mesoreactors: a new approach for gas-liquid contacting in multiphase microreaction technology. *Chemical Engineering Journal*, **169**(1–3), 239–246.

Aran H. C., Pacheco Benito S., Luiten-Olieman M. W. J., Er S., Wessling M., Lefferts L., Benes N. E. and Lammertink R. G. H. (2011b). Carbon nanofibers in catalytic membrane microreactors. *Journal of Membrane Science*, **381**(1–2), 244–250.

Aran H. C., Salamon D., Rijnaarts T., Mul G., Wessling M. and Lammertink R. G. H. (2011c). Porous photocatalytic membrane microreactor (P2M2): a new reactor concept for photochemistry. *Journal of Photochemistry and Photobiology A: Chemistry*, **225**(1), 36–41.

Baker R. T. K., Barber M. A., Harris P. S., Feates F. S. and Waite R. J. (1972). Nucleation and growth of carbon deposits from the nickel catalyzed decomposition of acetylene. *Journal of Catalysis*, **26**(1), 51–62.

Cheng I. F., Muftikian R., Fernando Q. and Korte N. (1997). Reduction of nitrate to ammonia by zero-valent iron. *Chemosphere*, **35**(11), 2689–2695.

Chinthaginjala J. K., Bitter J. H. and Lefferts L. (2010). Thin layer of carbon-nano-fibers (CNFs) as catalyst support for fast mass transfer in hydrogenation of nitrite. *Applied Catalysis A: General*, **383**(1–2), 24–32.

Chinthaginjala J. K. and Lefferts L. (2009). Influence of hydrogen on the formation of a thin layer of carbon nanofibers on Ni foam. *Carbon*, **47**(14), 3175–3183.

Chinthaginjala J. K. and Lefferts L. (2010). Support effect on selectivity of nitrite reduction in water. *Applied Catalysis B: Environmental*, **101**(1–2), 144–149.

Chinthaginjala J. K., Seshan K. and Lefferts L. (2007). Preparation and application of carbon-nanofiber based microstructured materials as catalyst supports. *Industrial & Engineering Chemistry Research*, **46**(12), 3968–3978.

Chinthaginjala J. K., Thakur D. B., Seshan K. and Lefferts L. (2008). How carbon-nano-fibers attach to Ni foam. *Carbon*, **46**(13), 1638–1647.

Coyle E. E. and Oelgemoller M. (2008). Micro-photochemistry: photochemistry in microstructured reactors. The new photochemistry of the future? *Photochemical & Photobiological Sciences*, **7**(11), 1313–1322.

Dutczak S. M., Luiten-Olieman M. W. J., Zwijnenberg H. J., Bolhuis-Versteeg L. A. M., Winnubst L., Hempenius M. A., Benes N. E., Wessling M. and Stamatialis D. (2011). Composite capillary membrane for solvent resistant nanofiltration. *Journal of Membrane Science*, **372**(1–2), 182–190.

Ebbesen S. D., Mojet B. L. and Lefferts L. (2008). *In situ* ATR-IR study of nitrite hydrogenation over Pd/Al_2O_3. *Journal of Catalysis*, **256**(1), 15–23.

Gavriilidis A., Angeli P., Cao E., Yeong K. K. and Wan Y. S. S. (2002). Technology and applications of microengineered reactors. *Chemical Engineering Research and Design*, **80**(1), 3–30.

Gorges R., Meyer S. and Kreisel G. (2004). Photocatalysis in microreactors. *Journal of Photochemistry and Photobiology A: Chemistry*, **167**(2–3), 95–99.

Hessel V., Angeli P., Gavriilidis A. and Löwe H. (2005a). Gas–liquid and Gas–liquid–solid microstructured reactors: contacting principles and applications. *Industrial & Engineering Chemistry Research*, **44**(25), 9750–9769.

Hessel V., Hardt S. and Löwe H. (2005b). *A Multi-Faceted, Hierarchic Analysis of Chemical Micro Process Technology: Sections 1.1–1.5*. Wiley–VCH Verlag GmbH & Co. KGaA, Weinheim.

Hessel V., Hardt S. and Löwe H. (2005c). *A Multi-Faceted, Hierarchic Analysis of Chemical Micro Process Technology: Sections 1.6–1.9*. Wiley–VCH Verlag GmbH & Co., KGaA, Weinheim.

Hu H.-Y., Goto N. and Fujie K. (2001). Effect of pH on the reduction of nitrite in water by metallic iron. *Water Research*, **35**(11), 2789–2793.

Huang Y. H. and Zhang T. C. (2006). Nitrite reduction and formation of corrosion coatings in zerovalent iron systems. *Chemosphere*, **64**(6), 937–943.

Jähnisch K., Hessel V., Löwe H. and Baerns M. (2004). Chemistry in microstructured reactors. *Angewandte Chemie International Edition*, **43**(4), 406–446.

Jarrah N. A., van Ommen J. G. and Lefferts L. (2006). Mechanistic aspects of the formation of carbon-nanofibers on the surface of Ni foam: a new microstructured catalyst support. *Journal of Catalysis*, **239**(2), 460–469.

Kashid M. N. and Kiwi-Minsker L. (2009). Microstructured reactors for multiphase reactions: state of the art. *Industrial & Engineering Chemistry Research*, **48**(14), 6465–6485.

Kiwi-Minsker L. and Renken A. (2005). Microstructured reactors for catalytic reactions. *Catalysis Today*, **110**(1–2), 2–14.

Liang F., Fan J., Guo Y., Fan M., Wang J. and Yang H. (2008). Reduction of nitrite by ultrasound-dispersed nanoscale zero-valent iron particles. *Industrial & Engineering Chemistry Research*, **47**(22), 8550–8554.

Lin Y.-T., Liang C. and Chen J.-H. (2011). Feasibility study of ultraviolet activated persulfate oxidation of phenol. *Chemosphere*, **82**(8), 1168–1172.

Lindstrom H., Wootton R. and Iles A. (2007). High surface area titania photocatalytic microfluidic reactors. *AIChE Journal*, **53**(3), 695–702.

Luiten-Olieman M. W. J., Winnubst L., Nijmeijer A., Wessling M. and Benes N. E. (2011). Porous stainless steel hollow fiber membranes via dry-wet spinning. *Journal of Membrane Science*, **370**(1–2), 124–130.

Matsushita Y., Iwasawa M., Ohba N., Kumada S., Suzuki T. and Ichimura T. (2007). Photocatalytic Oxidation and Alkylation Processes in Microreactors. Nano/Micro Engineered and Molecular Systems, 2007. NEMS '07. *2nd IEEE International Conference on*, pp. 851–854. doi: 10.1109/NEMS.2007.352152, IEEE, Washington DC.

Matsushita Y., Iwasawa M., Suzuki T. and Ichimura T. (2009). Multiphase photocatalytic oxidation in a microreactor. *Chemistry Letters*, **38**(8), 846–847.

Matsushita Y., Ohba N., Kumada S., Sakeda K., Suzuki T. and Ichimura T. (2008). Photocatalytic reactions in microreactors. *Chemical Engineering Journal*, **135** (Supplement 1), S303–S308.

Pacheco Benito S. and Lefferts L. (2010). The production of a homogeneous and well-attached layer of carbon nanofibers on metal foils. *Carbon*, **48**(10), 2862–2872.

Pardeshi S. K. and Patil A. B. (2008). A simple route for photocatalytic degradation of phenol in aqueous zinc oxide suspension using solar energy. *Solar Energy*, **82**(8), 700–705.

Pintar A., Berčič G. and Levec J. (1998). Catalytic liquid-phase nitrite reduction: kinetics and catalyst deactivation. *AIChE Journal*, **44**(10), 2280–2292.

Rodriguez N. M., Kim M.-S. and Baker R. T. K. (1994). Carbon nanofibers: a unique catalyst support medium. *The Journal of Physical Chemistry*, **98**(50), 13108–13111.

Roig B., Gonzalez C. and Thomas O. (2003). Monitoring of phenol photodegradation by ultraviolet spectroscopy. *Spectrochimica Acta Part A: Molecular and Biomolecular Spectroscopy*, **59**(2), 303–307.

Shukla A., Pande J., Bansiwal A., Osiceanu P. and Biniwale R. (2009). Catalytic hydrogenation of aqueous phase nitrate over Fe/C catalysts. *Catalysis Letters*, **131**(3), 451–457.

Subramanian M. and Kannan A. (2008). Effect of dissolved oxygen concentration and light intensity on photocatalytic degradation of phenol. *Korean Journal of Chemical Engineering*, **25**(6), 1300–1308.

Van Gerven T., Mul G., Moulijn J. and Stankiewicz A. (2007). A review of intensification of photocatalytic processes. *Chemical Engineering and Processing*, **46**(9), 781–789.

Vorlop K. D. and Tacke T. (1989). Erste Schritte auf dem Weg zur edelmetallkatalysierten Nitrat- und Nitrit-Entfernung aus Trinkwasser. *Chemie Ingenieur Technik*, **61**, 836–837.

Chapter 14

Biomimetic membranes for water separation applications

C. Hélix-Nielsen

14.1 INTRODUCTION

Nature provides an excellent palette of highly effective membranes capable of highly selective vectorial transport of a large number of molecular species. It is therefore striking that the membrane industry has developed synthetic separation membrane processes in a very different way (Rios *et al.*, 2007).

Traditional separation membranes are mostly dense polymeric films where advanced chemistry is used to control the surface properties of the films produced (Baker, 2004; Byrne, 2002). A wide range of polymers and production techniques are being used, resulting in a great diversity in structure and function of separation membranes tailored to a wide variety of applications. Separation is usually described in terms of pore/solute size, pore/solute charge and dielectric effects, coupled with diffusion or convective flow. Occasionally, more complex partitioning and transport mechanisms are used, however, most synthetic membranes may be broadly described as (supported) polymer sheets containing micro- to nanometer sized pores or interstitial cavities created by thermal motion of the polymer molecules.

This is in stark contrast to the bewildering complexity of biological membranes. Biological membranes are capable of performing intricate transport functions based on active and passive mechanisms. The complexity is reflected in the fact that 30% of the human genome codes for membrane proteins (Bernsel & Von Heijne, 2005) and typical mammalian cell membranes host several hundred lipid types (Dowhan, 1997). Despite dramatic progress over the last decades in our understanding of the molecular basis for biological membrane transport (Chung *et al.*, 2007; Hille, 2001) and protein-lipid interactions (Nielsen *et al.*, 1998; Nielsen, 2009b), this complexity remains a major obstacle in our molecular understanding of how living cells maintain their integrity and perform their function.

A bottom-up strategy leading to a better understanding of membrane design and membrane transport is to focus on a few components and features of biological membranes. By combining these components and features we may be able to exploit – or mimic – nature's tremendous capability for selective membrane transport. The term *biomimetic membranes,* is often used to describe such endeavors (Hélix-Nielsen, 2012; Nielsen, 2009a). Recent examples of bottom-up membrane biomimetics include low noise recording devices for ion channel research (Mayer *et al.*, 2003), free-standing lipid and triblock copolymer membranes (Gonzalez-Perez *et al.*, 2009; Hansen *et al.,* 2009a; Hansen *et al.*, 2009b; Nardin *et al.*, 2000; Rein *et al.*, 2011), enzyme-immobilization (Trau & Renneberg, 2003), and gas-extraction membranes (Romero *et al.*, 2006). In this type of development it is important to know the morphological descriptors such as the amount and intrinsic properties of amphiphiles (lipidic or block copolymeric types) forming the membrane, the equilibrium thickness, and the coverage. Also important is the stability against mechanical perturbations (e.g. viscoelastic responses to changes in hydrostatic or osmotic pressure differences), the rate of regeneration as well as the ease with which functional peptides or proteins can be adsorbed/incorporated (Pszon-Bartosz *et al.*, 2011; Rossi & Chopineau, 2007) and, once incorporated, how proteins interact with the amphiphilic matrix (Andersen *et al.*, 1998; Nielsen *et al.*, 1998; Nielsen, 2009b). Perhaps the most challenging part of bottom-up biomimetic membrane development is to understand the interaction between the <5 nm thick membrane and its support – in particular when this support also must be porous and able to support mass transport across the membrane without compromising osmotic/hydrostatic transport driving forces (Ibragimova *et al.*, 2011; Roerdink Lander *et al.*, 2011; Vogel *et al.*, 2009).

In any bottom-up based understanding of biomimetic membrane transport and how it relates to the functioning of its synthetic membrane counterparts, one must consider the framework by which we describe water and solute transport. The basic transport description of synthetic membranes is based on Darcy's law for pressure-driven convective flow in case of a pore model description and Fick's law for a solution-diffusion flow in case of a model with fluctuating volume elements appearing and disappearing on the same timescale as the motions of the permeant molecules (Baker, 2004) The transition between the Darcy and Fick regimes is in the range of 5–10Å in volume element effective diameter (Baker, 2004). Biological selective pore-forming proteins and transporters operate in the "low Darcy" regime, where the channel lumens or transporter cavities are finely tuned to maintain constriction dimensions of a few Å effectively limiting protein mediated transport of water and solute to a single file mechanism. This implies that the classical Goldman continuum electro-diffusion description (Goldman, 1943) and its Poisson-Nernst-Planck (PNP) extension (Chen

& Eisenberg, 1993a; Chen & Eisenberg, 1993b) may not be adequate to describe transport of ions and water through individual proteins (Miller, 1999). Descriptions based on microscopic Brownian dynamics (Chung & Kuyucak, 2002; Roux, 2005) have been presented to account for the increasing amount of experimentally observed single channel properties (Edwards *et al.*, 2002). Although the most realistic picture is likely to emerge from atomistic (molecular dynamics MD) simulations (Allen *et al.*, 2006), the close interactions between solute and protein in the pore region (which may require quantum mechanical descriptions of specific amino acid moieties) and the timescale of individual permeation (around 1 event per ns), make it a non-trivial issue to parameterize the systems and relate the behavior of dynamic proteins to macroscopic experimental observations.

In contrast to a bottom-up strategy, a top-down strategy for biomimetic membrane design would start at looking at entire barrier/transport systems. In higher organisms these could include the blood-brain barrier, the skin barrier, and intestinal epithelial walls. These structures share some general features, but are also developed to very specific tasks. In terms of water and salt transport the kidney stands out as *the* system dealing with fluid and electrolyte balance (Brenner, 2007). One of the key features of the kidney is its remarkable capacity in terms of water filtration. Under normal conditions the human kidney filters approximately 180 L blood plasma and excretes about 1.5 L per day (Rehberg, 1926). This implies that most of the water and solutes is re-circulated and that this transport is facilitated by the kidney membranes in a complex interplay between active and passive facilitators. Since the discovery of a specialized class of proteins – the aquaporins – facilitating passive water transport across lipid bilayers (Preston *et al.*, 1992), there has been considerable interest in using this protein as an "active layer" component in reverse osmosis (RO) type membranes, the reason being that with ideally selective nano-scale conduits (the aquaporin proteins) for water with a high osmotic permeability one would in principle be able to design semi-permeable membranes with 100% rejection of ions provided that the aquaporins can be embedded in an ion-impermeable stable matrix (Kumar *et al.*, 2007).

Shortly after their discovery (Preston *et al.*, 1992), the proteins were given the name aquaporins (Agre *et al.*, 1993) and this has fostered the canonical picture of aquaporins as conduits for water only. Thus at a first glance the spectacular properties of aquaporins *per se* make them the obvious choice as a biological component in a biomimetic application such as an RO based desalination applications. Another aspect of aquaporin based biomimetic is the fact that aquaporins in Nature do not perform RO but forward osmosis (FO). This has implications for the design of a water-selective membrane application based on aquaporins. Any encapsulation of a <5 nm thick biomimetic aquaporin containing membrane may potentially impede the osmotic gradient needed to drive the FO process. Thus, while aquaporins can be seen as the immediate bottom-up inspiration for designing "rejection layers", integrated organs such as the kidney may provide top-down inspiration with respect to membrane operation and module design.

From the biology of membrane transport, three themes may be identified as inspirational focal points for future biomimetic membrane design and operation:

(1) The use of FO as the operational principle for membrane processes
(2) Fabrication and use of nano-scale water conduits for selective water transport
(3) The use of aquaporins in membrane fabrication.

In the following each of these focal points will be discussed and concluded by briefly presenting the latest advances in aquaporin biomimetic membrane design and fabrication.

14.2 FORWARD OSMOSIS MEMBRANE TECHNOLOGY

Forward osmosis (FO) is a membrane process that is increasingly attracting international interest as a novel technology for treatment of a wide variety of aqueous solutions (Cath *et al.*, 2006; Teoh *et al.*, 2008) as well as energy production in pressure retarded osmosis (PRO) processes (Skilhagen *et al.*, 2008). Forward osmosis takes advantage of the osmotic pressure gradient across a semi-permeable membrane. This is in contrast to the hydraulic gradient applied in RO extraction of clean water from impaired water (Figure 14.1a).

FO operates by concentrating a feed solution (FS) and diluting a highly concentrated aqueous solution – the draw solution (DS). The larger osmotic pressure of the DS compared to the FS constitutes the driving force for the separation process. Thus, FO is an osmotic concentration process which in principle can operate at zero hydrostatic pressure providing a sustainable water treatment solution.

FO is not getting implemented at a large-scale for several reasons. There are at least three "must-win" battles to successfully overcome before FO becomes a large-scale technology comparable with established technologies such as RO. First, a new type of membrane is needed which has to be very thin in order to minimize internal concentration polarization (Figure 14.1b). Second good DS's are needed, the most studied currently are ammonium-bicarbonate-based DS (McCutcheon *et al.*, 2006) and their use poses several problems including unfavorable interactions with known (conventional) membrane materials. Third, a new system solution: an FO membrane cannot simply replace an RO membrane in conventional membrane modules (Gruber *et al.*, 2011; Gruber *et al.*, 2012). Until the prospect of winning battles one and two looks promising, no system-provider is likely to engage in the last challenge.

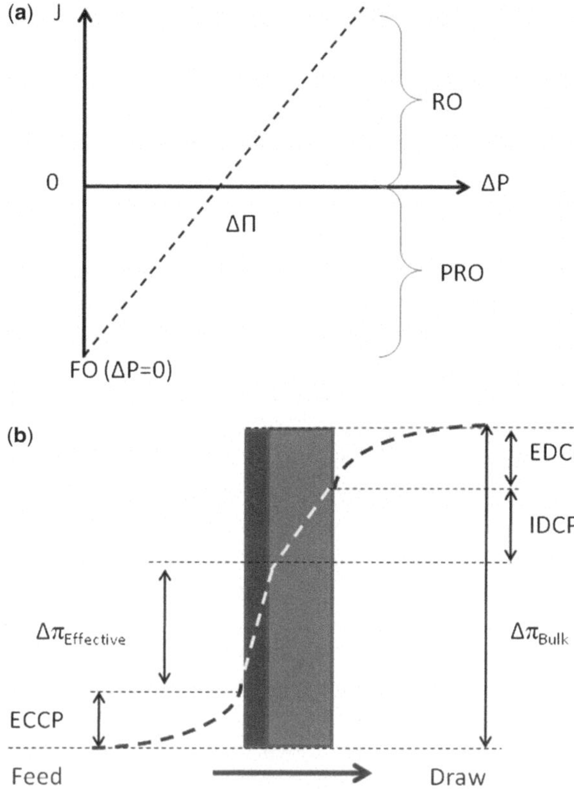

Figure 14.1 (a) Flux (J) *versus* hydraulic pressure (ΔP) in forward osmosis (FO), reverse osmosis (RO) and pressure retarded osmosis (PRO). For FO, there is no hydraulic pressure across the membrane. For RO, the applied hydraulic pressure must exceed the osmotic pressure ($\Delta\Pi$) and convectively drive the water from low to high water potential. For PRO, the hydraulic pressure retards the FO process but the hydraulic pressure applied is still less than the osmotic pressure. (b) Polarization effects in FO asymmetric membranes. The active (filter) layer (dark) rests on top of a porous support (light). The osmotic pressure difference between the two phases ($\Delta\pi_{Bulk}$) results in a driving force for water movement (red arrow) from one phase to the other. However, the effective FO driving force (arising from the effective osmotic pressure difference $\Delta\pi_{Effective}$) is smaller than the nominal maximal force due to polarization concentration profiles inside and outside the membranes. ECCP: External concentration polarization, IDCP: Internal dilutive concentration polarization, EDCP: External dilutive polarization.

Recently a water absorbant cellulose triacetate FO membrane with a thin ($<50\ \mu m$) support has been commercialized. So far the use of this membrane – the only one on the market – has been focusing on low-volume nutrient drink production (Salter, 2006). Thus a concentrated nutrient solution with a high osmotic pressure is used as DS where unsafe water as FS is purified through the FO membrane with a salt rejection between 93 and 95%. However, when operated in a high-volume application set-up, the flux is still low ($<6\ l\ m^{-2}\ h^{-1}$) compared to RO membranes where fluxes can reach $75\ l\ m^{-2}\ h^{-1}$ although typical values are around $30\ l\ m^{-2}\ h^{-1}$ (Xu *et al.*, 2010).

A common approach in large scale osmotic concentration processes is to use microporous hydrophobic membranes (Mengual *et al.*, 1993). Microporous hydrophobic membranes contain a very large number of small pores that penetrate completely through the membrane. Since the hydrophobic membrane is not wetted, water vapor passes through the membrane pores but the aqueous solution is prevented from passing through the pores, provided the pressure is maintained below a critical value. A net flux of water vapor across the membrane will result if there is a partial pressure gradient (due to a maintained temperature gradient) across the membrane (Drioli *et al.*, 1986).

Microporous hydrophobic membranes have the advantage that there is virtually no cross-contamination of salts (rejection levels $>99\%$). Since liquid does not transport across the hydrophobic membrane, dissolved ions (with virtually no vapor pressure) are completely rejected. Unfortunately, this high selectivity is coupled to a low flux across the membrane. Typical fluxes are below $3\ l\ m^{-2}\ h^{-1}$. The low mass transfer rates result from relatively small vapor pressure gradients coupled with small pore sizes and relatively thick membranes (Kunz *et al.*, 1999). In addition, only a portion of the membrane area is actually available for transport (40% porosity is typical). Another drawback is that the membranes tend to lose their hydrophobicity over time as they degrade (Petrotos *et al.*, 2001).

14.3 NANOSCALE WATER CONDUITS

A special case of water transport through nanoscale pores arises when water is confined to hydrophobic pores with diameters corresponding to the size of one or two water molecules. In this regime, phase transitions not observed in bulk systems may arise (Finkelstein, 1987) – which makes nanoscale conduits interesting as a physical system. But this area of research has also attracted considerable interest because biological ion and water selective channels obtain their high degree of selectivity combined with a high flux by constricting the flow to nanoscale pore geometries (Hille, 2001).

The well defined nanoscale and atomically smooth structure of single walled carbon nanotubes (CNTs) presents a system amendable to both experimental and theoretical approaches for studying water transport in hydrophobic pores (see Figure 14.2). This has attracted interest in the development of CNT isoporous membranes where the CNTs are aligned as cylindrical pores spanning an otherwise impermeable membrane. Other CNT membrane developments include the formation of Bucky-paper structures where the CNTs are randomly arranged in a non-woven paper-like structure, thereby creating a highly porous network with a large specific surface area. Also membranes with randomly dispersed CNTs (typically without alignment) in a polymer matrix, where the CNT act as nanofillers similar to zeolites, have been studied (Bakhtiari *et al.*, 2011). In the following, the discussion will be limited to CNTs as building blocks for isoporuos membranes; for further information about other CNT membranes the reader is referred to recent reviews (Sears *et al.*, 2010; van der Bruggen, 2012).

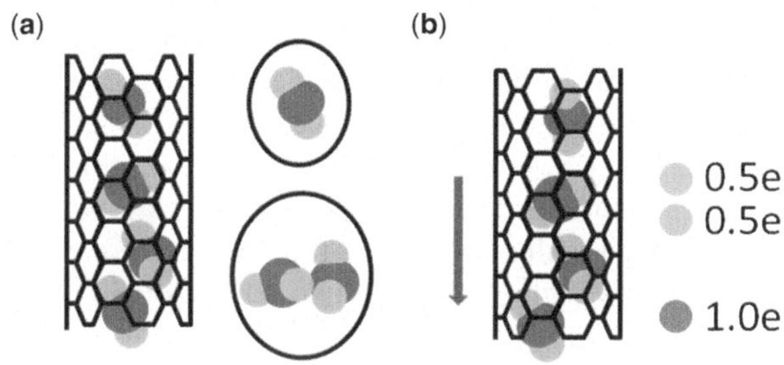

Figure 14.2 (a) Sketch of a water filled carbon nanotube (CNT). Increasing the diameter produces distinct density patterns (ordering) of water molecules. (b) Directional flow of water in a CNT with external charges. Modified from Gong *et al.* (2007).

Recent computer simulation studies have revealed interesting aspects of water interactions with narrow CNTs. These simulations have shown two remarkable features: first, in a water bath the CNT can fill with water despite the non-polar nature of their walls (Hummer *et al.*, 2001). Second, water flow through single wall CNTs can be orders of magnitude faster than that expected from continuum hydrodynamics (Holt *et al.*, 2006), where the velocity of a fluid approaches zero at the inner surfaces of a tube – essentially producing zero flow in narrow tubes. These observations seems to contradict the intuition that water filling and flow in confining structures depend on the ability to form a hydrogen-bonding (HB) network – an ability that is seemingly diminished in hydrophobic pores.

In CNTs with sub-nanometer diameters water forms hydrogen-bonded (HB) wires. In very narrow CNTs with inner lumen diameters of 3.2 Å single HB wires are formed along the lumen axis. Under certain circumstances the narrow CNT seem to hydrate and dehydrate in an all or none fashion (Beckstein & Sansom, 2003). As the inner diameter of the CNT is increased, double and quadruple HB chains have been observed in molecular dynamics (MD) simulations (Corry, 2008). Even in larger CNTs water form highly ordered structures (Kotsalis *et al.*, 2004). The key to the hydration and flow capabilities of the CNTs lies in the exact balance of the HB interactions. Outside the CNT each water molecule participates in a large number of fluctuating HB structures. The fluctuations imply that some interactions amongst the individual water molecules will indeed be strong but also that others will be less energetically favorable. In the CNT the higher degree of order induced by the nanoscale confinement results in less fluctuation for the HBs between water molecules. On average the interaction energies outside and inside the CNT are similar and the CNT thus does not present a significant energetic barrier for water transport (Corry, 2008).

The ordering of water molecules in narrow tubes is a direct result of the water dipole moments. Thus simple one dimensional (1D) dipole moments representations provide a powerful description of water in nanopores (Kofinger *et al.*, 2009). This implies that the HB dipole-dipole interactions are sensitive to external electrical fields. A dramatic example of this is found in CNT simulations where three discrete charges suitably arranged outside the CNT result in a unidirectional flow of water through the CNT (Gong *et al.*, 2007). Three positive charges with charge magnitudes of 1.0e, 0.5e and 0.5e are positioned along the channel, all with an axial radial distance of delta (and corresponding counter charges further away), see Figure 14.2b. When the two 0.5e charges are placed close to the centre of the channel and the position of the 1.0e charge is close to one of the CNT openings, a directional flow of water molecules is established conducting water into the CNT via the opening furthest away from the 1.0e charge and out via the opening closest to the 1.0e charge. The electrostatic forces exerted by water molecules on the three positive charges are substantial: 533 pN, 313 pN and 300 pN, from the lower to upper charges, respectively. If the charges were not constrained, they would be displaced and the net water flux would vanish (Gong *et al.*, 2007).

The fact that the exact hydration state of narrow tubes depends strongly on the van der Waals interactions between the water molecules and the tube wall also implies that in simulations it also depends on the *representation* of these interactions in simulation force fields. Representation of polarizability in MD simulations is a non-trivial issue and one should exercise caution when extrapolating results from nanoscopic simulations of water permeation to macroscopic flux estimations. An important tool in this respect is therefore to develop methods to experimentally probe water at the nanoscale (Holt, 2008) – so that theoretical and experimental data on water confinement may be compared.

The exciting theoretical and experimental work on water-filled CNTs and other pore forming molecules have spurred the idea of using such pore-molecules as building blocks for future high flux separation devices (e.g. water purification membranes and desalination systems) and nanofluidic sensor devices (Burykin & Warshel, 2006; Corry, 2008). The ideal is the perfect embodiment of the semipermeable membrane letting only water molecules pass, however in order to achieve this there are still major obstacles to overcome. One of the major issues to address is the requirement that the pore-formers can be packed in a solid non-porous material. Indeed dense arrays of CNTs have been successfully embedded into silicon nitride (Holt *et al.*, 2004) and polymer (Hinds *et al.*, 2004) substrates. However, the arrays consisted of multi-walled CNTs with outer diameters ranging from 20 to 50 nm and although the nanotubes diameter is determined by the catalyst particle, a precise tuning of the inner diameter (see above) requires a very well-determined catalyst synthesis with Å-scale diameter dispersion – not currently available. Also CNT may adopt exotic (but blocked) tubular structures (e.g. "bamboo" structures) (Holt *et al.*, 2004). Characterization of CNT morphology is often carried out using transmission electron microscopy (TEM) on small samples (100 nm length scale). Thus, based on TEM evidence a nominally hollow tube may in fact be blocked or otherwise not suitable for water permeation applications.

Regarding the use of CNTs in water purification in terms of constructing isoporous CNT membranes, blocking charge transport is crucial. Ion permeation though nanoscopic pores is directly related to hydration energies. Thus ions must be partly desolvated in order to pass though nano-scale channels and this remarkably exemplified in potassium selection ion channels. They let K^+ ions but not the smaller Na^+ ions pass due to the fact that fixed carbonyl oxygens in the "selectivity filter" exactly "replace" some of the K^+ hydration water molecules during ion passage; the Na^+ ion is too small for this "replacement" or solvent substitution to occur and they stay bulk hydrated outside the channel (Roux 2005). Simulations in CNT points to the possibility of perfect salt rejection (i.e. only bulk water ion hydration) provided that the inner diameter is less than 4.7 Å. At 5.9 Å inner CNT diameter the salt rejection is 96% and drops to 58% for a 7.5 Å inner radius (Corry, 2008). Thus if the CNT inner diameter can be controlled with Å precision, they could in principle provide perfect salt rejection properties.

While ions may be rejected by pore diameter constriction, rejection of protons may appear as a more difficult task to accomplish. Since water transport with a high flux is coupled to HB patterns between the water molecules in the confined tube geometry, it has been suggested that protons will engage in a Grotthuss-type "hopping" where protons perform nearest neighbour hopping from one hydronium (H_3O^+) molecule to another. This idea stems from proton mobility in bulk water where mobility likely involves periodic series of isomerizations between $H_9O_4{}^+$ (Eigen) and $H_5O_2^+$ (Zundel) ions (Agmon, 1995). In the Grotthuss mechanism only (electrostatic) barriers that can (transiently) disturb the HB pattern can stop protons in confined tube geometries. However, computational studies of the free energy barriers for transfer of proton and water molecules through a membrane assembled from double wall carbon nanotubes (3.2 Å inner diameter) show water transfer while blocking protons (Burykin & Warshel, 2006). This result goes against the Grotthuss mechanism and suggests a solvent substitution mechanism where the charged hydronium ion is prevented to enter the CNT simply because of the low internal dielectic environment.

Remarkably natural evolution has developed a water transport facilitating protein molecule that has the key features discussed above: a narrow pore confining geometry with a high flux and dipole/electrostatic modulation preventing ion and proton conduction. In addition the protein molecule is self-assembling and, under the right conditions, spontaneously inserts into a nanometer thick matrix with a very low ion intrinsic permeability. The naturally occurring protein molecules – aquaporins – may thus hold the key to understanding and designing osmotic biomimetic membranes.

14.4 AQUAPORINS: NANO-SCALE WATER CONDUITS

Aquaporins constitute a large family of membrane spanning proteins, for recent reviews see (Fu & Lu, 2007; Gonen & Walz, 2006; Ludewig & Dynowski, 2009). Much has been revealed about this class of proteins since the purification of a red blood cell membrane protein and subsequent heterolog expression of this protein revealing rapid water diffusion along osmotic gradients.

The primary sequence of aquaporin reveals its general unit structure in the form of two tandem repeats, each containing three transmembrane spanning segments. Biochemical analysis and later crystal structure analysis revealed an hour-glass structure with pseudo two-fold symmetry where the six transmembrane segments surround a central pore. Each tandem repeat contains a loop between TM2 and TM3 with an asparagine-proline-alanine (NPA) signature motif. Each unit functions as a pore and the predominant unit-assembly in biological membranes is a four-unit arrangement (see Figure 14.3).

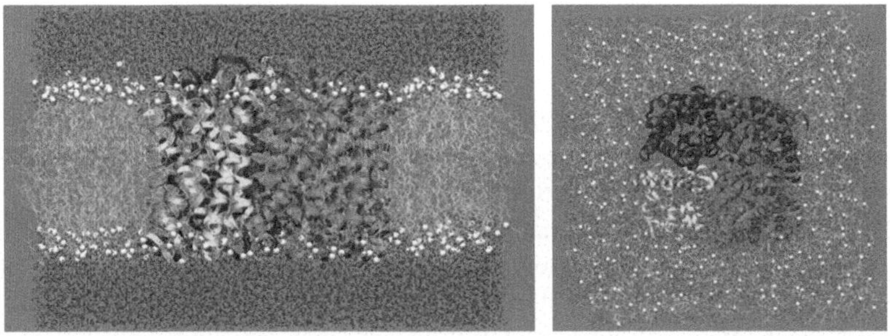

Figure 14.3 Spinach (SoPiP2;1) aquaporin tetramer in a lipid bilayer consisting of 1-palmitoyl-2-oleoyl-phosphatidylcholine surrounded by water on each side. Left: cross section; Right: top view. The four identical monomers are each represented by their amino acid backbone structure. The simulation snapshots were kindly provided by Dr. Himanshu Khandelia, University of Southern Denmark.

The simplicity of the aquaporin structure is apparent. The structure is the basis of the remarkable ability to transport water molecules and simultaneously reject charged species including protons. Molecular dynamics (MD) simulations based on the experimental structure of the mammalian AQP1 show single-file transport of water molecules through a narrow <3 Å pore in which steric and electrostatic factors prevent electrolyte passage. At the center of the membrane (close to the NPA region) the average water dipole orientation is reversed. This observation initially led to the suggestion of a proton rejection mechanism based on the disruption of a Grotthuss hopping along the water wire (Tajkhorshid *et al.*, 2002). This picture has now been challenged by simulations where the large barrier for proton transfer is due to the loss of salvation (Burykin & Warshel, 2003, 2006; Kato *et al.*, 2006) when moving a proton into the narrow partly hydrophobic part of the channel. The result is an impeded proton flux – even in the presence of an intact water wire (cf. the CNT proton permeability discussed above). The mechanism is not fully elucidated, yet the delicate balance between HB forces in a hydrophobic narrow pore combined with electrostatic barriers (Beitz *et al.*, 2006; Chen *et al.*, 2006) and helical dipolar moments (de Groot & Grubmuller, 2005) result in a very efficient and highly selective transport of water, although various conducting states are likely to exist (Xin *et al.*, 2011; Xin *et al.*, 2012).

However, it has become increasingly clear that aquaporins are not only "water channels" but a substantial group in fact have permeabilities for other solutes – most notably expressed in the class of aquaglyceroporins, where the name implies a permeability for both glycerol and water (Stroud *et al.*, 2003). Other more exotic solutes are also transported by members of the aquaporin family for example, metalloids (Nielsen, 2009c). Today close to 500 different isoforms of aquaporins are known (Sasaki, 2008) and functional roles beyond water transport are now being elucidated (Verkman, 2005).

14.5 DESIGNING BIOMIMETIC MEMBRANES

Biomimetic membranes in the form of lipid and polymer based membrane arrays have been investigated (Gonzalez-Perez *et al.*, 2009; Hansen *et al.*, 2009a; Hansen *et al.*, 2009b; Rein *et al.*, 2011; Roerdink Lander *et al.*, 2011; Vogel *et al.*, 2009). These membrane arrays with incorporated proteins were formed across apertures in a scaffold material allowing for upscaling. Osmotic applications require additional support which is sufficiently porous in order to allow substantial vectorial flow and at the same time sufficiently dense to support the membrane in the presence of transmembrane osmotic pressure gradients (Figure 14.4). Using hydrogels, we have demonstrated that functional lipid bilayer arrays with lifetimes of days can be constructed (Ibragimova *et al.*, 2011; Pelaez *et al.*, 2011).

However the main challenge with all planar designs is the inherent problem in securing the stability of a 3–4 nm thick film while up-scaling to square meter membrane areas. Several attempts have been described recently. These include charged lipid vesicle depositions onto commercially available nanofiltration membranes where the recipient surface was either cross-linked polyamide or sulfonated polysulfone both negatively charged at pH 7 (Kaufman *et al.*, 2010); rupture of aquaporin containing polymersomes on methacrylate functionalized cellulose acetate membranes (Zhong *et al.*, 2012); detergent-stabilized His-tagged aquaporin added to monolayers with nickel-chelating lipids (Sun *et al.*, 2012); and proteopolymersome deposition onto polycarbonate track-etched substrates coated with gold and functionalized with photo-active acrylate groups (Wang *et al.*, 2012).

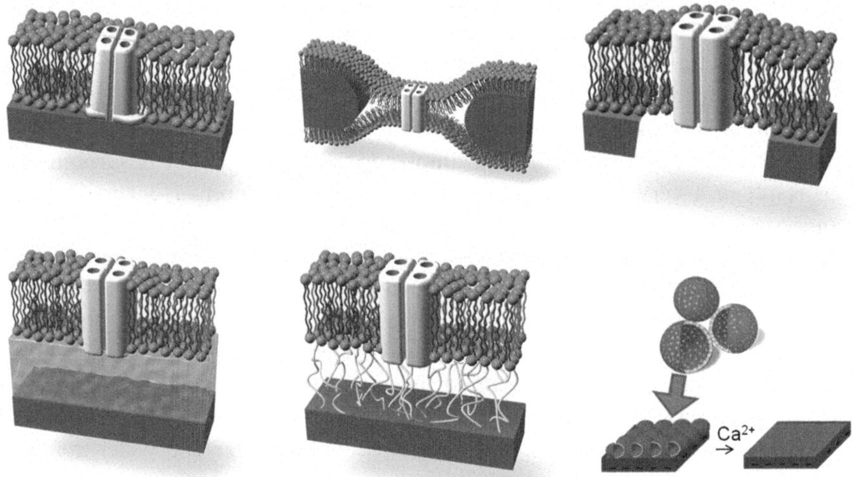

Figure 14.4 Example of biomimetic membrane design strategies. Top row from left to right: deposition of membranes with embedded aquaporin proteins directly onto a nano-scale flat solid surface may lead to protein denaturation (and consequently loss of function); aquaporin proteins embedded in a membrane established across a micron scale aperture in a scaffold; deposition of a protein containing membrane on a nanoporous material. Bottom row from left to right: deposition on a soft hydrogel cushion and deposition based on conjugated polymer anchors where the cushion/anchor material effectively separates the "soft" membrane from the "hard" support; deposition of lipid vesicles on a surface with subsequent induced vesicle fusion/collapse forming a continuous supported bilayer.

Recently, a design was developed where vesicles are embedded in thin (200 nm) films deposited on a porous support (Zhao *et al.*, 2012). Briefly, a microporous polysulfone substrate was soaked with an m-phenylene-diamine (MPD) aqueous solution that contained aquaporin vesicles. The soaked substrate was subsequently exposed to trimesoyl chloride (TMC) to form a three-dimensionally crosslinked polyamide layer in which the aquaporin vesicles are embedded. The resulting membranes are made in sheets up to 600 cm^2 and are sufficiently stable to allow normal handling and shipping procedures.

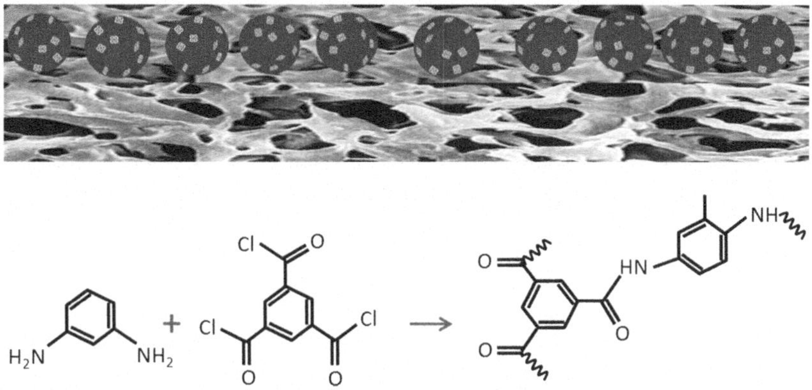

Figure 14.5 Top: Aquaporin TFC membrane design. Spherical vesicles with incorporated aquaporin proteins are embedded in a polyamide layer established on top of a porous support material. Bottom: The polyamide layer is made by interfacial polymerization of m-phenylene-diamine (MPD) (in the presence of aquaporin vesicles) and trimesoyl chloride (TMC) resulting in a three dimensional rejection layer with embedded aquaporin vesicles.

The resulting thin film composite (TFC) membrane has been characterized in cross-flow RO and FO setups. RO membranes were tested under applied pressures up to 14 bar, and benchmarked against commercially available RO membranes (Zhao *et al.*, 2012). Compared to the control membranes (membranes with protein free vesicles and membranes with protein vesicles where the protein was mutated to an inactive form), the aquaporin TFC membranes achieved significantly higher water permeability ($\sim 4\,l\,m^{-2}\,h^{-1}\,bar^{-1}$) with comparable NaCl rejection (~97%). Their permeability was ~40% higher compared to a commercial brackish water RO membrane (Dow Filmtec BW30) and an order of magnitude higher compared to a seawater RO membrane (Dow Filmtec SW30HR). In FO the aquaporin membranes had >90% rejection of urea with a water flux of $10\,l\,m^{-2}\,h^{-1}$ with 2M NaCl as draw solution. This clearly demonstrates the great potential of the aquaporin TFC membranes for industrial RO and FO applications.

REFERENCES

Agmon N. (1995). The Grotthuss mechanism. *Chem. Phys. Lett.*, **244**, 456–462.

Agre P., Sasaki S. and Chrispeels M. J. (1993). Aquaporins: a family of water channel proteins. *Am. J. Physiol.*, **265**(3 Pt 2), F461.

Allen T. W., Andersen O. S. and Roux B. (2006). Molecular dynamics – potential of mean force calculations as a tool for understanding ion permeation and selectivity in narrow channels. *Biophys. Chem.*, **124**(3), 251–267.

Andersen O. S., Nielsen C., Maer A. M., Lundbæk J. A., Goulian M. and Koeppe R. E. II. (1998). Gramicidin channels as molecular force transducers in lipid bilayers. *Biol. Skr. Dan. Vid. Selsk.*, **49**, 75–82.

Baker R. W. (2004). Membrane Technology and Applications, 2nd edn. John Wiley & Sons, Ldt., Chichester.

Bakhtiari O., Mosleh S., Khosravi T. and Mohammadi T. (2011). Synthesis and characterization of polyimide mixed matrix membranes. *Sep. Sci. Technol.*, **46**, 2138–2147.

Beckstein O. and Sansom M. S. (2003). Liquid-vapor oscillations of water in hydrophobic nanopores. *Proc. Natl. Acad. Sci. U S A*, **100**(12), 7063–7068.

Beitz E., Wu B., Holm L. M., Schultz J. E. and Zeuthen T. (2006). Point mutations in the aromatic/arginine region in aquaporin 1 allow passage of urea, glycerol, ammonia, and protons. *Proc. Natl. Acad. Sci. U S A*, **103**(2), 269–274.

Bernsel A. and Von Heijne G. (2005). Improved membrane protein topology prediction by domain assignments. *Protein Sci.*, **14**(7), 1723–1728.

Brenner B. M. (2007). Brenner and Rector's The Kidney, 8th edn. Saunders Elsevier, Maryland Heights, MO, USA.

Burykin A. and Warshel A. (2003). What really prevents proton transport through aquaporin? Charge self-energy versus proton wire proposals. *Biophys. J.*, **85**(6), 3696–3706.

Burykin A. and Warshel A. (2006). Membranes assembled from narrow carbon nanotubes block proton transport and can form effective nanofiltration devices. *J. Comput. Theor. Nanosci.*, **3**, 237–242.

Byrne W. (2002). Reverse Osmosis, 2nd edn. Tall Oaks Publishing Inc., Littleton, CO.

Cath T. Y., Childress A. E. and Elimelech M. (2006). Forward osmosis: principles, applications and recent developments. *J. Membr. Sci.*, **281**, 70–87.

Chen D. and Eisenberg R. (1993a). Charges, currents, and potentials in ionic channels of one conformation. *Biophys. J.*, **64**(5), 1405–1421.

Chen D. P. and Eisenberg R. S. (1993b). Flux, coupling, and selectivity in ionic channels of one conformation. *Biophys. J.*, **65**(2), 727–746.

Chen H., Wu Y. and Voth G. A. (2006). Origins of proton transport behavior from selectivity domain mutations of the aquaporin-1 channel. *Biophys. J.*, **90**(10), L73–L75.

Chung S. H., Andersen O. S. and Krishnamurthy V. (eds) (2007). Biological Membrane Ion Channels. Dynamics, Structure and Applications. Springer, New York.

Chung S. H. and Kuyucak S. (2002). Recent advances in ion channel research. *Biochim. Biophys. Acta*, **1565**(2), 267–286.

Corry B. (2008). Designing carbon nanotube membranes for efficient water desalination. *J. Phys. Chem. B*, **112**(5), 1427–1434.

de Groot B. L. and Grubmuller H. (2005). The dynamics and energetics of water permeation and proton exclusion in aquaporins. *Curr. Opin. Struct. Biol.*, **15**(2), 176–183.

Dowhan W. (1997). Molecular basis for membrane phospholipid diversity: why are there so many lipids? *Annu. Rev. Biochem.*, **66**, 199–232.

Drioli E., Calabro V. and Wu Y. (1986). Microporous membranes in membrane distillation. *Pure Appl. Chem.*, **58**, 1657–1662.

Edwards S., Corry B., Kuyucak S. and Chung S. H. (2002). Continuum electrostatics fails to describe ion permeation in the gramicidin channel. *Biophys. J.*, **83**(3), 1348–1360.

Finkelstein A. (ed.) (1987). Water Movement Through Lipid Bilayers, Pores, and Plasma Membranes. Theory and Reality. Wiley–Interscience, New York.

Fu D. and Lu M. (2007). The structural basis of water permeation and proton exclusion in aquaporins. *Mol. Membr. Biol.*, **24**(5–6), 366–374.

Goldman D. E. (1943). Potential, impedance and rectification in membranes. *J. Gen. Physiol.*, **27**, 37–60.

Gonen T. and Walz T. (2006). The structure of aquaporins. *Q Rev. Biophys.*, **39**(4), 361–396.

Gong X., Li J., Lu H., Wan R., Hu J. and Fang H. (2007). A charge-driven molecular water pump. *Nat. Nanotechnol.*, **2**(11), 709–712.

Gonzalez-Perez A., Stibius K. B., Vissing T., Nielsen C. H. and Mouritsen O. G. (2009). Biomimetic triblock copolymer membrane arrays: a stable template for functional membrane proteins. *Langmuir*, **25**, 10447–10450.

Gruber M., Johnson C. E., Tang C., Jensen M. H., Yde L. and Hélix-Nielsen C. (2012). Validation and analysis of forward osmosis CFD model in complex 3D geometries. *Membranes*, **2**, 764–782.

Gruber M. F., Johnson C., Tang C. Y., Jensen M. H., Yde L. and Hélix-Nielsen C. (2011). Computational fluid dynamics simulations of flow and concentration polarization in forward osmosis membrane systems. *J. Membr. Sci.*, **379**, 488–495.

Hansen J. S., Perry M., Vogel J., Groth J., Vissing T., Larsen M., Geschke O., Emneus J., Bohr H. and Nielsen C. H. (2009a). Large scale biomimetic membrane arrays. *Anal. Bioanal. Chem.*, **395**, 719–727.

Hansen J. S., Perry M. E., Vogel J., Vissing T., Geschke O., Emneus J. and Nielsen C. H. (2009b). Development of an automation technique for the establishment of functional lipid bilayer arrays. *J. Micromech. Microeng.*, **19**, 025014.

Hélix-Nielsen (ed.) (2012). Biomimetic Membranes for Sensor and Separation Applications. Springer Verlag, Dordrecht.

Hille B. (2001). Ionic Channels of Excitable Membranes. Sinauer, Sunderland.

Hinds B. J., Chopra N., Rantell T., Andrews R., Gavalas V. and Bachas L. G. (2004). Aligned multiwalled carbon nanotube membranes. *Science*, **303**(5654), 62–65.

Holt J. K. (2008). Methods for probing water at the nanoscale. *Microfluid. Nanofluid.*, **5**, 425–442.

Holt J. K., Noy A., Huser T., Eagelsham D. and Bakajin O. (2004). Fabrication of a carbon nanotube-embedded silicon nitride membrane for studies of nanometer-scale mass transport. *Nano Lett.*, **4**, 2245–2250.

Holt J. K., Park H. G., Wang Y., Stadermann M., Artyukhin A. B., Grigoropoulos C. P., Noy A. and Bakajin O. (2006). Fast mass transport through sub-2-nanometer carbon nanotubes. *Science*, **312**(5776), 1034–1037.

Hummer G., Rasaiah J. C. and Noworyta J. P. (2001). Water conduction through the hydrophobic channel of a carbon nanotube. *Nature*, **414**(6860), 188–190.

Ibragimova S., Stibius K. B., Szewczykowski P., Perry M., Bohr H. and Nielsen C. H. (2011). Hydrogels for *in situ* encapsulation of biomimetic membrane arrays. *Polym. Adv. Technol.*, **23**, 182–189.

Kato M., Pisliakov A. V. and Warshel A. (2006). The barrier for proton transport in aquaporins as a challenge for electrostatic models: the role of protein relaxation in mutational calculations. *Proteins*, **64**(4), 829–844.

Kaufman Y., Berman A. and Freger V. (2010). Supported lipid bilayer membranes for water purification by reverse osmosis. *Langmuir*, **26**(10), 7388–7395.

Kofinger J., Hummer G. and Dellago C. (2009). A one-dimensional dipole lattice model for water in narrow nanopores. *J. Chem. Phys.*, **130**(15), 154110.

Kotsalis E. M., Walther J. H. and Koumoutsakos P. (2004). Multiphase water flow inside carbon nanotubes. *Int. J. Multiph. Flow*, **30**, 995–1010.

Kumar M., Grzelakowski M., Zilles J., Clark M. and Meier W. (2007). Highly permeable polymeric membranes based on the incorporation of the functional water channel protein aquaporin Z. *Proc. Natl. Acad. Sci. U S A*, **104**(52), 20719–20724.

Kunz W., Benhabiles A. and Ben-Aim R. (1999). Osmotic evaporation through macroporous hydrophobic membranes: a survey of current research and applications. *J. Membr. Sci.*, **121**, 25–36.

Ludewig U. and Dynowski M. (2009). Plant aquaporin selectivity: where transport assays, computer simulations and physiology meet. *Cell Mol. Life Sci.*, **66**(19), 3161–3175.

Mayer M., Kriebel J. K., Tosteson M. T. and Whitesides G. M. (2003). Microfabricated teflon membranes for low-noise recordings of ion channels in planar lipid bilayers. *Biophys. J.*, **85**(4), 2684–2695.

McCutcheon J. R., McGinnis R. L. and Elimelech M. (2006). Desalination by a novel ammonia–carbon dioxide forward osmosis process: influence of draw and feed solution concentrations on process performance. *J. Membr. Sci.*, **278**, 114–123.

Mengual J. I., Ortiz de Zarate J. M., Pena L. and Velazquez A. (1993). Osmotic distillation through porous hydrophobic membranes. *J. Membr. Sci.*, **82**, 129–140.

Miller C. (1999). Ionic hopping defended. *J. Gen. Physiol.*, **113**(6), 783–787.

Nardin C., Winterhalter M. and Meier W. (2000). Giant free-standing ABA triblock copolymer membranes. *Langmuir*, **16**, 7708–7712.

Nielsen C., Goulian M. and Andersen O. S. (1998). Energetics of inclusion-induced bilayer deformations. *Biophys. J.*, **74**, 1966–1983.

Nielsen C. H. (2009a). Biomimetic membranes for sensor and separation applications. *Anal. Bioanal. Chem.*, **395**, 697–718.

Nielsen C. H. (2009b). Lipid-protein interactions in biomembranes. In: Handbook of Molecular Biophysics, H. G. Bohr (ed.), Wiley, Berlin, pp. 329–358.

Nielsen C. H. (2009c). Major intrinsic proteins in biomimetic membranes. *Adv. Exp. Med. Biol.*, **679**, 127–145.

Pelaez L., Romero V., Escalera S., Ibragimova S., Stibius K., Benavente J. and Nielsen C. H. (2011). Electrochemical characterization of hydrogels for biomimetic applications. *Polym. Adv. Technol.*, **22**, 1381–1388.

Petrotos K. B., Lazarides H. N., Wolf W., Behsnilian D. and Spiess W. (2001). Osmotic concentration of liquid foods. *J. Food. Eng.*, **49**, 201–206.

Preston G. M., Carroll T. P., Guggino W. B. and Agre P. (1992). Appearance of water channels in Xenopus oocytes expressing red cell CHIP28 protein. *Science*, **256**(5055), 385–387.

Pszon-Bartosz K., Hansen J. S., Stibius K. B., Groth J. S., Emneus J., Geschke O. and Helix-Nielsen C. (2011). Assessing the efficacy of vesicle fusion with planar membrane arrays using a mitochondrial porin as reporter. *Biochem. Biophys. Res. Commun.*, **406**(1), 96–100.

Rehberg P. B. (1926). Studies on kidney function. I. The rate of filtration and reabsorption in the human kidney. *Biochem. J.*, **20**, 447–460.

Rein C., Pszon-Bartosz K., Stibius K. B., Bjornholm T. and Helix-Nielsen C. (2011). Free-standing biomimetic polymer membrane imaged with atomic force microscopy. *Langmuir*, **27**(2), 499–503.

Rios G. M., Belleville M. P. and Paolucci-Jeanjean D. (2007). Membrane engineering in biotechnology: quo vamus? *Trends Biotechnol.*, **25**(6), 242–246.

Roerdink Lander M., Ibragimova S., Rein C., Vogel J., Stibius K., Geschke O., Perry M. and Helix-Nielsen C. (2011). Biomimetic membrane arrays on cast hydrogel supports. *Langmuir*, **27**(11), 7002–7007.

Romero J., Draga H., Belleville M. P., Sanchez J., Combe-James C., Dornier M. and Rios G. M. (2006). New hydrophobic membranes for contactor processes – applications to isothermal concentration of solutions. *Desalination*, **193**, 280–285.

Rossi C. and Chopineau J. (2007). Biomimetic tethered lipid membranes designed for membrane-protein interaction studies. *Eur. Biophys. J.*, **36**(8), 955–965.

Roux B. (2005). Ion conduction and selectivity in K(+) channels. *Annu. Rev. Biophys. Biomol. Struct.*, **34**, 153–171.

Salter R. (2006). Forward osmosis. *Water Cond. Purif.*, **48**, 36–38.

Sasaki S. (2008). Introduction for special issue for aquaporin: expanding the world of aquaporins: new members and new functions. *Pflugers Arch.*, **456**(4), 647–649.

Sears K., Dumée L., Schütz J., She M., Huynh C., Hawkins S., Duke M. and Gray S. (2010). Recent developments in carbon nanotube membranes for water purification and gas separation. *Materials*, **3**, 127–149.

Skilhagen S. E., Dugstad J. E. and Aaberg R. J. (2008). Osmotic power – power production based on the osmotic pressure difference between waters with varying salt gradients. *Desalination*, **220**, 476–482.

Stroud R. M., Savage D., Miercke L. J., Lee J. K., Khademi S. and Harries W. (2003). Selectivity and conductance among the glycerol and water conducting aquaporin family of channels. *FEBS Lett.*, **555**(1), 79–84.

Sun G., Zhou H., Li Y., Jeyaseelan K., Armugam A. and Chung T.-S. (2012). A novel method of aquaporin Z incorporation via binary-lipis Langmuir monolayers. *Colloids Surf. B: Biointerfaces*, **89**, 283–288.

Tajkhorshid E., Nollert P., Jensen M. O., Miercke L. J., O'Connell J., Stroud R. M. and Schulten K. (2002). Control of the selectivity of the aquaporin water channel family by global orientational tuning. *Science*, **296**(5567), 525–530.

Teoh M. M., Wang K., Bonyadi S. and Tai-Shung N. (2008). Forward osmosis and membrane distillation proceses for freshwater production. *Innovation*, **8**, 15–17.

Trau D. and Renneberg R. (2003). Encapsulation of glucose oxidase microparticles within a nanoscale layer-by-layer film: immobilization and biosensor applications. *Biosens. Bioelectron.*, **18**(12), 1491–1499.

van der Bruggen B. (2012). The separation power of nanotubes in membranes: a review. *ISRN Nanotechnol.*, **2012**, ID693485.

Verkman A. S. (2005). More than just water channels: unexpected cellular roles of aquaporins. *J. Cell Sci.*, **118**(Pt 15), 3225–3232.

Vogel J., Perry M. E., Hansen J. S., Bollinger P.-Y., Nielsen C. H. and Geschke O. (2009). Support structure for biomimetic applications. *J. Micromech. Microeng.*, **19**, 025026.

Wang H., Chung T. S., Tong Y. W., Jeyaseelan K., Armugam A., Chen Z., Hong M. and Meier W. (2012). Highly permeable and selective pore-spanning biomimetic membrane embedded with aquaporin Z. *Small*, **8**(8), 1185–1190.

Xin L., Helix-Nielsen C., Su H., Torres J., Tang C., Wang R., Fane A. G. and Mu Y. (2012). Population shift between the open and closed states changes the water permeability of an aquaporin Z mutant. *Biophys. J.*, **103**(2), 212–218.

Xin L., Su H., Nielsen C. H., Tang C., Torres J. and Mu Y. (2011). Water permeation dynamics of AqpZ: a tale of two states. *Biochim. Biophys. Acta*, **1808**(6), 1581–1586.

Xu Y., Peng X., Tang C., Shiang Fu Q. and Nie S. (2010). Effect of draw solution concentration and operating conditions on forward osmosis and pressure retarded osmosis performance in a spiral wound module. *J. Membr. Sci.*, **348**, 298–309.

Zhao Y., Qiu C., Li X., Vararattanavech A., Shen W., Torres J., Hélix-Nielsen C., Wang R., Hu X., Fane A. G. and Tang C. Y. (2012). Synthesizing robust and high-performance aquaporin based biomimetic membrane by interfacial polymerization – membrane preparation and RO performance characterization. *J. Membr. Sci.*, **423–424**, 422–428.

Zhong P. S., Chung T.-S., Jeyaseelan K. and Armugam A. (2012). Aquaporin-embedded biomimetic membranes for nanofiltration. *J. Membr. Sci.*, **407–408**, 27–33.

Chapter 15

Functionalised graphene: a novel platform for biosensors

W. Yang and C. J. Barrow

15.1 INTRODUCTION

One of key goals of biomolecular science today is to monitor biomolecular interactions with high sensitivity in real-time. Biosensors are ever more important to all the aspects of modern life, from medical tests for disease diagnosis to detecting threats from bioterrorism. The ideal biosensor would offer the sensitivity and selectivity to detect single molecules in real-time. The ability to detect and study a specific single molecule (Fan & Bard, 1995; Lu *et al.*, 1998; Nie & Emery, 1997) represents the ultimate challenge for biosensors in particular and for analytical chemistry in general. Today, a number of biomolecule-detection methods exist that can sense molecules such as DNA and proteins. These methods can be divided into two general categories: label-based detection and label-free detection. Typical label-based methods involve detecting the fluorescence, chemiluminescence or radioactivity from a specific label after it has become bound to a target on the biomolecule. Examples of these methods include the enzyme-linked immunosorbent assay (ELISA) (Cui *et al.*, 2007; Hou *et al.*, 2007) and nanoparticle-labelling assays (Mirkin *et al.*, 1996). The highest sensitivity reported for a labelled-detection scheme is in the order of *atto-molar* (10^{-18} M) (Nam *et al.*, 2003), but these assays require two antibody receptors, labelling and other manipulation steps that can potentially increase the time and cost of the assay and reduce quantitative accuracy. Furthermore, this detection limit is still far from the *zepto-molar* concentration (10^{-21} mol/L or approximately one molecule per millilitre) needed for single-molecule detection. In addition, almost all of these assays are unsuitable for the real-time detection and/or kinetic analysis. Researchers have developed a number of label-free detection approaches, including optical resonator (Armani *et al.*, 2007), microcantilevers (Wu *et al.*,

2001) and nanowire or carbon-nanotube devices (Besteman *et al.*, 2003; Cui *et al.*, 2001; Patolsky *et al.*, 2004) which are able to detect changes in resonance shift, deflection and electrical conductivity, respectively. Specific advances have been made with each of these methods; however, typically, the sensitivity achieved was lower than that reported by the label-based methods.

One promising approach for the direct electrical detection of biological macromolecules uses semiconducting nanowires or carbon nanotubes configured as field-effect transistors, which change conductance when charged macromolecules bind to receptors on the device surface (Cui *et al.*, 2001). This is because the current flows close to the surface of a nanowire or nanotube and, therefore, is highly sensitive to small changes in the surface's charge state. Thus, any biological macromolecules that bind to the surface can be sensed electrically as they perturb the current flow near the surface of the nanowire or nanotube. By using this principle, Dekker's (Besteman *et al.*, 2003) and Star's (Star *et al.*, 2003) research groups have investigated the application of carbon-nanotube devices as electrical biosensors for various biomolecules including enzymes and proteins. Similarly, Lieber's group (Hahm & Lieber, 2004) has shown that semiconductor nanowires can be functionalised with biological macromolecules and incorporated into the electrical circuits, and that, in this configuration, the current that flows in the nanowire is very sensitive to binding events of the macromolecules. They achieved detection limits as low as 10×10^{-15} M for a wild-type DNA, from the cystic-fibrosis-transmembrane-receptor gene, when binding to peptide-nucleic-acid receptors.

Recently, graphene (Figure 15.1) has emerged as a new advanced carbon material with enormous potential for numerous applications in electronics (Chen *et al.*, 2008; Geim & Novoselov, 2007; Li & Kaner, 2008; Novoselov *et al.*, 2004; Novoselov *et al.*, 2007; Ratinac *et al.*, 2010). In 2004, Novoselov, Geim and their colleagues demonstrated that few-layer, and even single-layer graphene could be isolated from graphite and definitively identified by microscopy (Novoselov *et al.*, 2004). They also measured the electrical properties of few-layer graphene and found some remarkable behaviour, such as the fact that graphene behaved as a zero-gap semiconductor with ballistic transport of charge carriers and linear current-voltage activities, and that it displayed the ambi-polar field effect. These, and the other intriguing properties observed, were considered to offer exciting prospects for graphene to be used in creating nanometre-scale metallic transistors as well as other types of devices.

Since then, of course, the number of experimental and theoretical studies of graphene has increased dramatically. Some of this volume has been generated by the condensed-matter physicists who are having a field day with a material that had long been modelled in physics, but, until recently, had not been believed to exist in a free state. Other research has occurred in fields such as chemistry and nanotechnology, where scientists and engineers are trying to come to grips with making graphene in a controllable way, understanding its properties and then exploiting its potential for practical devices such as field-effect transistors and biosensors.

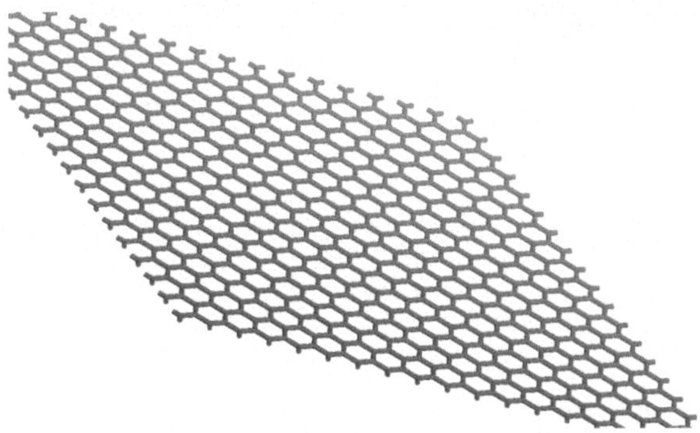

Figure 15.1 Schematic image of graphene with a hexagonal crystal structure.

It was suggested that graphene would offer far superior performance in detecting biomolecules than nanomaterials-based sensors, for several reasons. Firstly, a graphene sheet is only one atomic layer thick, with every atom at the surface of the sheet, so that adsorption of a molecule onto the graphene will change the electrical properties of the entire graphene sheet; it also offers a larger detection area compared to carbon nanotubes. This means that graphene-based sensors will be capable of extremely high sensitivity. Secondly, graphene is a truly two-dimensional material and, as such, has its whole volume exposed to surface adsorbates, which maximises their effects on its conductivity. Furthermore, well-prepared graphene has few crystal defects (Choucair & Stride, 2009; Geim & Novoselov, 2007) which ensures a low level of excess noise caused by thermal switching. Add to this the robust nature of carbon chemistry (Wang *et al.*, 2009) – graphene was recently demonstrated to be the strongest material ever measured (Lee *et al.*, 2008) – and there is obvious potential for developing reliable, long-lived and exquisitely sensitive devices (Ao *et al.*, 2008; Yang *et al.*, 2010). Furthermore, graphene is highly amenable to microfabrication. Indeed, ever since the first experimental study on graphene, researchers have used traditional microfabrication approaches, such as masking and lithography, to manufacture graphene devices. Finally, graphene (Figure 15.2) which consists of a single atomic sheet of conjugated sp^2 carbon atoms has a wide open double-sided surface that can undergo a broad class of organic reactions analogous to unsaturated systems in organic molecules. This creates a unique structural motif with double-sided decoration of functional groups on an extended sheet. In principle these functional groups allow layer-by-layer coordination assembly in a "supramolecular" fashion, schematically shown in Figure 15.2. These unique constructs will help to develop the high performance chemical and biosensor devices.

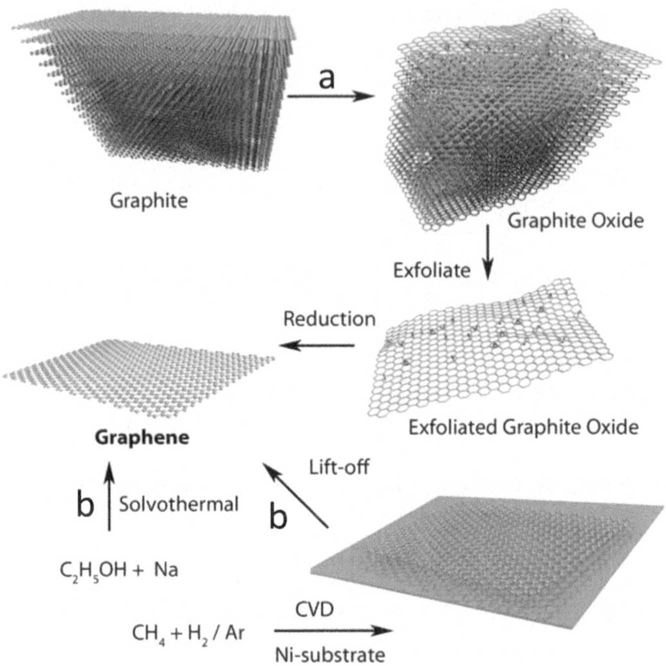

Figure 15.2 (a) Schematic representation of the oxidation-exfoliation-reduction process used to generate individual sheets of reduced graphene oxide from graphite. (b) Schematic representation of the approaches to produce graphene by chemical vapor deposition or by solvothermal reaction (Yang *et al.*, 2010).

15.2 SYNTHESIS AND FUNCTIONALISATION OF GRAPHENE

Graphene-based biosensors require methods for reproducibly fabricating quality graphene in large volumes. Experimentally, graphene is an extremely young material, so no such methods exist yet. There is, however, a flurry of research worldwide trying to find methods to make industrial fabrication of graphene a reality. Many of these efforts are drawing inspiration from the more mature synthetic approaches that have been developed for fabricating carbon nanotubes during the last 20 years, which, no doubt, is accelerating the rate of progress. Still, there is some way to go for developing large-volume, high-purity methods to make graphene, as will be illustrated in this overview. At present, there are four main approaches to 'making' graphene (Yang *et al.*, 2010) (Figure 15.2). Clearly, each of the four classes of graphene-production methods has its limitations, and none have yet reached the point needed for commercial manufacture of graphene-based biosensors. It is also apparent that a variety of different forms of graphene-like materials can be produced via these methods.

This offers an even larger diversity of properties for designing biosensors, but also endless opportunities for confusion if the materials are poorly characterized or inadequately, or simply incorrectly, described when the work is published.

The coupling of a sensing element (such as an organic molecule, enzyme, cell or antibody) with a transducer is the basis of an electrochemical bio/chemical sensor. The two crucial components of a bio/chemical sensor are usually integrated by the immobilisation of the recognition molecule onto the surface of the transducer. The integration of functional sensing molecule-nanoparticle hybrid systems with surfaces paves the way for the generation of ordered architectures with new functionalities. Such hybrid nanostructured materials, which exhibit unique electronic, electrochemical properties with sensing molecules, display unique recognition for analytical applications and can be used for the construction of the sensing components. The rich chemistry of graphene provides great potential for building up nanostructured sensing interfaces for electrochemical bio/chemical sensors. Surface modification at the molecular level can potentially provide a reproducible and robust method of fabricating electrochemical sensors where some control over the orientation and distribution of the sensing elements is afforded. In other words, immobilisation of the sensing elements (molecules) onto the graphene surface is a particularly demanding aspect of the fabrication of bio/chemical sensors (Gooding *et al.*, 2003).

There are two basic approaches to immobilise sensing elements onto graphene electrodes: non-covalent and covalent binding. Direct covalent attachment of sensing elements onto the graphene surface has the greatest potential for the development of commercial sensors due to the stability of the resultant bond. One of the most common methods for the covalent functionalization of graphene oxides or graphene involves reaction with carboxylic acids (–COOH) residues on graphene oxide by using carbodiimide coupling, which couples amines to carboxylic acids but involved low efficiency and tedious processes. The click chemistry, especially the Cu(I)-catalyzed 1,3-dipolar azide-alkyne cycloaddition reaction, has received enormous attention in polymer and surface science since its reinvention in 2001. The success of the click cycloaddition on the graphene surface with high yield was reported recently (Salvio *et al.*, 2009; Sun *et al.*, 2010). The most common non-covalent approach to immobilising sensing elements to the graphene surface has been via π-π stacking interactions. The π–π stacking interactions usually occur between two relatively non-polar aromatic rings having overlapping π orbitals. Pyrene-functional precursors have been successfully attached onto aromatic macromolecules such as carbon nanotubes (Liu *et al.*, 2009; Liu *et al.*, 2007), graphene oxide (Liu *et al.*, 2010; Liu *et al.*, 2008), and fullerene and related materials (Sygula *et al.*, 2007). The modification of graphene using π–π stacking should have advantages over previous attachment strategies, because of the conjugated structure of graphene, therefore electrical conductivity will be retained after modification, as the covalent bonds have the potential of disrupting the conjugated structure of grapheme and leading to compromised physical properties.

The surface control of graphene provided over molecular architecture of the sensing interface fabricated has seen some highly elegant sensing interfaces developed recently. More exciting developments are expected in the near future with regard to reproducibility of sensors, sensor arrays and the ability to switch sensing interfaces "on" or "off". If one puts functionalized graphene sheet after sheet on top of one another and thus created a gradual transition from a 2D sheet to a (thin) 3D layered material, this would open up the attractive possibility to systematically explore highly elegant 3D complex sensor concepts.

We expect to see exciting developments in electrochemical sensors made by stacking carefully functionalised graphene, sheet after sheet, on top of one another. Such approaches could be used to move from essentially two-dimensional electrodes to three-dimensional layered electrodes, unlocking options for systematic exploration of sophisticated sensors based on three-dimensional electrode architectures (Figure 15.3).

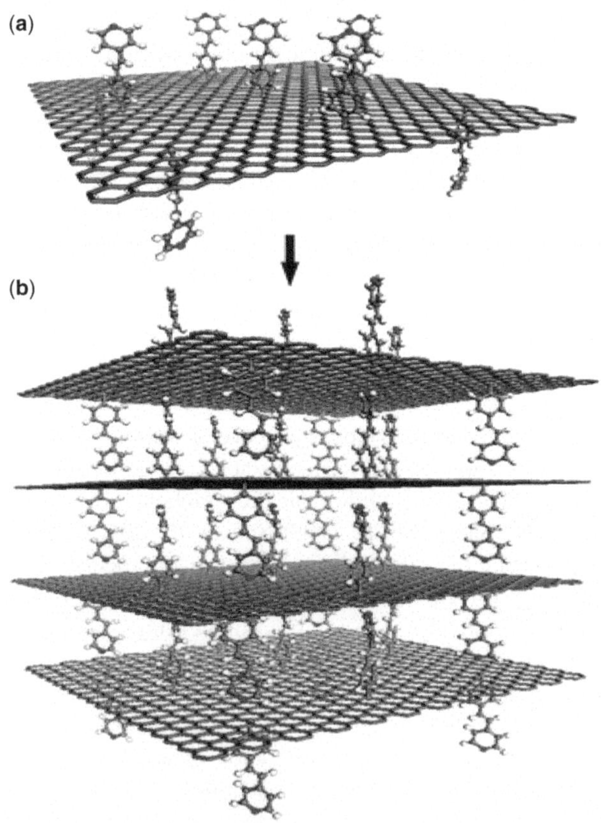

Figure 15.3 Schematic structure of a double-sided decoration of functional groups on graphene (Feng *et al.*, 2010).

It is quite clear that unique properties of functionalized graphene have much to contribute to the construction of relevant nanostructured surfaces. Apart from their use as building block, they have many useful properties which can be tuned by control over size, thickness of the functional graphene sheet and can be utilized in the fabrication of nanostructured sensing devices. The chemical functionalities associated with graphene enable the assembly of 3D nanoparticle architectures on surfaces. Graphene functionalized with various molecular and biomolecular units can be assembled into complex hybrid systems. It is very encouraging that functional multilayer graphene has been constructed by using layer by layer techniques (Liu *et al.*, 2010; Wang *et al.*, 2010). The rapid synthesis of three-dimensional graphene superstructures with tuneable properties will surly continue to expand. Eventually the integration of graphene with sensing elements, which display unique recognition properties, can yield novel 3-D hybrid electrochemical sensing systems.

15.3 GRAPHENE-BASED GAS SENSORS

Demonstrating the potential of graphene as a sensor material (Schedin *et al.*, 2007), first reported their results in 2007 for graphene-based molecular sensors that detected gas molecules, such as NO_2, NH_3, and O_2, with fast response time and high sensitivity at room temperature. The sensing mechanism involves measuring the change in conductance of graphene, caused by charge transfer from gas molecules adsorbed on its surface. A number of theoretical studies in recent years have examined the interactions between gases and graphene layers and substantiated the foregoing findings. Simulations, for example, confirmed that H_2O and NO_2 were acceptors and CO and NH_3 were donors. In addition, these revealed that charge transfer was largely independent on binding sites on pristine graphene, but strongly depended on molecular orientation (Leenaerts *et al.*, 2008). Supporting the need to functionalize graphene for good sensitivity, the other modeling verified that gas molecules were adsorbed only weakly on pristine graphene, however, adsorbed more strongly on doped graphene or defective graphene that contained a vacancy (Ao *et al.*, 2008; Zhang *et al.*, 2009). Specifically, defective graphene had the greatest sensitivity for CO, NO and NO_2, whereas boron-doped (p-type) graphene was most sensitive to NH_3. An *ab-initio* study of graphene nanoribbons also implicated dangling edge bonds in strong gas adsorption (Huang *et al.*, 2008). Stronger adsorption, whether from defects, doping or edge sites, generally resulted in more substantial changes in the density of states and the conductivity of graphene, as were required for effective sensors. The main problem with these strong interactions, however, is that adsorption is effectively irreversible on realistic timeframes and for realistic regeneration temperatures (Zhang *et al.*, 2009). So, there must be a balance between sensitivity and reversibility of the sensing reactions, unless we intend to use graphene only in "one off" sensors.

Several recent papers have focused on making reduced graphene oxide into gas sensors (Arsat *et al.*, 2009; Fowler *et al.*, 2009; Lu *et al.*, 2009; Robinson *et al.*, 2008). The basic reasoning behind these studies is that graphene oxide, besides being easy to process, offers the ability to tailor the amount of functional groups, albeit in a rather coarse way, by varying the degree of reduction. This in turn alters the response of sensors based on the graphene oxide to the adsorption of gas molecules because, consistent with the aforementioned theoretical studies, defects or functional groups in graphene oxide offer higher energy adsorption sites and so can increase the selectivity and sensitivity to gas adsorption. Of course, these defects and sp^3 regions also severely decrease the conductivity of graphene oxide, and thus a balance must be struck. In their 2008 publication, Robinson *et al.* (2008) concluded that gas molecules adsorbed on both, regions of pristine graphene and on functionalised and defective areas in the reduced graphene oxide. Such conclusion was reached because the electrical signal displayed two parts. The first part was a rapid response due to gas adsorption onto low-energy sites, that is, sp^2-bonded graphene areas, and it was recoverable simply by removing the gas. The second part was a slow response due to gas interactions with functional groups and defects, which was only recovered after a moderate heating. Overall, the reduced graphene oxide was able to detect four types of gases at parts-per-billion levels. Similar types of results were obtained in the other studies (Arsat *et al.*, 2009; Fowler *et al.*, 2009; Lu *et al.*, 2009).

15.4 GRAPHENE-BASED BIOSENSORS

In a departure from gas detection, by exploiting quenching of fluorescent tags during adsorption of biomolecules on graphene oxide, Lu and co-workers (Lu *et al.*, 2010; Lu *et al.*, 2009) have developed an optical approach to sensing DNA fragments and proteins in solution (Figure 15.4). Recognizing that nucleobases of DNA weakly bind to graphene oxide and reduced graphene oxide, probably via van der Waals forces (Varghese *et al.*, 2009), the team speculated that fluorescently tagged olignucleotides would bind to graphene-oxide flakes and, in doing so, would have their fluorescence quenched. This concept was confirmed when the presence of graphene oxide quenched 97% of the fluorescent signal from the single-stranded DNA. As the group hypothesized, the subsequent addition of the complementary strands of DNA caused hybridisation, displacing the tagged DNA from the graphene-oxide surfaces and restoring approximately 77% of the original fluorescent signal. To demonstrate that this approach could be applied more generally, they also attached a fluorescently tagged apatamer for the human-thrombin protein to graphene oxide. Once again, the fluoresence almost entirely disappeared, but returned strongly upon the addition of human thrombin to the system. Additions of other types of proteins, including bovine thrombin, to apatamer quenched on graphene oxide produced only small increases in fluorescent signal, indicating the reasonable specificity of this approach.

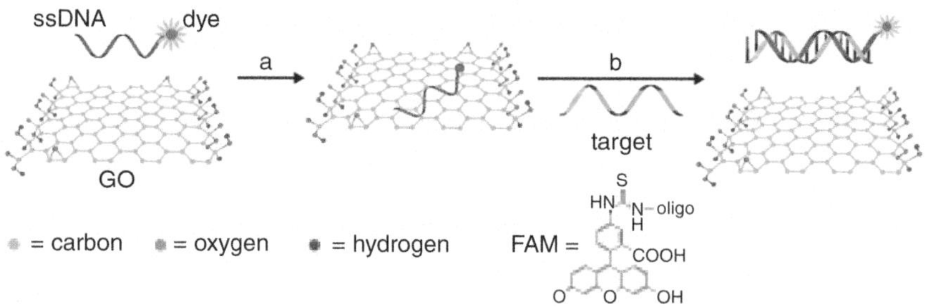

Figure 15.4 Schematic representation of the target-induced fluorescence change of the ssDNA-FAM-GO complex (see text for details). FAM is the fluorescein-based fluorescent dye (adopted from Lu *et al.*, 2010).

Subsequently, the same group developed a simple and highly sensitive aptamer-based assay that relied on the ability of nanomaterials to protect DNA from nuclease cleavages, the aptamer was released from the nanomaterial when challenged with a target. The nuclease then cleaved the free aptamer and released the target, which could initiate a new cycle. Recycling of the target in this way led to a significant signal amplification (Lu *et al.*, 2010).

Ang *et al.* (2010) examined the response of expitaxial few-layer graphene to changes in hydrogen-ion concentration in solution. This was done by fabricating a "solution gated" field effect transistor (FET) made of the expitaxial graphene. Solution gating is used to change graphene's conductance by applying a gate potential to the FET through a reference electrode placed above the FET, with the intervening electrolyte solution acting as the gate dielectric. By studying the changes in capacitance and conductance of this device, the team found that graphene was highly sensitive to changes in pH, even showing supra-Nernstian response for negative gate voltages. These findings indicated that graphene might make an extremely responsive pH sensor for chemical or biological applications.

There have been a few reports of using graphene-based materials in electrical (i.e., field-effect type) biosensors to date (Huang *et al.*, 2010; Mohanty & Berry, 2008; Ohno *et al.*, 2010). Mohanty and Berry (2008) used graphene oxide heavily functionalised with covalent bonds, rather than pristine graphene. The functionalisation allowed them to ensure specificity of detecting the hybridisation of DNA strands or bacteria, as measured by small changes in the low conductivity of the graphene oxide. Consequently, although making effective biosensors, they were unable to exploit the unique electronic properties and low electrical noise of graphene sheets to achieve ultrahigh sensitivity. Subsequently, Ohno *et al.* (2010) built an electrolyte-gated field-effect transistor from micromechanically exfoliated graphene. This showed high selective detection of IgE proteins as measured by changes in conductivity. They also attempted to detect surface-protein adsorption and showed that the conductance of the graphene device increased with exposure to a single type of protein at a concentration of several hundred pico-molar, although the device lacked selectivity for specific proteins.

The ability to pattern large-area ultrathin rGO films with high yield and reproducibility will open up an avenue to widely incorporate graphene into electronics or bioelectronics. He and co-workers (2010) demonstrated the patterning of centimetre long, micrometer wide, ultrathin (1–3 nm), continuous rGO – network films as the conducting channels of FETs on various substrates. Their rGO FETs exhibit remarkable and bending-insensitive field-effects, which are able to detect the presence and dynamic cellular secretion of biomolecules (He *et al.*, 2010). Recently, Ang *et al.* (2010) demonstrated the effect of charged lipid membranes and the interplay of charged impurity and ionic screening mechanisms on the control of electrical response in a biomimetic membrane-coated graphene biosensor. A charged biomimetic membrane imposes an impurity potential which causes voltage shifts in the charge neutrality point and increases minimum conductivity. The impurity potential is screened by mobile ions upon thinning or disrupting of the membrane by a membrane-disrupting agent, leading to a recovery of the voltage-shift initially induced by the uncoated graphene plane. The sensing concept developed here based on the dynamics of membrane diffusion and changes in ionic screening on graphene may be extended to other biorecognition systems, such as ligand–receptor binding and gated control of ion channels embedded in membranes (Figure 15.5).

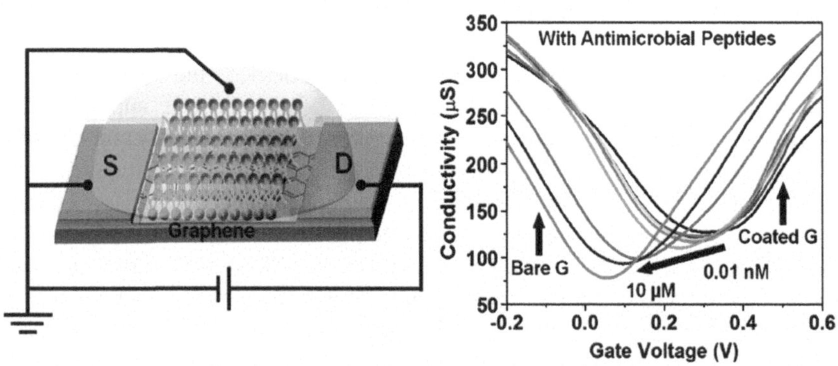

Figure 15.5 The electronic properties of graphene can be modulated by a charged lipid bilayer adsorbing on the surface. Biorecognition events which lead to changes in membrane integrity can be monitored electrically using an electrolyte-gated biomimetic membrane–graphene (adopted from Ang *et al.*, 2010).

The remainder of the biosensor papers to date has focused on detecting biomolecules by means of electrochemical reactions involving various types of "graphenes". For instance, Lu and co-workers (Lu *et al.*, 2008; Lu *et al.*, 2007) have created novel electrode materials for measuring the glucose concentration, a process that typically involves the oxidation of glucose by the enzyme glucose oxidase and then measurement of the resulting hydrogen peroxide (H_2O_2) concentration. The new electrodes were nanocomposite films made of thermally exfoliated graphite "nanoplatelets" – effectively micrometre-scale flakes of multilayer graphene – dispersed in the conductive polymer Nafion, with or without the addition of glucose oxidase. The team found that the nanocomposites of Nafion and multilayer graphene exhibited a rapid response and significant oxidation and reduction currents, starting at about 0.3 V and −0.15 V, respectively, in H_2O_2 solution, whereas a Nafion-modified gold electrode had negligible electrochemical response. These results showed that the graphite nanoplatelets catalysed the oxidation and reduction of the H_2O_2 and lowered the over potential to detect the peroxide, much like carbon nanotubes do. However, the addition of glucose oxidase to the nanocomposites produced electrodes that had an excellent response to glucose, performing up to three times better than most sensors made with carbon nanotubes (Lu *et al.*, 2007). In the subsequent publication, this group precipitated catalytic nanoparticles of platinum or palladium onto the graphite nanoplatelets before incorporating them into Nafion nanocomposites (Lu *et al.*, 2008). In this way, the nanoplatelets acted as a substrate for the metallic particles, keeping them extremely small and, at least in the case of platinum, evenly distributed. Thus, the electrodes had exceptionally high surface area-to-volume ratio for these costly metals, producing low-cost, high-performance electrodes that markedly decreased the over potential needed to detect H_2O_2.

A different approach to glucose detection was attempted by Shan *et al.* (2009), who reported the use of electrodes based on reduced graphene oxide "protected" with polyvinylpyrrolidone and incorporated into a film of polyethylenimine-functionalised ionic liquid, doped with glucose oxidase. A key point of their study was to exploit the high surface area and moderate electrical conductivity of reduced graphene oxide to attempt direct electron transfer between the glucose oxidase and the electrode. Like many studies in this field, however, the question remains as to whether the observed redox peaks of the electrochemistry observed were genuinely from flavin adenine dinucleotide sitting in its native configuration within glucose oxidase or merely from a denatured form of the enzyme. It is well-known that the active redox center of glucose oxidase, the flavin adenine dinucleotide, is deeply embedded in a protective protein shell, which makes direct electron transport to electrodes extremely difficult. So it seems far more likely that the observed redox peaks came from denatured glucose oxidase, leaving it unclear if or how the reduced graphene oxide might have mediated direct electrochemical communications between the redox-active enzyme and the electrode used in this study.

Furthermore, Long *et al.* (2011) reported a disposable tyrosinase electrochemical biosensor based on the immobilization of tyrosinase-protected gold nanoparticles onto the 1-pyrenebutanoic acid, succinimidyl ester functional graphene oxide for the determination of catechol. The bi-functional molecule PASE was assembled onto grapheme oxide (GO) sheets through noncovalent interaction, which was confirmed by UV–VIS, FT-IR and AFM. Subsequently, the Tyr–Au was immobilized on the PASE–GO sheets forming a biocompatible nanocomposite. TEM indicated that Tyr–Au successfully assembled onto the functionalized GO sheets. The disposable tyrosinase biosensor was fabricated by modifying the screen printed electrode with the Tyr–Au/PASE–GO bionanocomposite. The biosensor provides a suitable microenvironment for retaining bioactivity of tyrosinase and exhibits a good analytical performance for the amperometric detection of catechol. In addition, this biosensor exhibits high sensitivity, good reproducibility and acceptable stability, showing a great potential for rapid, cost-effective and on-field analysis of phenolic compounds (Figure 15.6).

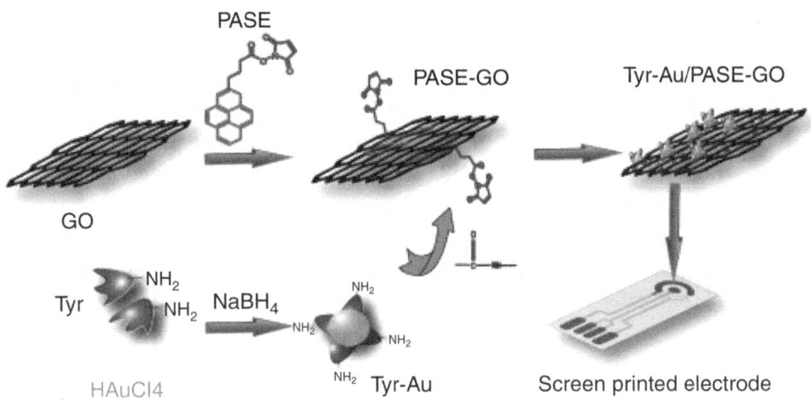

Figure 15.6 Assembling process of Tyr-Au/PASE-GO bionanocomposite on a screen-printed electrode transducer (adopted from Long *et al.*, 2011).

Based on the first clinical trial used aptamer AS1411 and functionalized graphen, other researchers developed a reusable electrochemical sensor that can realize label-free cancer cell detection (Feng *et al.*, 2011). Generally, AS1411 and its complementary DNA are used as a nanoscale anchorage substrate to capture/release cells. The constructed graphene surface does not show cytotoxicity under the experimental conditions. Furthermore, AS1411 forms G-quadruplex and has high binding affinity and specificity to the over-expressed nucleolin on the cancer cell surface compared with normal cells. Also, a designed electrochemical aptasensor has the ability to differentiate cancer cells and normal ones. The sensor can be regenerated using AS1411 cDNA and reusable for cancer cell detection. This graphene/aptamer-based design can be adaptable for detection of protein, small molecules, and nucleic acid targets by using different DNA sequence and specificity (Feng *et al.*, 2011).

The anodized epitaxial grapheme (EG) electrode demonstrates high level of performance for electrochemical impedance as well as differential pulse voltammetry detection of immobilized DNA and free DNA, respectively, at solid–liquid interfaces. On the anodized EG surface, because of the presence of oxygen functionalities as well as π conjugated domains, the anchoring of the DNA probe can be achieved by either covalent grafting or noncovalent π–π stacking (Dubuisson *et al.*, 2011; Loh *et al.*, 2010). It was found that anodized EG also allowed the direct voltametric sensing of single-nucleotide mismatch in the DNA hybridization assay (Figure 15.7).

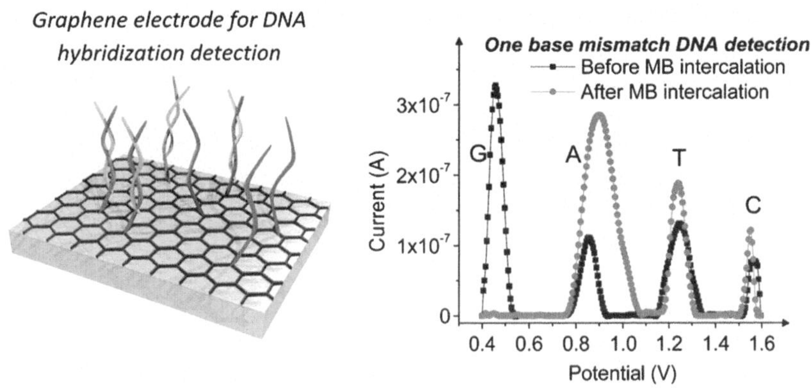

Figure 15.7 The anodized epitaxial graphene (EG) electrode demonstrates a high level of performance for differential pulse voltammetry detection of immobilized DNA and free DNA using methylene blue (MB) (Dubuisson *et al.*, 2011).

15.5 FUTURE OUTLOOK

The rapidly growing activity in graphene-based biosensors is due to the promising properties of these materials and their obvious success in electrical and optical sensors to date. Of course, commercial production of graphene-based sensors is still some way off, and numerous important challenges must be met along the way. Fabrication is one of these challenges, as researchers seek cost-effective production methods that retain graphene's essential properties and can be scaled up. If high-purity graphene is the primary material needed for sensor applications, then, at this stage, chemical vapor deposition or epitaxial growth seem to be the synthetic routes that will likely come to dominate. However, if reduced graphene oxide continues to build on its early promise for gas sensing, then chemical-exfoliation of graphite and subsequent wet processing, such as spin casting, presumably will be scaled up. Either way, or perhaps with entirely new approaches, there is currently so much research directed towards fabricating graphene that it appears the challenge of cost-effective, large-scale fabrication cannot remain unconquered for too much longer.

The other major challenge is avoiding undesirable contaminants on the sensors. Graphene is lipophilic, so it adsorbs contaminants such as hydrocarbons, and water vapor tends to be prevalent on graphene and/or its supporting substrates. Furthermore, some processing methods add other contaminants, such as the polymer residues used in lithography or the solvents used for casting graphene films. Such impurities can be removed by annealing of the graphene under vacuum or reducing atmosphere, but the key is to avoid the subsequent re-contamination during the use of graphene devices. The most likely solution at this stage would seem to be exploiting the controlled functionalization of clean graphene that is essential for producing sensors with high selectivity. With appropriate molecular engineering, the functionalization layer can also be used to protect graphene's high surface area from adsorption of impurities, thereby minimizing these issues while obtaining the greatest sensitivity. Well-designed functionalization of graphene sensors could also help balance the strength of adsorption of analytes, which is needed for good sensitivity, against the reversibility of the detection process, which is a fundamental requirement for gas sensors that can take continuous measurements.

REFERENCES

Ang P. K., Jaiswal M., Lim C. H. Y. X., Wang Y., Sankaran J., Li A., Lim H. K., Wohland T., Barbaros O. and Loh K. P. (2010). A bioelectronic platform using a graphene-lipid bilayer interface. *ACS Nano*, **4**(12), 7387–7394.

Ao Z. M., Yang J., Li S. and Jiang Q. (2008). Enhancement of CO detection in Al doped graphene. *Chemical Physics Letters*, **461**(4–6), 276–279.

Armani A. M., Kulkarni R. P., Fraser S. E., Flagan R. C. and Vahala K. J. (2007). Label-free, single-molecule detection with optical microcavities. *Science*, **317**(5839), 783–787.

Arsat R., Breedon M., Shafiei M., Spizziri P. G., Gilje S., Kaner R. B., Kalantar-zadeha K. and Wlodarskia W. (2009). Graphene-like nano-sheets for surface acoustic wave gas sensor applications. *Chemical Physics Letters*, **467**(4–6), 344–347.

Besteman K., Lee J. O., Wiertz F. G. M., Heering H. A. and Dekker C. (2003). Enzyme-coated carbon nanotubes as single-molecule biosensors. *Nano Letters*, **3**(6), 727–730.

Chen H., Muller M. B., Gilmore K. J., Wallace G. G. and Li D. (2008). Mechanically strong, electrically conductive, and biocompatible graphene paper. *Advanced Materials*, **20**(18), 3557–3561.

Choucair M. and Stride J. (2009). Gram-scale production of graphene based on solvothermal synthesis and sonication. *Nature Nanotechnology*, **4**(1), 40–44.

Cui R., Pan H. C., Zhu J. J. and Chen H. Y. (2007). Versatile immunosensor using CdTe quantum dots as electrochemical and fluorescent labels. *Analytical Chemistry*, **79**(22), 8494–8501.

Cui Y., Wei Q. Q., Park H. K. and Lieber C. M. (2001). Nanowire nanosensors for highly sensitive and selective detection of biological and chemical species. *Science*, **293** (5533), 1289–1292.

Dubuisson E., Yang Z. Y. and Loh K. P. (2011). Optimizing label-free DNA electrical detection on graphene platform. *Analytical Chemistry*, **83**(7), 2452–2460.

Fan F. R. F. and Bard A. J. (1995). Electrochemical detection of single molecules. *Science*, **267**(5199), 871–874.

Feng J. C., Sun S. T., Cao Y. W. and Wu P. Y. (2010). Click chemistry as a route for the immobilization of well-defined polystyrene onto graphene sheets. *Journal of Materials Chemistry*, **20**(27), 5605–5607.

Feng L. Y., Chen Y., Ren J. S. and Qu X. G. (2011). A graphene functionalized electrochemical aptasensor for selective label-free detection of cancer cells. *Biomaterials*, **32**(11), 2930–2937.

Fowler J. D., Allen M. J., Tung V. C., Yang Y., Kaner R. B. and Weiller B. H. (2009). Practical chemical sensors from chemically derived graphene. *ACS Nano*, **3**(2), 301–306.

Geim A. K. and Novoselov K. S. (2007). The rise of graphene. *Nature Materials*, **6**(3), 183–191.

Gooding J. J., Mearns F., Yang W. and Liu J. (2003). Self-assembled monolayers into the 21st century: recent advances and applications. *Electroanalysis*, **15**, 81–96.

Hahm J. and Lieber C. M. (2004). Direct ultrasensitive electrical detection of DNA and DNA sequence variations using nanowire nanosensors. *Nano Letters*, **4**(1), 51–54.

He Q. Y., Sudibya H. G., Yin Z. Y., Wu S. X., Li H., Boey F., Huang W., Chen P. and Zhang H. (2010). Centimeter-long and large-scale micropatterns of reduced graphene oxide films: fabrication and sensing applications. *ACS Nano*, **4**(6), 3201–3208.

Hou S. Y., Chen H. K., Cheng H. C. and Huang C. Y. (2007). Development of zeptomole and attomolar detection sensitivity of biotin-peptide using a dot-blot gold nanoparticle immunoassay. *Analytical Chemistry*, **79**(3), 980–985.

Huang B., Li Z. Y., Liu Z. R., Zhou G., Hao S. G., Wu J., Gu B. L. and Duan W. H. (2008). Adsorption of gas molecules on graphene nanoribbons and its implication for nanoscale molecule sensor. *Journal of Physical Chemistry C*, **112**(35), 13442–13446.

Huang Y. X., Dong X. C., Shi Y. M., Li C. M., Li L. J. and Chen P. (2010). Nanoelectronic biosensors based on CVD grown graphene. *Nanoscale*, **2**(8), 1485–1488.

Lee C., Wei X. D., Kysar J. W. and Hone J. (2008). Measurement of the elastic properties and intrinsic strength of monolayer graphene. *Science*, **321**(5887), 385–388.

Leenaerts O., Partoens B. and Peeters F. M. (2008). Adsorption of H_2O, NH_3, CO, NO_2, and NO on graphene: a first-principles study. *Physical Review B*, **77**(12), 125416.

Li D. and Kaner R. B. (2008). Materials science – graphene-based materials. *Science*, **320**(5880), 1170–1171.

Liu J. Q., Tao L., Yang W. R., Li D., Boyer C., Wuhrer R., Braet F. and Davis T. P. (2010). Synthesis, characterization, and multilayer assembly of pH sensitive graphene-polymer nanocomposites. *Langmuir*, **26**(12), 10068–10075.

Liu Z., Fan A. C., Rakhra K., Sherlock S., Goodwin A., Chen X. Y., Yang Q. W., Felsher D. W. and Dai H. J. (2009). Supramolecular stacking of doxorubicin on carbon nanotubes for *in vivo* cancer therapy. *Angewandte Chemie-International Edition*, **48**(41), 7668–7672.

Liu Z., Robinson J. T., Sun X. M. and Dai H. J. (2008). PEGylated nanographene oxide for delivery of water-insoluble cancer drugs. *Journal of the American Chemical Society*, **130**(33), 10876–10877.

Liu Z., Sun X. M., Nakayama-Ratchford N. and Dai H. J. (2007). Supramolecular chemistry on water-soluble carbon nanotubes for drug loading and delivery. *ACS Nano*, **1**(1), 50–56.

Loh K. P., Lim C. X., Hoh H. Y. and Ang P. K. (2010). Direct voltammetric detection of DNA and pH sensing on epitaxial graphene: an insight into the role of oxygenated defects. *Analytical Chemistry*, **82**(17), 7387–7393.

Long Y. T., Song W., Li D. W., Li Y. T. and Li Y. (2011). Disposable biosensor based on graphene oxide conjugated with tyrosinase assembled gold nanoparticles. *Biosensors & Bioelectronics*, **26**(7), 3181–3186.

Lu C. H., Li J. A., Lin M. H., Wang Y. W., Yang H. H., Chen X. and Chen G. N. (2010). Amplified aptamer-based assay through catalytic recycling of the analyte. *Angewandte Chemie-International Edition*, **49**(45), 8454–8457.

Lu C. H., Yang H. H., Zhu C. L., Chen X. and Chen G. N. (2009). A graphene platform for sensing biomolecules. *Angewandte Chemie-International Edition*, **48**(26), 4785–4787.

Lu G. H., Ocola L. E. and Chen J. H. (2009). Gas detection using low-temperature reduced graphene oxide sheets. *Applied Physics Letters*, **94**(8), 083111.

Lu H. P., Xun L. Y. and Xie X. S. (1998). Single-molecule enzymatic dynamics. *Science*, **282** (5395), 1877–1882.

Lu J., Do I., Drzal L. T., Worden R. M. and Lee I. (2008). Nanometal-decorated exfoliated graphite nanoplatelet based glucose biosensors with high sensitivity and fast response. *ACS Nano*, **2**(9), 1825–1832.

Lu J., Drzal L. T., Worden R. M. and Lee I. (2007). Simple fabrication of a highly sensitive glucose biosensor using enzymes immobilized in exfoliated graphite nanoplatelets Nafion membrane. *Chemistry of Materials*, **19**(25), 6240–6246.

Mirkin C. A., Letsinger R. L., Mucic R. C. and Storhoff J. J. (1996). A DNA-based method for rationally assembling nanoparticles into macroscopic materials. *Nature*, **382**(6592), 607–609.

Mohanty N. and Berry V. (2008). Graphene-based single-bacterium resolution biodevice and DNA transistor: interfacing graphene derivatives with nanoscale and microscale biocomponents. *Nano Letters*, **8**(12), 4469–4476.

Nam J. M., Thaxton C. S. and Mirkin C. A. (2003). Nanoparticle-based bio-bar codes for the ultrasensitive detection of proteins. *Science*, **301**(5641), 1884–1886.

Nie S. M. and Emery S. R. (1997). Probing single molecules and single nanoparticles by surface-enhanced Raman scattering. *Science*, **275**(5303), 1102–1106.

Novoselov K. S., Geim A. K., Morozov S. V., Jiang D., Zhang Y., Dubonos S. V., Grigorieva I. V. and Firsov A. A. (2004). Electric field effect in atomically thin carbon films. *Science*, **306**(5296), 666–669.

Novoselov K. S., Jiang Z., Zhang Y., Morozov S. V., Stormer H. L., Zeitler U., Maan J. C., Boebinger G. S., Kim P. and Geim A. K. (2007). Room-temperature quantum hall effect in graphene. *Science*, **315**(5817), 1379–1379.

Ohno Y., Maehashi K. and Matsumoto K. (2010). Label-free biosensors based on aptamer-modified graphene field-effect transistors. *Journal of the American Chemical Society*, **132**(51), 18012–18013.

Patolsky F., Zheng G. F., Hayden O., Lakadamyali M., Zhuang X. W. and Lieber C. M. (2004). Electrical detection of single viruses. *Proceedings of the National Academy of Sciences of the United States of America*, **101**(39), 14017–14022.

Ratinac K. R., Yang W. R., Ringer S. P. and Braet F. (2010). Towards ubiquitous environmental gas sensors – capitalising on the promise of graphene. *Environmental Science and Technology*, **44**(4), 1167–1176.

Robinson J. T., Perkins F. K., Snow E. S., Wei Z. Q. and Sheehan P. E. (2008). Reduced graphene oxide molecular sensors. *Nano Letters*, **8**(10), 3137–3140.

Salvio R., Krabbenborg S., Naber W. J. M., Velders A. H., Reinhoudt D. N. and van der Wiel W. G. (2009). The formation of large-area conducting graphene-like platelets. *Chemistry-A European Journal*, **15**(33), 8235–8240.

Schedin F., Geim A. K., Morozov S. V., Hill E. W., Blake P., Katsnelson M. I. and Novoselov K. S. (2007). Detection of individual gas molecules adsorbed on graphene. *Nature Materials*, **6**(9), 652–655.

Shan C. S., Yang H. F., Song J. F., Han D. X., Ivaska A. and Niu L. (2009). Direct electrochemistry of glucose oxidase and biosensing for glucose based on graphene. *Analytical Chemistry*, **81**(6), 2378–2382.

Star A., Gabriel J. C. P., Bradley K. and Gruner G. (2003). Electronic detection of specific protein binding using nanotube FET devices. *Nano Letters*, **3**(4), 459–463.

Sun S. T., Cao Y. W., Feng J. C. and Wu P. Y. (2010). Click chemistry as a route for the immobilization of well-defined polystyrene onto graphene sheets. *Journal of Materials Chemistry*, **20**(27), 5605–5607.

Sygula A., Fronczek F. R., Sygula R., Rabideau P. W. and Olmstead M. M. (2007). A double concave hydrocarbon buckycatcher. *Journal of the American Chemical Society*, **129** (13), 3842–3843.

Varghese N., Mogera U., Govindaraj A., Das A., Maiti P. K., Sood A. K. and Rao C. N. R. (2009). Binding of DNA nucleobases and nucleosides with graphene. *Chemphyschem*, **10**(1), 206–210.

Wang G. X., Wang B., Park J., Yang J., Shen X. P. and Yao J. (2009). Synthesis of enhanced hydrophilic and hydrophobic graphene oxide nanosheets by a solvothermal method. *Carbon*, **47**(1), 68–72.

Wang Z. Q., Zeng G. H., Xing Y. B., Gao J. A. and Zhang X. (2010). Unconventional layer-by-layer assembly of graphene multilayer films for enzyme-based glucose and maltose biosensing. *Langmuir*, **26**(18), 15022–15026.

Wu G. H., Datar R. H., Hansen K. M., Thundat T., Cote R. J. and Majumdar A. (2001). Bioassay of prostate-specific antigen (PSA) using microcantilevers. *Nature Biotechnology*, **19**(9), 856–860.

Yang W. R., Ratinac K. R., Ringer S. P., Thordarson P., Gooding J. J. and Braet F. (2010). Carbon nanomaterials in biosensors: should you use nanotubes or graphene? *Angewandte Chemie-International Edition*, **49**(12), 2114–2138.

Zhang Y. H., Chen Y. B., Zhou K. G., Liu C. H., Zeng J., Zhang H. L. and Peng Y. (2009). Improving gas sensing properties of graphene by introducing dopants and defects: a first-principles study. *Nanotechnology*, **20**(18), 185504.

Chapter 16

Nanoparticle based sensors for water quality testing

P. Pati, R. Kent and P. Vikesland

16.1 INTRODUCTION

Recurring outbreaks of waterborne disease in developed countries are a continual challenge to public health and economic productivity (Hrudey & Hrudey, 2007; Straub & Chandler, 2003). Improved pathogen detection strategies would considerably improve the water industry's capacity to rapidly respond to and prevent future outbreaks. The burgeoning field of nanomaterial-enabled biosensors holds significant promise for pathogen detection in drinking water systems.

Nanomaterial-enabled biosensors generally include three components—a recognition element, a nanomaterial, and a signal transduction technique (Vikesland & Wigginton, 2010). Antibodies are the most widely used recognition element, with nucleic acid aptamers, carbohydrates, and antimicrobial peptides used less frequently (Ngundi *et al.*, 2006). Recognition elements conjugated to a nanomaterial provide the sensor with its specificity, and a recognition event produces a quantifiable signal that is used to detect the presence of target organisms. Commonly employed signal transduction techniques rely upon optical and electrochemical methods. The novel optical, and electrical properties of various nanomaterials make them uniquely adapted for integration with these signal transduction methods (Alivisatos, 2004). This chapter discusses the application of nanomaterials that are appropriate for different signal transduction techniques for pathogen detection in water treatment.

16.2 OPTICAL SIGNAL TRANSDUCTION

Optical techniques are the most widespread signal transduction mechanisms used for biosensing. The exceptional optical characteristics of many nanomaterials have naturally led to their use in biosensor applications. In this section the use of fluorescent nanomaterials and noble metal nanomaterials for pathogen detection is discussed.

16.2.1 Quantum dot nanomaterials

Fluorescence measurements take advantage of the intense emission of light from a fluorophore that occurs following absorption of light at a different wavelength. Pathogens stained with fluorescent labels can be enumerated directly under a fluorescence microscope or indirectly by correlating the intensity of the emitted light to the pathogen concentration. Such measurements are widely employed in the detection of waterborne pathogens. US EPA method 1623, for example, uses antibodies labelled with fluorescent dyes to detect *Crytposporidium parvum* oocysts and *Giardia lamblia* cysts. Fluorescent nanomaterials, however, offer a number of advantages over the organic dyes traditionally used in these detection schemes.

Quantum dots (QDs) are a prominent nanomaterial in fluorescent biosensor applications (Figure 16.1). They have been used for detecting diverse pathogens ranging from enteric and respiratory viruses to pathogenic bacteria and infectious protozoa (Bentzen *et al.*, 2005; Lee *et al.*, 2004; Rawsthorne *et al.*, 2009; Xue *et al.*, 2009). QDs are quasi-spherical semiconductor nanocrystals with a core-shell structure. The core is often CdSe or CdTe, with typical shell materials including CdSe, CdTe, CdS, ZnS, ZnSe, PbS, and PbTe. The emission wavelengths of QDs can be tuned simply by varying their size, shell thickness, and chemical composition (Alivisatos, 2004; Smith & Nie, 2010).

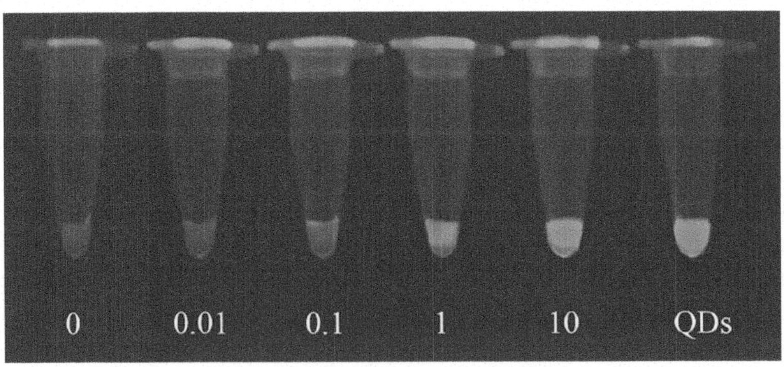

Figure 16.1 Applications of QDs in pathogen detection. Visual illustration of the correlation between fluorescence intensity and concentration of human enterovirus 71 (EV71) in a QD-based detection scheme. In this scheme, QD fluorescence is initially quenched in dye-antibody-QD conjugates and subsequently restored upon binding of the antibody to the target organism. Concentrations of 0, 0.01, 0.1, 1, and 10 μg EV71/mL and a pure QD suspension are shown (Chen *et al.*, 2011; Adapted with permission from *Anal. Chem.* Vol. 83 (19), pp. 1147–1152. Copyright 2011, American Chemical Society).

The emission spectra of QDs are characteristically narrow and symmetric, and their absorption spectra are broad (Hahn *et al.*, 2008; Resch-Genger *et al.*, 2008). These features are desirable for multiplex detection because a single excitation source can be used to produce multiple distinguishable signals. Multiplex detection enabled by QDs has been accomplished for important pairs of pathogens, such as *C. parvum* and *G. lamblia*, *Bacillus anthracis* and *Yersinia pestus*, and *Escherichia coli* and *Salmonella typhimurium* (Yang & Li, 2006; Zahavy *et al.*, 2010; Zhu *et al.*, 2004). In contrast, organic dyes generally have narrower absorption bands, so multiple excitation sources are needed in order to simultaneously detect different pathogens. Moreover, the emission spectra of traditional fluorescent dyes are broad and asymmetric, which can lead to spectral overlap that complicates multicolour analyses (Hahn *et al.*, 2008). These drawbacks necessitate pathogen specific methodologies that can be costly and time-consuming. Since QDs largely overcome these difficulties they promise to substantially improve the efficiency of waterborne pathogen monitoring since they enable detection of multiple pathogens in a single analysis. The multiplexing ability of QDs may one day make automated platforms, such as flow cytometry (FCM), feasible for routine water quality testing (Hahn *et al.*, 2008; Zahavy *et al.*, 2010).

Beyond their potential for multiplex analyses, detection limits reported for QD-enabled assays rival or surpass those of established protocols (Lee *et al.*, 2004; Sanvicens *et al.*, 2011; Su & Li, 2004). QDs impart added sensitivity because of their relatively long fluorescence lifetimes, photostability, and bright fluorescence intensity. Fluorescence lifetime is the average time that a fluorophore will remain in its excited state prior to emitting light to return to its ground state. Common fluorescence lifetimes are 1–10 ns for organic dyes and 10–100 ns for QDs (Resch-Genger *et al.*, 2008). The autofluorescent background typically has a fluorescence lifetime comparable to organic dyes. Time gating can thus be used to take advantage of the longer fluorescence lifetime of QDs by removing the short-lived autofluorescence signal (Dahan *et al.*, 2001; Grecco *et al.*, 2004). Upon removal of the background signal in this manner, the signal-to-noise ratio reportedly can be increased by more than an order of magnitude (Dahan *et al.*, 2001). The excellent photostability of QDs further enhances their sensitivity compared to dye molecules (Lee *et al.*, 2004; Zhu *et al.*, 2004). Organic dyes are highly susceptible to photobleaching, so the diminished sensitivity of traditional fluorescent dyes is expected when high energy or extended illumination are required (Alivisatos, 2004). Finally, QDs generally fluoresce more brightly than dye molecules because they have higher molar extinction coefficients and quantum yields due to their crystallinity and precisely engineered structure (Smith & Nie, 2010). Blinking (i.e. intermittent dark periods experienced by individual QDs) is sometimes reported to diminish the brightness and reproducibility of QD-based biosensors (Alivisatos, 2004; Resch-Genger *et al.*, 2008; Smith & Nie, 2010). Nevertheless, QD fluorescence is consistently reported to be brighter than that of dyes in pathogen detection (Hahn *et al.*, 2008; Su & Li, 2004).

Specificity is as important as sensitivity in pathogen detection, and QD-based assays have demonstrated adequate specificity for environmental analyses. Interference from reservoir water, treated wastewater effluent, membrane bioreactor permeate, and tap water has been shown to be 50% lower for QDs than for organic dyes (Lee *et al.*, 2008). Detection of 1% pathogenic *E. coli* in 99% non-pathogenic *E. coli* was achieved in another study (Hahn *et al.*, 2008). Others have also reported that immunostaining with QDs has resulted in highly specific binding to target pathogens in mixtures of viruses or bacteria, although some nonspecific binding has been noted, as expected for any technique employing antibodies as a recognition element (Agrawal *et al.*, 2005; Ikanovic *et al.*, 2007; Liu *et al.*, 2011; Su & Li, 2004; Xue *et al.*, 2009; Yang & Li, 2006). At least one study has found that non-specific binding and QD aggregation reduced QD specificity relative to common dyes (Ferrari & Bergquist, 2007). Particle aggregation can be decreased by using an appropriate surface coating (Resch-Genger *et al.*, 2008). Specificity of QD labels was improved in an FCM study by staining a target pathogen with two sizes of QDs (Zahavy *et al.*, 2010).

16.2.2 Silica and polymeric nanomaterials

Although the focus here has been on QDs, much of the previous discussion holds for dye-doped silica and polymeric nanoparticles as well. In general, silica nanoparticles are affected by aggregation and dye leakage to a lesser extent than polymeric nanoparticles (Wang & Tan, 2006). These nanomaterials contain thousands of luminescent molecules encapsulated within a silica or polymer matrix. The high concentration of dye molecules within an individual nanoparticle results in intense fluorescent emissions that enable sensitive pathogen detection down to 1 cell/mL (Zhao *et al.*, 2004). Importantly, the silica or polymer matrix stabilizes the dye molecules against photobleaching, so the fluorescence signal remains stable over prolonged periods of illumination (Yan *et al.*, 2007). As shown in Figure 16.2, multiple dyes with overlapping excitation wavelengths can be incorporated into these nanomaterials at different ratios to fine tune their emission spectra and enable multiplex detection (Jiang *et al.*, 2010; Wang *et al.*, 2005; Tan *et al.*, 2006; Wang & Tan, 2006). Simultaneous detection of *E. coli*, *S. typhimurium* and *Bacillus cereus* has been reported (Zhao *et al.*, 2004). Time-gating is also possible for silica and polymeric nanoparticles if inorganic dyes with sufficiently long fluorescence lifetimes are used (Wu *et al.*, 2009). The versatility of silica nanoparticles is further extended by the ability to readily functionalize their surface using established silane chemistry, which can be exploited to decrease aggregation and nonspecific binding (Bagwe *et al.*, 2006).

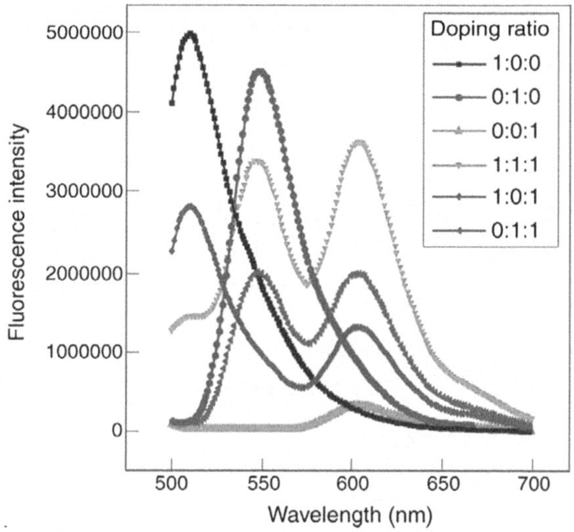

Figure 16.2 Fluorescence emission spectra of silica nanoparticles doped with variable ratios of FITC, rhodamine 6G, and 6-carboxyl-X-rhodamine (X:Y:Z) under 300 nm ultraviolet illumination (Wang & Tan, 2006; Adapted with permission from *Nano Letters* Vol. 6 (1), pp. 84–88. Copyright 2006, American Chemical Society).

16.2.3 Noble metal nanomaterials

Noble metal nanomaterials exhibit interesting optical properties due to the localized surface plasmon resonance (LSPR) that results from the collective oscillation of their conduction electrons upon light absorption. One useful outcome of this behaviour is the surface enhanced Raman scattering (SERS) effect. Raman spectroscopy takes advantage of inelastically scattered light from the molecular vibrations of polarisable molecules to yield fingerprint spectra that can be used to identify molecules and organisms. The intensity of Raman scattered light is too low to be used for environmental analyses; however, the strong electromagnetic field produced near noble metal nanomaterial surfaces due to the LSPR can enhance the Raman signal by orders of magnitude. Further enhancements can be achieved by a charge transfer mechanism that occurs when the analyte specifically binds to the metal surface and by a resonance effect that occurs when the incident laser light overlaps an absorption peak of the analyte (Halvorson & Vikesland, 2010). These combined effects can result in enhancements as high as 10^{14}, enabling single molecule detection under special conditions; thus, SERS generates signals that provide sufficient sensitivity to compete with fluorescence techniques (Kneipp *et al.*, 1997; Nie & Emery, 1997). Furthermore, SERS is well suited for multiplex analyses since Raman bands are 10–100 times narrower than fluorescence bands.

Either the intrinsic SERS signal of an organism or the signal from a SERS label can be used for pathogen detection (Figure 16.3). Directly measuring the SERS

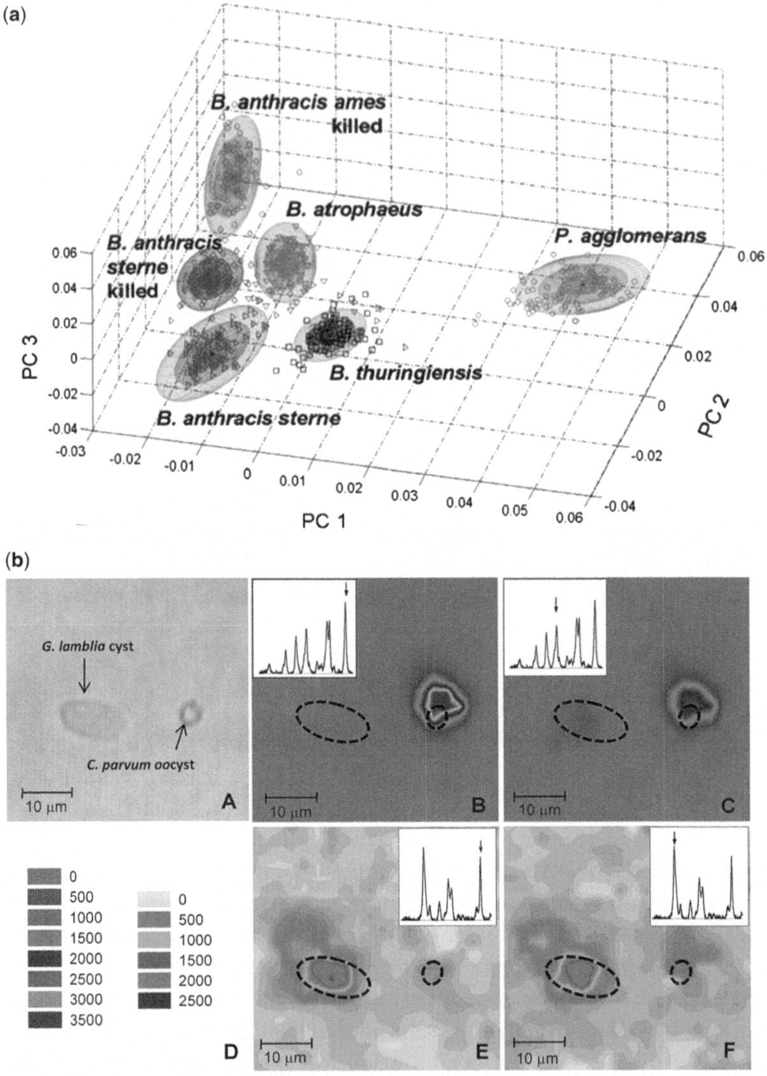

Figure 16.3 Pathogen detection by SERS. (a) Differentiation of closely related bacteria by applying principle components analysis (PCA) to the intrinsic SERS spectra of the organisms (Guicheteau *et al.*, 2008; Adapted with permission from *Appl. Spect.* Vol. 62 (3), pp. 267–272. Copyright 2008, Society for Applied Spectroscopy). (b) Multiplex detection of *C. parvum* oocysts (top row) and *G. lamblia* cysts (bottom row) using two different dyes (RBITC and MGITC, respectively) with gold nanoparticle SERS labels (Rule & Vikesland, 2009; Adapted with permission from *Env. Sci. Technol.* Vol. 43 (4), pp 1147–1152. Copyright 2009, American Chemical Society).

spectra of a microorganism eliminates the need for additional recognition elements along with the false positives associated with nonspecific binding. To obtain an organism's intrinsic SERS signal, the target must be brought into close proximity to a noble metal (typically gold or silver) nanostructure. This step can be accomplished by depositing organisms directly onto a SERS-active substrate, by coating the organisms with noble metal nanoparticles in solution, or by reducing noble metal ions to metallic nanoparticles directly on the organism's surface (Chu *et al.*, 2008; Kudo *et al.*, 2011). Multivariate statistical tools can be used to differentiate organisms based on variations in their SERS spectra, and it is even possible to obtain strain and mutant specificity and to distinguish between viable and nonviable cells (Chu *et al.*, 2008; Guicheteau *et al.*, 2008; Premasiri *et al.*, 2005).

The need for advanced statistical methods to analyze the intrinsic SERS spectra of different organisms could limit the routine application of this technique in water quality testing. Also, complex environmental samples can complicate collection and interpretation of SERS spectra. For example, chloride and proteins can bind strongly to SERS substrates and adversely affect detection limits (Golightly *et al.*, 2009).

An alternate method for water quality monitoring by SERS is to tag pathogens with SERS probes. The probes could be, for example, gold nanoparticles conjugated to antibodies and Raman-active dye molecules. The antibodies enable the nanoparticles to selectively bind to target organisms, and the fingerprint SERS spectrum of the dye can be used to identify the organism following the binding event. An immunoassay using SERS labels for feline calicivirus has a reported detection limit of 10^6 PFU/mL, which is competitive with other viral assays (Driskell *et al.*, 2005).

Because numerous dyes with unique Raman spectra are available, this technique has significant potential for multiplexing. In fact, this technique enabled the simultaneous detection of eight different probes without any complicated statistical processing of the SERS spectra (Sun *et al.*, 2007; Sun *et al.*, 2008). Pathogenic *E. coli*, *S. typhimurium*, and *Staphylococcus aureus* have been simultaneously identified to a detection limit of 10^2 CFU/mL in a mixture with non-pathogenic organisms using this labelling strategy, and multiplex detection of *C. parvum* and *G. lamblia* has also been demonstrated (Ravindranath *et al.*, 2011; Rule & Vikesland, 2009). These studies demonstrate the specificity and sensitivity of this method; however, information about the organisms of interest is lost by using antibodies and reporter molecules (e.g. viability cannot be established, non-specific binding will result in false positives, and mutant strains may not be detected).

16.3 ELECTROCHEMICAL SIGNAL TRANSDUCTION

Electrochemical techniques are a second widely used category of methods employed for the development of nano-enabled biosensors. Nanoparticles, owing to their large surface area, tunable electronic properties and rapid electrochemical activity (Alivisatos, 2004; Sanvicens *et al.*, 2009; Sadik *et al.*, 2009), are desirable candidates for use in electrochemical biosensors. Electrochemical techniques in nano-biosensors primarily rely on two features: (i) large signal amplification and (ii) improved biomolecule immobilization that reduces background "noise" (current signal) and takes advantage of rapid electrochemistry (Sanvicens *et al.*, 2009; Sadik *et al.*, 2009).

In most electrochemical biosensors, biorecognition occurs due the pairing of the sensor and the target molecule (Drummond *et al.*, 2003). The primary mechanism of "electrochemical sensing" involves measurement of changes in the conductivity of the nanostructure following a biorecongnition element. These conductivity changes in the biorecognition element can occur upon interaction with an analyte containing the biogenic species of interest or due to the electro-chemical stripping of the metal after its immobilization onto the analyte (Sanvicens *et al.*, 2009; Sadik *et al.*, 2009; Zhang *et al.*, 2009). Besides metallic nanoparticles, polymer nanowires (Pal *et al.*, 2007; Cho & Lee, 2008) and carbon nanotubes are also used in selective electrochemical detection of biological binding activity. In the following paragraphs, the applications of nano-based electrochemical techniques for pathogen detection are discussed.

Thevenot *et al.* (2001) defined electrochemical biosensors as follows: "An electrochemical biosensor is a self-contained integrated device, which is capable of providing specific quantitative or semi-quantitative analytical information using a biological recognition element (biochemical receptor) which is retained in direct spatial contact with an electrochemical transduction element." Carbon nanotubes (CNTs) are one of the most widely used nanomaterials for electrochemical signal transduction owing to their large specific surface area, high tensile strength, greatly enhanced thermal and electrical conductivity, and their amenability towards self-assembly (Dai, 2002; Richard *et al.*, 2003; Ren & Chou, 2001).

Carbon-based nanomaterials functionalized with metallic nanoparticles and biological recognition elements (e.g. antibodies or aptamers) have been used effectively as electrochemical biosensors for detection of pathogens and chemical contaminants. Villamizar *et al.* (2008). devised a label-free technique to selectively detect *Salmonella infantis* using CNT field effect transistors (FETs). The FETs consisted of a network of single-walled carbon nanotubes (SWCNTs) acting as the conductor channel, with the SWCNT walls doped with anti-*Salmonella* antibodies (Figure 16.4). This method combined the electrochemical transduction properties of the SWCNT-FET with antigen-antibody based biorecognition. The authors reported a detection limit of 100 CFU/mL in 1 hour. The biosensor also selectively detected *Salmonella infantis* in the presence of *Streptococcus pyogenes* and *Shigella sonnei* at concentrations of 500 CFU/mL. Aptamer-functionalized SWCNT-FETs have also been used to selectively detect *E. coli* in the presence of *Salmonella typhimurium*, with a detection time of less than 20 minutes (So *et al.*, 2008).

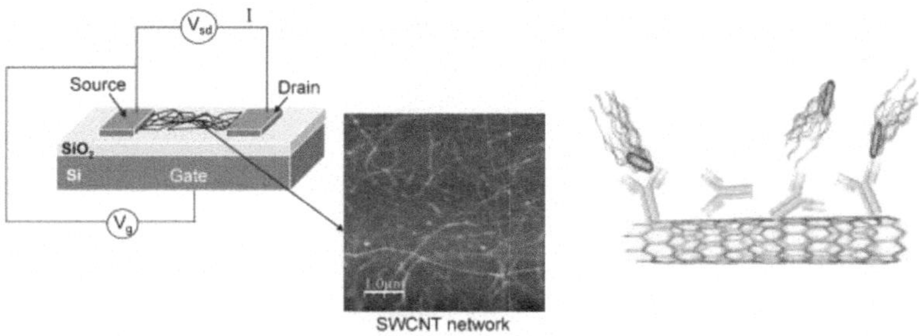

Figure 16.4 Experimental set up for the detection of *Salmonella infantis* using anti-*Salmonella*-antibody functionalized CNT-FETs (Villamizar *et al.*, 2008; Licensed with permission from *Biosensors and Bioelectronics* Vol. 24 (2), pp. 279–283. Copyright 2008, Elsevier).

In addition to the electrochemical properties of the sensing element, the size and morphology of the biomolecules or pathogens targeted also play an important role in detection efficiency. Garcia-Aljaro *et al.* (2010) proposed a CNT-based immunosening method to detect *E. coli* O157:H7 and the bacteriophage T7, using antibody-functionalized SWCNTs as the transduction element. *E. coli* K12 and the bacteriophage MS2 were used to test the selectivity of the biosensor. The authors reported detection limits of 10^5 CFU/mL for whole bacterial cells, 10^3 CFU/mL for lysed bacterial cells and 10^3 PFU/mL for the bacteriophage T7. The sensor response-time was ≈60 minutes for the bacteria and ≈5 minutes for the bacteriophage. The improved detection limit with the lysed cells was attributed to the higher diffusivity and lower steric hindrance of the lysates because of their lower particle size. Similarly, the bacteriophage T7, due to its nearly spherical shape and much smaller size (hence, faster diffusion), had a markedly lower detection time. Impressively, SWCNTs have been shown to have ≈27 times higher adsorption affinity for *Bacillus subtillis* spores relative to activated carbon, despite activated carbon having similar surface properties and a larger BET surface area relative to SWCNTs (Upadhyayula *et al.*, 2009). The authors attributed this result to the large portion of mesopore and macropore volume accessible to *B. subtilis* spores and the SWCNTs' fibrous and highly ordered graphite sheet structure.

Nanocomposites of CNTs, metallic nanoparticles and biological receptors have also been successfully used for pathogen detection in aqueous and food matrices. In such nanocomposites, the metallic nanoparticles help enhance the signal triggered by the electrochemical reaction occurring at the receptor's surface. Cheng *et al.* (2009) used CNTs coated with Fe_3O_4 magnetic nanoparticles (MNPs) to enhance the current signal in a tyrosinase (Tyr) biosensor designed to detect *E. coli* (Figure 16.5). Tyr catalyzes the oxidation of phenolic compounds in the presence of oxygen. In a two-step reaction, phenol was hydroxylated to form catechol, which then further reacted to form o-quinone. O-quinone was reduced electrochemically back to catechol at the electrode, without any mediator, resulting in a change (in this case, a decrease) in the signal (i.e. the current).

$$\text{phenol} + O_2 \xrightarrow{\text{tyrosinase}} \text{catechol}$$

$$\text{catechol} + O_2 \xrightarrow{\text{tyrosinase}} o - quinone + H_2O$$

At the electrode:

$$\text{o} - \text{quinone} + H^+ + 2e \xrightarrow{\text{yields}} \text{catechol}$$

The presence of CNTs in the nanocomposite was found to improve the performance of the Tyr biosensor by increasing the conductivity and the specific area, allowing more Tyr to be immobilized on the nanocomposite (Cheng *et al.*, 2009).

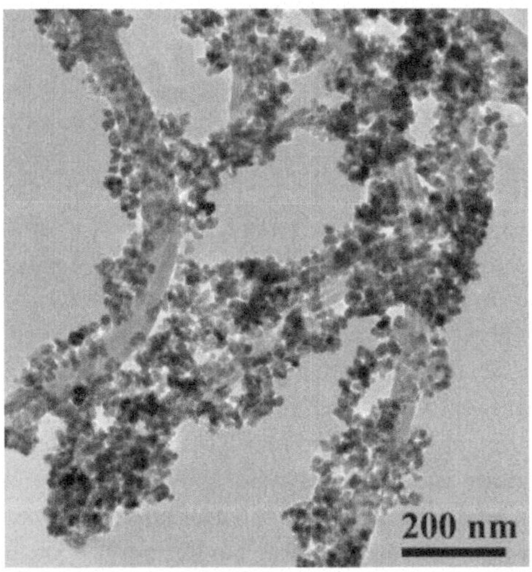

Figure 16.5 TEM image of Fe_3O_4 MNPs-coated CNTs nanocomposite (Cheng *et al.*, 2009; Licensed with permission from *Electrochimica Acta* Vol. 54 (9), pp. 2588–2594. Copyright 2009, Elsevier).

In electrochemical biosensing techniques, the targeted biomolecules undergo electrochemical transformations due to the redox reactions occurring at the electrode. In amperometric biosensors (where the detection is based on the measurement of current), the measured signal is proportional to the concentration of the species at the electrode's surface (Sadik *et al.*, 2009; Sanvicens *et al.*, 2009). Nanoparticles and nanocomposites that increase the velocity of a desired reaction at the electrode can therefore be extremely useful in amplifying the signal and improving detection. For example, in a study by Tang *et al.* (2007), gold nanoparticles were found to improve the sensitivity of antigen-antibody interactions for mammary the cancer cell CA15-3. The authors attached anti-CA15-3 to immobilized red blood cells (RBCs), functionalized with gold nanoparticles and found that gold nanoparticles and RBC provided a "congenial microenvironment for adsorbed biomolecules" that helped reduce the electron transfer impedance.

This concept has also been applied to amperometric biosensors for pathogen detection. Lin *et al.* (2008) prepared disposable immunosensing strips to detect *E. coli.* O15:H7 using gold nanoparticles (AuNPs) for signal amplification (Figure 16.6). The authors employed 13 nm diameter AuNP and ferrocenedicarboxylic acid (FeDC) to modify disposable screen-printed carbon electrodes (SPCEs) using two *E. coli* O15:H7-specific antibodies for specificity. The authors suggested that the AuNP in the immunosensing strips increased the effective area of the electrodes and provided a compatible microenvironment that facilitated the formation of complexes between the antibodies and the bacterial cells. The authors tested the specificity of the AuNPs/FeDC–SPCE immunosensing strips using 5×10^7 CFU/mL samples of *E. coli* O157:H7, *E. coli* K12, *L. monocytogens*, *S. choleraesuis* and *V. parahaemolyticus* cultures. The detection limit for *E. coli* O157:H7 was reported to be 5×10^3 CFU/mL) in milk with an analysis time of approximately 1 hour.

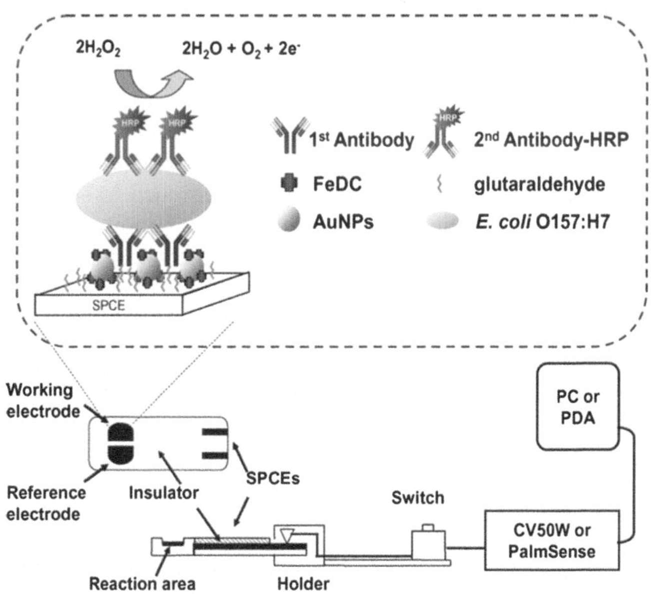

Figure 16.6 Schematic of a disposable immunosensing strip using AuNPs/FeDC–SPCE immunosensor system to detect *E. coli* O157:H7 detection (Lin *et al.*, 2008; Licensed with permission from *Biosensors and Bioelectronics* Vol. 23 (12), pp. 1832–1837. Copyright 2008, Elsevier).

In a recent study, Cheng *et al.* (2011), developed a novel electrochemical nanobiosensor based on a nanoporous alumina membrane modified electrode for label-free detection of *E. coli* cells and the analysis of viable but non-culturable (VBNC) *E. coli*. Anti-*E. coli* antibody was attached to the alumina membrane surface and bovine serum albumin (BSA) was employed to coat any remaining surface sites to prevent nonspecific binding. The formation of immune complexes on the membrane surface hindered the diffusive mass-transfer of a redox probe, thus resulting in the reduction of the signal response of the sensor in the presence of the biomolecule. The authors reported high specificity towards *E. coli* in the presence of *Serratia marcescens* and *Salmonella typhimurium*, with these interferences contributing to 8.2–12.4% of the total signal at 10^3 CFU/mL.

In addition to pathogens and biogenic species, electrochemical nanobiosensors can also be used for the detection of dissolved chemicals. Yao *et al.* (2006) developed a biosensor to detect sterigmatocystin (ST), a carcinogenic mycotoxin, using gold electrodes modified with multi-wall carbon nanotubes (MWCNTs), onto which aflatoxin–detoxifizyme (ADTZ) was covalently bound. The Au/MWNTs–ADTZ electrodes had a detection limit of 0.13 μM (Yao *et al.*, 2006). In another study, Chen (2008) presented a conductometric immunosensing technique for the detection of staphylococcal enterotoxin B (SEB) by immobilizing horseradish peroxidase (HRP)-labeled SEB antibody (HRP-anti-SEB) onto AuNP/chitosan-multi-walled carbon nanotube composites (Au/CTS-MWNT). The detection mechanism relied upon the variation in the measured conductivity due to the formation of antibody-antigen binding complexes.

Rahman *et al.* (2006) fabricated an enzymatic biosensor to amperometrically detect phosphate ions by covalently immobilizing pyruvate oxidase (PyO) onto nanoparticles comprised of poly-5, 2:5′,2″-terthiophene-3′-carboxylic acid, poly-TTCA (nano-CP). The detection limit was determined to be about 0.3 μM in 20 mM HEPES buffer solution, with a response time of ≈6 s. The authors reported the sensor to be very selective towards phosphate ions with no interference observed at 0.1 μM concentrations of other biologically common solutes such as Cl^-, SO_4^{2-}, NO_3^-, ClO_4^-, ascorbate, glutamate, or urate (Rahman *et al.*, 2006).

Acetylcholinesterase (AChE) based renewable biosensors have been fabricated for organophosphate pesticides (OPs). Gong *et al.* (2009) devised such a biosensor by immobilizing AChE onto AuNP–polypyrrole (AuNP-PPy) nanowires composites, with the PPy deposited onto a glass carbon electrode (GCE). The biosensor had a detection limit of 2 ng/mL and the AChE was reactivated to 95% of the original activity after immersing in 5.0 mM pralidoxime iodide. The authors mentioned that the Au NP-PPy nanocomposite improved the detection sensitivity by providing a biocompatible microenvironment (Gong *et al.*, 2009).

As described in the studies mentioned above, electrochemical techniques in nanobiosensors can be used to fabricate highly sensitive and selective detectors of pathogens and chemical contaminants in both aqueous and food matrices. Miniaturized nanobiosensors hold promise as hand-held devices for rapid or real time pathogen detection. Besides the electrochemical properties of the transduction element, the physicochemical properties of the species of interest (bacterial cells, spores, proteins) and the Debye screening effects (Sorgenfrei *et al.*, 2011) during the interaction between the biomolecule and the recognition element also determine the effectiveness of electrochemical biosensors.

16.4 CONCLUSIONS AND FUTURE OUTLOOK

Nanomaterial based sensors are continually being developed both within the water treatment community and within the broader nanotechnology field. At present, a majority of the existing sensors are solely laboratory based, however, with continued investment many of the most promising sensors will ultimately become commercially available and thus their merits relative to existing detection schemes should be carefully considered. This chapter has highlighted only a select few areas where great strides have been made in recent years. Numerous other types of nanomaterial based sensors exist and even more will be developed in the coming years. The capacity for rapid, multiplex detection of numerous pathogens and chemical contaminants ensures that interest in this field will continue to grow.

ACKNOWLEDGEMENTS

This work was supported by grants from the Virginia Tech Institute of Critical Technology and Applied Science (ICTAS) and NSF Awards CBET-0547342 & CBET-1133746.

REFERENCES

Agrawal A., Tripp R. A., Anderson L. J. and Nie S. M. (2005). Real-time detection of virus particles and viral protein expression with two-color nanoparticle probes. *Journal of Virology*, **79**(13), 8625–8628.

Alivisatos P. (2004). The use of nanocrystals in biological detection. *Nature Biotechnology*, **22**(1), 47–52.

Bagwe R. P., Hilliard L. R. and Tan W. H. (2006). Surface modification of silica nanoparticles to reduce aggregation and nonspecific binding. *Langmuir*, **22**(9), 4357–4362.

Bentzen E. L., House F., Utley T. J., Crowe J. E. and Wright D. W. (2005). Progression of respiratory syncytial virus infection monitored by fluorescent quantum dot probes. *Nano Letters*, **5**(4), 591–595.

Chen L., Zhang X. W., Zhang C. L., Zhou G. H., Zhang W. P., Xiang D. S., He Z. K. and Wang H. Z. (2011). Dual-color fluorescence and homogeneous immunoassay for the determination of human enterovirus 71. *Analytical Chemistry*, **83**(19), 7316–7322.

Cheng Y., Liu Y., Huang J., Li K., Xian Y., Zhang W. and Jin L. (2009). Amperometric tyrosinase biosensor based on Fe_3O_4 nanoparticles-coated carbon nanotubes nanocomposite for rapid detection of coliforms. *Electrochimica Acta*, **54**(9), 2588–2594.

Cheng M. S., Lau S. H., Chow V. T. and Toh C. S. (2011). Membrane-based electrochemical nanobiosensor for *Escherichia coli* detection and analysis of cells viability. *Environmental Science and Technology*, **45**(15), 6453–6459.

Cho S. I. and Lee S. B. (2008). Fast electrochemistry of conductive polymer nanotubes, synthesis, mechanism, and application. *Accounts of Chemical Research*, **41**(6), 699–707.

Chu H. Y., Huang Y. W. and Zhao Y. P. (2008). Silver nanorod arrays as a surface-enhanced Raman scattering substrate for foodborne pathogenic bacteria detection. *Applied Spectroscopy*, **62**(8), 922–931.

Dahan M., Laurence T., Pinaud F., Chemla D. S., Alivisatos A. P., Sauer M. and Weiss S. (2001). Time-gated biological imaging by use of colloidal quantum dots. *Optics Letters*, **26**(11), 825–827.

Dai H. (2002). Carbon nanotubes, synthesis, integration, and properties. *Accounts of Chemical Research*, **35**(12), 1035–1044.

Driskell J. D., Kwarta K. M., Lipert R. J., Porter M. D., Neill J. D. and Ridpath J. F. (2005). Low-level detection of viral pathogens by a surface-enhanced Raman scattering based immunoassay. *Analytical Chemistry*, **77**(19), 6147–6154.

Drummond T. G., Hill M. G. and Barton J. K. (2003). Electrochemical DNA sensors. *Nature Biotechnology*, **21**(10), 1192–1199.

Ferrari B. C. and Bergquist P. L. (2007). Quantum dots as alternatives to organic fluorophores for *Cryptosporidium* detection using conventional flow cytometry and specific monoclonal antibodies, lessons learned. *Cytometry Part A*, **71A**(4), 265–271.

Garcia-Aljaro C., Cella L. N., Shirale D. J., Park M., Munoz F. J., Yates M. V. and Mulchandani A. (2010). Carbon nanotubes-based chemiresistive biosensors for detection of microorganisms. *Biosensors & Bioelectronics*, **26**(4), 1437–1441.

Golightly R. S., Doering W. E. and Natan M. J. (2009). Surface-enhanced Raman spectroscopy and homeland aecurity, a perfect match? *ACS Nano*, **3**(10), 2859–2869.

Gong J., Wang L. and Zhang L. (2009). Electrochemical biosensing of methyl parathion pesticide based on acetylcholinesterase immobilized onto Au-polypyrrole interlaced network-like nanocomposite. *Biosensors & Bioelectronics*, **24**(7), 2285–2288.

Grecco H. E., Lidke K. A., Heintzmann R., Lidke D. S., Spagnuolo C., Martinez O. E., Jares-Erijman E. A. and Jovin T. M. (2004). Ensemble and single particle photophysical properties (two-photon excitation, anisotropy, FRET, lifetime, spectral conversion) of commercial quantum dots in solution and in live cells. *Microscopy Research and Technique*, **65**(4–5), 169–179.

Guicheteau J., Argue L., Emge D., Hyre A., Jacobson M. and Christesen S. (2008). *Bacillus* spore classification via surface-enhanced Raman spectroscopy and principal component analysis. *Applied Spectroscopy*, **62**(3), 267–272.

Hahn M. A., Keng P. C. and Krauss T. D. (2008). Flow cytometric analysis to detect pathogens in bacterial cell mixtures using semiconductor quantum dots. *Analytical Chemistry*, **80**(3), 864–872.

Halvorson R. A. and Vikesland P. J. (2010). Surface-enhanced Raman spectroscopy (SERS) for environmental analyses. *Environmental Science & Technology*, **44**(20), 7749–7755.

Hrudey S. E. and Hrudey E. J. (2007). Published case studies of waterborne disease outbreaks – Evidence of a recurrent threat. *Water Environment Research*, **79**(3), 233–245.

Ikanovic M., Rudzinski W. E., Bruno J. G., Allman A., Carrillo M. P., Dwarakanath S., Bhahdigadi S., Rao P., Kiel J. L. and Andrews C. J. (2007). Fluorescence assay based on aptamer-quantum dot binding to *Bacillus thuringiensis* spores. *Journal of Fluorescence*, **17**(2), 193–199.

Jiang H. F., Wang G. L., Zhang W. Z., Liu X. Y., Ye Z. Q., Jin D. Y., Yuan J. L. and Liu Z. G. (2010). Preparation and time-resolved luminescence bioassay application of multicolor luminescent lanthanide nanoparticles. *Journal of Fluorescence*, **20**(1), 321–328.

Kneipp K., Wang Y., Kneipp H., Perelman L. T., Itzkan I., Dasari R. and Feld M. S. (1997). Single molecule detection using surface-enhanced Raman scattering (SERS). *Physical Review Letters*, **78**(9), 1667–1670.

Kudo H., Itoh T., Kashiwagi T., Ishikawa M., Takeuchi H. and Ukeda H. (2011). Surface enhanced Raman scattering spectroscopy of Ag nanoparticle aggregates directly photo-reduced on pathogenic bacterium (*Helicobacter pylori*). *Journal of Photochemistry and Photobiology A-Chemistry*, **221**(2–3), 181–186.

Lee L. Y., Hu J. Y., Ong S. L., Ng H. Y., Wong S. W., Feng Y. and Tan X. (2008). Alternative immunofluorescent labeling of *Cryptosporidium parvum* in water samples using semiconductor quantum dots. *Water Environment Research*, **80**(8), 725–731.

Lee L. Y., Ong S. L., Hu J. Y., Ng W. J., Feng Y. Y., Tan X. L. and Wong S. W. (2004). Use of semiconductor quantum dots for photostable immunofluorescence labeling of *Cryptosporidium parvum*. *Applied and Environmental Microbiology*, **70**(10), 5732–5736.

Lin Y. H., Chen S. H., Chuang Y. C., Lu Y. C., Shen T. Y., Chang C. A. and Lin C. S. (2008). Disposable amperometric immunosensing strips fabricated by Au nanoparticles-modified screen-printed carbon electrodes for the detection of foodborne pathogen *Escherichia coli* O157:H7. *Biosensors & Bioelectronics*, **23**(12), 1832–1837.

Liu X. H., Luo J. P., Tian Q., Liu C. X. and Cai X. X. (2011). Detection of *Bacillus anthracis* using fluorescence immunoassay with quantum dots labels. *Chinese Journal of Analytical Chemistry*, **39**(2), 163–167.

Ngundi M. M., Kulagina N. V., Anderson G. P. and Taitt C. R. (2006). Nonantibody-based recognition: alternative molecules for detection of pathogens. *Expert Review of Proteomics*, **3**(5), 511–524.

Nie S. M. and Emery S. R. (1997). Probing single molecules and single nanoparticles by surface-enhanced Raman scattering. *Science*, **275**(5303), 1102–1106.

Pal S., Alocilja E. C. and Downes F. P. (2007). Nanowire labeled direct-charge transfer biosensor for detecting *Bacillus* species. *Biosensors & Bioelectronics*, **22**(9–10), 2329–2336.

Premasiri W. R., Moir D. T., Klempner M. S., Krieger N., Jones G. and Ziegler L. D. (2005). Characterization of the surface enhanced Raman scattering (SERS) of bacteria. *Journal of Physical Chemistry B*, **109**(1), 312–320.

Rahman M. A., Park D. S., Chang S. C., McNeil C. J. and Shim Y. B. (2006). The biosensor based on the pyruvate oxidase modified conducting polymer for phosphate ions determinations. *Biosensors & Bioelectronics*, **21**(7), 1116–1124.

Ravindranath S. P., Wang Y. L. and Irudayaraj J. (2011). SERS driven cross-platform based multiplex pathogen detection. *Sensors and Actuators B-Chemical*, **152**(2), 183–190.

Rawsthorne H., Phister T. G. and Jaykus L. A. (2009). Development of a fluorescent *in situ* method for visualization of enteric viruses. *Applied and Environmental Microbiology*, **75**(24), 7822–7827.

Ren Z. F. and Chou T. W. (2001). Advances in the science and technology of carbon nanotubes and their composites: a review. *Composites Science and Technology*, **61**(13), 1899–1912.

Resch-Genger U., Grabolle M., Cavaliere-Jaricot S., Nitschke R. and Nann T. (2008). Quantum dots versus organic dyes as fluorescent labels. *Nature Methods*, **5**(9), 763–775.

Richard C., Balavoine F., Schultz P., Ebbesen T. W. and Mioskowski C. (2003). Supramolecular self-assembly of lipid derivatives on carbon nanotubes. *Science*, **300**(5620), 775–778.

Rule K. L. and Vikesland P. J. (2009). Surface-enhanced resonance Raman spectroscopy for the rapid detection of *Cryptosporidium parvum* and *Giardia lamblia*. *Environmental Science & Technology*, **43**(4), 1147–1152.

Sadik O. A., Aluoch A. O. and Zhou A. (2009). Status of biomolecular recognition using electrochemical techniques. *Biosensors and Bioelectronics*, **24**(9), 2749–2765.

Sanvicens N., Pascual N., Fernandez-Arguelles M. T., Adrian J., Costa-Fernandez J. M., Sanchez-Baeza F., Sanz-Medel A. and Marco M. P. (2011). Quantum dot-based array for sensitive detection of *Escherichia coli*. *Analytical and Bioanalytical Chemistry*, **399**(8), 2755–2762.

Sanvicens N., Pastells C., Pascual N. and Marco M. P. (2009). Nanoparticle-based biosensors for detection of pathogenic bacteria. *TrAC Trends in Analytical Chemistry*, **28**(11), 1243–1252.

Smith A. M. and Nie S. M. (2010). Semiconductor nanocrystals: structure, properties, and band gap engineering. *Accounts of Chemical Research*, **43**(2), 190–200.

So H.-M., Park D.-W., Jeon E.-K., Kim Y.-H., Kim B. S., Lee C.-K., Choi S. Y., Kim S. C., Chang H. and Lee J.-O. (2008). Detection and titer estimation of *Escherichia coli* using aptamer-functionalized single-walled carbon-nanotube field-effect transistors. *Small*, **4**(2), 197–201.

Sorgenfrei S., Chiu C. Y., Johnston M., Nuckolls C. and Shepard K. L. (2011). Debye screening in single-molecule carbon nanotube field-effect sensors. *Nano Letters*, **11**(9), 3739–3743.

Straub T. M. and Chandler D. P. (2003). Towards a unified system for detecting waterborne pathogens. *Journal of Microbiological Methods*, **53**(2), 185–197.

Su X. L. and Li Y. B. (2004). Quantum dot biolabeling coupled with immunomagnetic separation for detection of *Escherichia coli* O157:H7. *Analytical Chemistry*, **76**(16), 4806–4810.

Sun L., Yu C. X. and Irudayaraj J. (2007). Surface-enhanced Raman scattering based nonfluorescent probe for multiplex DNA detection. *Analytical Chemistry*, **79**(11), 3981–3988.

Sun L., Yu C. X. and Irudayaraj J. (2008). Raman multiplexers for alternative gene splicing. *Analytical Chemistry*, **80**(9), 3342–3349.

Tan W. H., Wang L. and O'Donoghue M. B. (2006). Nanoparticles for multiplex diagnostics and imaging. *Nanomedicine*, **1**(4), 413–426.

Tang D., Yuan R. and Chai Y. (2007). Biochemical and immunochemical characterization of the antigen-antibody reaction on a non-toxic biomimetic interface immobilized red blood cells of crucian carp and gold nanoparticles. *Biosensors & Bioelectronics*, **22**(6), 1116–1120.

Thevenot D. R., Toth K., Durst R. A. and Wilson G. S. (2001). Electrochemical biosensors: recommended definitions and classification. *Biosensors & Bioelectronics*, **16**(1–2), 121–131.

Upadhyayula V. K., Deng S., Smith G. B. and Mitchell M. C. (2009). Adsorption of *Bacillus subtilis* on single-walled carbon nanotube aggregates, activated carbon and NanoCeram. *Water Research*, **43**(1), 148–156.

Vikesland P. J. and Wigginton K. R. (2010). Nanomaterial enabled biosensors for pathogen monitoring – A review. *Environmental Science & Technology*, **44**(10), 3656–3669.

Villamizar R. A., Maroto A., Rius F. X., Inza I. and Figueras M. J. (2008). Fast detection of *Salmonella infantis* with carbon nanotube field effect transistors. *Biosensors & Bioelectronics*, **24**(2), 279–283.

Wang L. and Tan W. H. (2006). Multicolor FRET silica nanoparticles by single wavelength excitation. *Nano Letters*, **6**(1), 84–88.

Wang L., Yang C. Y. and Tan W. H. (2005). Dual-luminophore-doped silica nanoparticles for multiplexed signaling. *Nano Letters*, **5**(1), 37–43.

Wu J., Ye Z. Q., Wang G. L., Jin D. Y., Yuan J. L., Guan Y. F. and Piper J. (2009). Visible-light-sensitized highly luminescent europium nanoparticles: preparation and application for time-gated luminescence bioimaging. *Journal of Materials Chemistry*, **19**(9), 1258–1264.

Xue X. H., Pan J., Xie H. M., Wang J. H. and Zhang S. (2009). Fluorescence detection of total count of *Escherichia coli* and *Staphylococcus aureus* on water-soluble CdSe quantum dots coupled with bacteria. *Talanta*, **77**(5), 1808–1813.

Yan J. L., Estevez M. C., Smith J. E., Wang K. M., He X. X., Wang L. and Tan W. H. (2007). Dye-doped nanoparticles for bioanalysis. *Nano Today*, **2**(3), 44–50.

Yang L. J. and Li Y. B. (2006). Simultaneous detection of *Escherichia coli* O157:H7 and *Salmonella Typhimurium* using quantum dots as fluorescence labels. *Analyst*, **131**(3), 394–401.

Yao D. S., Cao H., Wen S., Liu D. L., Bai Y. and Zheng W. J. (2006). A novel biosensor for sterigmatocystin constructed by multi-walled carbon nanotubes (MWNT) modified with aflatoxin-detoxifizyme (ADTZ). *Bioelectrochemistry*, **68**(2), 126–133.

Zahavy E., Heleg-Shabtai V., Zafrani Y., Marciano D. and Yitzhaki S. (2010). Application of fluorescent nanocrystals (q-dots) for the detection of pathogenic bacteria by flow-cytometry. *Journal of Fluorescence*, **20**(1), 389–399.

Zhang X., Geng P., Liu H., Teng Y., Liu Y., Wang Q., Zhang W., Jin L. and Jiang L. (2009). Development of an electrochemical immunoassay for rapid detection of *E. coli* using anodic stripping voltammetry based on Cu@Au nanoparticles as antibody labels. *Biosensors & Bioelectronics*, **24**(7), 2155–2159.

Zhao X. J., Hilliard L. R., Mechery S. J., Wang Y. P., Bagwe R. P., Jin S. G. and Tan W. H. (2004). A rapid bioassay for single bacterial cell quantitation using bioconjugated nanoparticles. *Proceedings of the National Academy of Sciences of the United States of America*, **101**(42), 15027–15032.

Zhu L., Ang S. and Liu W. T. (2004). Quantum dots as a novel immunofluorescent detection system for *Cryptosporidium parvum* and *Giardia lamblia*. *Applied and Environmental Microbiology*, **70**(1), 597–598.

Chapter 17

Green synthesis of nanoparticles and nanomaterials

J. Virkutyte and R. S. Varma

17.1 INTRODUCTION

17.1.1 Importance of nanomaterials/nanoparticles

Commercial and research interest in nanotechnology significantly increased in the past several years translating into more than US$9 billion in investment from public and private sources (Eckelman *et al.*, 2008). Nanoparticles are generally defined as particulate matter with at least one dimension below 100 nm (Figure 17.1) (Christian *et al.*, 2008). They also have an exceptionally large surface area to volume ratio, which is the most important characteristic responsible for the widespread use of nanomaterials in a great variety of industries. Unfortunately, high surface area often results into numerous problems associated with surface phenomena, for instance the outer layer atoms in the particle may have a different composition and therefore, chemistry from the rest of the particle. Furthermore, nanoparticle surface will be prone to environmental changes such as redox conditions, pH, ionic strength, adhesion to natural organic matter (NOM) and microorganisms.

17.1.2 Composition of nanomaterials/nanoparticles

In most cases, the exact composition of the target nanomaterial and, most importantly, nanoparticle surface is closely related to their intended application. For instance, if the nanoparticle is manufactured to interact with biological systems, it will have suitable functional groups attached to its surface, such as short peptides and some surfactants to prevent from aggregation and/or agglomeration (Christian *et al.*, 2008). Therefore, most of the scientific efforts are currently directed towards the synthesis of nanoparticles and nanomaterials that suspend in the target media. It involves introduction of coatings or surface active materials that form transient van-der-Waals interactions with the surface of the nanoparticle and exist in equilibrium with the free surfactant molecule.

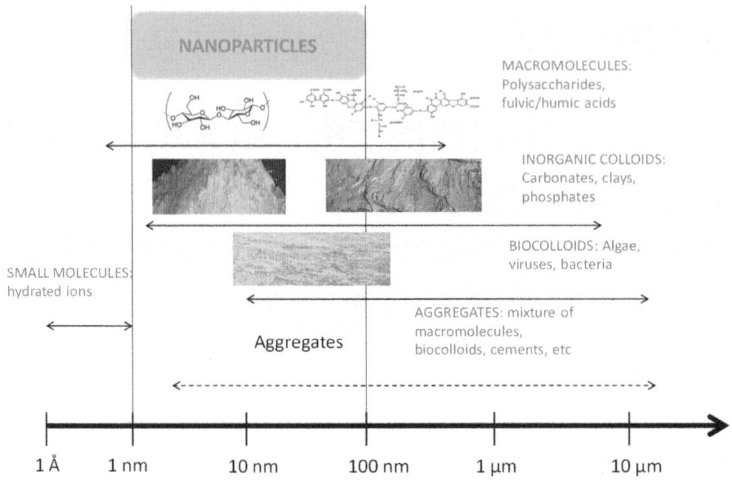

Figure 17.1 Size domains and main representatives.

Other size-dependent significance of synthesized nanoparticles includes manufacturing nanomaterials that enhance the strength and hardness of various solids, modifications of electrical properties by controlling the arrangements within the constituents' nanoclusters and of their assembly to have materials that can be used in the electronics industry. Additionally, the control of chemical reactivity by the attachment of functional side groups to for example, manufacture biocompatible nanocatalysts or the control of optical properties by variation of the size and microstructure to produce for example, visible light active catalysts (Huczko, 2000).

Nanoparticles consist of three different layers: the surface that can be functionalized, a shell material that may be added according to the application needs and the core material that can be synthesized utilizing various methods, reaction conditions and precursors (Christian *et al.*, 2008). The surface of nanoparticles can be functionalized with various metals and metal oxides, small molecules, surfactants and/or polymers. Moreover, due to functionalization, the nanoparticle surface can be charged (e.g. base catalyzed hydrolysis of tetraethyl orthosilicate, SiO^-M^+) or uncharged (citrate, sodium dodecylsulfate (SDS), polyethylene glycol (PEG) depending on the application. Generally, the shell is made of chemically different inorganic material and has a completely different structure than the core, for example iron oxide on iron nanoparticle, quantum dots (zinc sulphide on cadmium selenide) and polystyrene-polyaniline nanoparticles (Christian *et al.*, 2008). The core is usually referred to as the nanoparticle itself. Importantly, physicochemical properties of nanoparticles are nearly always governed by the properties of the core. However, environmental fate and transport most likely will be dominated by the core and shell properties rather than core alone.

17.1.3 Nanomaterials/nanoparticle synthesis

There are two main methods to synthesize nanomaterials: "top down" and "bottom up" approaches. The "top-down" approach means that nanoparticles are prepared by lithographic techniques, etching, grinding in a ball mill – decreasing the size of large pieces of a material until only a nanoparticle remains. The most commonly used method is, however, the "bottom up" approach, where nanoparticle are "grown" from more simple molecules – reaction precursors. It is possible to control the size and shape of nanoparticles depending on the subsequent application through variation in precursor's concentrations, reaction conditions (temperature, pH, etc), functionalizing of the nanoparticles surface, and so on.

Functionalization of the surface is one of the most important aspects of manufacturing nanoparticles and nanomaterials for desired applications. The high chemical reactivity of nanoparticles with a large surface is usually the main reason for undesirable and most often irreversible processes such as aggregation (Pomogailo & Kestelman, 2005). Aggregation reduces the specific surface area and the interfacial free energy, thereby diminishing particle reactivity (Figure 17.2). Therefore, to avoid aggregation, nanoparticles must be coated or treated with certain chemicals to increase their stability during storage, transportation and overall life cycle (Figure 17.2).

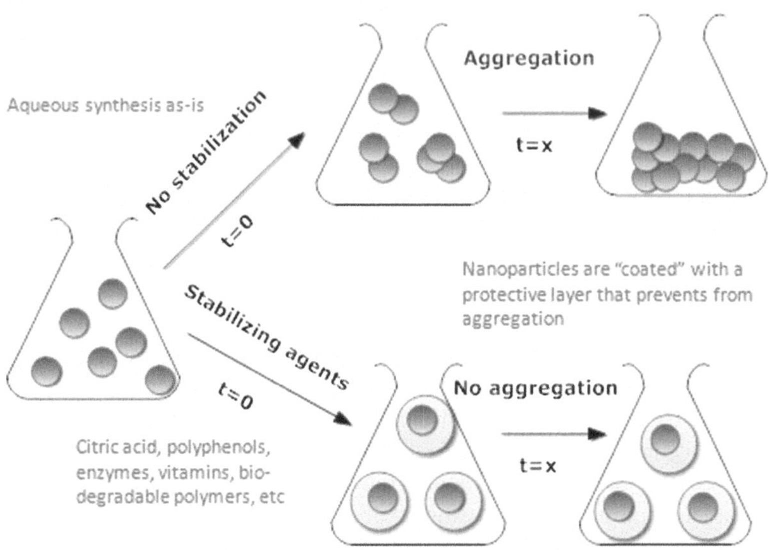

Figure 17.2 Schematized aggregation of nanoparticles in the presence and absence of stabilizing agents (adapted from Virkutyte and Varma, 2011a).

The majority of the stabilization methods involve dispersant molecules such as surfactants or polyeletrolytes (Stubbs & Gilman, 2007). These not only alter the chemistry of the nanoparticle surface, but also produce a huge amount of waste material, because they occupy a significant (>50%) mass fraction of a nanoparticle system (Eckelman *et al.*, 2008). Thus, in order to avoid producing wastes and a subsequent contamination of the environment, there is an urge to search for environmentally benign stabilization and functionalization pathways as well as biocompatible (i.e. non-immunogenic, non-toxic and hydrophilic) stabilizing agents.

Over the past couple of years, nanoparticles functionalization and thus stabilization research has significantly advanced, identifying new functionalization pathways and uses of benign stabilizers such as polyphenols, citric acid, vitamins (B, C, D, K), biodegradable polymers and silica (Baruwati & Varma, 2009; Nadagouda *et al.*, 2009; Nadagouda & Varma, 2006, 2007a, 2007b, 2008b).

Regardless of the synthesis approach, synthesized nanomaterials are expected to (Dahl *et al.*, 2007): (i) exhibit new size-based properties (both beneficial and detrimental) that are intermediate between molecular and particulate, (ii) incorporate a wide range of elemental and material compositions, including organics, inorganics, and hybrid structures, and iii) possess a high degree of surface functionality.

Green chemistry has already been successfully employed in the preparation of highly functionalized pharmaceuticals that exhibit similar characteristics to functionalized materials. Therefore it is expected that green chemistry will be increasingly used in the synthesis of nanomaterials as well. So, the efficient development and utilization of green nanotechnology, where green chemistry and green engineering principles are combined, has a potential to transform many currently available processes and products that enhance environmental quality, reduce pollution and conserve natural and non-renewable resources (Virkutyte & Varma, 2011a).

Green chemistry is a set of principles that encourages the design of products and processes that reduce or completely eliminate the need to utilize and generate hazardous substances (Anastas & Warner, 1998). 12 principles along with the corresponding green nanotechnology approach are summarized in Table 17.1 (Virkutyte & Varma, 2011a).

These 12 principles include the most important aspects of nanotechnology and chemistry, including design of syntheses methods that utilize renewable and benign precursors and mild reaction conditions, the use of negligible amounts or non-toxic solvents, determination of fate and transport of nanomaterials after their life cycle as well as real time monitoring and process control development to avoid or at least minimize the possibility of accidents.

As summarized by Dahl *et al.* (2007), the application of green chemistry principles and approaches has already significantly reduced the use of hazardous solvents and other toxic reaction ingredients, optimized reaction conditions, improved materials and energy efficiency of numerous chemical processes and enhanced the design of products. Essentially, the main focal point of green nanotechnology is processes and products. Processes include synthesis of nanomaterials without threatening the environment and human health, whereas products include synthesis of nanomaterials that provide sustainable solutions to various environmental problems (Karn, 2008). Therefore, green chemistry and nanotechnology have the same purpose, which includes the ability of chemical processes to function more like ecosystems or cells, in which benign materials are used at room temperature, chemical reactions are conducted in water, wastes are recycled and energy is used efficiently.

Table 17.1. Principles of green chemistry along with corresponding green nanotechnology approach (adapted from Virkutyte & Varma, 2011a).

Green chemistry principles	Green nanotechnology approach
1. Prevent waste	Develop new 'greener' chemical syntheses pathways of nanomaterials that minimize or completely eliminate waste streams and products by designing pollution prevention at the source.
2. Manufacture safer chemicals and products	Synthesize nanomaterials that possess the desired physicochemical properties and that have little or no toxicity.
3. Design less hazardous chemical syntheses	Discover and develop syntheses that utilizes benign precursors and solvents and, thus, generate substances with little or no toxicity to the environment and humans.
4. Use renewable feedstocks	Use raw materials and feedstock that are renewable and benign.
5. Utilize catalysts	Minimize waste by utilizing nanocatalysts synthesized in a green manner. Develop catalytic methods that can greatly enhance reaction selectivity and overall process efficiency.
6. Avoid chemical derivatives	Avoid using blocking or protecting groups or any temporary modifications if possible. Eliminate solvent-intensive syntheses by utilizing selective nanosyntheses.
7. Maximize atom economy	Design and develop syntheses so that the final product contains the maximum proportion of the starting materials.
8. Use safer solvents and reaction conditions	Avoid using hazardous solvents, separation agents, or other auxiliary chemicals. Utilize bottom-up approach to enhance materials efficiency and eliminate steps.

(*Continued*)

Table 17.1. Principles of green chemistry along with corresponding green nanotechnology approach (adapted from Virkutyte & Varma, 2011a) (*Continued*).

Green chemistry principles	Green nanotechnology approach
9. Increase energy efficiency	Develop synthesis protocols so that they are carried out at ambient temperature and pressure whenever possible.
10. Design chemicals and products to degrade after use	Assess and quantify nanomaterial behavior in the environment, study their fate and major chemical interactions with other particles or organisms. Design nanomaterials to degrade to innocuous substances after use so that they do not accumulate and harm the environment.
11. Real-time monitoring and process control to prevent pollution	Utilize real-time monitoring to optimize reaction chemistry, minimize energy costs and to minimize or eliminate the formation of by-products. Develop Life Cycle Assessment (LCA) of nanoscience, which is an essential tool for managing risks in research, development and commercialization of this field.
12. Minimize the potential for accidents	Produce chemicals and their forms (solid, liquid, or gas) to minimize the potential for chemical accidents including explosions, fires, and release to the environment.

Effective use of nanomaterials in environmental applications, electronics, catalysis, biology and biomedical applications, physics, material science and numerous interdisciplinary fields as well as understanding their fate, transport and toxicity, critically depend on the synthesis method, structural features such as size, shape, composition and surface structure/chemistry (core/shell, etc) and surface functionalization of the nanomaterials (Buhro & Colvin, 2003; Dahl *et al.*, 2007). Therefore, to extend the array of properties, it is necessary to (i) develop various environmentally-friendly synthesis pathways through manufacturing materials that can be effectively extracted after the life cycle (e.g. magnetic nanoparticles) and (ii) produce nanoparticles coated with protective layers that prevent harmful effects (e.g. consumer products), or are biocompatible (e.g. bionanoparticles) (Mohanpuria *et al.*, 2008).

This chapter summarizes the "state of the art" in the exploitation of various synthesis methods, reaction precursors and conditions to manufacture magnetic, metal, metal oxide and bionanoparticles for a vast variety of purposes.

17.2 MAGNETIC NANOPARTICLES

It is of vital importance to functionalize and therefore stabilize nanoparticles and nanomaterials for a great array of applications. However, easy and relatively inexpensive recoverability and thus reusability of these particles is an essential part of the life cycle of nanomaterials (Behrens, 2011). For several years, magnetic nanoparticles have been extensively used in metal ions and dye recovery, drug delivery, enzyme immobilization, protein and cell separation, magnetic fluids, catalysis, biotechnology/biomedicine, magnetic resonance imaging, data storage and environmental remediation because magnetic separation offers unique advantages of high efficiency, cost effectiveness and rapidity in comparison to non-magnetic nanoparticles (Chang *et al.*, 2011; Lu *et al.*, 2007; Virkutyte & Varma, 2011a). Currently, magnetic nanoparticles are mostly used as heterogeneous supports (e.g. magnetic nanocores) in various catalytic transformations providing added benefits of easy recoverability with a sample magnet, eliminating the need of solvent swelling before or catalyst filtration after the process (Abu-Reziq *et al.*, 2008; Baruwati *et al.*, 2009; Polshettiwar & Varma, 2009a; Polshettiwar & Varma, 2009b).

Iron oxides exist in many forms in nature, with magnetite (Fe_3O_4), maghemite (γ-Fe_2O_3) and hematite (α-Fe_2O_3) among the most common ones (Teja & Koh, 2009). Hematite is one of the most widespread and known iron oxides in rocks and soils. It is sometimes called ferric oxide, iron sesquioxide, red ochre or martite. It has blood-red colour when grinded and black or grey if coarsely crystalline. Hematite is extremely stable at ambient conditions and is often the end product of the transformation of other iron oxides. On the other hand, magnetite is called black iron oxide, magnetic iron ore or Hercules stone. It is one of the strongest magnetic materials of any transition metal oxide (Majewski & Thierry, 2007). Maghemite is usually found in soils as a weathering product of magnetite or as a by-product of heating of other iron oxides. It is rather meta-stable with respect to hematite and forms continuous solid solutions with magnetite (Teja & Koh, 2009).

The crystal structure of all these iron oxides can be described in terms of close-packed planes of oxygen anions with iron cations in octahedral or tetrahedral interstitial sites (Cornell & Schwertmann, 2003). Thus, in hematite, oxygen ions are in hexagonal close-packed arrangements, where Fe(III) ions occupy octahedral sites, whereas in magnetite and maghemite, the oxygen ions are in cubic close-packed arrangements. Furthermore, magnetite has an inverse spinel structure with Fe(III) ions distributed randomly between octahedral and tetrahedral sites and Fe(II) ions in octahedral sites (Klotz *et al.*, 2008). On the contrary, maghemite exhibits spinel structure that is somewhat similar to that of magnetite but with vacancies in the cation sub-lattice.

Unfortunately, one of the most pronounced problems is their intrinsic instability over long periods of time. As magnetic nanoparticle sizes are only <20 nm, they tend to aggregate and agglomerate, which significantly reduces the energy associated with the high surface area to volume ratio. Also, as all naked metal nanoparticles, magnetic nanomaterials are highly active and thus can be easily oxidized in air, resulting in the loss of magnetism and dispersability. Therefore, in order to manufacture stable and effective nanoparticles, they must not only be synthesized using green methods and sustainable pathways, but also functionalized, for example, coated with certain species, such as surfactants, polymers, carbon or silica species.

Despite the fact that there are numerous methods currently available for the synthesis of magnetic nanoparticles with various compositions, shapes and sizes, effective application of these particles in different industries highly depends on their stability under a range of conditions and stresses. In most applications, magnetic nanoparticles perform the best if the size is below a critical value, which strongly depends on the material, but is typically around 10–20 nm (Lu *et al.*, 2007). When the size is around 10–20 nm, each nanoparticle becomes a single magnetic domain and exhibits superparamagnetic behaviour when the temperature is above the so called blocking temperature. Such individual nanoparticles have a large constant magnetic moment and behave like a giant paramagnetic atom with a fast response to applied magnetic fields with negligible residual magnetism and coercivity, which is the field required to bring the magnetization to zero (Lu *et al.*, 2007).

17.2.1 Liquid phase methods

Liquid phase methods include co-precipitation, sonochemical approaches and sol-gel techniques. They are usually inexpensive and offer better yields of products and ease of surface treatment (Teja & Koh, 2009). Co-precipitation is a facile and appropriate way to synthesize iron oxides from aqueous Fe^{2+}/Fe^{3+} salts by adding calculated amounts of base under inert atmosphere at room or elevated temperature. The size, shape and composition of as-prepared magnetic nanoparticles depend on the type of precursor salt used (e.g. chlorides, nitrates, sulphates, etc), the Fe^{2+}/Fe^{3+} ratio, the reaction temperature, the pH of the system and ionic strength of the medium (Lu *et al.*, 2007). Importantly, magnetite nanoparticles are not very stable under ambient conditions, and therefore, can be easily oxidized to maghemite, whereas maghemite being a ferrimagnet, is resistant to oxidation. Also, nanoparticles synthesized by co-precipitation tend to be polydisperse.

17.2.1.1 Iron oxide particles

To prepare monodisperse magnetic nanoparticles, various organic additives such as stabilizers and/or reducing agents must be added. These include anionic surfactants, dispersing agents, proteins, starches and polyelectrolytes. Resulting nanomaterials and nanocomposites can effectively be utilized for a number of chemical, biological and physical applications. For instance, to be used as contrast agents for nuclear magnetic resonance (NMR), iron oxide magnetic nanoparticles were first coated with a fluorescent silica shell, which would allow their visualization by optical means, and then grafted with hyperbranched polyglycerol (HPG) to attain water dispersibility and macrophage-evading properties. Subsequently, folic acid was coupled onto the surface of polyglycerol (HPG)-grafted Fe_3O_4–SiO_2 by an effective 'thiol' click reaction at room temperature to achieve cancer targeting capability (Wang *et al.*, 2011a). The nanocomposites were spherical with the distinct core-shell structure and the size determined by transmission electron microscopy (TEM) was ca. 30 nm (i.e. silica shell of ca. 8 nm and Fe_3O_4 core of ca. 13 nm) and hydrodynamic diameters of 46 nm without and 54 nm with the addition of folic acid. Also, synthesized nanoparticles were not aggregated, exhibited excellent dispersability and mostly existed as single particles in solution. Saturation magnetization was 80 emu g^{-1} (electromagnetic units) along with the absence of hysteresis in their magnetic profiles, indicating that nanocomposites were superparamagnetic. Moreover, the uptake of FA-HPG-grafted Fe_3O_4–SiO_2 nanoparticles by SKOV-3 cells was about four and seven times that by macrophages and 3T3 fibroblasts, respectively, suggesting their potential to be used as high contrast agents for diagnostic imaging.

17.2.1.2 Bimetallic nanoparticles

Furthermore, bifunctional precious metal-magnetic nanoparticles are constantly gaining an increased attention due to their potential applications in novel electrical, optical, magnetic, catalytic and sensing technologies. For instance, the Fe_2O_3/Ag core–shell magnetic nanocomposite was successfully prepared via a simple method at room temperature in the presence of sodium hydroxide (NaOH), TSA-doped polyaniline, dimethylsulfoxide (DMSO) and dimethylformamide (DMF) without calcination (Sun *et al.*, 2011a). The average particle size reported by both XRD and TEM was in the range of 30 to 60 nm. Magnetization of the nanocomposite was 0.96 emu g^{-1} and coercivity 104.0 Oe (Oersted, the unit of magnetizing field). The rather low specific saturation magnetization and coercivity were attributed to the nonmagnetic Ag, which was introduced into the composite. However, the magnetization could be saturated at a low applied field, indicating that the as-prepared Fe_2O_3/Ag nanocomposites exhibited typical ferromagnetic behavior (Sun *et al.*, 2011a).

Metal and metal oxide-magnetic nanocomposites can also be used for sensors and biosensors fabrication. For example, the bifunctional Fe_3O_4–Al_2O_3 core–shell nanocomposite, synthesized by a simple sonochemical approach (ca. 3 h sonication, vacuum-drying at 80°C overnight), can be deposited on the electrode surface by applying a constant magnetic field, and then for example, heme proteins could be immobilized on the Fe_3O_4–Al_2O_3 nanocomposite surface via electrostatic interactions (Peng *et al.*, 2011a). This way synthesized nanoparticles were monodispersed, with an average particle size of 40 nm to 50 nm (i.e. ca. 30 nm core and ca. 20 nm shell) and provided direct electron transfer for different heme proteins, that is, established effective electronic communication between proteins and the electrode. Also, such a biosensor exhibited excellent electrocatalytic ability to reduce H_2O_2.

17.2.1.3 Spinel ferrites

In recent years, spinel ferrites (Adam *et al.*, 2002; Li *et al.*, 2011a; Naidek *et al.*, 2011; Özgür *et al.*, 2009) – magnetic $CoFe_2O_4$ nanoparticles in particular – became one of the most important constituents in synthetic and biological chemistry as multiferroic materials. As well as the magnetic labels of biological systems due to their unusual properties such as doping or strain enhanced coercivity, photoinduced magnetic effects among a few (Senapati *et al.*, 2011). Spinel structure materials are closed packed cubic in the form of AB_2O_4 where A are divalent cations and B are trivalent cations. Furthermore, spinel ferrites are ferromagnetic materials with the general chemical composition $MeO{\cdot}Fe_2O_3$ where Me is a divalent metal such as iron, manganese, magnesium, nickel, zinc, cadmium, cobalt, copper, or a combination of these. The Fe ions are the trivalent variety, Fe^{3+}. $MeO{\cdot}Fe_2O_3$ has a structure identical to naturally occurring $MgAl_2O_4$ which is called "spinel" (Özgür *et al.*, 2009).

Synthesis of extremely stable (no agglomeration was observed during storage for several months) spinel cobalt ferrite magnetic nanoparticles with average sizes of 40–50 nm has been achieved using a combined sonochemical and co-precipitation technique in aqueous medium (appropriate amounts of $FeCl_3$, $CoCl_2{\cdot}6H_2O$ and KOH were mixed in an ultrasonic bath applied at a frequency of 40 KHz and power of 40 KW) without any surfactant or organic capping agent at 60°C (Senapati *et al.*, 2011). Prepared magnetic nanoparticles were used as quasi-homogeneous catalysts in the Knoevenagel condensation reaction, which is one of the most widely used synthetic pathways for carbon-carbon formation in organic synthesis. Another approach to synthesize spinel magnetic $CoFe_2O_4$ nanoparticles ($Co^{2+}Fe^{2+}Fe^{3+}$-CO_3^{2-}-layered double hydroxides (LDHs), in particular) is to use porous degreased popped corn with subsequent calcinations at 900°C for 5 h in air. Mesoporous Co-spinel ferrites with pore sizes ranging from 2.7 nm to 4.2 nm were obtained that can be potentially utilized as catalysts and as a vessel for oxidation reactions of small organic molecules (Ji *et al.*, 2011).

It is also possible to use other metals (e.g. Co, Zn, Mn, Ni, etc) for the synthesis of spinel ferrites for a wide array of applications. Thus, preparation of $Co_{1-x}Zn_xFe_{2-x}Cr_xO_4$ ferrites with variable composition ($x = 0.0$–0.5) utilizing chemical co-precipitation techniques with appropriate metal sulphates at 60°C with NaOH followed by calcinations at 600°C for 12 h in air was demonstrated (Alone *et al.*, 2011). These nanoparticles had an average size in the range of 15 to 30 nm depending on the composition (x). The values of saturation magnetization (Ms) and coercivity (Hc) were in the range of 46.6–17.6 emu g^{-1} and 1453–135 Oe, respectively depending on x. The $Mn_{0.5}Zn_{0.5-x}Cd_xFe_2O_4$ ($x =$ 0.0, 0.1, 0.2 and 0.3) nanoparticles were prepared by a chemical co-precipitation method for 60 min at 100°C followed by pelletelization using a hydraulic press and calcination at 500°C for 5 h (Venkataraju *et al.*, 2011). Furthermore, thermally stable nickel ferrite ($NiFe_2O_4$) nanosheet synthesis using a polymer (polyvinyl pyrrolidone, PVP) assisted co-precipitation method followed by 4 h drying in hot air at 100°C was demonstrated (Sivakumar *et al.*, 2011). The observed value of saturation magnetization (Ms) and coercivity (Hc) were 35 emu g^{-1} and 8.6 Oe respectively. They argued that significantly lower magnetization in comparison to reported bulk particles (47.5 emu g^{-1}) may be attributed to the spin canting at the surface of ferrite nanoparticles.

Preparation of stable amine functionalized magnetic nanoparticles, using ammonia aqueous solution, tetraethylorthosilicate (TEOS) and γ-aminopropyltriethoxysilane (APTES) at room temperature under constant mixing was reported (Huang *et al.*, 2011). These nanoparticles were able to kill 94% of *E.coli* and *S. lutea* after 10 min of incubation time (at a nanoparticle concentration of 0.5 mg ml^{-1}), which is an important addition to the degradation of various organic contaminants. In addition to amine-functionalization, enzyme-conjugated magnetic nanoparticles can also be synthesized utilizing green chemistry approaches (Park *et al.*, 2011). To summarize, glucose oxidase was covalently bound to the magnetic nanoparticles by modifying the surfaces with 3-(aminopropyl)triethoxysilane (APTES) and the common protein crosslinking with the agent glutaraldehyde at room temperature. This way, nanoparticles with average particle sizes of 5 nm (magnetization 48.5 emu g^{-1}), 26 nm (magnetization 72.8 emu g^{-1}) and 51 nm (magnetization 84.5 emu g^{-1}), were synthesized depending on the reaction conditions. They were applied in the field of drug delivery, biosensors, environmental remediation and biocatalysis. The enzyme-stabilized nanoparticles retained their activity and stability even after 10 consecutive cycles.

17.2.1.4 Hydroxyapatites

Other materials such as hydroxyapatites can also be used to synthesize effective and inexpensive magnetic nanocomposites for various industrial applications, including remediation of soils, water and wastewater. Hydroxyapatite (HAP) is the main constituent of bones and teeth and is widely used in such fields as bone repairs, bone implants, bioactive materials, purification and separation of biological molecules. It is a photocatalyst active under the UV irradiation. Fe_3O_4–HAP nanocomposite synthesis was reported in details, via simple homogeneous precipitation of $FeCl_2{\cdot}4H_2O$, $FeCl_3{\cdot}6H_2O$ and NH_4OH solution at 90°C for ca. 3 h (Yang *et al.*, 2011). After overnight drying at room temperature, the as-prepared samples were calcined at 400°C for 3 h in air. Nanoparticles were nearly spherical in shape and monodispersed with unique size of about 25 nm in diameter. According to the authors, the saturation magnetization (Ms) and coercive force (Hc) were estimated to be 5.2 emu g^{-1} and 0 Oe, respectively. Magnetic properties were significantly lower than those of bulk Fe_3O_4 nanoparticles (84 emu g^{-1} and 500 – 800 Oe), which can be attributed to the diamagnetic contribution of HAP, that would significantly weaken the magnetic moment, whereas the low Hc could be due to the size of Fe_3O_4 particles embedded into the HAP matrix. However, the remanence of the nanocomposite was zero once the applied magnetic field was removed, indicating superparamagnetic behavior of the Fe_3O_4–HAP nanocomposites (Yang *et al.*, 2011). The utilization of as-prepared nanocomposite as UV-activated catalyst proved to be effective in removing 75% of insecticide diazinon (10 mg L^{-1}) in 60 min.

In addition to hydroxyapatite, magnetic nanoparticles can also be functionalized with β-cyclodextrin and pluronic polymer (F127) at room temperature under a nitrogen atmosphere for successful drug delivery applications (Yallapu *et al.*, 2011). Utilizing this co-precipitation approach, stable nanoparticles with average sizes from 90 to 256 nm were produced depending on formulations.

17.2.2 Gas phase methods

Gas phase methods include thermal, solvothermal and hydrothermal decomposition methods. Thermal decomposition of maghemite γ-Fe_2O_3 without the addition of any dopants, solvents or surfactants after calcinations at 500°C for 1 h in air, resulted in hematite (α-Fe_2O_3) nanoparticles with an average size of 18 ± 2 nm (Darezereshki, 2011). The saturation magnetization of hematite nanoparticles was 0.00654 emu g^{-1} and a coercivity of 101 Oe, indicating that these nanoparticles exhibited weak ferromagnetic behaviour at room temperature.

Usually, most authors that report their findings in literature synthesize small amounts of nanoparticles and use those for laboratory scale experiments. In contrast, a facile and effective sonochemical approach (ca. 1500 W and 20 kHz) for the large scale synthesis of extremely stable iron oxide magnetic nanoparticles using inexpensive and non-toxic metal salts as reactants as well as relatively short synthesis duration at room temperature (90 min) was demonstrated (Nazrul Islam *et al.*, 2011). Thus, as-prepared nanoparticles were nearly spherical with an average diameter of 11 nm and had a very low coercivity of 10 Oe, which indicated superparamagnetic behavior. In addition, the magnetization value at the maximum applied field was 80 emu g^{-1}. The authors also determined that nanoparticles had a relatively high susceptibility ($Xg = 5.71 \times 10^{-4}$m3 kg^{-1}) and could be potentially used for various biomedical applications.

17.2.2.1 Carbonaceous nanoparticles

According to numerous research efforts, it is also possible to synthesize non-metal, metal and metal oxide/salt doped and/or coated magnetic nanocomposites and nanoparticles for numerous applications. For instance, synthesis of magnetic carbonaceous nanoparticles (MNP@C) utilizing a facile two-step solvothermal synthesis approach for adsorption of Pb ions from solution was attempted (Nata *et al.*, 2011). During this synthesis, nanoparticles with 100 nm were synthesized. However, after the surface functionalization with 1,6-hexanediamine, the nanoparticle size reduced to ca. 30 nm. These nanoparticles were effectively encapsulated in the carbonaceous shell using 0.5 M glucose solution. Moreover, the saturation magnetization decreased significantly from 70 to 25 emu g^{-1} after the carbonaceous shell was formed. Furthermore, carbonaceous nanoparticles carried a negative charge (− 30 mV) at neutral pH and had a point of zero charge at pH 2. MNP@C did not show any superior adsorption capacity to other readily available materials, but superparamagnetic properties of MNP@C could facilitate the retrieval of the spent adsorbent from the waste stream.

Synthesis (1 h in air) of recyclable and effective Fe_3O_4@Ag composite utilizing Ag salt and (3-aminopropyl)triethoxysilane (APTES) at room temperature without the addition of reducing agents or other potentially hazardous chemicals was reported (Zhang *et al.*, 2011b). Thus, as-synthesized particles were quasi-spheres and ca. 10 nm in diameter and were effective in removing rhodamine blue from the solution.

17.2.2.2 Spinels

Magnetic $CoFe_2O_4$ nanoparticles (spinel ferrites) are very important for various synthetic and biological chemistry applications and can be synthesized utilizing liquid and gas phase methods. Solid state ball-milling of the reaction precursors (cobalt acetate tetrahydrates ($Co(CH_3COO)_2{\cdot}4H_2O$), ferric acetate anhydrous (Fe $(CH_3COO)_2$) and citric acid ($C_6H_8O_7$) and subsequent calcination (250 to 350°C for 3 h in air) in the presence of citric acid can produce closely packed nanoparticles (4–5 nm) with almost spherical shape (Eshraghi & Kameli, 2011). In addition, by increasing the particle size from 3.5 to 7.2 nm, the room temperature coercivity and magnetization increased from 0 to 152 Oe and from 8 to 30 emu g^{-1}, respectively with the mean value of the effective anisotropy constant $K_{eff} = 9.2 \times 10^6$ erg cm^{-3}. To enhance properties and extend the array of applications, $CoFe_2O_4$ nanoparticles can be doped with rare earth-elements such as gadolinium and samarium to effectively lower the Curie temperature (Tc) as compared to pure ferrites, making them more promising for magneto-optical recording applications (Peng *et al.*, 2011b). Thus, a gadolinium-doped $CoFe_2O_4$ nanocomposite was synthesized from stoichiometric amounts of $Fe(NO_3)_3{\cdot}9H_2O$, $Co(NO_3)_2{\cdot}6H_2O$ and $Gd(NO_3)_3$, which was prepared via dissolving Gd_2O_3 in 10 mL dilute nitric acid and added drop wise into the solution with vigorous stirring. The reaction mixture was autoclaved at 200°C for 16 h prior usage. As observed from TEM images, as-prepared nanomaterials were crystals, roughly spherical, uniformly distributed and not highly agglomerated. According to the authors, Gd^{3+} had no influence on the nanocomposite morphology but significantly affected the average crystallite size, that is, average crystallite size increased with an increase in dopant amount. The reactions occurring during the hydrothermal treatment are (Peng *et al.*, 2011b):

$$Co(NO_3)_2 + 2\text{-}xFe(NO_3)_3 + xGd(NO_3)_3 + 8NaOH \longrightarrow Co(OH)_2 + 2\text{-}x\,FeO(OH) + xGdO(OH) + 8NaNO_3 + 2H_2O \qquad \text{(Eq. 17.1)}$$

$$Co(OH)_2 + 2\text{-}xFeO(OH) + xGdO(OH) \text{ hydrothermal treatment (200°C for 16h)} \rightarrow CoFe_{2}\text{-}xGd_xO_4\downarrow + 2H_2O \quad \text{(Eq. 17.2)}$$

The room temperature magnetic field versus magnetization (M–H) measurements confirmed the strong influence of rare-earth metal doping on the saturation magnetization (highest of 58.2 emu g^{-1}) and coercivity (highest of 996 Oe, depending on the composition) due to large lattice distortion and grain growth of small particles.

In addition, other compositions of spinel ferrites MFe_2O_4 (where M = Fe, Co, Ni) can also be synthesized via solvothermal decomposition of trimetallic acetate complex precursors in benzyl ether in the presence of oleic acid and oleylamine, using 1,2-dodecanediol as the reducing agent with particle sizes in the range of 9 nm to 25 nm (Naidek *et al.*, 2011). Moreover, Mn-doped $BiFeO_3$ ($BiFe_{1-x}Mn_xO_3$ x = 0, 0.05, 0.1) nanoparticles (average size of 55 nm) using a hydrothermal method in the autoclave for 3 hours at 256°C were also synthesized (Basu *et al.*, 2011). They found that below 93°C, all samples exhibited ferromagnetic behavior and the magnetization value at −12°C increased from 0.0027 to 2.75 emu gm^{-1} when doped with 5% Mn. The cubic structure Mn ferrite nanoparticles with average sizes from 12 to 22 nm in the presence of polyvinyl pyrrolidon (PVP) and various calcination temperatures up to 467°C were manufactured (Goodarz Naseri *et al.*, 2011). The cubic structure Mn ferrite nanoparticles with average sizes from 12 to 22 nm in the presence of polyvinyl pyrrolidon (PVP) and various calcination temperatures up to 467°C were manufactured. Magnetic properties were demonstrated by vibrating sample magnetometers which indicated superparamagnetic behavior of all samples and also increasing saturation magnetization from 3.06 to 15.78 emu g^{-1} with increasing calcination temperatures. Furthermore, novel and extremely fast (ca. 15 s) metallic iron (α–Fe)/manganese–zinc ferrite ($Fe_{3-x-y}Mn_xZn_yO_4$) nanocomposite synthesis was demonstrated using microwave hydrothermal treatment of alcoholic solutions of chloride precursors and sodium ethoxide (Caillot *et al.*, 2011). They found three variations of crystallites depending on the calcination duration: (i) spherical grains with a diameter close to 10 nm, (ii) particles in form of lozenge with a size between 50 and 200 nm and (iii) grains without particular form with a size between 50 and 200 nm. The formation of ferrite can be expressed as (Caillot *et al.*, 2011):

$$1 - x - yMn(OEt)_2 + xZn(OEt)_2 + 2Fe(OEt)_3 + yFe(OEt)_2 + 4H_2O \longrightarrow Fe_{2+y}Mn_{1-x-y}Zn_xO_4 + 8EtOH \quad \text{(Eq. 17.3)}$$

17.2.2.3 Nanocomposites

In the last several years, bi-functional nanomaterials with unique magnetic and luminescent properties became attractive for a great number of applications, including magnetic resonance imaging as contrast agents, drug delivery carriers, cell sorting and labelling, bioseparations due to unique magnetic responsivity, visible luminescence, low cytotoxicity and good biocompatibility (Sun *et al.*, 2011b). Thus, a hybrid material with magnetic and luminescent properties was hydrothermally synthesized from Fe_3O_4 nanoparticles encapsulated with SiO_2 via the Stöber method and further functionalized by the deposition of YVO_4: Eu^{3+} phosphors in the presence of citric acid as a reducing and capping agent and subsequently calcined at 500°C for 3 hours. The average particle size was ca. 50 nm and the magnetization of the bifunctional nanocomposite was 25.2 emu g^{-1}.

The preparation of a new kind of magnetic and luminescent magnetic Fe_3O_4/CdTe (Quantum dots) nanocomposite modified with tetramethylammonium hydroxide (TMAOH) and chemically activated with aspartic acid at room temperature, using hydrothermal co-precipitation that can be used for bioassays, drug delivery, enhanced magnetic nuclear resonance, fluorescent detection and information storage was reported (Wang *et al.*, 2011b). The as-prepared nanocomposites had a structure with an average diameter of 40–50 nm, yellow-green emission feature and room temperature ferro-magnetism. Importantly, the nanocomposites retained the ferromagnetic property with a saturation magnetization of 8.9 emu g^{-1}.

To use surface-functionalized magnetic nanoparticles for heavy metal removal from aqueous solutions, treated amine-functionalized Fe_3O_4 nanoparticles with ethylenediaminetetraacetic acid (EDTA) and polyethyleneimine (PEI) and autoclaved at 200°C for 6 h or heated at 80°C for 30 min in air depending on the target nanoparticle were used (Zhang *et al.*, 2011a). These nanoparticles had a narrow size distribution (60–80 nm) and exhibited a very good dispresability without evident agglomeration for several months. Removal (i.e. adsorption) of Pb^{2+}, Cu^{2+} and Cd^{2+} with initial concentration of 10 mg l^{-1} was 98%, 95% and 50%, respectively in 5 min of the reaction.

One of the interesting approaches to synthesize metal-magnetic nanoparticles is to use extremely fast (synthesis of only several minutes) Carbon Combustion Synthesis of Oxides (CCSO). During CCSO the exothermic oxidation of carbon $C + O_2 = CO_2$ $\Delta H_{CO2}{}^{298} = -393.5$ kJ mol^{-1} generates a steep thermal reaction wave (temperature gradient of up to 500°C cm^{-1}) that propagates through a solid reactant mixture converting it to a desired complex oxide product (Martirosyan *et al.*, 2011). Thus, carbon combustion synthesis was applied to rapid and energy efficient fabrication of crystalline barium hexaferrite nanoparticles that yielded an average particle size of 50–100 nm.

$$BaCO_3 + 6Fe_2O_3 + \alpha(C + O_2) = BaFe_{12}O_{19} + (\alpha + 1)CO_2 \qquad \text{(Eq. 17.4)}$$

The saturation magnetization (M_s) and coercivity (H_c) value of 11 wt.% were relatively high (Ms ~58 emu g^{-1}, 900 Oe), indicating good magnetic properties.

The sintered samples had low AC-conductivity of the order of 10^{-7} S m^{-1} at 10 MHz, which is favorable for high frequency applications of these magnetic nanomaterials.

17.3 CLAY SUPPORTED NANOPARTICLES

Metal and metal oxide nanoparticles can be effectively utilized for a great variety of purposes. Unfortunately, one of the main drawbacks in wide spread use of nanomaterials is potential agglomeration or aggregation into larger particles and clumps, because of the loss of the initial activity and the surface area. Thus, to avoid agglomeration, different porous solids such as clays are commonly utilized as active or passive supporting media to disperse various nanoparticles of active phases. Importantly, porous materials provide high surface areas and easily accessible channels for molecules and improve their diffusion and catalytic properties (Garrido-Ramírez *et al.*, 2011; Li *et al.*, 2008). Clay supported nanomaterials are prepared by exchanging the interlamellar alkali cations with particular metallic polyoxocations in aqueous solution (Li *et al.*, 2008). Subsequently, calcinations convert these polyoxocations to robust nanoparticles of corresponding metals or metal oxides anchored on the silicate platelets. Importantly, the clay layers partition metal or metal oxide nanoparticles preventing them from sintering, whereas a wide interlayer gallery prevents pores from being blocked by the metal oxide particles. In such a way, clay supported or pillared nanoparticles can be used for numerous industrial applications due to large surface areas, good thermal stability and special surface acidity.

Partial hydrolysis to synthesize iron, chromium, cobalt, manganese and cerium pillared clay (laponite) with the surface areas of 510 to 640 m^2 g^{-1} and total pore volume of 0.43 to 0.51 cm^3 g^{-1} in the presence of stabilizer was utilized in the presence of acetic acid which was used to degrade benzene (Li *et al.*, 2008). According to the authors, metal precursors substituted the sodium ions in laponite clay forming mesoporous solids with well dispersed metal oxides and high surface area. Also, a significant increase in the porosity after the intercalation process was attributed to delamination of the laponite clay caused by the reaction between the acidic metal precursor solutions and the silicate platelets. A complete conversion of benzene was achieved at 300°C with Co-pillared laponite. Unfortunately, other catalyst formulations were not that effective despite a significant increase in temperature.

Ag is one of the most common and mostly used metals to synthesize clay supported nanomaterials. Thus, Ag nanoparticles using UV irradiation and $NaBH_4$ in the interlamellar space of a layered kaolinite were synthesized (Patakfalvi & Dékány, 2004). Disaggregation of the lamellae of the nonswelling kaolinite was achieved by the intercalation of dimethyl sulfoxide (DMSO), which is less toxic than other similar solvents. Interestingly, nanoparticles synthesized using UV irradiation were larger (8.3 to 14.8 nm) in comparison to $NaBH_4$ (7.1 to 10.5 nm) reduced ones.

A very simple and facile synthesis approach without addition of reducing agents or extensive heat treatment was used by applying the UV light (irradiation intensity of 0.362 mW cm^{-2} for 1 h, 3 h, 18 h, 46 h and 93 h) to manufacture stable Ag nanoparticles in layered laponite suspensions at room temperature was used (Huang & Yang, 2008). The average particle size was from ca. 20 nm to 110 nm depending on irradiation time with particles being the largest after 3 h of UV treatment.

In addition to the conversion and degradation of various organics, clay supported nanomaterials can also be used to eliminate bacterial contamination. For instance, highly stable Ag nanoparticles (average size of 2.5 $\pm$ 0.6 nm) were prepared on montmorillonite clay in n-hexanol with an intense yellow surface plasmon band observed at 397 to 422 nm at Ag concentration of 0.01 to 1 M (Miyoshi *et al.*, 2011). Unfortunately, reduction of Ag^+ to Ag^0 was performed with the addition of $NaBH_4$, which is considered dangerous and is not under the green chemical synthesis guidelines. The authors found that as-prepared Ag-montmorillonite nanoparticles exhibited antibacterial activity against *E. coli* even after 12 years. This synthesis route of Ag nanoparticles requires significant improvement.

Combinations of two metals incorporated into the clay matrix may prove to be effective antimicrobial agents. For instance, nanostructure ensembles of silver nanoparticles/zinc-substituted anionic clay matrices (layered double hydroxides) (Ag/ZnLDH) were synthesized via a simple single step route at room temperature (Carja *et al.*, 2009). These nanomaterials consisted of silver nanoparticles (average size 7 nm) deposited on the larger nanoparticles of clay (average size 85 nm). Ag/ZnLDH exhibited significant antimicrobial activity against the Gram-positive bacterium *S. aureus* and the Gram-negative bacterium *E. coli*, which was higher than the unsupported nanosized silver.

Other metals and/or metal oxides can be intercalated into the clay structure rendering clay supported nanoparticles for various uses. Semiconductor-clay organocomplexes can be synthesized in the presence of hexadecylpyridinium and hexadecylammonium, cyclohexane and ethanol along with relevant salts, metals and oxides (CdS, TiO_2 and Pd) and Wyoming bentonite (Dékány *et al.*, 1999). Nanoparticles are intercalated into the interlayer space of the bentonite clay. Also, CeO_2, which is one of the most important solid electrolytes in solid oxide fuel cells and catalyst, can be intercalated into the montmorillonite utilizing simple synthesis approach (Gorobinskii *et al.*, 2007). Thus, clay was subjected to $(NH_4)_2Ce(NO_3)_6$ at room temperature for about 24 h and then calcined at 550°C for 2 h at a heating rate of 10°C min^{-1} rendering CeO_2 nanoparticles with a mean size of 3.5 nm. The resulting nanocomposite was an amorphous solid with high permanent porosity ($S_{BET} = 239$ m^2 g^{-1}, micropore volume = 0.1839 cm^3 g^{-1} and average pore diameter = 3.07 nm) and excellent stability at high temperatures.

Simple synthesis of Au nanoparticles in clay were also reported (Zhu *et al.*, 2009). Supported gold nanoparticles were prepared from the smectite and sepiolite-palygorskite groups (i.e. montmorillonite and sepiolite) by cation adsorption from the cationic gold precursor $Au(en)_2Cl_3$ (en = ethylenediamine). The novelty of this approach lies in the use of this specific precursor. Thus, most of the research on gold nanoparticles uses $HAuCl_4{\cdot}3H_2O$ as precursor to manufacture Au nanoparticles. However, the mean diameter of these nanoparticles is around 20 to 50 nm, which is larger than the small nano-size range (2 to 5 nm), which is considered to be the most effective in Au catalytic applications (Zhu *et al.*, 2009). Indeed, the authors synthesized Au nanoparticles on montmorillonite with average sizes below 5 nm after calcinations at 350 to 450°C with a gold loading lower than 2 wt%. Unfortunately, the increase in Au loading significantly increased agglomeration. Furthermore, Au nanoparticles on sepiolite had a narrow size distribution and an average size of 2.7 nm.

TiO_2 may also be used to prepare intercalated clays (laponite, montmorillonite, iron-rich smectite and vermiculite) (Manova *et al.*, 2011; Zhu *et al.*, 2005). For instance, it was reported that an acidic sol solution was able to react with the clay platelets and leach out part of the magnesium or aluminum in clay, while TiO_2 nanoparticles (average grain size 3 to 9 nm) were formed and condensed on the fragmentized pieces of the leached silicate (Zhu *et al.*, 2005). Also, the use of polyethylene oxide (PEO) surfactant significantly increased the porosity and surface area (from 217 to as high as 635 $m^2\ g^{-1}$) of the nanocomposite rendering an effective and stable catalyst capable of degrading phenol and dyes in aqueous solutions in 5 h. Furthermore, TiO_2 intercalated clay in the presence of the surfactant CTAB, calcined at 500°C for 1 h with nanoparticle size ranging from 9 to 15 nm were synthesized (Manova *et al.*, 2011). The as-prepared catalyst was effective in degrading dichlorophenol in 12 hours.

Another successful approach to synthesize homogeneous gold nanoparticles, with a narrow particle size distribution, on acidified sepiolite was introduced (Letaief *et al.*, 2011). During this synthesis approach, very well dispersed Au nanoparticles (average size of 2 to 5 nm) were prepared from a gold hydrosol and deposited on the external surface of sepiolite, previously functionalized by aminopropyltriethoxysilane (APTES). The applicability of these nanomaterials were tested for acetylation of phenol derivatives under soft and organic solvent-free conditions with acetic anhydride as acetyl donor and in the presence of conventional or microwave irradiation heating. Typical yields of 90+% were achieved regardless of the heating source. However, a rather dramatic reduction of the reaction time was observed when microwave irradiation was used.

Iron is another promising metal to be intercalated into clays for a large array of applications. Synthesis of iron nanoparticles in the presence of K10 bentonite was reported (Shahwan *et al.*, 2011). Unfortunately, the synthesis approach using $NaBH_4$ is not green and safe, but still remains as one of the most effective in fabricating stable nanoparticles regardless of the precursor used. Notably, the introduction of K10 during synthesis resulted in a partial decrease in the aggregation of iron nanoparticles. The dispersed nanoparticles had a typical core-shell structure and were in the range of 10 to 60 nm. In order to test the effectiveness of such nanocomposite as an adsorbent for environmental remediation, Co^{2+} was targeted to adsorb from aqueous solutions at various contact times, pH, concentrations, and so on, The highest Co^{2+} removal (up to 88%) was achieved when the pH was increased to 10.

Ni can also be incorporated into the clay structure to produce highly active, selective and stable catalysts for various organic transformations. Thus, several saponites under microwave irradiation (pH = 7 and lower) and subsequently impregnated and ion exchanged $Ni^{2+}(Ni(NO_3)_2)$ precursors (15 to 20 wt%) at room temperature, followed by reduction in pure hydrogen at 400°C for 6 h calcinations at 400°C were synthesized (Vicente *et al.*, 2011). The average Ni nanoparticle size varied from 10 nm to 25 nm depending on Ni content in the saponites. This catalyst was used to hydrogenate styrene oxide to 2-phenylethanol, with conversions up to 70%. Also, green synthesis of the Ni nanoparticles in interlamellar spaces of K10-montmorillonite by chemical reduction by hydrazine in ethanol at mild reaction temperatures (60°C for 5 hours) was reported (Dhakshinamoorthy & Pitchumani, 2008). The Ni content of 2.84 wt% was effective in hydrogenation of alkenes and alkynes with high yields (ranging from 50 to 91%). Resulting nanoparticles were essentially mono-dispersed with an average diameter of 15–20 nm.

Rare metals such as Pd, Pt and Rh nanoparticles can also be supported on or in the clay structure. However, the vast majority of researchers utilize $NaHB_4$ to reduce metals, which is not a "green" approach according to the 12 principles of Green Chemistry (Manikandan *et al.*, 2007a; Manikandan *et al.*, 2007b; Mastalir *et al.*, 2002; Sidhpuria *et al.*, 2009). Therefore, Pd nanoparticles were synthesized via reduction by ethanol at 65°C for 24 h in the presence and absence of non-ionic poly(*N*-vinyl-2-pyrrolidone (PVP) and the cationic poly-diallyldimethylammonium chloride (PDDA) stabilizers on montmorillonite layer silicates (Papp *et al.*, 2001). The growth of Pd^0 nanoparticles was mostly in the internal, interlamellar space and on the external surface of the lamellae. The average diameter of nanoparticles was 2 to 3 nm but aggregation without the stabilizing agent in the aqueous medium was also observed. Another attempt to reduce Pd^{2+} in ethanol at 65°C for 2 h on aluminum hydroxide pillared montmorillonite and saponite produced Pd nanocrystals with an average diameter of 5 to 15 nm predominantly in the interlamellar space of clays (Szucs *et al.*, 2000). The Al content in the supports varied between 0.72 and 2.24% (w:w).

The preparation of Pt (1 and 2 wt.%) nanoparticles reduced with $NaHB_4$ at room temperature in the presence of a cationic surfactant hexadecyl trimethylammonium bromide (CTAB) in intercalated (Na^+) montmorillonite utilizing the swellability and ion exchange capability of the clay mineral was reported (Manikandan *et al.* 2007a; Manikandan *et al.*, 2007b). The effectiveness of the catalyst was tested in selective dehydrogentation reactions of α, β unsaturated aldehydes (cinnamaldehyde). A maximum of 91% cinnamaldehyde conversion with 78% selectivity towards cinnamyl alcohol was achieved with 0.5 g of 1% Pt intercalated montmorillonite catalyst at the end of 5 h. In addition, Rh nanoparticles (average particle size 25.6 ± 2.5 nm) were intercalated into montmorillonite and used for hydrogenation of benzene (Sidhpuria *et al.*, 2009). The benzene was completely hydrogenated with 100% selectivity to cyclohexane at 180°C and 6.2 MPa. However, the catalyst was poisoned by thiophene, which resulted in the decrease in benzene hydrogenation with increasing thiophene concentrations.

To increase the activity of clay supported nanoparticles, it is viable to use surfactants to modify clay properties prior the subjection to metal ions solutions. For instance an environment-friendly montmorillonite-hexadecyl trimethylammonium/iron nanocomposite ($S_{BET} = 28.7 - 38.1\ m^2\ g^{-1}$, particle size ranging from 20 to 100 nm) was synthesized and used to reduce (up to 80% in ca. 6 hours) highly toxic Cr (VI) (Li *et al.*, 2011b). In addition to only iron or iron oxides, vanadium can also be incorporated into the clay matrix. Thus, nanostructured catalysts of iron and vanadium oxides supported on the matrices of iron substituted hydrocalcite-like anionic clays (LDHs) were synthesized using a simple and facile coprecipitation method followed by calcinations in air at 450°C for 7 h. Nanocomposites had an average particle size of 75 nm (clays) with nanoparticles in the range of 7 to 11 nm and were effectively used for ethylbenzene dehydrogenation to styrene (Carja *et al.*, 2008).

Clay supported bimetallic nanoparticles (e.g. Bi^{3+} and Cu^{2+}) can also be used as visible light activated catalysts to degrade various organic contaminants (Virkutyte & Varma, 2011b). Bi/Cu-pillared montmorillonite was activated by the visible light and was able to nearly completely degrade methyl orange and dichlorophenol in 4 hours without the addition of any oxidizing agents.

17.4 GREEN SYNTHESIS ROUTES

17.4.1 Green reducing agents

One of the most used methods to prepare metal and metal oxide nanoparticles is their treatment with appropriate reducing agents such as borohydrides, citrates, and so on. Unfortunately, many of these reducing agents are associated with adverse effects on the environment. Therefore, it is of utmost importance to develop synthesis protocols and processes that are in line with the green chemistry principles. Biosynthesis of metal nanoparticles utilizing fruits, plants and derived benign natural products (that contain vast amounts of polyphenols, limonoids, carbohydrates, polyols, proteins, etc) that are responsible for the reduction of metal salts most likely will prove to be effective in the synthesis of metal and metal oxide nanoparticles and/or nanomaterials utilizing green chemistry approaches.

Ag is one of the most important nanoparticles used in a wide array of applications including medical, biomedical, engineering and environmental fields. There are different approaches and ways to synthesize it. One of the ways to synthesize stable Ag nanoparticles (6 to 8 nm in diameter) is to use hydroxypropyl starch as both reducing and stabilizing agents, which eliminates the need for separate or additional stabilizers or reducing agents (El-Rafie *et al.*, 2011). Also, it is possible to synthesize Ag nanoparticles with a relatively narrow size distribution (average particle size of 25 nm), varying ammonia as the reaction precursor concentration (0.2 to 0.005 mol l^{-1}), using saccharides, such as monosaccharides (glucose and galactose) and disaccharides (maltose and lactose) (Panácek *et al.*, 2006). These nanoparticles (a concentration as low as 1.69 μlg ml^{-1} of Ag) showed high antimicrobial and bactericidal activity against Gram-positive and Gram-negative bacteria, including highly multi-resistant strains such as methicillin-resistant *Staphylococcus aureus*. On the other hand honey can be used to prepare Ag nanoparticles as well. For instance, the pH controlled synthesis of Ag nanoparticles in the presence of honey as a reducing and stabilizing agent was demonstrated (Philip, 2010b). The author discovered that by adjusting the pH, it is possible to synthesize highly crystalline nanoparticles of various sizes at room temperature. For instance, at pH 8.5 predominantly monodispersed and nearly spherical colloidal Ag nanoparticle size of ca. 4 nm. According to the FTIR results, nanoparticles were bound to protein through the carboxylate ion group. Ag nanoparticles utilizing polyphenols present in tea extract and epicatechin were synthesized and the benign nature of nanoparticles was demonstrated when in contact with human keratinocytes (HaCaTs) used as an *in vitro* model of exposure (Moulton *et al.*, 2010). During a typical synthesis, the relevant Ag salt was subjected to the treatment with tea extract and/or epicatechin at room temperature. Synthesized Ag nanoparticles sizes ranged from 4 nm to 100 nm depending on the ratio of water to tea extract used for the synthesis and from 15 to 26 nm when various ratios of epicatechin were used. In addition to Ag nanoparticles, Pd nanoparticles utilizing polyphenols from tea and coffee were also fabricated (Nadagouda & Varma, 2008b). The nanoparticles were in the size range of 20–60 nm, depending on the ratios of the reaction precursors and crystallized in face centered cubic symmetry.

Vitamin B_2 is also reported to be an effective reducing and capping agent due to its high water solubility, biodegradability and low-toxicity in comparison to other available reducing agents (Nadagouda & Varma, 2006, 2008a). Also, the addition of a solvent along with B_2 may alter the sizes of as-synthesized nanoparticles. Indeed, the average particle sizes of Ag and Pd were 6.1 ± 0.1 nm and 4.1 ± 0.1 nm in ethylene glycol as well as 5.9 ± 0.1 nm and 6.1 ± 0.1 in acetic acid and *N*-methylpyrrolidinone (NMP), respectively. In addition, when water was used as a solvent medium, Ag and Pd nanoparticles self-assembled into rod-like structures. However, in iso-propanol nanoparticles yielded wire-like structures with a thickness in the range of 10 to 20 nm and several hundred microns in length. Finally, in acetone and acetonitrile medium, the Ag and Pd nanoparticles were self-assembled into a regular pattern making nanorod structures with thicknesses ranging from 100 to 200 nm and lengths of a few microns (Nadagouda & Varma, 2008a).

There are numerous plants and fruits that can be used to reduce and stabilize both single and several metal nanoparticles in a "one-pot" synthesis. For instance, *Citrus sinensis* peel was used in a novel, simple and green chemistry procedure at room temperature and 60°C to prepare Ag nanoparticles (Kaviya *et al.*, 2011). *Citrus sinensis* reduced and stabilized nanoparticles were effective against *E. coli*, *Ps. aeruginosa* (Gram-negative) and *S. aureus* (Gram-positive). Dried leaves of mahogany (*Swietenia mahogany JACQ.)* that contain various polyhydroxy limonoids, were used to in a rapid and benign Ag, Au and Ag/Au alloy synthesis protocol (Mondal *et al.*, 2011). Interestingly, it was observed that when metal salts were exposed to aqueous mahogany leaf extract, competitive reduction of Au (II) and Ag (I) occurred producing bimetallic Au/Ag alloy nanoparticles. Also, *Hibiscus rosa sinensis* was reported to produce stable Au and Ag nanoparticles at room temperature (Philip, 2010a). Indian propolis that contained hydroxyflavonoids and its chemical constituents pinocembrin and galangin in ethanol and water extracts were also used to manufacture stable Au and Ag nanoparticles with various morphologies under alkaline conditions ($pH \sim 10.62$) (Roy *et al.*, 2010). *Anacardium occidentale* leaf extract containing polyols can also be used to synthesize Au, Ag, Au-Ag alloy and Au core-Ag shell nanoparticles at room temperature without the use of hazardous reducing and stabilizing agents (Sheny *et al.*, 2011). According to the XRD results, these nanoparticles were highly crystalline and predominantly in the cubic phase. *Camellia sinensis* extract that contains polyphenols was used to synthesize colloidal Ag and Au nanoparticles with average sizes of 40 nm that exhibited highly efficient single photon-induced luminescence that could be manipulated by changing concentrations of metal ions and the quantity of the reducing agent (Vilchis-Nestor *et al.*, 2008).

The use of stem latex of a medicinally important plant, *Euphorbia nivulia* that contains peptides and terpenoids, to synthesize Ag and Cu nanoparticles was demonstrated (Valodkar *et al.*, 2011). During a synthesis protocol, nanoparticles (with average diameter of 5 to 10 nm) were produced at room temperature and in the presence of microwave irradiation (30 sec.). These synthesized nanoparticles were exceptionally effective against gram-negative and gram-positive bacteria. In addition, benign synthesis of noble metal nanoparticles at 45 to 75°C in water in the presence of gum kondagogu (*Cochlospermum gossypium*), specifically, –OH groups present in the gum matrix was demonstrated (Vinod *et al.*, 2011). According to their findings, Au, Ag and Pt colloidal (crystalline in nature with face centered cubic geometry) nanoparticles (7.8 ± 2.3 nm, 5.5 ± 2.5 nm and 2.4 ± 0.7 nm, respectively) were extremely stable without undergoing any oxidation.

Starch, an effective reducing and morphology-directing agent can not only be used to manufacture Ag, Au and similar widely used nanoparticles but also to synthesize tellurium nanowires (Lu *et al.*, 2005). During a typical synthesis, tellurium salt was kept in the autoclave at 160°C for 15 h, and then cooled to room temperature. Synthesized nanowires were ca. 25 nm in diameter and ca. 10 μm in length. According to the authors, a possible synthetic mechanism involved the chain-shaped bioorganic molecule acting as a template for the one-dimensional growth of inorganic tellurium. Also wood materials such as cellulose and lignin can also be employed in spherical and cubic Pt nanoparticle synthesis without the addition of reducing, capping or dispersing agents that can be utilized for the degradation of *p*-nitrophenols (Lin *et al.*, 2011).

Interestingly, biocompatible hyperbranched polyglycidol (HBP) also demonstrated properties of an effective reducing and stabilizing agent for the synthesis of highly water-soluble monometallic (Au, Ag, Pt, Pd, and Ru) and bimetallic (Au/Pt, Au/Pd, and Au/Ru) nanoparticles at room temperatures (Li *et al.*, 2010). It is reported that the HBP-assisted reduction of metal ions follows an analogous polyol process and obtained average diameters were 6.8 ± 1.3 nm, 11.3 ± 4.2 nm, 1.8, 2.4, 1.4 nm, 4.5 ± 1.5 nm for Au, Ag, Pt, Pd, Ru and Au/Pt nanoparticles, respectively.

To further advance the use of plant derivatives and potential wastes in the bulk synthesis of metal nanoparticles, aqueous hybrid sorghum (*Sorghum spp.*) that contains high levels of diverse phenolic compounds that can function as both reducing and capping agents in the synthesis of nanoparticles to produce stable Fe and Ag nanocolloids at room temperature was used (Njagi *et al.*, 2010). Hybrid sorghuyms (phenolic compounds) are water-soluble, nontoxic, and biodegradable, affording a green synthesis process. Also, most sorghum varieties are drought and heat tolerant, making sorghum economical to produce. Sorghum-reduced and stabilized Ag and Fe nanoparticles were largely uniform with a narrow size distribution (average diameter of spherical Ag and Fe nanoparticles was 10 nm and 50 nm, respectively).

Not only metals but also metal oxides can be synthesized utilizing green synthesis principles. For instance, mesoporous carbon and nitrogen doped TiO_2 nanomaterials with a predominantly anatase phase can be synthesized by a one-step "green" synthesis approach with low cost Ti salt and glucose, guanidine as precursors at mild temperatures and without the formation of undesirable by-products (Dong *et al.*, 2009; Virkutyte *et al.*, 2010; Virkutyte & Varma, 2010). According to the results, oxygen sites in the TiO_2 lattice are mostly substituted by dopant atoms (C, N, etc.) forming O-Ti-C(N, etc.) bonds. It was also found that new electronic states above the valence band edge were directly responsible for the electronic origin of the band gap narrowing and visible light photoactivity of doped TiO_2. On the other hand, magnetite nanoparticles (Fe_3O_4) with an average diameter of 27.2 nm can be synthesized using a biopolymer (sodium alginate) via the hydrothermal route (autoclave at 180°C for 24 h) in the presence of urea (0.5 g of urea added to 10 mL of aqueous solution containing 1.5 mmol of $FeCl_3$, followed by the drop-wise addition of 25 mL of a 25.4 mM sodium alginate aqueous solution) as a reaction precursor (Gao *et al.*, 2008).

17.4.2 Biocatalysis

An array of physical, chemical and biological methods has been used to synthesize metal nanoparticles and other nanomaterials. However, it is vitally important to develop a reliable "green" chemistry process for the synthesis of bionanoparticles. Therefore: (i) the use of fungi and bacteria emerges as an eco-friendly and exciting approach that reduce waste products (ultimately leading to atomically precise molecular manufacturing with zero waste); (ii) the use of nanomaterials as catalysts for greater efficiency in current manufacturing processes by minimizing or eliminating the use of toxic materials (green chemistry principles); (iii) the use of nanomaterials and nanodevices to reduce pollution (e.g. water and air filters); and (iv) the use of nanomaterials for more efficient alternative energy production (e.g. solar and fuel cells) (Jebali *et al.*, 2011).

Microbial (bacteria, actinomycetes, fungi, yeasts and viruses) synthesis of metal nanoparticles is a green chemistry approach that interconnects nanotechnology and microbial biotechnology (Mandal *et al.*, 2006). Despite the stability, biological nanoparticles are not monodispersed and the rate of synthesis is slow. To overcome these problems, several factors such as microbial cultivation methods and the extraction techniques need to be optimized (Narayanan & Sakthivel, 2010).

Fungi have numerous advantages for nanoparticle synthesis compared with other organisms. Thus, the fungus *Trichoderma viride* was used to synthesize polydispersed Ag nanoparticles (size ranging from 5 to 40 nm) at near-room temperature (27°C) that showed maximum absorbance at 420 nm on ultraviolet-visible spectra (Fayaz *et al.*, 2010). Antibactericidal properties of as-synthesized Ag nanoparticles were tested against four bacterial strains—namely, *Salmonella typhi* (gram-negative rods), *Escherichia coli* (gram-negative rods), *Staphylococcus aureus* (gram-positive cocci), and *Micrococcus luteus* (gram-positive cocci). It was also found that antibacterial activities of ampicilin, kanamycin, erythromycin and chloramphenicol were significantly increased in the presence of Ag nanoparticles. In addition, *Geotricum sp.* was found to successfully produce Ag nanoparticles with particle sizes ranging from 30 to 50 nm (Jebali *et al.*, 2011). According to FTIR analyses, the presence of amide (I) and (II) bands of protein were identified and acted as capping and stabilizing agent on the surface of nanoparticles. Also, the fungus *Verticillium* can be used to synthesize Ag nanoparticles with an average size of 25 ± 12 nm at room temperature (Mukherjee *et al.*, 2001). According to the authors, silver particles were formed below the cell wall surface, possibly due to reduction of the metal ions by enzymes present in the cell wall. It is noteworthy to mention that Ag ions were not toxic to the fungal cells and the cells continued to multiply after biosynthesis of the silver nanoparticles.

Rice husk is a cheap agro-based waste material, which harbors a substantial amount of silica in the form of amorphous hydrated silica grains. Therefore, it would be ideal to harness the amorphous silica that can be bio-transformed at room temperature to crystalline silica nanoparticles for numerous applications. It was reported that the fungus *Fusarium oxysporum* rapidly bio-transformed amorphous plant biosilica into crystalline silica at room temperature and leached out silica extracellularly in the form of 2-6 nm quasi-spherical, highly crystalline silica nanoparticles capped by stabilizing proteins (Bansal *et al.*, 2006).

Bacteria and actinomycetes can also be effectively utilized for the synthesis of metal nanoparticles. The novel extracellular synthesis of well-dispersed Au nanoparticles (average size of 8 nm) using the prokaryotic microorganism *Thermomonospora* sp. was demonstrated (Ahmad *et al.*, 2003). Also, the bacterium *Brevibacterium casei* can be used for the "green" synthesis of Ag (sizes from 10 to 50 nm) and Au (sizes from 10 to 50 nm) nanoparticles (Kalishwaralal *et al.*, 2010). Surface plasmon absorbance was observed at 420 nm and 540 nm for Ag and Au nanoparticles, respectively. According to FTIR observations, the presence of proteins in fungi was responsible for the reduction and capping of nanoparticles. Figure 17.3 shows the synthesis pathway during microorganisms-induced synthesis of metal nanoparticles.

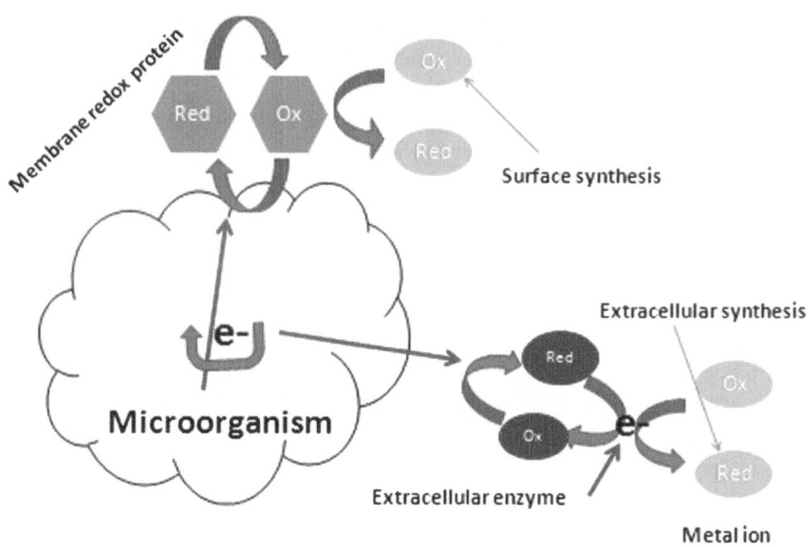

Figure 17.3 Use of microorganisms for metal nanoparticle synthesis (adapted from Kalishwaralal *et al.*, 2010).

Pd supported on ferrimagnetic particles have a diverse range of uses in catalysis (Chapters 9 and 10; Coker *et al.*, 2010). However, synthesis utilizing conventional synthetic methods results in potentially high environmental and economic costs. Nonetheless, the use of the Fe(III)-reducing bacterium *Geobacter sulfurreducens* may significantly reduce synthesis costs; facilitate easy recovery of the catalyst

(Pd nanoparticles on biomagnetite with 10 mol% of Pd) with superior performance due to the reduced agglomeration and particle size ranging from 20 to 30 nm. Such catalyst was highly effective in the Heck reaction coupling iodobenzene to ethyl acrylate or styrene with a complete conversion to ethyl cinnamate or stilbene within 90 and 180 min, respectively.

TiO_2 nanoparticles with sizes ranging from 8 to 35 nm can also be synthesized using *Lactobacillus* sp. and *Sachharomyces cerevisae* at room temperature (Jha *et al.*, 2009). According to the authors, the synthesis of *n*-TiO_2 might have resulted because of pH-sensitive membrane bound oxido-reductases and carbon source dependent rH2 in the culture medium.

17.5 OUTLOOK

Currently, green nanosynthesis faces extensive challenges in order to achieve the maximum efficiency while minimizing or eliminating the adverse impacts on human health and the environment. Only by adapting green chemistry and engineering principles, we will be able to provide a background for greener nanomaterials through greener nanosynthesis methods. In the nearest future, synthesis of nanomaterials will finally emerge from the laboratory scale attempts to the synthesis of large quantities of structurally well-defined, functionalized, stable and effective nanoparticles applicable for a great array of applications. This chapter focused on the importance and future of "green" synthesis of nanomaterials, that is utilization of "green" reaction precursors and mild reaction conditions to manufacture stable, safe, high-performance and effective as well as functionalized nanomaterials for targeted processes. Advances in "greener" synthesis pathways, surface functionalization and shape control that presented the state-of-the-art and identified future challenges of nanoparticle preparation approaches were described. Importantly, detailed examination of examples and case studies presented in this chapter suggest that more advances are necessary in developing the ultimate "green" synthesis methods. The key findings in the area of synthesis of nanomaterials described in this chapter are as following: developed green synthesis methods are methods that: (1) eliminated toxic reagents and organic solvents, (2) had very little or no by-product formation during synthesis, (3) had better control of shapes, particle sizes and dispersivity of nanoparticles and (4) had no need for purification of the final product, nanoparticle, which eliminates the use of organic solvents.

Despite the substantial progress in the area of nanomaterial synthesis, there are still some issues that have to be addressed in the future: (1) identification of risks associated with biological, ecological and human exposure to nanomaterials during all life cycle stages including synthesis, use and degradation, (2) development of even safer and "greener" synthesis approaches, eliminating energy intensive processes, the use of solvents and valuable materials that can be substituted with various wastes, and (3) development and optimization of analytical techniques that can perform real-time and routine analysis of the nanoparticles for morphology, composition, structure and purity, regardless the experimental matrix. These issues present a significant challenge that will require profound understanding of nanoparticle formation and transformations, reactivity of reaction precursors, and so on.

Reaction precursors and possible nanoparticle transformations must be amenable to scale up for the production of large quantities of nanomaterials. Despite the encouraging findings and developed methods, further research is highly warranted to improve current approaches and processes that will provide opportunities and challenges for the scientific community and general public for the years to come.

REFERENCES

Abu-Reziq R., Wang D., Post M. and Alper H. (2008). Separable catalysts in one-pot syntheses for greener chemistry. *Chem. Mater.*, **20**(7), 2544–2550.

Adam J. D., Davis L. E., Dionne G. F., Schloemann E. F. and Stitzer S. N. (2002). Ferrite devices and materials. *Microw. Theory Technol.*, **50**(3), 721–737.

Ahmad A., Senapati S., Khan M. I., Kumar R. and Sastry M. (2003). Extracellular biosynthesis of monodisperse gold nanoparticles by a novel extremophilic actinomycete, *Thermomonospora* sp. *Langmuir*, **19**(8), 3550–3553.

Alone S. T., Shirsath S. E., Kadam R. H. and Jadhav K. M. (2011). Chemical synthesis, structural and magnetic properties of nano-structured Co-Zn-Fe-Cr ferrite. *J. Alloys Comp.*, **509**(16), 5055–5060.

Anastas P. and Warner J. (1998). *Green Chemistry: Theory and Practice*. Oxford University Press, USA.

Bansal V., Ahmad A. and Sastry M. (2006). Fungus-mediated biotransformation of amorphous silica in rice husk to nanocrystalline silica. *J. Am. Chem. Soc.*, **128**(43), 14059–14066.

Baruwati B., Polshettiwar V. and Varma R. S. (2009). Magnetically recoverable supported ruthenium catalyst for hydrogenation of alkynes and transfer hydrogenation of carbonyl compounds. *Tetrahedron Lett.*, **50**(11), 1215–1218.

Baruwati B. and Varma R. S. (2009). High value products from waste: grape pomace extract – a three-in-one package for the synthesis of metal nanoparticles. *ChemSusChem*, **2**(11), 1041–1044.

Basu S., Hossain S. M., Chakravorty D. and Pal M. (2011). Enhanced magnetic properties in hydrothermally synthesized Mn-doped $BiFeO_3$ nanoparticles. *Curr. Appl. Phys.*, **11**(4), 976–980.

Behrens S. (2011). Preparation of functional magnetic nanocomposites and hybrid materials: recent progress and future directions. *Nanoscale*, **3**(3), 877–892.

Buhro W. E. and Colvin V. L. (2003). Semiconductor nanocrystals: shape matters. *Nat. Mater.*, **2**(3), 138–139.

Caillot T., Pourroy G. and Stuerga D. (2011). Novel metallic iron/manganese-zinc ferrite nanocomposites prepared by microwave hydrothermal flash synthesis. *J. Alloys Comp.*, **509**(8), 3493–3496.

Carja G., Kameshima Y., Nakajima A., Dranca C. and Okada K. (2009). Nanosized silver-anionic clay matrix as nanostructured ensembles with antimicrobial activity. *Int. J. Antimicr. Agents*, **34**(6), 534–539.

Carja G., Kameshima Y. and Okada K. (2008). Nanoparticles of iron and vanadium oxides supported on iron substituted LDHs: synthesis, textural characterization and their catalytic behavior in ethylbenzene dehydrogenation. *Micropor. Mesopor. Mater.*, **115** (3), 541–547.

Chang P. R., Yu J., Ma X. and Anderson D. P. (2011). Polysaccharides as stabilizers for the synthesis of magnetic nanoparticles. *Carbohydr. Polym.*, **83**(2), 640–644.

Christian P., Von der Kammer F., Baalousha M. and Hofmann T. (2008). Nanoparticles: structure, properties, preparation and behaviour in environmental media. *Ecotoxicology*, **17**(5), 326–343.

Coker V. S., Bennett J. A., Telling N. D., Henkel T., Charnock J. M., van der Laan G., Pattrick R. A. D., Pearce C. I., Cutting R. S., Shannon I. J., Wood J., Arenholz E., Lyon I. C. and Lloyd J. R. (2010). Microbial engineering of nanoheterostructures: biological synthesis of a magnetically recoverable palladium nanocatalyst. *ACS Nano*, **4**(5), 2577–2584.

Cornell R. M. and Schwertmann U. (2003). *The Iron Oxides: Structure, Properties, Reactions, Occurrences and Uses*, 2nd edn. Wiley–VCH, Weinheim.

Dahl J. A., Maddux B. L. S. and Hutchison J. E. (2007). Toward greener nanosynthesis. *Chem. Rev.*, **107**(6), 2228–2269.

Darezereshki E. (2011). One-step synthesis of hematite ([alpha]-Fe_2O_3) nano-particles by direct thermal-decomposition of maghemite. *Mater. Lett.*, **65**(4), 642–645.

Dékány I., Turi L. and Király Z. (1999). CdS, TiO_2 and Pd^o nanoparticles growing in the interlamellar space of montmorillonite in binary liquids. *Appl. Clay Sci.*, **15**(1–2), 221–239.

Dhakshinamoorthy A. and Pitchumani K. (2008). Clay entrapped nickel nanoparticles as efficient and recyclable catalysts for hydrogenation of olefins. *Tetrahedron Lett.*, **49** (11), 1818–1823.

Dong F., Wang H. and Wu Z. (2009). One-step "Green" synthetic approach for mesoporous C-doped titanium dioxide with efficient visible light photocatalytic activity. *J. Phys. Chem. C*, **113**(38), 16717–16723.

Eckelman M. J., Zimmerman J. B. and Anastas P. T. (2008). Toward green nano. *J. Ind. Ecol.*, **12**(3), 316–328.

El-Rafie M. H., El-Naggar M. E., Ramadan M. A., Fouda M. M. G., Al-Deyab S. S. and Hebeish A. (2011). Environmental synthesis of silver nanoparticles using hydroxypropyl starch and their characterization. *Carbohydr. Polym.*, **86**(2), 630–635.

Eshraghi M. and Kameli P. (2011). Magnetic properties of $CoFe_2O_4$ nanoparticles prepared by thermal treatment of ball-milled precursors. *Curr. Appl. Phys.*, **11**(3), 476–481.

Fayaz A. M., Balaji K., Girilal M., Yadav R., Kalaichelvan P. T. and Venketesan R. (2010). Biogenic synthesis of silver nanoparticles and their synergistic effect with antibiotics: a study against gram-positive and gram-negative bacteria. *Nanomed. Nanotechnol. Biol. Med.*, **6**(1), 103–109.

Gao S., Shi Y., Zhang S., Jiang K., Yang S., Li Z. and Takayama-Muromachi E. (2008). Biopolymer-assisted green synthesis of iron oxide nanoparticles and their magnetic properties. *J. Phys. Chem. C*, **112**(28), 10398–10401.

Garrido-Ramírez E. G., Theng B. K. G. and Mora M. L. (2011). Clays and oxide minerals as catalysts and nanocatalysts in Fenton-like reactions – a review. *Appl. Clay Sci.*, **47**(3–4), 182–192.

Goodarz Naseri M., Saion E. B., Ahangar H. A., Hashim M. and Shaari A. H. (2011). Synthesis and characterization of manganese ferrite nanoparticles by thermal treatment method. *J. Magn. Magn. Mater.*, **323**(13), 1745–1749.

Gorobinskii L. V., Yurkov G. Y. and Baranov D. A. (2007). Production of high porosity nanoparticles of cerium oxide in clay. *Micropor. Mesopor. Mater.*, **100**(1–3), 134–138.

Huang H. and Yang Y. (2008). Preparation of silver nanoparticles in inorganic clay suspensions. *Comp. Sci. Technol.*, **68**(14), 2948–2953.

Huang Y.-F., Wang Y.-F. and Yan X.-P. (2011). Amine-functionalized magnetic nanoparticles for rapid capture and removal of bacterial pathogens. *Environ. Sci. Technol.*, **44**(20), 7908–7913.

Huczko A. (2000). Template-based synthesis of nanomaterials. *Appl. Phys. A*, **70**(4), 365–376.

Jebali A., Ramezani F. and Kazemi B. (2011). Biosynthesis of silver nanoparticles by *Geotricum* sp. *J. Cluster Sci.*, **22**(2), 225–232.

Jha A. K., Prasad K. and Kulkarni A. R. (2009). Synthesis of TiO_2 nanoparticles using microorganisms. *Colloid Surf. B*, **71**(2), 226–229.

Ji X., Liu Y., Li M., Evans D. G. and Xu Q. (2011). Synthesis of mesoporous Co-spinel ferrite using a hard template and gaseous diffusion. *Solid State Sci.*, **13**(2), 404–408.

Kalishwaralal K., Deepak V., Ram Kumar Pandian S., Kottaisamy M., BarathManiKanth S., Kartikeyan B. and Gurunathan S. (2010). Biosynthesis of silver and gold nanoparticles using *Brevibacterium casei*. *Colloid Surf. B*, **77**(2), 257–262.

Karn B. (2008). *Research on Nanotechnology Applications: Green Nanotechnology for Past, Present, and Preventing Future Problems*. NanoECO Monte Verità, Switzerland.

Kaviya S., Santhanalakshmi J., Viswanathan B., Muthumary J. and Srinivasan K. (2011). Biosynthesis of silver nanoparticles using citrus sinensis peel extract and its antibacterial activity. *Spectrochim. Acta A*, **79**(3), 594–598.

Klotz S., Steinle-Neumann G., Straumlssle T., Philippe J., Hansen T. and Wenzel M. J. (2008). Magnetism and the Verwey transition in Fe_3O_4 under pressure. *Phys. Rev. B*, **77**(1), 012411.

Letaief S., Grant S. and Detellier C. (2011). Phenol acetylation under mild conditions catalyzed by gold nanoparticles supported on functional pre-acidified sepiolite. *Appl. Clay Sci.*, **53**(2), 236–243.

Li H., Jo J. K., Zhang L. D., Ha C.-S., Suh H. and Kim I. (2010). Hyperbranched polyglycidol assisted green synthetic protocols for the preparation of multifunctional metal nanoparticles. *Langmuir*, **26**(23), 18442–18453.

Li J. J., Mu Z., Xu X. Y., Tian H., Duan M. H., Li L. D., Hao Z. P., Qiao S. Z. and Lu G. Q. (2008). A new and generic preparation method of mesoporous clay composites containing dispersed metal oxide nanoparticles. *Micropor. Mesopor. Mater.*, **114**(1–3), 214–221.

Li N.-H., Lo S.-L., Hu C.-Y., Hsieh C.-H. and Chen C.-L. (2011a). Stabilization and phase transformation of $CuFe_2O_4$ sintered from simulated copper-laden sludge. *J. Hazard. Mater.*, **190**(1–3), 597–603.

Li S., Wu P., Li H., Zhu N., Li P., Wu J., Wang X. and Dang Z. (2011b). Synthesis and characterization of organo-montmorillonite supported iron nanoparticles. *Appl. Clay Sci.*, **50**(3), 330–336.

Lin X., Wu M., Wu D., Kuga S., Endo T. and Huang Y. (2011). Platinum nanoparticles using wood nanomaterials: eco-friendly synthesis, shape control and catalytic activity for p-nitrophenol reduction. *Green Chem.*, **13**(2), 283–287.

Lu A.-H., Salabas E. L. and Schüth F. (2007). Magnetic nanoparticles: synthesis, protection, functionalization, and application. *Ang. Chemie Int. Ed.*, **46**(8), 1222–1244.

Lu Q., Gao F. and Komarneni S. (2005). A green chemical approach to the synthesis of tellurium nanowires. *Langmuir*, **21**(13), 6002–6005.

Majewski P. and Thierry B. (2007). Functionalized magnetite nanoparticles – synthesis, properties, and bio-applications critical reviews in solid state and materials sciences. *Crit. Rev. Solid State Mater. Sci.*, **32**(3–4), 203–215.

Mandal D., Bolander M., Mukhopadhyay D., Sarkar G. and Mukherjee P. (2006). The use of microorganisms for the formation of metal nanoparticles and their application. *Appl. Microbiol. Biotechnol.*, **69**(5), 485–492.

Manikandan D., Divakar D., Rupa A. V., Revathi S., Preethi M. E. L. and Sivakumar T. (2007a). Synthesis of platinum nanoparticles in montmorillonite and their catalytic behaviour. *Appl. Clay Sci.*, **37**(1–2), 193–200.

Manikandan D., Divakar D. and Sivakumar T. (2007b). Utilization of clay minerals for developing Pt nanoparticles and their catalytic activity in the selective hydrogenation of cinnamaldehyde. *Catal. Comm.*, **8**(11), 1781–1786.

Manova E., Aranda P., Angeles Martín-Luengo M., Letaïef S. and Ruiz-Hitzky E. (2011). New titania-clay nanostructured porous materials. *Micropor. Mesopor. Mater.*, **131** (1–3), 252–260.

Martirosyan K. S., Galstyan E., Hossain S. M., Wang Y.-J. and Litvinov D. (2011). Barium hexaferrite nanoparticles: synthesis and magnetic properties. *Mater. Sci. Eng. B*, **176**(1), 8–13.

Mastalir Á., Szöllösi G., Király Z. and Rázga Z. (2002). Preparation and characterization of platinum nanoparticles immobilized in dihydrocinchonidine-modified montmorillonite and hectorite. *Appl. Clay Sci.*, **22**(1–2), 9–16.

Miyoshi H., Ohno H., Sakai K., Okamura N. and Kourai H. (2011). Characterization and photochemical and antibacterial properties of highly stable silver nanoparticles prepared on montmorillonite clay in *n*-hexanol. *J. Colloid Interface Sci.*, **345**(2), 433–441.

Mohanpuria P., Rana N. and Yadav S. (2008). Biosynthesis of nanoparticles: technological concepts and future applications. *J. Nanopart. Res.*, **10**(3), 507–517.

Mondal S., Roy N., Laskar R. A., Sk I., Basu S., Mandal D. and Begum N. A. (2011). Biogenic synthesis of Ag, Au and bimetallic Au/Ag alloy nanoparticles using aqueous extract of mahogany (Swietenia mahogani JACQ.) leaves. *Colloid Surf. B*, **82**(2), 497–504.

Moulton M. C., Braydich-Stolle L. K., Nadagouda M. N., Kunzelman S., Hussain S. M. and Varma R. S. (2010). Synthesis, characterization and biocompatibility of "green" synthesized silver nanoparticles using tea polyphenols. *Nanoscale*, **2** (5), 763–770.

Mukherjee P., Ahmad A., Mandal D., Senapati S., Sainkar S. R., Khan M. I., Parishcha R., Ajaykumar P. V., Alam M., Kumar R. and Sastry M. (2001). Fungus-mediated synthesis of silver nanoparticles and their immobilization in the mycelial matrix: a novel biological approach to nanoparticle synthesis. *Nano Lett.*, **1**(10), 515–519.

Nadagouda M. N., Polshettiwar V. and Varma R. S. (2009). Self-assembly of palladium nanoparticles: synthesis of nanobelts, nanoplates and nanotrees using vitamin B1, and their application in carbon-carbon coupling reactions. *J. Mater. Chem.*, **19**(14), 2026–2031.

Nadagouda M. N. and Varma R. S. (2006). Green and controlled synthesis of gold and platinum nanomaterials using vitamin B_2: density-assisted self-assembly of nanospheres, wires and rods. *Green Chem.*, **8**(6), 516–518.

Nadagouda M. N. and Varma R. S. (2007a). Microwave-assisted shape-controlled bulk synthesis of Ag and Fe nanorods in poly(ethylene glycol) solutions. *Cryst. Growth Des.*, **8**(1), 291–295.

Nadagouda M. N. and Varma R. S. (2007b). Synthesis of thermally stable carboxymethyl cellulose/metal biodegradable nanocomposites for potential biological applications. *Biomacromolecules*, **8**(9), 2762–2767.

Nadagouda M. N. and Varma R. S. (2008a). Green synthesis of Ag and Pd nanospheres, nanowires, and nanorods using vitamin B_2: catalytic polymerisation of aniline and pyrrole. *J. Nanomater.*, **2008**, 1–8.

Nadagouda M. N. and Varma R. S. (2008b). Green synthesis of silver and palladium nanoparticles at room temperature using coffee and tea extract. *Green Chem.*, **10**(8), 859–862.

Naidek K. P., Bianconi F., da Rocha T. C. R., Zanchet D., Bonacin J. A., Novak M. A., das Graças Fialho Vaz M. and Winnischofer H. (2011). Structure and morphology of spinel MFe_2O_4 (M = Fe, Co, Ni) nanoparticles chemically synthesized from heterometallic complexes. *J. Colloid Interface Sci.*, **358**(1), 39–46.

Narayanan K. B. and Sakthivel N. (2010). Biological synthesis of metal nanoparticles by microbes. *Adv. Colloid Interface Sci.*, **156**(1–2), 1–13.

Nata I. F., Salim G. W. and Lee C.-K. (2011). Facile preparation of magnetic carbonaceous nanoparticles for Pb^{2+} ions removal. *J. Hazard. Mater.*, **183**(1–3), 853–858.

Nazrul Islam M., Van Phong L., Jeong J.-R. and Kim C. (2011). A facile route to sonochemical synthesis of magnetic iron oxide (Fe_3O_4) nanoparticles. *Thin Solid Films*, **519**(23), 8277–8279.

Njagi E. C., Huang H., Stafford L., Genuino H., Galindo H. M., Collins J. B., Hoag G. E. and Suib S. L. (2010). Biosynthesis of iron and silver nanoparticles at room temperature using aqueous sorghum bran extracts. *Langmuir*, **27**(1), 264–271.

Özgür Ü., Alivov Y. and Morkoç H. (2009). Microwave ferrites, Part 1: Fundamental properties. *J. Mater. Sci.*, **20**(9), 789–834.

Panácek A., Kvítek L., Prucek R., Kolář M., Večeřová R., Pizúrová N., Sharma V. K., Nevěčná T. J. and Zbořil R. (2006). Silver colloid nanoparticles: synthesis, characterization, and their antibacterial activity. *J. Phys. Chem. B*, **110(33)**, 16248–16253.

Papp S., Szücs A. and Dékány I. (2001). Preparation of Pd^0 nanoparticles stabilized by polymers and layered silicate. *Appl. Clay Sci.*, **19**(1–6), 155–172.

Park H. J., McConnell J. T., Boddohi S., Kipper M. J. and Johnson P. A. (2011). Synthesis and characterization of enzyme-magnetic nanoparticle complexes: effect of size on activity and recovery. *Colloid Surf. B*, **83**(2), 198–203.

Patakfalvi R. and Dékány I. (2004). Synthesis and intercalation of silver nanoparticles in kaolinite/DMSO complexes. *Appl. Clay Sci.*, **25**(3–4), 149–159.

Peng H.-P., Liang R.-P. and Qiu J.-D. (2011a). Facile synthesis of Fe_3O_4@Al_2O_3 core-shell nanoparticles and their application to the highly specific capture of heme proteins for direct electrochemistry. *Biosens. Bioelectron.*, **26**(6), 3005–3011.

Peng J., Hojamberdiev M., Xu Y., Cao B., Wang J. and Wu H. (2011b). Hydrothermal synthesis and magnetic properties of gadolinium-doped $CoFe_2O_4$ nanoparticles. *J. Magn. Magn. Mater.*, **323**(1), 133–137.

Philip D. (2010a). Green synthesis of gold and silver nanoparticles using *Hibiscus rosa sinensis*. *Physica E*, **42**(5), 1417–1424.

Philip D. (2010b). Honey mediated green synthesis of silver nanoparticles. *Spectrochim. Acta A*, **75**(3), 1078–1081.

Polshettiwar V. and Varma R. S. (2009a). Nanoparticle-supported and magnetically recoverable palladium (Pd) catalyst: a selective and sustainable oxidation protocol with high turnover number. *Org. Biomol. Chem.*, **7**(1), 37–40.

Polshettiwar V. and Varma R. S. (2009b). Nanoparticle-supported and magnetically recoverable ruthenium hydroxide catalyst: efficient hydration of nitriles to amides in aqueous medium. *Chem. Eur. J.*, **15**(7), 1582–1586.

Pomogailo A. D. and Kestelman V. N. (eds) (2005). *Metallopolymer Nanocomposites*. Springer, Berlin, Heidelberg, New York.

Roy N., Mondal S., Laskar R. A., Basu S., Mandal D. and Begum N. A. (2010). Biogenic synthesis of Au and Ag nanoparticles by Indian propolis and its constituents. *Colloid Surf. B*, **76**(1), 317–325.

Senapati K. K., Borgohain C. and Phukan P. (2011). Synthesis of highly stable $CoFe_2O_4$ nanoparticles and their use as magnetically separable catalyst for Knoevenagel reaction in aqueous medium. *J. Mol. Catal. A – Chem.*, **339**(1–2), 24–31.

Shahwan T., Üzüm Ç., Eroglu A. E. and Lieberwirth I. (2011). Synthesis and characterization of bentonite/iron nanoparticles and their application as adsorbent of cobalt ions. *Appl. Clay Sci.*, **47**(3–4), 257–262.

Sheny D. S., Mathew J. and Philip D. (2011). Phytosynthesis of Au, Ag and Au-Ag bimetallic nanoparticles using aqueous extract and dried leaf of *Anacardium occidentale*. *Spectrochim. Acta A*, **79**(1), 254–262.

Sidhpuria K. B., Patel H. A., Parikh P. A., Bahadur P., Bajaj H. C. and Jasra R. V. (2009). Rhodium nanoparticles intercalated into montmorillonite for hydrogenation of aromatic compounds in the presence of thiophene. *Appl. Clay Sci.*, **42**(3–4), 386–390.

Sivakumar P., Ramesh R., Ramanand A., Ponnusamy S. and Muthamizhchelvan C. (2011). Synthesis and characterization of $NiFe_2O_4$ nanosheet via polymer assisted co-precipitation method. *Mater. Lett.*, **65**(3), 483–485.

Stubbs D. and Gilman P. (2007). *Nanotechnology Applications in Environmental Health: Big Plans for Little Particles*. Oak Ridge Center for Advanced Studies, Oak Ridge, TN, USA.

Sun Y., Guo G., Yang B., Zhou X., Liu Y. and Zhao G. (2011a). One-step fabrication of Fe_2O_3/Ag core-shell composite nanoparticles at low temperature. *J. Non-Cryst. Solids*, **357**(3), 1085–1089.

Sun Z., Liu D., Tong L., Shi J., Yang X., Yu L., Tao Y. and Yang H. (2011b). Synthesis and properties of magnetic and luminescent $Fe_3O_4/SiO_2/YVO_4$: Eu^{3+} nanocomposites. *Solid State Sci.*, **13**(2), 361–365.

Szucs A., Berger F. and Dékány I. (2000). Preparation and structural properties of Pd nanoparticles in layered silicate. *Colloid Surf. A*, **174**(3), 387–402.

Teja A. S. and Koh P.-Y. (2009). Synthesis, properties, and applications of magnetic iron oxide nanoparticles. *Prog. Cryst. Growth Char. Mater.*, **55**(1–2), 22–45.

Valodkar M., Nagar P. S., Jadeja R. N., Thounaojam M. C., Devkar R. V. and Thakore S. (2011). Euphorbiaceae latex induced green synthesis of non-cytotoxic metallic nanoparticle solutions: a rational approach to antimicrobial applications. *Colloid Surf. A*, **384**(1–3), 337–344.

Venkataraju C., Sathishkumar G. and Sivakumar K. (2011). Effect of Cd on the structural, magnetic and electrical properties of nanostructured Mn-Zn ferrite. *J. Magn. Magn. Mater.*, **323**(13), 1817–1822.

Vicente I., Salagre P. and Cesteros Y. (2011). Ni nanoparticles supported on microwave-synthesised saponite for the hydrogenation of styrene oxide. *Appl. Clay Sci.*, **53**(2), 212–219.

Vilchis-Nestor A. R., Sánchez-Mendieta V., Camacho-López M. A., Gómez-Espinosa R. M., Camacho-López M. A. and Arenas-Alatorre J. A. (2008). Solventless synthesis and optical properties of Au and Ag nanoparticles using Camellia sinensis extract. *Mater. Lett.*, **62**(17–18), 3103–3105.

Vinod V. T. P., Saravanan P., Sreedhar B., Devi D. K. and Sashidhar R. B. (2011). A facile synthesis and characterization of Ag, Au and Pt nanoparticles using a natural hydrocolloid gum kondagogu (*Cochlospermum gossypium*). *Colloid Surf. B*, **83**(2), 291–298.

Virkutyte J., Baruwati B. and Varma R. S. (2010). Visible light induced photobleaching of methylene blue over melamine-doped TiO_2 nanocatalyst. *Nanoscale*, **2**(7), 1109–1111.

Virkutyte J. and Varma R. S. (2010). Fabrication and visible light photocatalytic activity of a novel $Ag/TiO_{2-x}N_x$ nanocatalyst. *New J. Chem.*, **34**(6), 1094–1096.

Virkutyte J. and Varma R. S. (2011a). Green synthesis of metal nanoparticles: biodegradable polymers and enzymes in stabilization and surface functionalization. *Chem. Sci.*, **2**, 837–846.

Virkutyte J. and Varma R. S. (2011b). Photoinduced catalytic adsorption of model contaminants on Bi/Cu pillared montmorillonite in the visible light range. *Sep. Purif. Technol.*, **78**(2), 201–207.

Wang L., Neoh K. G., Kang E.-T. and Shuter B. (2011a). Multifunctional polyglycerol-grafted $Fe_3O_4@SiO_2$ nanoparticles for targeting ovarian cancer cells. *Biomaterials*, **32**(8), 2166–2173.

Wang X. L., Wei L., Tao G. H. and Huang M. Q. (2011b). Synthesis and characterization of magnetic and luminescent Fe_3O_4/CdTe nanocomposites using aspartic acid as linker. *Chin. Chem. Lett.*, **22**(2), 233–236.

Yallapu M. M., Othman S. F., Curtis E. T., Gupta B. K., Jaggi M. and Chauhan S. C. (2011). Multi-functional magnetic nanoparticles for magnetic resonance imaging and cancer therapy. *Biomaterials*, **32**(7), 1890–1905.

Yang Z.-P., Gong X.-Y. and Zhang C.-J. (2011). Recyclable Fe_3O_4/hydroxyapatite composite nanoparticles for photocatalytic applications. *Chem. Eng. J.*, **165**(1), 117–121.

Zhang F., Zhu Z., Dong Z., Cui Z., Wang H., Hu W., Zhao P., Wang P., Wei S., Li R. and Ma J. (2011a). Magnetically recoverable facile nanomaterials: synthesis, characterization and application in remediation of heavy metals. *Microchem. J.*, **98**(2), 328–333.

Zhang X., Jiang W., Gong X. and Zhang Z. (2011b). Sonochemical synthesis and characterization of magnetic separable Fe_3O_4/Ag composites and its catalytic properties. *J. Alloys Comp.*, **508**(2), 400–405.

Zhu H. Y., Li J. Y., Zhao J. C. and Churchman G. J. (2005). Photocatalysts prepared from layered clays and titanium hydrate for degradation of organic pollutants in water. *Appl. Clay Sci.*, **28**(1–4), 79–88.

Zhu L., Letaief S., Liu Y., Gervais F. and Detellier C. (2009). Clay mineral-supported gold nanoparticles. *Appl. Clay Sci.*, **43**(3–4), 439–446.

Chapter 18

Plant-based nanoparticle manufacturing

J. Hong, J. R. Peralta-Videa and J. L. Gardea-Torresdey

18.1 INTRODUCTION

Nanotechnology is "the understanding and control of matter at dimensions of roughly 1 to 100 nanometers (a nanometer is one billionth of a meter), where unique phenomena enable novel applications" (NNI, 2007). The research on nanotechnology is still in its infancy and will be enormously expanded in the near future. Although this is new technology, nanotechnology is currently been applied in electronics, engineering, energy, material science, pharmaceutical, biomedical, cosmetics, textiles, food industry, and agriculture (Biswas & Wu 2005; Chen *et al.*, 2006; Maynard *et al.*, 2006).

The building blocks of nanotechnology are nanoparticles (NPs), which have at least two dimensions between 1 and 100 nm. Currently, NPs are produced by chemical and physical methods. Physical methods include chemical vapor deposition, laser ablation, sonication, ultraviolet irradiation and aerosol technology (Kim *et al.*, 2007; Yu *et al.*, 2007; Zhang *et al.*, 2007; Kammler *et al.*, 2001). Chemical methods involve reducing agents (sodium borohydride, hydroxylamine methoxypolyethylene glycol, and potassium bitartrate), protecting agent or surfactant (polyvinyl pyrrolidone), stabilizers (2-vinylpyridine telomers, citric acid, and alginate), (Meltzer *et al.*, 2001; Esumi *et al.*, 2001; Mallick *et al.*, 2004; Tan *et al.*, 2003; Shimazaki *et al.*, 2005; Liu *et al.*, 2009; Kimling *et al.*, 2006; Fukuoka *et al.*, 2004). However, these methods may use toxic or dangerous solvents and to some extent are sources of health and environmental hazards. For this reason, eco-friendly methods have been developed for the synthesis of NPs of potential industrial use. In this attempt, biological resources including fungi, bacteria, yeast, plants and plant extracts have been used for the synthesis of metal NPs (Shanker *et al.*, 2003; Mohanpuria *et al.*, 2008).

The research on the formation of NPs by microorganisms is relatively recent. The presence of Ag NPs ranging between 35 to 46 nm in the periplasmic space of prokaryotic bacteria was reported in the 1990s (Slawson *et al.*, 1992). At the end of the 1990s, other reports described the formation of 200 nm Ag NPs by *Pseudomonas stutzeri* AG259 treated in the dark with 50 mM $AgNO_3$ at 30°C for 48 h (Klaus *et al.*, 1999). It has been reported that bacteria can also form CdS nanocrystals (Sweeney *et al.*, 2004), FeS, magnetite, and Au NPs (Watson *et al.*, 1999; Bharde *et al.*, 2005; Konishi *et al.*, 2004; He *et al.*, 2007). Besides bacteria, fungi such as *Verticillium*, *Aspergillus fumigates*, and *Trichoderma asperellum* have also been found to synthesizing Ag NPs (Mukherjee *et al.*, 2001; Bhainsa & D'Souza, 2006; Mukherjee *et al.*, 2008). Marine algae *Sargassum wightii*, have proven to synthesize highly stable extracellular Au NPs in a relatively short time compared with other biological procedures (Singaravelu *et al.*, 2007).

At the end of the 1990s-beginning of 2000s, plant biomass was identified as another resource for the green synthesis of metal NPs. Several studies reported the formation of Au NPs (Armendariz *et al.*, 2004; Das & Pramanik, 2004). Other studies demonstrated that living plants can also form NPs. Gardea-Torresdey *et al.*, 2002, 2003) reported for the first time that Au and Ag NPs can be formed by living alfalfa (*Medicago sativa*) plants. They found that alfalfa plants grown in Au(III)-enriched medium can form face-centered-cubic (FCC) configuration Au NPs varying from 2 nm to 20 nm in diameter. More recently, plant extracts have been used as reducing agents for the production of NPs. Plant extracts have some advantages over the biomass because the metal reduction process require less time and energy, and there is no presence of strange materials (Shankar *et al.*, 2003, 2004). One of the problems faced by the green synthesis of NPs is the lack of uniformity. To overcome this problem, attempts have been made to find methods for size controlled NPs production. One of the first attempts to control the size of green synthesized NPs was reported in 2004 by the Gardea-Torresdey research group (Armendariz *et al.*, 2004). Recently, Montes *et al.* (2011) reported that alfalfa biomass extracts prepared in water produced mainly Au triangles varying from 500 nm to 4 μm, while isopropanol extract produced two types of Au NPs: 30 nm decahedra and 15 nm icosahedra. Nanotechnology is under continuous development and there are more and more researches about synthesizing NPs with controlled shape and size. This chapter will focus on the green synthesis of NPs using plant extracts and plant derived materials.

18.2 FORMATION OF METALLIC NPs BY PLANT EXTRACTS

18.2.1 Formation of Ag NPs by leaf extract

Silver NPs have special prosperities such as catalysis, magnetic and optical polarizability, electrical conductivity and antimicrobial activity (Vaseeharan *et al.*, 2010). These NPs can be synthesized by chemical reduction of Ag ions in aqueous solution, thermal decomposition in organic solvents, photoreduction in reverse micelles and radiation chemical reduction (Forough *et al.*, 2010). However, most of these techniques are expensive, based on toxic or hazardous chemicals, and may cause environmental contamination. Recently, the green synthesis of Ag NPs has been reported in numerous researches. Table 18.1 shows the most recent publications on synthesis of Ag NPs by plant leaf extracts.

Table 18.1 Studies about the synthesis of Ag NPs by plant leaf extract.

Plant	Nanoparticle size (nm)	Reaction time	Reaction temperature (°C)	$AgNO_3$ concentration (mM)	Reference
Tinospora cordifolia	55–80	1 h	25	1	Jayaseedan *et al.* (2011)
Bryophyllum sp., *Cyperus* sp., *Hydrila* sp.	2–5	4 h	90	25	Jha *et al.* (2009)
Cacumen platycladi	13.8–23		30, 60, 90	1, 3, 5	Huang *et al.* (2011)
Acanthe phylum bracteatum	29–68	13 min	86	3	Foroug and Farhadi (2010)
Lippia citriodora	15–30	15 min–3 h	25–90	1	Cruz *et al.* (2010)
Pelargonium peltatum		over night	25	1	Bunghez *et al.* (2011)
Rhizophora apiculata	19–42		30–90	1	Antony *et al.* (2011)
Camellia sinensis		30 min	40	10	Vaseeharan *et al.* (2010)
Azadirachta indica	20	4 h		10	Tripathy *et al.* (2010)
Artocarpus heterophyllus		1–2 h	25	1	Thirumurugan *et al.* (2010)
Pinus desiflora, Diospyros kaki, Ginkgo biloba, Magnolia kobus, Platanus orientalis	15–500	11 min	95	1	Song *et al.* (2009a)
Citrullus colocynthis	31	24 h	35	1	Satyavani *et al.* (2011)

(*Continued*)

Table 18.1 Studies about the synthesis of Ag NPs by plant leaf extract (*Continued*).

Plant	Nanoparticle size (nm)	Reaction time	Reaction temperature (°C)	$AgNO_3$ concentration (mM)	Reference
Pungamia pinnata, Hemidesmus indicus, Syzygium cumini	24–55	14 min–12 h	25–95	1	Panda *et al.* (2011)
Magifera indica	20	30 min	25	0.5	Philip (2011)
Ocimum sanctum	3–20			1	Mallikarjuna *et al.* (2011)
Acalypha indica	20–30	30 min	37	1	Krishnaraj *et al.* (2010)
Polyalthia longifolia	15, 20, 35, 50		25, 60	0.1, 1	Kaviya *et al.* (2011)
Nelumbo nucifera		10 min	25	1	Santhoshkumar *et al.* (2011)
Catharanthus roseus	45–70	24 h, 15 min	27 , 80	1	Kannan *et al.* (2011)
Vitex negundo L.	10–30	48 h	25	100	Zargar *et al.* (2011)
Phyllanthus amarus	30		25	1	Kasthuri *et al.* (2009)
Murraya koenigii	10	10 min	27	1	Philip *et al.* (2011)
Memecylon edule	50–90	3 h		1	Elavazhangan and Arunachalam (2011)

Although plant materials used in the synthesis of NPs are different, the processes are rather similar. The leaf extracts for most of these studies are prepared by distilled water with or without boiling, followed by centrifugation and filtration. Then, the leaf extract is mixed with an $AgNO_3$ solution and monitored by UV/vis spectra. At the end of the reaction, the colour of the mixture changes from light green to dark brown or black. The key mechanism is:

$$Ag(I) \xrightarrow{\text{Leaf Extract}} Ag(0)$$

The plant extract can be used as the reducing agent in the synthesis of Ag NPs, because the extract contains a lot of bio-molecules such as reducing sugars, flavonoids, saccharine, and proteins. Jha *et al.* (2009) used the leaf extract of three plants (xerophytes-*Bryophyllum* sp., mesophytes-*Cyperus* sp. and hydrophytes-*Hydrilla* sp.) to synthesize Ag NPs (Figure 18.1). According to the X-ray diffraction profile and TEM micrographs, the Ag NPs were FCC and the unit cell edge was 4.062 Å, 4.065 Å and 4.067 Å with sizes in the range of 2–5 nm (Jha *et al.*, 2009). The mechanisms proposed by the authors are as follows:

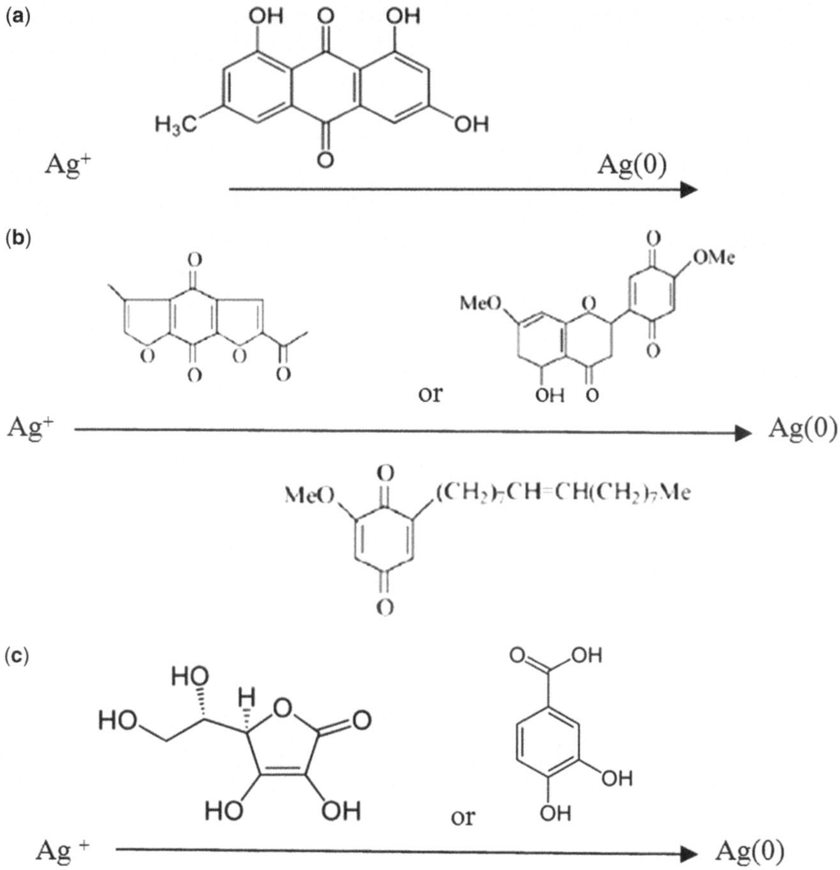

Figure 18.1 Mechanism of biosynthesis of Ag NPs using plant extracts of (a) xerophytes, (b) mesophytes, and (c) hydrophytes (Modified from Jha *et al.*, 2009).

Changes in the reaction time, temperature, and the concentration of leaf extract affect the size and shape of the formed NPs. Cruz *et al.* (2010) used *Lippia citriodora* leaves aqueous extract as the reducing agent to synthesize Ag NPs. They found that the reaction of $AgNO_3$ with the leaf extract occurred rapidly during the first 6 h. Then, the rate was lowered and saturated after 24 h. The TEM images clearly showed that the number and size of Ag NPs increased with the reaction time. The average diameter of the Ag NPs was 15 nm after 3 h and increased to 20 nm after 24 h, which proved that reaction time can affect the size of the NPs (Cruz *et al.*, 2010).

Huang *et al.* (2011) synthesized Ag NPs by using *Cacumen platycladi* leaves extract. In the experiment, the synthesis reaction was performed at 30, 60, and 90°C and 1, 3, and 5 mM $AgNO_3$. At 30°C, the conversions of Ag ions were 80.8%, 27.8%, and 19.1% within 1 h for 1, 3, and 5 mM $AgNO_3$, respectively. The rate increased to 94.1%, 89.1% and 83.6% when the reaction time was prolonged to 24 h. When the temperature increased to 60°C, the conversion rate changed to 83.6%, 52.6%, and 32.6% for 1, 3, and 5 mM $AgNO_3$, respectively, within 10 min, and increased to 98.2%, 91.3% and 82% in 6 h. But for 90°C, the rates were 91.5%, 84.4%, and 46.4% after 10 min and at all silver nitrate concentrations, the rate were 100% after 3 h. The TEM images showed that higher temperatures lead to the production of monodisperse Ag NPs (Huang *et al.*, 2011). Kasthuri *et al.* (2009) reported that the phyllanthin extract from *Phyllanthus amarus* leaves at 2 mM, reacted with the $AgNO_3$ solution at 1 mM produced quasispherical Ag NPs with average size ~30 nm. When the concentration of phyllanthin increased to 4 mM, the formed Ag NPs were ellipsoidal, with low density dispersion, and average size ~60 nm.

18.2.2 Formation of Ag NPs by other plant extracts

The synthesis of Ag NPs has also been achieved using extracts of callus from plant cells obtained by tissue culture, seeds, stems, roots, and fruit peels. These different parts of plants have different biochemical compounds, which determine the rate, size, and shape of the synthesized NPs.

18.2.2.1 Callus extracts

It has been reported that Ag NPs can be synthesized using papaya (*Carica papaya*) callus extract obtained by tissue culture. One month old, greenish callus was used to prepare the extract. During the process, 20 g of callus powder was mixed with 10 mL of distilled water and filtered. Then, the extract was added to 90 mL of $AgNO_3$ (1 mM) solution at room temperature for 24 h. The reaction started in the first hour, and the formation of NPs was confirmed by the brown color in the reaction mixture. The SEM images showed Ag NPs ranging between 60 and 80 nm. The difference in size was explained by the capping of the NPs by papaya proteins (Mude *et al.*, 2009).

Another study showed that extracts from callus and leaf of the saltmarsh plant *Sesuvium portulacastrum* formed different Ag NPs. To prepare both extracts, 20 g of a fresh material were boiled for 5 min and centrifuged at 3000 rpm. Subsequently, 5 mL of extracts were mixed, separately, with 45 mL of a 10^{-3} M $AgNO_3$ solution. The Ag NPs formed by callus and leaf extracts were mostly spherical varying from 5 to 20 nm in diameter. The TEM analysis showed that the NPs produced by callus extract were more distinct and scattered in distribution than those formed by the leaf extract (Nabikhan *et al.*, 2010).

18.2.2.2 Seed extracts

The synthesis of metallic NPs by plant seed extracts was reported in the early 20th century (Miller & Hibbard, 1926), but the NPs were not morphologically characterized due to the scarcity of appropriate analytical techniques at that time. Plant seeds can release a variety of metabolites during the germination process, such as carbohydrates, vitamins, amino acids and other organic compounds. Some of these can be used as reducing and stabilizing agents during the synthesis of NPs. Extracts of seeds from *Medicago sativa* were used to synthesize Ag NPs (Lukman *et al.*, 2011). The reaction occurred instantly during the first minutes when the extract was mixed with the $AgNO_3$ solution. About 90% of the Ag^+ ions were reduced in 50 min. Spherical NPs (5–51 nm) were formed when the Ag concentration was 0.01 M whereas flower-like particle clusters (104 nm) were formed when the Ag concentration increased to 0.1 M. Pre-dilution of the seed extract generated single-crystalline Ag nanoplates, hexagonal particles and nanotriangles with edge lengths of 86–100 nm. At pH 11, monodisperse Ag NPs with an average size of 12 nm were formed.

Kumar *et al.* (2010) synthesized Ag NPs by leaf and seed extract of *Syzygium cumini*. The extracts were mixed with 1 mM $AgNO_3$ solution for 24 h at room temperature. According to TEM and AFM results, the leaf extract formed Ag NPs with a size around 30 nm, and the seed extract produced 92 nm Ag NPs. The possible reason of this difference may be that the seed extract contains more polyphenols (40 mg g^{-1}) than the leaf extract (23 mg g^{-1}). Such compounds have been reported to play an important role in the synthesis of NPs in *Aloe vera* plant extract (Chandran *et al.*, 2006). This is evidence that the extracts of different parts of a plant can form NPs of different size.

18.2.2.3 Fruit extracts

Prathna *et al.* (2011) used lemon juice as the bio-reducing agent to synthesize Ag NPs. These authors mixed $AgNO_3$ solutions at 10^{-2} M, 10^{-3} M and 10^{-4} M with lemon juice at the mixing ratios of 1:4, 2:3, 3:2, 1:1 and 4:1 for 4 h at 30°C in a shaker at 120 rpm. The TEM images showed spherical and spheroidal Ag NPs less than 50 nm in diameter after 4 h of the reaction time. They demonstrated that citric acid was the major reducing and stabilizing agent for the nanosynthesis process (Prathna *et al.*, 2011). The suggested synthesis mechanism was:

$$4Ag^+ + C_6H_5O_7 + X + 2H_2O \longrightarrow 4Ag^0 + C_6H_5O_7H_3 + H^+ + Y + O_2$$

where X stands for the unknown bio-organic substance in lemon juice, other than citric acid, participating in the reduction of the Ag^+ ions, whereas Y stands for the oxidized products of X.

The extract of orange peel has also been used to synthesize Ag NPs (Konwarh *et al.*, 2011). Orange peel contains abundant pectin, flavonoids, ascorbic acid, sugars, carotenoids and a myriad of flavones, which may be effective reducing agents. In this experiment, the orange peel extract was mixed with 1 M $AgNO_3$ solution under basic condition adjusted by aqueous 2 M KOH. The size of the synthesized Ag NPs ranged from 3 to 12 nm. The author proposed that the ascorbic acid was the reducing agent as shown in the following reaction:

$$2Ag^+ + C_6H_8O_6 \longleftrightarrow 2Ag^0 + C_6H_6O_6 + 2H^+$$

Ahamed *et al.* (2011) used garlic clove extract as the reducing and stabilizing agent for the synthesis of Ag NPs. During a typical synthesis, 5 mL of garlic extract was added to 50 mL aqueous solution of $AgNO_3$ (1 mM) and the reaction was kept for 30 min. Then, a golden yellow color suspension was formed, indicating Ag reduction. XRD data showed crystalline in nature and spherical shaped Ag NPs with an average size of 12 nm. Garlic contains antioxidants such as vitamin C, vitamin B6, beta-carotene, thiamine, riboflavin, and niacin, among others that can be effective reducing agents for the Ag ions (Ahamed *et al.*, 2011).

18.2.2.4 Rhizome extracts

Nagajyothi and Lee (2011) reported the synthesis of Ag NPs by *Dioscorea batatas* rhizome extract at 25°C and 80°C. The color of the rhizome extract $AgNO_3$ (1 mM) mixture changed from colorless to brownish in 2 h at 25°C and SEM images showed circular and flower shaped Ag NPs. However, when the temperature was increased to 80°C, the color changed and monodispersed Ag NPs were formed in 50 min. They also studied the antimicrobial activity of these synthesized Ag NPs and found that Ag NPs produced at 25°C had higher inhibitory activity on microbial community compared to Ag NPs produced at 80°C.

18.2.3 Formation of Au NPs by plant leaf extracts

Au NPs are of paramount importance due to their application in bio-labeling, biosensors, drug delivery, tissue imaging, photothermal therapy, and immune-chromatographic identification of pathogens in clinical specimens (Huang, 2006). Extracts from *Murraya koenigii* (Philip *et al.*, 2011), *Memcylon edule* (Elavazhagan & Arunachalam, 2011), *Mentha piperita* (Mubarakali *et al.*, 2011), *Cinnamomum camphora* (Huang *et al.*, 2007), *Swietenia mahogany* JACQ (Mondal *et al.*, 2011), *Azadirachta indica* (Ramezani *et al.*, 2008), *Tridax procumbens* L., *Jatropa curcas* L., *Calotropis gigantean* L., *Solanum melongena* L., *Datura metel* L., *Carica papaya* L. and *Citrus aurantium* L. (Rajasekharreddy *et al.*, 2010) have shown different degrees of efficiency in the formation of Au NPs. The size and shape of the Au NPs formed depend on factors such as pH, concentration, temperature, boiling and drying of leaves.

The major mechanism of these reactions is as follows:

$$HAuCl_4 \xrightarrow{\text{Leaf Extract}} Au(0)$$

The size and shape of Au NPs can be affected by many factors during the synthesis process including pH, concentration, temperature, boiling, and drying of leaves.

18.2.3.1 Effect of pH

Khalil *et al.* (2010) reported the synthesis of Au NPs using hot water olive leaf extracts at pH values of 3.3 and 9.6. Various shapes of Au NPs (triangle, hexagonal, and spherical) were obtained after 20 min of the reaction time at room temperature. Surface plasmon resonance spectra results showed that the absorbance increased with increasing pH from 2.3 to 9.6 with a blue shift in the spectra. The size of Au particles (60 nm) at pH 3.3 was larger while forming hexagonal structures, and these particles were unstable and precipitated within 12 h. At pH 9.6, the size of synthesized Au NPs was about 20 nm. Figure 18.2 shows the TEM images of formed Au NPs at pH 3.3 and 9.6.

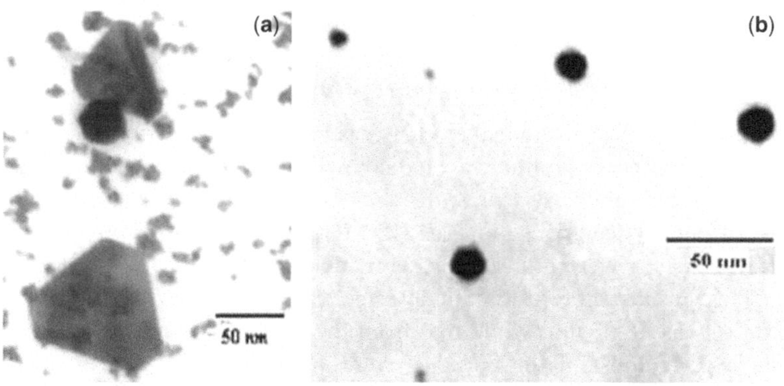

Figure 18.2 TEM images of Au synthesized nanoparticles at pH 3.3 (a) and pH 9.6 (b) (Reprinted from Khalil *et al.*, 2010 with permission from the Arabian Journal of Chemistry).

18.2.3.2 Effect of the reaction temperature

Au NPs have been synthesized using *Magnolia kobus* and *Diopyros kaki* leaf extracts under different temperatures and concentrations of the leaf extracts (Song *et al.*, 2009b, 2010). Song and co-workers (2009b) found that the reaction temperature significantly affected the reaction rate and the size of Au NPs. At 25°C, the conversion rate was only 60% after 6 h; while at 60°C, it needed 1.5 h to reach 100%. However, it required only 5 min for 100% conversion to Au NPs at 95°C. It is clear that at higher temperature, higher conversion occur. SEM and TEM images also showed that the particles size decreased from 110 nm at 25°C to 40 nm at 95°C, and a mixture of plates (triangles, pentagons, and hexagons) and spheres were formed at 25°C and 60°C, but only spherical shapes were obtained at 95°C.

18.2.3.3 Effect of leaf extracts concentration

Smitha and co-workers (2009) reported that the particle size was observed to decrease with an increase in the leaf extract concentration. They put 24 g of thoroughly washed and finely cut *Cinnamomum zeylanicum* leaves into 300 mL of deionized water and boiled the mixture to make the leaf broth. Subsequently, different volumes (4, 6, 8, 10, 12, 15, 17 and 20 mL) of the extract were mixed with 0.2 mM of aqueous $HAuCl_4$ solution at room temperature. The results showed that the size and form of the NPs changed as the concentration of leaf broth increased. At lower quantity (4 mL) of the leaf extract, most of the Au NPs were large spheres and prisms, while at higher leaf broth quantity (17 mL) almost only small sized spherical NPs were observed. Thus, results confirmed that the shape and size of synthesized Au NPs could be controlled by the concentration of the leaf extract (Song *et al.*, 2009b).

18.2.3.4 Effect of boiling and drying of the extracts

The synthesis of Au NPs using leaf extracts has been reported with and without the boiling step. The effect of boiling was studied by Raju *et al.* (2011). During a typical synthesis, 30 g of *Semecarpus anacardium* L. leaves were extracted in 100 mL deionized water by centrifugation at 10,000 rpm for 15 min "green extract." Another 30 g leaves were boiled in 100 mL deionized water for 30 min "boiled extract." TEM results showed that the average size of Au NPs synthesized by the green extract was 18.63 nm, while that from the boiled extract was 19.22 nm. In the green extract, most of the particles were 15 and 16 nm, and in the boiled extract these were in the range of 16 and 18 nm. According to ICP-AES results, the amount of Au NPs in green extract (20.82 ppm) was higher than in the boiled extract (17.99 ppm). The reason may be that the synthesis occurs by active proteins and microorganisms in the leaf extract what are destroyed by the boiling process (Raju *et al.*, 2011).

Philip (2010) synthesized Au NPs using fresh and dry leaf extract of *Mangifera indica*. He found that the Au NPs could be formed in 2 min when the fresh or dried leaf extract were mixed with $HAuCl_4 \cdot 3H_2O$. These colloids were stable for more than 5 months. The author also reported that the drying process did affect the synthesis process, because Au NPs synthesized by dry leaf extract were smaller (17 nm) and more uniform in size than those formed with fresh leaf extract (20 nm).

18.2.4 Formation of Au NPs by seeds, fruits, flowers, and rhizomes

Au NPs have also been synthesized using extracts from other plant parts. Recently, Montes *et al.* (2011) synthesized Au NPs by alfalfa shoots extract prepared in water and isopropanol separately. They found that the size of Au NPs synthesized by isopropanol extract ranged from 30–60 nm lying onto the (111) lattice plane and the morphologies included 30 nm decahedra and 15 nm icosahedra. For water alfalfa shoot extracts, the Au NPs were mainly triangular with a thickness of 15–30 nm, ranging from 500 nm to 4 μm in size.

Mucuna pruriens seeds extracts were used to biosynthesize Au NPs (Arulkumar & Sabesan, 2010). The dried seeds were ground to powder and extracted with methanol by a Soxhlet apparatus. One hundred mL of *M. pruriens* seeds extract were mixed with 100 mL of 1 mM $HAuCl_4$ solution at room temperature. After 10 min, 95% of the Au(III) ions were reduced to Au(0) and Au NPs formed ranging from 6 to 17.7 nm; most of them around 12.5 nm in diameter. These Au NPs have been proved to be stable in solution over one month at room temperature (Arulkumar & Sabesan, 2010).

The extract of the ginger (*Zingiber officinale*) rhizome has also been used to synthesize Au NPs. One mL of ginger rhizome extract was added to 25 mL 1 mM $HAuCl_4$ solution and boiled for 20 min. The TEM analysis indicated NPs varying from 5 nm to 15 nm, with a large fraction at around 4 nm. The stability of the synthesized Au NPs was tested at different temperature for one month at pH 7.4. The results revealed that these NPs were very stable even at 50°C (Kumar *et al.*, 2011).

Extracts of flowers of *Nyctanthes arbortristis* have also been used for the synthesis of Au NPs (Das *et al.*, 2011). Shade dried flowers were grinded and extracted by ethanol at room temperature for 72 h. Then, the extract was added to 1 mM $HAuCl_4$ solution, and stirred at 200 rpm at 25°C and 80°C for 30 min. The formed Au NPs were triangular, pentagonal, rod shaped and spherical and had an average diameter of 19.8 ± 5.0 nm and were stable for 6 months. The authors explained that the variation in shape was due to the susceptibility of the nucleation process of metallic NPs to the reaction temperature.

Fruit extracts have also been used for the synthesis of Au NPs. Extracts of *Tanacetum vulgare* (Dubey *et al.*, 2010), fruit banana peel (Bankar *et al.*, 2010), orange, papaya, peach, and lemon (Tai *et al.*, 2011) have been effectively used to synthesize Au NPs. Dubey *et al.* (2010) used fresh fruit of *T. vulgare* to synthesize Au NPs. 50 mL aqueous solution (1 mM) of auric acid were reduced using 1.8 mL of fruit extract at room temperature for 10 min. This resulted in the formation of Au NPs with an average size of 11 nm. In this experiment, factors such as quantity of fruit extract, Au concentration, reaction temperature, and time were studied. They observed a decrease in particles size with an increase in fruit extract concentration (Dubey *et al.*, 2010). However, an increase in particle size was observed with an increase in Au ion concentration. The higher temperature produced a smaller particle size, which has been also reported by other researches (Song *et al.*, 2009b). The extract of fruit banana peel also showed ability to reduce Au ions to form Au NPs (Bankar *et al.*, 2010). According to SEM and dynamic light scattering (DLS) results, the average size of synthesized Au NPs was around 300 nm. The XRD analysis demonstrated that the synthesized Au NPs were FCC, based on the peaks (111), (200), (220) and (311). The authors proved that these Au NPs exhibited antifungal and antibacterial activity. In another study, extracts of orange fruits (*Citrus sinensis*), papaya (*Carica papaya*) peach (*Prunus persica*), and lemon (*Cirtus limon*) without peel were successfully used to synthesize Au NPs of various shapes and sizes (Tai *et al.*, 2011). The peel of the fruits was removed and the flesh was extracted by dionised water. Then, each extract was reacted with 1 mM $HAuCl_4$ solution for 24 h. Results showed that the orange extract formed spherical Au NPs of 10–30 nm. The extract of the papaya fruit formed aggregates of 150 ± 50 nm. Peach and lemon extract produced multi-shaped NPs, such as triangular, hexagonal, decahedra, pentagon, full-grown triangle and diamond shapes. Noteworthy to mention, lemon extract produced more shapes in comparison to the peach extract. The biocompatibility of these Au NPs was studied by haemolysis tests and the results showed non-toxic effects in blood cells (Tai *et al.*, 2011).

18.2.5 Formation of other NPs by plant extracts

18.2.5.1 Formation of Cu NPs

The stem latex of *Euphorbia nivulia*, a medicinal plant, was successfully used to synthesize Cu and Ag NPs (Valodkar *et al.*, 2011). In this experiment, 10 mL of the diluted latex was mixed with 0.6 mL of 0.01 M $AgNO_3$ solution and heated in a domestic microwave oven for 30 s at about 95°C. Results indicated that the latex alone did not reduce the Cu ions but it reduced the Ag ions. Thus, the authors added ascorbic acid as the reducing agent to copper nitrate ($Cu(NO_3)_2 \cdot 3H_2O$) solution together with stem latex. According to TEM analysis, spherical Cu NPs of about 5–10 nm size were clearly observed and from the XRD, the peaks observed at 43.25°, 50.59° and 74.29° represented (111), (200) and (220) planes of FCC crystal structures of bulk copper. Although the stem latex of *E. nivulia* was not a powerful reducing agent for Cu, it was an effective stabilizing agent. Without the latex, the synthesized Cu NPs aggregated into a black precipitates immediately after the formation due to the tendency of copper to get oxidized in aqueous medium.

18.2.5.2 Formation of Pd NPs

Grape pomace extract has been used by Baruwati and Varma (2009) to synthesize Pd NPs. They mixed 5 mL of the wine pomace with 2 mL Na_2PdCl_4 (10 mmol/l) and put the mixture under 50 W microwave power for 60 s. The size of the synthesized NPs was around 5–10 nm. In this study, the grape pomace extract was used as the solvent, reducing and capping agent.

18.2.5.3 Formation of Pt NPs

Leaf extract of *Diospyros kaki* was reacted with an aqueous $H_2PtCl_6 \cdot 6H_2O$ solution at various temperatures between 25°C and 95°C to form Pt NPs with sizes ranging from 2 to 12 nm (Song *et al.*, 2010). Authors proved that the size of Pt NPs could be controlled by the temperature or the composition of the reaction mixture. The average NPs size decreased from 12 nm at 25°C to 5 nm at 95°C, and decreased from 11 nm at 5% to 3 nm at 50% the leaf broth concentration. Also, the size of Pt NPs increased from 3 nm to 9 nm as $PtCl_6$ ion concentration increased from 0.1 mM to 2 mM. Pt NPs were also synthesized by mixing wine pomace extract with a $HPtCl_4$ solution (Baruwati & Varma, 2009).

A natural hydrocolloid kondagogu gum (*Cochlospermum gossypium*) was used to fabricate Au, Ag, and Pt NPs (Vinod *et al.*, 2011). The reduction and formation of Pt NPs from H_2PtCl_6 was achieved by autoclaving the Pt-gum solution (pH 8.0) for 15 min at 120°C. After the reaction, the mixture changed from colorless to black indicating the formation of NPs. The average size of Pt NPs was 2.4 ± 0.7 nm, which was smaller compared to Ag and Au NPs synthesized by the same kondagogu gum (Vinod *et al.*, 2011). The solution was observed over six months with no sign of aggregation or precipitation of the Pt NPs. This proved that kondagogu gum acts as both, the reducing and stabilizing agent for the target NPs.

18.2.5.4 Formation of Fe NPs

Fe NPs have been successfully synthesized by aqueous sorghum bran extract (Njagi *et al.*, 2010). In this study, 0.1 M $FeCl_3$ solution was added to the sorghum bran extract in a 2:1 volume ratio. The mixture was hand shaken for 1 min and kept at room temperature for 1 h. According to the Field Emission Transmission Electron Microscopy (FE-TEM) images, the size of the thus synthesized Fe NPs was about 50 nm and these particles agglomerated to form irregular clusters. The XRD patterns of Fe NPs lacked distinct diffraction peaks, indicating they were amorphous. Selected area electron diffraction (SAED) micrographs also confirmed the amorphous nature of the bioformed Fe NPs, because there were no diffraction rings or spot patterns but a diffuse ring in the micrographs. Generally, Fe NPs application in environmental remediation and hazardous waste treatment is limited by their stability, because NPs aggregate easily in an aquatic environment. However, in this experiment, no noticeable flocculation or sedimentation was observed after one week, suggesting that the sorghum extract was a good stabilizing agent for Fe NPs.

18.2.5.5 Formation of In_2O_3 and TiO_2 NPs

Aloe vera plant extract has been used to synthesize In_2O_3 NPs with sizes of 5–50 nm using indium acetylacetonate (Maensirei *et al.*, 2008). This study gave a potential avenue for the production of oxide NPs. The NPs were formed with/without calcination. During the synthesis process, 3 g indium(III) acetyulacetonate was added to the leaf extract and the mixture was stirred at 60°C until dry. Then, the dried precursor was grounded and calcined in a box-furnace at 400, 500 and 600°C for 2 h. The TEM analysis showed that the calcination temperature determined the morphology and size of the In_2O_3 NPs. When calcined at 400°C, In_2O_3 NPs were of 5–10 nm and increased to 10–25 nm when calcined at 500°C. At 600°C, the NPs had a size of 30–50 nm.

TiO_2 NPs have also been synthesized using plant extracts with a subsequent calcination step (Sundrarajan & Gowri, 2011). For instance, 0.4 M of titanium tetraisoproxide were mixed with *Ethanolic nyctanthes* leaf extract and stirred at 50°C for 4 h. After centrifugation, these particles were washed with ethanol and dried. Then, the dried and grinded particles were calcinated at 500°C in a muffle furnace for 3 h. The XRD results showed six peaks at 23.5°, 37.8°, 47.9°, 54.5°, 62.8° and 75.1°, which represented the planes (101), (004), (200), (105), (204), (116), and (215) of TiO_2 NPs, respectively. The SEM analysis demonstrated that the average size of these NPs ranged from 100–150 nm with uniformed spherical shapes. These two studies proved that plant extracts can be used for the synthesis of metal and metal oxide NPs.

18.2.5.6 Formation of bimetallic NPs

Bimetallic Au-Pd NPs have also been produced by using plant extracts (Zhan *et al.*, 2011). Powder of dry leaves of *Cacumen platycladi* was dispersed in deionized water by shaking for four h followed by filtration. The extract was then mixed with $HAuCl_4$ (40 mL, 0.25 mM) and $PdCl_2$ (40 mL, 0.25 mM). After 2 h of the reaction, about 7.5 nm Au-Pd bimetallic NPs were observed. These NPs were dark brown in color, while pure Au NPs were ruby red, the pure Pd NPs were light brown, and the physical mixture of the two NPs were ruby red. The XRD results of the bimetallic Au-Pd NPs showed a broad band at 39°, which was different from the pure Au (38.3°) and pure Pd (40.0°). All of these results indicated that the newly formed bimetallic NPs were in fact metal alloys.

18.3 FORMATION OF METAL NPs BY INTACT PLANTS

18.3.1 Formation of metal NPs by living plants

The first report on the synthesis of metal NPs by living plants appeared in 2002 (Gardea-Torresdey *et al.*, 2002). In this study, alfalfa plants (*Medicago sativa*) were grown for two weeks in an Au(III)-enriched agar system and the roots as well as shoots were analyzed by TEM and XAS (X-ray absorption spectroscopy).

Figure 18.3 shows the EXAFS of the Au and the tetrachloroaurate samples. The nearest neighbor in the Au(0) foil was observed at the same distance as the nearest neighbor in the Au plant and agar sample, and this confirmed that shoot, roots, and agar samples contained Au(0). This proved that alfalfa can actively uptake Au(0) from a solid media. The TEM images showed that the alfalfa synthesized Au NPs between 2 and 20 nm, with FCC configuration. Also, the NPs were multitwinned and the shortest lattice parameter calculated was 0.23 nm, which was the interplanar spacing between the (111) plane.

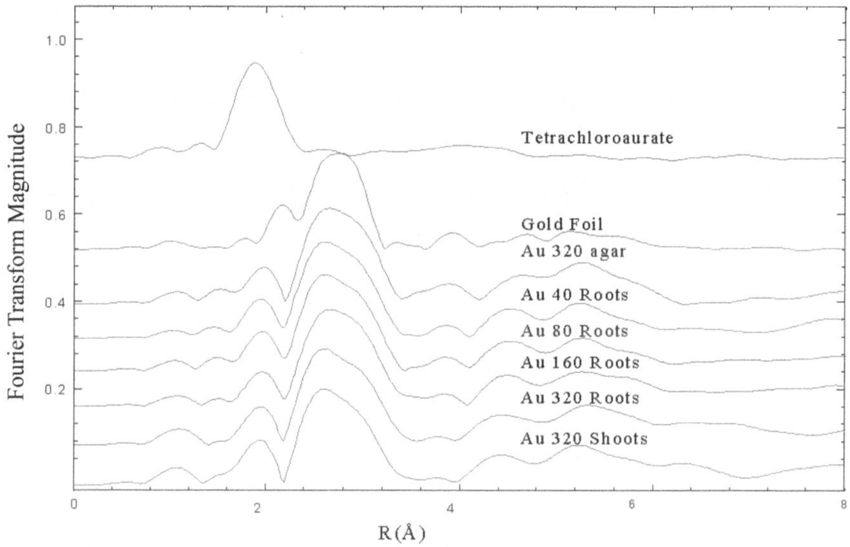

Figure 18.3 EXAFS of the gold in alfalfa roots and shoots, gold (0) foil, and gold-enriched agar samples (Reprinted from Gardea-Torresdey *et al.*, 2002 with permission from American Chemical Society).

The same research group also used living alfalfa to produced Ag NPs (Gardea-Torresdey *et al.*, 2003). Alfalfa seedlings were germinated and grown in agar-based medium containing 0, 40, 80, 160, 320 mg/L $AgNO_3$. After 15 days of treatment, roots and stems were examined with TEM. The TEM micrographs exhibited Ag NPs varying in diameter from 2 to 20 nm. The NPs were dispersed throughout the plant in the small assemblies like nanowires. In addition, it has been reported that Desert willow (*Chilopsis linearis*) and Indian mustard (*Brassica juncea*) can be used to form Au NPs in their living tissues (Rodriguez *et al.*, 2007; Marshall *et al.*, 2007; Sharma *et al.*, 2007). The average size of the NPs can be affected by the Au concentration in the tissues and the location in the plant. For example, larger NPs were found in the stems and smaller in the roots of *B. juncea* (Marshall *et al.*, 2007). Morning glory (*Ipomoea lacunosa*) and alfalfa (*Medicago sativa*) produced titanium NPs in their tissue through bio-precipitation and/or bio-mineralization when they were exposed to a titanium-enriched aqueous solution (Yates *et al.*, 2005).

More recently, the formation of Au and Ag NPs by living alfalfa (*M. sativa*) (Bali & Harris, 2010) and Indian mustard (*B. juncea*) (Shekhawat & Arya, 2009) was reported. Both *B. juncea* and *M. sativa* plants were exposed to 5, 40, 1000 ppm of $KAuCl_4$ aqueous solution for 24, 48 and 72 h. At all concentrations and exposure times, the color of roots changed from white to purple, which indicated the reduction of Au(III) to Au(0). The Au(0) particles were present in the epidermal layer, cortex, and vascular cylinder (Figure 18.4). The TEM images showed that Au NPs were distributed extracellularly and intracellularly; in particular within the epidermal cell walls, xylem cell walls, cell membranes, and vacuoles. Moreover, the particle size ranged from 2 nm to 2 μm in *B. juncea* and from 2 nm to 1 μm in *M. sativa*. The TEM results showed that the small NPs formed larger nanoplates with spherical, hexagonal, triangular and irregularly shapes.

The production of NPs by living plants is more difficult and has less control on size and shapes compared to the synthesis using plant extracts. Another crucial factor is the time. Plant extracts must be used immediately after the extraction, while living plants need time to growth. So far, although the mechanisms of NPs formation by living plants is not completely clear, it is hypothesized that many plant metabolites, such as amino acid, histidines and petides, phytochelatins and glutathione can effectively act as the reducing agents (Sharma & Dietz, 2006).

18.3.2 Formation of metal NPs by plant biomass

Plant biomass is a renewable resource obtained from living or dead plants. A variety of biomasses have been used to synthesize Au and Ag NPs (Parsons *et al.*, 2009). Biomass of alfalfa (Parsons *et al.*, 2007), hops (Lopez *et al.*, 2005), wheat (Armendariz *et al.*, 2004), *Cinnamomum camphora* (Huang *et al.*, 2007), and algal biomass (Cai *et al.* 2011) has been reported as effective precursors for the synthesis of various NPs. Figure 18.5 shows the different shapes of Au NPs synthesized by wheat and oat biomass (Armendariz *et al.*, 2004). Sathishkumar *et al.* (2009) compared the difference between Ag NPs synthesized by *Cinnamon zeylanicum* bark extract and *C. zeylanicum* biomass. They found that the bark extract produced more NPs than the bark biomass and the reaction rate for the bark extract was much more rapid. The difference was attributed to a higher concentration of reducing agents found in the extract. The average size of the synthesized Ag NPs ranged from 31 to 40 nm for both, bark extract and the biomass, and these NPs remained stable in the solution for three months. XRD pattern results indicated the cubic and hexagonal phase of the Ag nanocrystals. Also, it confirmed that the pH is able to control the size and shape of the biosynthesized NPs. Indeed, large NPs were formed at lower pH and smaller ones were produced at high pH.

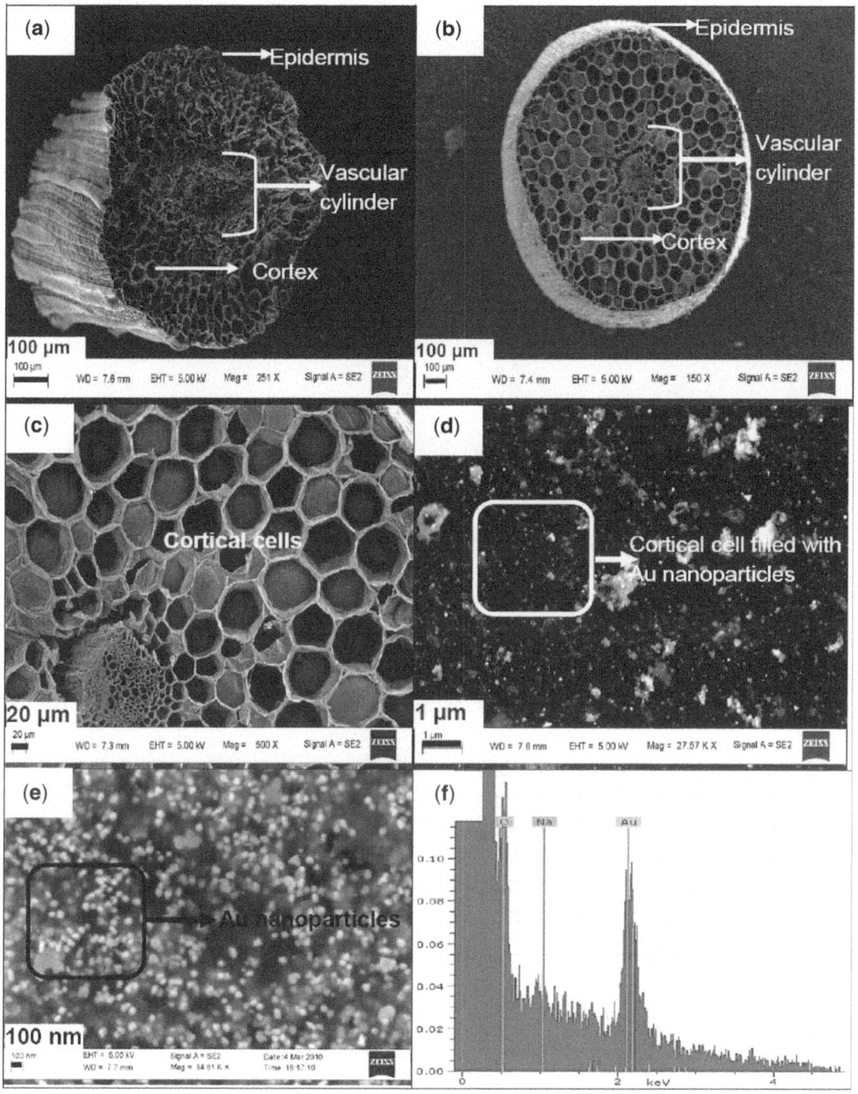

Figure 18.4 (a) SEM image of *B. juncea* root exposed to 1000 ppm of Au solution for 24 h, (b) SEM image of root cross section of *M. sativa* exposed to 40 ppm of Au solution, (c) higher magnification of parts of Figure b, (d) single cortical cell filled with Au particles, (e) Au nanoparticles of varying morphologies, and (f) EDS image confirming that the NPs contained Au. (Reprinted from Bali and Harris 2010 with permission from American Chemical Society).

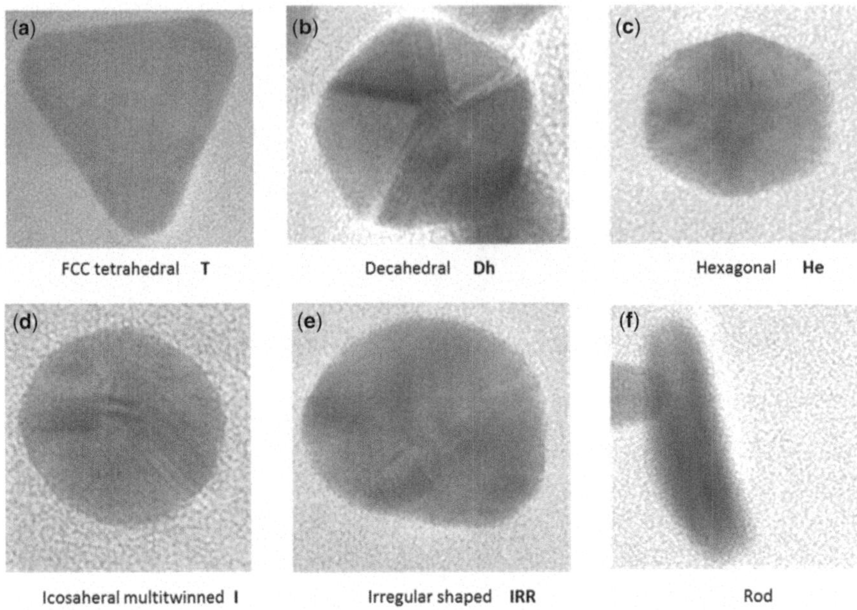

Figure 18.5 Classification of gold nanoparticles obtained with biomass (Reprinted from Armendariz *et al.*, 2004 with authorization from Revista Mexicana de Fisica).

Plant biomass can also be used as a valuable resource for metal ions. Qu *et al.* (2011a, 2011b) successfully synthesized ZnO NPs by the Zn-hyperaccumulator plants *Physalis alkekengi* L. and *Sedum alfredii* Hance. These plants were grown in soil with high Zn level (1080–2030 mg kg^{-1}) and dried after they were harvested. Then, the biomass was extracted by ethanol and water, and mixed with chemicals. The mechanisms are shown in the following equations:

(1) $C_{32}H_{30}ON_4Mg(CO_2CH)(CO_2C_{20}H_{39}) + 2NaOH \longrightarrow C_{32}H_{30}ON_4Mg\ (CO_2Na)_2 + CH_3OH + C_{20}H_{39}OH$

(2) $C_{32}H_{30}ON_4Mg(CO_2Na)(CO_2Na) + 4HCl \longrightarrow C_{32}H_{30}ON_4H_2(CO_2H)_2 + 2NaCl + MgCl_2$

(3) $C_{32}H_{30}ON_4H_2\ (CO_2H)(CO_2H) + ZnSO_4 \longrightarrow C_{32}H_{30}ON_4Zn(CO_2H)_2 + H_2SO_4$

(4) $NH_3 \cdot H_2O \longrightarrow NH_4^+ + OH^+$

(5) $Zn^{2+} + 2OH \longrightarrow Zn(OH)_2$

(6) $Zn(OH)_2 \longrightarrow ZnO + H_2O$

The results showed that *ca.* 211 mg of ZnO NPs were synthesized from 30 g shoot of *P. alkekengi*. TEM images showed polydispersed NPs with sizes ranging from 50 to 200 nm. This study opens a potential avenue to recycle metals derived from phytoremediation projects.

Agricultural wastes have also attracted attention as a source for nanomaterial production. Espindola-Gonzalez *et al.* (2010) synthesized silica oxide NPs from waste rice husk, coffee husk and sugarcane bagasse by vermicomposting the wastes with *Eisenia foetida*. The process includes stabilization, elimination of organic matter, acid treatment and neutralization (Espindola-Gonzalez *et al.*, 2010). In another study, rice husk ash was used to synthesize nanozeolite sodalite by a hydrothermal method without organic additives (Ghasemi *et al.*, 2011). The wheat straw has also been used to produce nano-silica materials through combustion and acid leaching (Chen *et al.*, 2010). Moreover, wheat straw and other agricultural wastes like kernel shell and coconut have been used to produce carbon nanofibers and biocellulose nanofibers (Chen *et al.*, 2008; Dahman *et al.*, 2010).

18.4 CONCLUSIONS

The review of current literature on the green synthesis of metal NPs has shown that plants and plant derived materials represent a real possibility for the synthesis of monodispersed monosized metallic NPs. The analyzed literature highlighted the advantages of using plant extracts over the synthesis by live plants or plant biomass. The biosynthesis of NPs using plant extracts suggests that natural compounds found in the biomass act as reducing agents and can be used to develop green methods for the production of NPs as an alternative to traditional and more harmful methods (Montes *et al.*, 2011; Mubarakali *et al.*, 2011). In this sense, it is foreseen that agricultural wastes can play an important role in the production of nanomaterials. However, problems related to the amounts of bioproduced NPs still need to be solved. Future research should be aimed to characterize plant extracts and to test pure compounds to achieve more control over the size and shape of the bioproduced NPs, while using a green method for their production. In addition, more studies are needed to determine whether the biosynthesized NPs are industrially used in the same way as the chemically produced NPs.

ACKNOWLEDGEMENT

This material is based upon work supported by the National Science Foundation and the Environmental Protection Agency under Cooperative Agreement Number DBI-0830117. Any opinions, findings, and conclusions or recommendations expressed in this material are those of the author(s) and do not necessarily reflect the views of the National Science Foundation or the Environmental Protection Agency. This work has not been subjected to EPA review and no official endorsement should be inferred. The authors also acknowledge the USDA grant numbers 2011-38422-30835 and the NSF Grant # CHE-0840525. J. L. Gardea-Torresdey acknowledges the Dudley family for the Endowed Research Professorship in Chemistry.

REFERENCES

Ahamed M., Majeed-Khan M. A., Siddiqui M. K. J., Alsalhi M. S. and Alrokayan S. A. (2011). Green synthesis, characterization and evaluation of biocompatibility of silver nanoparticles. *Physica E*, **43**, 1266–1271.

Antony J. J., Sivalingam P., Siva D., Kamalakkannan S., Anbarasu K., Sukirtha R., Krishnan M. and Achiraman S. (2011). Comparative evaluation of antibacterial activity of silver nanoparticles synthesized using *Rhizophora apiculata* and glucose. *Colloids and Surfaces B: Biointerfaces*, **88**, 134–140.

Armendariz V., Jose-Yacaman M., Duarte-Moller A., Peralta-Videa J. R., Troiani H., Herrera I. and Gardea-Torresdey J. L. (2004). HRTEM characterization of gold nanoparticles produced by wheat biomass. *Revista Mexicana de Fisica*, **50**, 7–11.

Arulkumar S. and Sabesan M. (2010). Biosynthesis and characterization of gold nanoparticles using antiparkinsonian drug *Mucuna pruiens* plant extract. *International Journal of Research in Pharmaceutical Science*, **1**(4), 417–420.

Bali R. and Harris A. T. (2010). Biogenic synthesis of Au nanoparticles using vascular plants. *Industrial and Engineering Chemistry Research*, **49**, 12762–12772.

Bankar A., Joshi B., Kumar A. R. and Zinjarde S. (2010). Banana peel extract mediated synthesis of gold nanoparticles. *Colloids and Surfaces B: Biointerfaces*, **80**, 45–50.

Baruwati B. and Varma R. S. (2009). High value products from waste: grape pomace extract-a three-in-one package for the synthesis of metal nanoparticles. *ChemSusChem*, **2**, 1041–1044.

Bhainsa K. C. and D'Souza S. F. (2006). Extracellular biosynthesis of silver nanoparticle using the fungus *Aspergillus fumigates*. *Colloids and Surfaces B: Biointerfaces*, **47**, 160–164.

Bharde A., Wani A., Shouche Y., Pattayil A., Bhagavatula L. and Sastry M. (2005). Bacterial aerobic synthesis of nanocrystalline magnetite. *Journal of the American Chemical Society*, **127**, 9326–9327.

Biswas P. and Wu C. (2005). Nanoparticles and the environment. *Air & Waste Management Association*, **55**, 708–746.

Bunghez I. R., Ghiurea M., Faraon V. and Ion R. M. (2011). Green synthesis of silver nanoparticles obtained from plant extracts and their antimicrobial activities. *Journal of Optoelectronics and Advanced Materials*, **13**(7), 870–873.

Cai F., Li J., Sun J. and Ji Y. (2011). Biosynthesis of gold nanoparticles by biosorption using *Magnetospirillum gryphiswaldense* MSR-1. *Chemistry Engineering Journal*, **175**, 70–75.

Chandran S. P., Chaudhary M., Pasricha R., Ahmad A. and Sastry M. (2006). Synthesis of gold nanoparticles and silver nanoparticles using *Aloe vera* plant extract. *Biotechnology Progress*, **22**, 577–583.

Chen H. D., Weiss J. C. and Shahidi F. (2006). Nanotechnology in nutraceuticals and functional foods. *Food Technology*, **25**(3), 30.

Chen X., Timpe O., Hamid S. B. A., Schlogl R. and Su D. S. (2008). Direct synthesis of carbon nanofibers on modified biomass-derived activated carbon. *Carbon*, **47**, 340–343.

Chen H., Wang F., Zhang C., Shi Y., Jin G. and Yuan S. (2010). Preparation of nano-silica materials: the concept from wheat straw. *Journal of Non-crystalline Solids*, **356**, 2781–2785.

Cruz D., Fale P. L., Mourato A., Vaz P. D., Serralheiro M. L. and Lino A. R. L. (2010). Preparation and physicochemical characterization of Ag nanoparticles biosynthesized by *Lippia citriodora* (Lemon Verbena). *Colloid Surface B: Biointerfaces*, **81**, 67–73.

Dahman Y., Jayasuriya K. E. and Kalis M. (2010). Potential of biocellulose nanofibers production from agricultural renewable resources: preliminary study. *Applied Biochemistry and Biotechnology*, **162**, 1647–1659.

Das R. K., Gogoi N. and Bora U. (2011). Green synthesis of gold nanoparticles using *Nyctanthes arbortristis* flower extract. *Bioprocess and Biosystems Engineering*, **34**, 615–619.

Das R. N. and Pramanik P. (2004). Preparation of nanocrystalline α-Al_2O_3 using plant fiber. *Journal of Nanoscience and Nanotechnology*, **4**, 94–97.

Dubey S. P., Lahtinen M. and Sillanpaa M. (2010). Tansy fruit mediated greener synthesis of silver and gold nanoparticles. *Process Biochemistry*, **45**, 1065–1071.

Elavazhagan T. and Arunachalam K. D. (2011). *Memecylon edule* leaf extract mediated green synthesis of silver and gold nanoparticles. *International Journal of Nanomedicine*, **6**, 1265–1278.

Espindola-Gonzalez A., Martinez-Hernandez A. L., Angeles-Chavez C., Castano V. M. and Elasco-Santos C. (2010). Novel crystalline SiO_2 nanoparticles via annelids bioprocessing of agro-industrial wastes. *Nanoscale Research Letters*, **5**, 1408–1417.

Esumi K., Kameo A., Suzuki A., Torigoe K., Yoshimura T., Koide Y. and Shosenji H. (2001). Preparation of gold nanoparticles using 2-vinylpyridine telomers possessing multi-hydrocarbon chains as stabilizer. *Colloids and Surfaces A*, **176**, 233–237.

Forough M. and Farhadi K. (2010). Biological and green synthesis of silver nanoparticles. *Turkish Journal of Engineering & Environmental Sciences*, **34**, 281–287.

Fukuoka A., Araki H., Kimura J., Sakamoto Y., Higuchi T., Sugimoto N., Inagaki S. and Ichikawa M. (2004). Template synthesis of nanoparticle arrays of gold, platinum and palladium in mesoporous silica films and powders. *Journal of Materials Chemistry*, **14**, 752–756.

Gardea-Torresdey J. L., Parsons J. G., Gomez E., Peralta-Videa J., Troiani H., Santiago P. and Jose-Yacaman M. (2002). Formation and growth of Au nanoparticles inside live alfalfa plants. *Nano Letters*, **2**(4), 397–401.

Gardea-Torresdey J. L., Gomez E., Peralta-Videa J. R., Parsons J. G., Troiani H. and Jose-Yacaman M. (2003). Alfalfa sprouts: a natural source for the synthesis of silver nanoparticles. *Langmuir*, **19**(4), 1357–1361.

Ghasemi Z., Younesi H. and Kazemian H. (2011). Synthesis of nanozeolite sodalite from rice husk ash without organic additives. *The Canadian Journal of Chemical Engineering*, **89**, 601–608.

He S., Guo Z., Zhang Y., Zhang S., Wang J. and Gu N. (2007). Biosynthesis of gold nanoparticles using the bacteria *Rhodopseudomonas capsulata*. *Materials Letters*, **61**(18), 3984–3987.

Huang J., Li Q., Sun D., Lu Y., Su Y., Yang X., Wang H., Wang Y., Shao W., He N., Hong J. and Chen C. (2007). Biosynthesis of silver and gold nanoparticles by novel sundried *Cinnamomum camphora* leaf. *Nanotechnology*, **18**, 105104–1105115.

Huang J., Zhan G., Zheng B., Sun D., Lu F., Lin Y., Chen H., Zheng Z., Zheng Y. and Li Q. (2011). Biogenic silver nanoparticles by *Cacumen Platycladi* extract: synthesis, formation mechanism, and antibacterial activity. *Industrial and Engineering Chemistry Research*, **50**, 9095–9106.

Huang S. H. (2006). Gold nanoparticle-based immunochromatographic text for identification of gold nanoparticles using *Staphylococcus aureus* from clinical specimens. *Clinica Chimica Acta*, **373**, 139–143.

Jayaseelan C., Rahuman A. A., Rajakumar G., Kirthi A. V., Santhoshkumar T., Marimuthu S., Bagavan A., Kamaraj C., Zahir A. A. and Elango G. (2011). Synthesis of pediculocidal and larvicidal silver nanoparticles by leaf extract from heartleaf moonseed plant, *Tinospora cordifolia* Miers. *Parasitology Research*, **109**, 185–194.

Jha A. K., Prasad K., Prasad K. and Kulkarni A. R. (2009). Plant system: natures's nanofactory. *Colloids and Surfaces B: Biointerfaces*, **73**, 219–223.

Kammler H. K., Madler L. and Pratsinis S. E. (2001). Flame synthesis of nanoparticles. *Chemical Engineering Technology*, **24**, 583–596.

Kannan N., Mukunthan K. S. and Balaji S. (2011). A comparative study of morphology, reactivity and stability of synthesized silver nanoparticles using *Bacillus subtilis* and *Catharanthus roseus* (L.) G. Don. *Colloids and Surfaces B: Biointerfaces*, **86**, 378–383.

Kasthuri J., Kathiravan K. and Rajendiran N. (2009). *Phyllanthin*-assisted biosynthesis of silver and gold nanoparticles: a novel biological approach. *Journal of Nanoparticle Research*, **11**, 1075–1085.

Kaviya S., Santhanalakshmi J. and Viswanathan B. (2011). Green Synthesis of silver nanoparticles using *Polyalthia longifolia* leaf extract along with D-sorbitol: study of antibacterial activity. *Journal of Nanotechnology*, vol. 2011, Article ID 152970, 5 pages. doi:10.1155/ 2011/152970.

Khalil M. M. H., Ismail E. H. and El-Magdoub F. (2010). Biosynthesis of Au nanoparticles using olive leaf extract. *Arabian Journal of Chemistry*, **5**, 431–437.

Kim H., Lee S. J. and Jung N. J. (2007). Chemical vapor deposition of Pt nanoparticles on carbon nanotubes. *Solid State Phenomena*, **124**(6), 1769–1772.

Kimling J., Maier M., Okenve B., Kotaidis V., Ballot H. and Plech A. (2006). Turkevich method for gold nanoparticle synthesis revisited. *The Journal of Physical Chemistry A*, **110**(32), 15700–15707.

Klaus T., Joerger R., Olsson E. and Granqvist C. G. (1999). Silver-based crystalline nanoparticles, microbially fabricated. *Proceedings of the National Academy of Sciences*, **96**, 13611–13614.

Konishi Y., Nomura T., Tsukiyama T. and Saitoh N. (2004). Microbial preparation of gold nanoparticles by anaerobic bacterium. *Transactions of the Materials Research Society of Japan*, **29**(5), 2341–2343.

Konwarh R., Gogoi B., Philip R., Laskar M. A. and Kark N. (2011). Biomimetic preparation of polymer-supported free radical scavenging, cytocompatible and antimicrobial 'green' silver nanoparticles using aqueous extract of *Citrus sinensis* peel. *Colloids and Surfaces B: Biointerfaces*, **84**, 338–345.

Krishnaraj C., Jagan E. G., Rajasekar S., Selvakumar P., Kalaichelvan P. T. and Mohan N. (2010). Synthesis of silver nanoparticles using *Acalypha indica* leaf extracts and its antibacterial activity against water borne pathogens. *Colloids and Surfaces B: Biointerfaces*, **76**, 50–56.

Kumar K. P., Paul W. and Sharma C. P. (2011). Green synthesis of gold nanoparticles with *Zingiber officinale* extract: characterization and blood compatibility. *Process Biochemistry*, **46**, 2007–2013.

Kumar V., Yadav S. C. and Yadav S. K. (2010). *Syzygium cumini* leaf and seed extract mediated biosynthesis of silver nanoparticles and their characterization. *Journal of Chemistry Technology Biotechnology*, **85**, 1301–1309.

Liu Y., Chen S., Zhong L. and Wu G. (2009). Preparation of high-stable silver nanoparticle dispersion by using sodium alginate as a stabilizer under gamma radiation. *Radiation Physics and Chemistry*, **78**(4), 251–255.

Lopez M. L., Parsons J. G., Peralta-Videa J. R. and Gardea-Torresdey J. L. (2005). An XAS study of the binding and reduction of Au(III) by hop biomass. *Microchemical Journal*, **81**(1), 50–56.

Lukman A. I., Gong B., Marjo C. E., Roessner U. and Harris A. T. (2011). Facile synthesis, stabilization, and anti-bacterial performance of discrete Ag nanoparticles using *Medicago sativa* seed exudates. *Journal of Colloid and Interface Science*, **353**, 433–444.

Maensirei S., Laokul P., Klinkaewnarong J., Phokha S., Promarak V. and Seraphin S. (2008). Indium oxide (In_2O_3) nanoparticles using *Aloe vera* plant extract: synthesis and optical properties. *Journal of Optoelectronics and Advanced Materials*, **10**(3), 161–165.

Mallick K., Witcomb M. J. and Scurell M. S. (2004). Polymer stabilized silver nanoparticles: a photochemical synthesis route. *Journal of Materials Science*, **39**, 4459–4463.

Mallikarjuna K., Narasimha G., Dillip G. R., Praveen B., Shreedhar B., Lakshmi C. S., Reddy B. V. S. and Raju B. D. P. (2011). Green synthesis of silver nanoparticles using *Ocimum* leaf extract and their characterization. *Digest Journal of Nanomaterials and Biostructures*, **6**(1), 181–186.

Marshall A. T., Haverkamp R. G., Davies C. E., Parsons J. G., Gardea-Torresdey J. L. and Agterveld D. (2007). Accumulated of gold nanoparticles in *Brassicajuncea*. *International Journal of Phytoremediation*, **9**(3), 197–206.

Maynard A. D., Aitken R. J., Butz T., Colvin V., Donaldson K., Oberdörster G., Philbert M. A., Ryan J., Seaton A., Stone V., Tinkle S. S., Tran L., Walker N. J. and Warheit D. B. (2006). Safe handling of nanotechnology. *Nature*, **444**, 267–269.

Meltzer S., Resch R., Koel B. E., Thompson M. E., Madhukar A., Requicha A. A. G. and Will P. (2001). Fabrication of nanostructures by hydroxylamine seeding of gold nanoparticle templates. *Langmuir*, **17**(5), 1713–1718.

Miller E. V. and Hibbard R. P. (1926). Aqueous extracts of seeds as agents in the preparation of silver sols. *Plant Physiology*, **1**(4), 409–413.

Mohanpuria P., Rana N. K. and Yadav S. K. (2008). Biosynthesis of nanoparticles: technological concepts and future applications. *Journal of Nanoparticle Research*, **10**, 507–517.

Mondal S., Roy N., Laskar R. A., Sk I., Basu S., Mandal D. and Begum N. A. (2011). Biogenic synthesis of Ag, Au and bimetallic Au/Ag alloy nanoparticles using aqueous extract of mahogany (*Swietenia mahogany* JACQ.) leaves. *Colloids and Surfaces B: Biointerfaces*, **82**, 497–504.

Montes M. O., Mayoral A., Deepak F. L., Parsons J. G., Jose-Yacamán M., Peralta-Videa J. R. and Gardea-Torresdey J. L. (2011). Anisotropic gold nanoparticles and gold plates biosynthesis using alfalfa extracts. *Journal of Nanoparticle Research*, **13**(8), 3113–3121.

Mubarakali D., Thajuddin N., Jeganathan K. and Gunasekaran M. (2011). Plant extract mediated synthesis of silver and gold nanoparticles and its antibacterial activity against clinically isolated pathogens. *Colloids and Surfaces B: Biointerfaces*, **85**, 360–365.

Mude N., Ingle A., Gade A. and Ral M. (2009). Synthesis of silver nanoparticles using callus extract of *Carica papaya* – a first report. *Journal of Plant Biochemistry and Biotechnology*, **18**(1), 83–86.

Mukherjee P., Ahmad A., Mandal D., Senapati S., Sainkar S. R., Khan M. I. *et al.* (2001). Fungus-mediated synthesis of silver nanoparticles and their immobilization in the mycelial matrix: a novel biological approach to nanoparticle synthesis. *Nano Letters*, **1**, 515–519.

Mukherjee P., Roy M., Mandal B., Dey G., Mukherjee P. and Ghatak J. (2008). Green synthesis of highly stabilized nanocrystalline silver particles by a non-pathogenic and agriculturally important fungus *T. asperellum*. *Nanotechnology*, **19**, 75103–75110.

Nabikhan A., Kandasamy K., Raj A. and Alikunhi N. M. (2010). Synthesis of antimicrobial silver nanoparticles by callus and leaf extracts from saltmarsh plant, *Sesuvium portulacastrum* L. *Colloids and Surfaces B: Biointerfaces*, **79**, 488–493.

Nagajyothi P. C. and Lee K. D. (2011). Synthesis of plant-mediated silver nanoparticles using *Dioscorea batatas* rhizome extract and evaluation of their antimicrobial activities. *Journal of Nanomaterials*, vol. 2011, Article ID 573429, 7 pages. doi: 10.1155/2011/573429.

Njagi E. C., Huang H., Stafford L., Genuino H., Galindo H. M., Collins J. B., Hoag G. E. and Suib S. L. (2010). Biosynthesis of iron and silver nanoparticles at room temperature using aqueous sorghum bran extracts. *Langmuir*, **27**, 264–271.

National Nanotechnology Initiative (NNI). (2007). National Nanotechnology Initiative: what is nanotechnology? Available at http://www.nano.gov/nanotech-101/what (accessed December 2012).

Panda K. K., Achary V. M. M., Krishnaveni R., Padhi B. K., Sarangi S. N., Sahu S. N. and Panda B. B. (2011). *In vitro* biosynthesis and genotoxicity bioassay of silver nanoparticles using plants. *Toxicology In Vitro*, **25**, 1097–1105.

Parsons J. G., Peralta-Videa J. R., Dokken K. M. and Gardea-Torresdey J. L. (2009). Biological and biomaterials-assisted synthesis of precious metal nanoparticles. In: *Metallic and Metal Oxide Nanomaterials (Nanomaterials for Life Science Vol. 1)*, C. S. S. R. Kumar (ed.), Wiley-VCH, Verlag, GmbH, KGaA, Weinheim, pp. 461–491.

Parsons J. G., Peralta-Videa J. R. and Gardea-Torresdey J. L. (2007). Use of plants in biotechnology: synthesis of metal nanoparticles by inactivated plant tissues, plant extracts, and living plants. In: *Concepts and Applications in Environmental Geochemistry*, D. Sarkar, R. Datta and R. Hannigan (eds), Vol. 5, Elsevier, Amsterdam, The Netherlands, pp. 463–486.

Philip D. (2010). Rapid green synthesis of spherical gold nanoparticles using *Mangifera indica* leaf. *Spectrochimica Acta Part A*, **77**, 807–810.

Philip D. (2011). *Mangifera indica* leaf-assisted biosynthesis of well-dispersed silver nanoparticles. *Spectrochimica Acta A*, **78**, 327–331.

Philip D., Unni C., Aromal S. A. and Vidhu V. K. (2011). *Murraya koenigii* leaf-assisted rapid green synthesis of silver and gold nanoparticles. *Spectrochimica Acta Part A*, **78**, 899–904.

Prathna T. C., Chandrasekaran N., Raichur A. M. and Mukherjee A. (2011). Biomimetic synthesis of silver nanoparticles by *Citrus limon* (lemon) aqueous extract and theoretical prediction of particle size. *Colloid Surface B: Biointerfaces*, **82**, 152–159.

Qu J., Luo C. and Hou J. (2011a). Synthesis of ZnO nanoparticles from Zn-hyperaccumulator (*Sedum alfredii* Hance) plants. *Micro & Nano Letters*, **6**(3), 174–176.

Qu J., Yuan X., Wang X. and Shap P. (2011b). Zinc accumulation and synthesis of ZnO nanoparticles using *Physalis alkekengi* L. *Environmental Pollution*, **159**, 1783–1788.

Rajasekharreddy P., Rani P. U. and Sreedhar B. (2010). Qualitative assessment of silver and gold nanoparticle synthesis in various plants: a photobiological approach. *Journal of Nanoparticle Research*, **12**, 1711–1721.

Raju D., Metha U. J. and Hazra S. (2011). Synthesis of gold nanoparticles by various laef fractions of *Semecarpus anacardium* L. tree. *Trees*, **25**, 145–151.

Ramezani N., Ehsanfar Z., Shamsa F., Amin G., Shahverdi H. R., Esfahani H. R. M., Shamsaie A., Bazaz R. D. and Shahverdi R. (2008). Screening of medicinal plant methanol extracts for the synthesis of gold nanoparticles by their reducing potential. *Zeitschrift für Naturforschung*, **63b**, 903–908.

Rodriguez E., Parsons J. G., Peralta-Videa J. R., Cruz-Jimenez G., Romero-Gonzalez J., Sanchez-Salcido B. E., Saupe G. B., Duarte-Gardea M. and Gardea-Torresdey J. L. (2007). Potential of *Chilopsis linearis* for gold phytomining: using XAS to determine gold reduction and nanoparticle formation within plant tissues. *International Journal of Phytoremediation*, **9**(2), 133–147.

Santhoshkumar T., Rahuman A. A., Rajakumar G., Marimuthu S., Bagavan A., Jayaseelan C., Zahir A. A., Elango G. and Kamaraj C. (2011). Synthesis of silver nanoparticles using *Nelumbo nucifera* leaf extract and its larvicidal activity against malaria and filariasis vectors. *Parasitology Research*, **108**, 693–702.

Sathishkumar M., Sneha K., Won S. W., Cho C. W., Kim S. and Yun Y. S. (2009). *Cinnamon zeylanicum* bark extract and powder mediated green synthesis of nano-crystalline silver particles and its bactericidal activity. *Colloid Surface B: Biointerfaces*, **73**, 332–338.

Satyavani K., Ramanathan T. and Gurudeeban S. (2011). Plant mediated synthesis of biomedical silver nanoparticles by using leaf extract of *Citrullus colocynthis*. *Research Journal of Nanoscience and Nanotechnology*, **1**(2), 95–101.

Shankar S., Ahmad A. and Sastry M. (2003). Geranium leaf assisted biosynthesis of silver nanoparticles. *Biotechnology Progress*, **19**(6), 1627–31.

Shankar S., Rai A., Ankamwar B., Singh A., Ahmad A. and Sastry M. (2004). Biological synthesis of triangular gold nanoprisms. *Nature Materials*, **3**, 482–8.

Sharma N. C., Sahi S. V., Nath S., Parsons J. G., Gardea-Torresdey J. L. and Pal T. (2007). Synthesis of plant-mediated gold nanoparticles and catalytic role of biomatrix-embedded nanomaterials. *Environmental Science and Technology*, **41**(14), 5137–5142.

Sharma S. S. and Dietz K. J. (2006). The signifance of amino acids and amino acid derived molecules in plant responses and adaptation to heavy metal stress. *Journal of Experimental Botany*, **57**, 711–726.

Shekhawat G. S. and Arya V. (2009). Biological synthesis of Ag nanoparticles through *in vitro* cultures of *Brassica juncea* C. Zern. *Advanced Materials Research*, **67**, 295–299.

Shimazaki Y., Kobayashi Y., Yamada S., Miwa T. and Konno M. (2005). Preparation and characterization of aqueous colloids of Pt–Ru nanoparticles. *Journal of Colloid and Interface Science*, **292**(1), 122–126.

Singaravelu G., Arockiamary J., Ganesh K. and Govindaraju K. (2007). A novel extracellular synthesis of monodisperse gold nanoparticles using marine alga, *Sargassum wightii* Greville. *Colloids and Surfaces B: Biointerfaces*, **57**, 97–101.

Slawson R. M., Van-Dyke M. I., Lee H. and Trevor J. T. (1992). Germanium and silver resistance, accumulation and toxicity in microorganisms. *Plasmid*, **27**, 73–79.

Smitha S. L., Philip D. and Gophchandran K. G. (2009). Green synthesis of gold nanoparticles using *Cinnamomum zeylanicum* leaf broth. *Spectrochimica Acta Part A*, **74**, 735–739.

Song J. Y. and Kim B. S. (2009a). Rapid biological synthesis of silver nanoparticles using plant leaf extracts. *Bioprocess Biosystem Engineering*, **32**, 79–84.

Song J. Y., Jang H. and Kim B. S. (2009b). Biological synthesis of gold nanoparticles using *Magnolia kobus* and *Diopyros kaki* leaf extract. *Process Biochemistry*, **44**, 1133–1138.

Song J. Y., Kwon E. and Kim B. S. (2010). Biological synthesis of platinum nanoparticles using *Diopyros kaki* leaf extract. *Bioprocess Biosystem Engineering*, **33**, 159–164.

Sundrarajan M. and Gowri S. (2011). Green synthesis of titanium dioxide nanoparticles by *Nyctanthes arbor-tristis* leaves extract. *Chalcogenide Letters*, **8**(8), 447–451.

Sweeney R. Y., Mao C., Gao X., Burt J. L., Belcher A. M., Georgiou G. and Iverson B. (2004). Bacterial biosynthesis of cadmium sulfide nanocrystals. *Chemistry & Biology*, **11**, 1553–1559.

Tai Y., Tran N. T. T., Tsai Y. C., Fang J. Y. and Chang L. W. (2011). One-step synthesis of highly biocompatible multi-shaped gold nanostructures with fruit extract. *The Institution of Engineering and Technology*, **5**(2), 52–59.

Tan Y., Dai Y., Li Y. and Zhua D. (2003). Preparation of gold, platinum, palladium and silver nanoparticles by the reduction of their salts with a weak reductant-potassium bitartrate. *Journal of Materials and Chemistry*, **13**, 1069–1075.

Thirumurugan A., Tomy N. A., Jaiganesh R. and Gobikrishnan S. (2010). Biological reduction of silver nanoparticles using plant leaf extracts and its effect on increased antimicrobial activity against clinically isolated organism. *Der Pharma Chemica*, **2**(6), 279–284.

Tripathy A., Raichur A. M., Chandrasekaran N., Prathna T. C. and Mukherjee A. (2010). Process variables in biomimetic synthesis of silver nanoparticles by aqueous extract of *Azadirachta indica* (Neem) leaves. *Journal of Nanoparticle Research*, **12**, 237–246.

Valodkar M., Nagar P. S., Jadeja R. N., Thounaojam M. C., Devkar R. V. and Thakore S. (2011). Euphorbiaceae latex induced green synthesis of non-cytotoxic metallic nanoparticle solutions: a rational approach to antimicrobial applications. *Colloids and Surfaces A Physicochemical and Engineering Aspects*, **384**(1–3), 337–344.

Vaseeharan B., Ramasamy P. and Chen J. C. (2010). Antibacterial activity of silver nanoparticles (AgNPs) synthesized by tea leaf extracts against pathogenic *Vibrio harveyi* and its protective efficacy on juvenile *Feneropenaeus indicus*. *Letters in Applied Microbiology*, **50**, 352–356.

Vinod V. T. P., Saravanan P., Sreedhar B., Devi D. K. and Sashidhar R. B. (2011). A facile synthesis and characterization of Ag, Au and Pt nanoparticles using a natural hydrocolloid gum kondagogu (*Cochlospermum gossypium*). *Colloids and Surfaces B: Biointerfaces*, **83**, 291–298.

Watson J. H. P., Ellwood D. C., Soper A. K. and Charnock J. (1999). Nanosized strongly magnetic bacterially-produced iron sulfide materials. *Journal of Magnetism and Magnetic Materials*, **203**, 69–72.

Yates B. J., Myre E., Breetz D. and Dionysiou D. D. (2005). Biotemplating of nanoparticles for environmental applications using phytomining techniques. Abstracts of Papers, 230th ACS National Meeting, Washington, DC, United States, August 28 – September 1, 2005.

Yu L., Yao J. and Fu Z. W. (2007). Laser ablated preparation of noble metal nanoparticles in liquid. *Wuli Huaxue XueBao*, **23**(6), 945–949.

Zargar M., Hamid A. A., Bakar F. A., Shamsudin M. N., Shameli K., Jahanshiri F. and Farahani F. (2011). Green synthesis and antibacterial effect of silver nanoparticles using *Vitex Negundo* L. *Molecules*, **16**, 6667–6676.

Zhan G., Huang J., Du M., Abdul-Rauf I., Ma Y. and Li Q. (2011). Green synthesis of Au-Pd bimetallic nanoparticles: single-step bioreduction method with plant extract. *Materials Letter*, **65**, 2989–2991.

Zhang G. H., Guo W. L. and Wang X. K. (2007). Sonochemical formation of nanocrystalline gold in aqueous solution. *Materials and Research Innovations*, **11**(4), 201–203.

Index